XXII International Conference "Physical Interpretations of Relativity Theory - 2021" (PIRT 2021)

Journal of Physics: Conference Series Volume 2081

Moscow, Russia
5 – 9 July 2021

ISBN: 978-1-7138-5052-6
ISSN: 1742-6588

Printed from e-media with permission by:

Curran Associates, Inc.
57 Morehouse Lane
Red Hook, NY 12571

Some format issues inherent in the e-media version may also appear in this print version.

Institute of Physics
Dirac House, Temple Back
Bristol BS1 6BE UK

Phone: 44 1 17 929 7481
Fax: 44 1 17 920 0979

techtracking@iop.org

Additional copies of this publication are available from:

Curran Associates, Inc.
57 Morehouse Lane
Red Hook, NY 12571 USA
Phone: 845-758-0400
Fax: 845-758-2633
Email: curran@proceedings.com
Web: www.proceedings.com

TABLE OF CONTENTS

XXII International Scientific Conference "Physical Interpretations of the Relativity Theory PIRT-2021"

V O Gladyshev[1], **P Rowlands**[2] **and A N Morozov**[1]

[1] Bauman Moscow State Technical University, 5, 2-nd Baumanskaya St., Moscow, 105005, Russia
[2] University of Liverpool, L69 3BX, UK

E-mail: gladyshev@bmstu.ru, p.rowlands@liverpool.ac.uk, amor@bmstu.ru

July 5 - 8, 2021, Bauman Moscow State Technical University hosted the XXII International Conference on Modern Problems of the Theory of Relativity, Cosmology, and Astrophysics "Physical Interpretations of the Relativity Theory PIRT-2021".

The PIRT Conference is an international scientific event that was first organized in 1988 and held at Imperial College in London once every two years, and since 2003 it has been held at Bauman University in Moscow. The main topics of the conference are the physical, geometric, and mathematical interpretation of the theory of relativity, its experimental verification and alternative theories of gravity, optical methods for detecting gravitational waves, new effects of moving-media optics and relativistic electrodynamics, as well as astrophysical observations and space experiments.

The Conference Program of the Moscow PIRT-2021 included papers dealing with the following major themes:

- Gravitation, cosmology and large-scale structure
- Gravitational waves and experimental tests of the relativity theory
- Relativistic electrodynamics
- High energy astrophysics
- Modern problems of classical and quantum field theory

The working languages of the XXII Conference were Russian and English; all reports were simultaneously interpreted.

The PIRT-2021 program was dedicated to the latest advances in the theory of gravity, cosmology, astrophysics; more than 120 researchers from 21 countries took part in the conference.

The conference welcomed well-known scientists, such as the President of the Russian Gravitational Society, Academician of the Russian Academy of Sciences, the world-renowned scientist in the field of cosmology and gravity Alexey Alexandrovich Starobinsky, Professor of Electromagnetism at the University of Naples Federico II, Member of the Optical Society of America Innocenzo M. Pinto and others.

The Proceedings of PIRT-2019 conference were published in the Journal of Physics: Conference Series https://iopscience.iop.org/issue/1742-6596/1557/1

The organizing and editorial committee of the PIRT-2021 conference hopes you enjoy the reading of this volume of the JPCS. We wish to thank all authors and participants for providing their valuable contributions for the PIRT proceeding as well as the reviewers for their constructive recommendations and criticism aiding to improve the presented papers.

International Advisory Council

- Prof. Aroonkumar Beesham, University of Zululand, South Africa
- Prof. Bivudutta Mishra, BITS-Pilani, Hyderabad Campus, India
- Prof. Carlos Romero, Universidade Federal da Paraiba, Brazil
- Prof. John B. Dainton, Liverpool University, Great Britain
- Prof. Barry C. Barish, California Institute of Technology, United States
- Prof. David Blair, University of South Western Australia, Australia
- Prof. Naresh Dadhich, IUCAA, PUNE, India
- Dr. Nicola N. Napolitano, INAF Astronomical Observatory of Capodimonte, Italy
- Prof. Alexander D. Dolgov, Institute of Theoretical and Experimental Physics of RAS, & INFN, Ferrara, Italy
- Prof. Carlo Bradaschia, Istituto Nazionale di Fisica Nucleare INFN, Universita di Pisa, Italy
- Prof. David Shoemaker, MIT LIGO Laboratory, Massachusetts Institute of Technology, Cambridge, U.S.A.
- Prof. Guido Pizzella, Physics Department, University of Rome, Rome, Italy
- Prof. James Hough, Department of Physics and Astronomy, University of Glasgow, Glasgow, U.K.
- Prof. Mairi Sakellariadou, King's College London, University of London, London, Great Britain
- Prof. Nathalie Deruelle, Institut des Hautes Etudes Scientifiques, Bures sur Yvette, France
- Prof. Peter Rowlands, University of Liverpool, Liverpool, United Kingdom
- Prof. Roland Triay, Centre de Physique Theorique CNRS – University de Provence, Marseille, France
- Prof. Rainer Weiss, Massachusetts Institute of Technology, Cambridge, U.S.A.
- Prof. Stanley E. Whitcomb, Physics/LIGO laboratory, California Institute of Technology, Pasadena, U.S.A.
- Dr. Richard Kerner, University Pierre et Marie Curie, France
- Prof. Yee Jack Ng, University of North Carolina – UNC, USA
- Prof. Dadhich Naresh, IUCAA, PUNE, India
- Prof. Mahim R. Adhikari, University of Calcutta, IMBIC, India
- Prof. H.P. Mazumdar, Physics and Earth Sciences Division, Indian Statistical Institute, Kolkata, India

Conference Committee

- Prof. Anatoly A. Aleksandrov, Bauman MSTU, Moscow, Russia (University rector)
- Prof. Vladimir O. Gladyshev, Bauman MSTU, Moscow, Russia (Chairman)
- Member of RAS, Prof. Alexei A. Starobinsky, Landau Institute for Theoretical Physics, Russia (Co-Chairman)
- Member of RAS, Prof. Vladislav I. Pustovoit, Science Technological Center of Unique Instruments of RAS, Moscow, Russia (Co-Chairman)
- Corr.-Member of RAS, Prof. Andrei N. Morozov, Bauman MSTU, Moscow, Russia (Co-Chairman)

Local Committee

- Member of RAS, Prof. Anatoly M. Cherepashchuk, Sternberg Astronomical Institute Moscow Russia
- Corr.-Member of RAS, Prof. Nikolay N. Kolachevsky, P.N. Lebedev Physical Institute of RAS, Russia
- Corr.-Member of RAS, Prof. Nikolay N. Rozanov, Saint Petersburg National Research University of Information Technologies, Mechanics and Optics; Vavilov State Optical Institute, Russia

- Prof. Valentin N. Rudenko, Sternberg State Astronomical Institute, Lomonosov Moscow State University, Moscow, Russia
- Prof. Sergey V. Sushkov, Kazan Federal University, Russia
- Prof. Igor V. Fomin, Bauman MSTU, Moscow, Russia
- Prof. Sergey V. Chervon, Ulyanovsk State University, Ulyanovsk, Russia
- Prof. Konstantin A. Postnov, Sternberg State Astronomical Institute Lomonosov Moscow State University, Russia
- Prof. Mikhail V. Sazhin, Sternberg State Astronomical Institute Lomonosov Moscow State University, Russia
- Prof. Boris M. Bolotovskii, Lebedev Physical Institute of RAS, Moscow, Russia
- Prof. Georgy N. Izmailov, Moscow Aviation Institute, Moscow, Russia

Invited Speakers

- Alexei Starobinsky, Landau Institute for Theoretical Physics, Russia
- Vladislav Pustovoit, Russian Academy of Science, Bauman University, Russia
- Innocenzo M. Pinto, University of Sannio at Benevento, Italy
- Peter Rowlands, University of Liverpool, United Kingdom
- Antonio C. Gutierrez-Pineres, Universidad Industrial de Santander, Colombia
- Bivudutta Mishra, BITS-Pilani, Hyderabad Campus, India
- Carlos Romero, Universidade Federal da Paraiba, Departamento de Fisica, Brazil
- Andrei Lebed, University of Arizona, USA
- Godwin Ibeh, Nigerian Defence Academy, Faculty of Science, Department of Physics, Nigeria
- Gogfrey Akpojotor, Delta State University, Faculty of Science, Physics Department, Nigeria
- Joann Sydney Rowlands, University of Liverpool, United Kingdom
- Muhammad Sharif, University of the Punjab, Pakistan
- Muhammad Zubair, COMSATS University Islamabad, Pakistan
- Pradyumn Kumar Sahoo, BITS-Pilani, Hyderabad Campus, India
- Sergey Chervon, Ulyanovsk State University, Russia
- Sunil Kumar Tripathy, Indira Gandhi Institute of Technology, Sarang, India
- Vladimir Vargashkin, Oryol State University, Russia
- Alexander Balakin, Kazan (Volga region) Federal University, Russia
- Alcides Garat, Universidad dela Republica, Uruguay

Support to the conference

- The State Duma Committee on Education of the Russian Federation
- Department of Physics, University of Liverpool (Great Britain)
- Russian Gravitational Society
- Moscow Physical Society
- The International Society on General Relativity & Gravitation
- British Society for the Philosophy of Science

Peer Review Declaration Form Guidance

On the following page you will find the declaration form.

- Please answer each question.
- You should submit the form along with the rest of your submission files.
- The deadline is the submission date written in your publishing agreement.

All conference organisers/editors are required to declare details about their peer review.

We will published the information you provide as part of your proceedings.

Peer review declaration

All papers published in this volume of Journal of Physics: Conference Series have been peer reviewed through processes administered by the Editors. Reviews were conducted by expert referees to the professional and scientific standards expected of a proceedings journal published by IOP Publishing.

- **Type of peer review:** Double-blind
- **Conference submission management system:**
  To participate in the PIRT-2021 Conference, Participants had to register on the web-site http://www.pirt.info/?lang=eng#reg_form
  Abstracts and papers had to be sent to the PIRT-2021 Organizing Committee by e-mail: dekan-fn@mail.ru
  All information about the format of abstracts and papers was on the web-site: http://www.pirt.info/?lang=eng
- **Number of submissions received:** 61
- **Number of submissions sent for review:** 51
- **Number of submissions accepted:** 38
- **Acceptance Rate (Number of Submissions Accepted / Number of Submissions Received X 100):** 62,29
- **Average number of reviews per paper:** 2
- **Total number of reviewers involved:**12
- **Any additional info on review process:**
  Plagiarism check system: antiplagiat.ru
  Authors could resubmit the paper with the necessary revisions.
- **Contact person for queries:**
  Name : Professor Vladimir Olegovich Gladyshev
  Affiliation: Head of the Faculty of Fundamental Sciences,
  Bauman Moscow State Technical University,
  5, 2-nd Baumanskaya St., Moscow, 105005, Russian Federation
  Email : gladyshev@bmstu.ru

PIRT 2021 IOP Publishing
Journal of Physics: Conference Series 2081 (2021) 012001 doi:10.1088/1742-6596/2081/1/012001

Reconstruction of some cosmological models from the deceleration parameter

Aroonkumar Beesham

Department of Mathematical Sciences, University of Zululand, P Bag X1001, Kwa-Dlangezwa 3886, South Africa; and
Faculty of Natural Sciences, Mangosuthu University of Technology, P O Box 12363, Jacobs 4026, South Africa

E-mail: abeesham@yahoo.com

Abstract. Since the discovery of the late-time acceleration of the universe, researchers are still trying to find an explanation for it. This is regarded as the most important unsolved problem in cosmology today. The most favoured explanation is dark energy, an unknown or exotic form of matter with negative pressure. One may argue that particle physics may provide the answer in time. Currently, the LambdaCDM model is regarded as the best model. Although this model is reasonably successful and widely accepted, there is growing interest in looking at alternatives. Some of the reasons for this are the fine-tuning, coincidence, inflationary paradigm and cosmological constant problems, and whether general relativity is valid on large scales. One focus in trying to understand dark energy is to assume some form of the scale, Hubble or deceleration parameter (or some other reasonable assumption), and then to see how well the model fits in with current observations. This approach is broadly called reconstruction. In this talk, we focus on the deceleration parameter. We provide a brief review of the various forms of the deceleration parameter that have been employed in the past in cosmology, and then focus on some particular forms of interest which have drawn some attention. We note that it is most worthwhile to study alternative dark energy and dark gravity models in order to fully understand the entire space of possibilities.

1. Introduction

Einstein's field equations with cosmological constant are:

$$R_{ab} - \frac{1}{2}Rg_{ab} + \Lambda g_{ab} = T_{ab}. \tag{1}$$

For the Friedmann-Lemaitre-Robertson-Walker metric, which represents a homogeneous and isotropic universe,

$$ds^2 = -dt^2 + a^2(t)\left\{\frac{dr^2}{1-kr^2} + r^2(d\theta^2 + sin^2\phi d\phi^2)\right\}, \tag{2}$$

where a is the scale factor, we get the following Friedmann and Raychaudhuri-type equations:

$$3\frac{\dot{a}^2}{a^2} + \frac{3k}{a^2} - \Lambda = \rho, \tag{3}$$

$$2\frac{\ddot{a}}{a} + \frac{\dot{a}^2}{a^2} + \frac{k}{a^2} - \Lambda = -p. \tag{4}$$

Now, the Hubble parameter H determines the expansion rate of the universe:

$$H = \frac{\dot{a}}{a}, \tag{5}$$

whereas the deceleration parameter q implies a small correction as the matter slows down the expansion

$$q = -\frac{\ddot{a}a}{\dot{a}^2}. \tag{6}$$

The importance of q is that it relates to the non-linear part of the time dependence of the scale factor. In terms of H and q, equations (3) and (4) can be written as:

$$3H^2 + \frac{3k}{a^2} - \Lambda = \rho, \tag{7}$$

$$H^2(1-2q) + \frac{k}{a^2} - \Lambda = -p. \tag{8}$$

Two other parameters that have proven useful in the analysis of dark energy models are the jerk and snap parameters. The jerk parameter j is effectively a derivative of q:

$$j = \frac{\dddot{a}}{a}\left(\frac{\dot{a}}{a}\right)^{-3}, \tag{9}$$

and the snap parameter s, which is effectively a derivative of j, is defined by:

$$s = \frac{\ddddot{a}}{a}\left(\frac{\dot{a}}{a}\right)^{-4}. \tag{10}$$

2. History of cosmology

If we take a look briefly at the history of cosmology, we notice that there have been three major revolutions. The first was the discovery of the expansion of the universe, largely due to Hubble [1]. What do the solutions to equations (3) and (4) look like for $\Lambda = 0,\ p = 0$?

Figure 1 [2] illustrates these solutions for the cases $k = -1, 0, +1$ corresponding to open, flat and closed spatial sections, respectively. Observations at that time indicated that the universe was expanding ($H > 0$), but that the expansion was decreasing, i.e., a decelerating universe ($q > 0$).

In the early 1970's, Sandage defined cosmology as simply the search for H_o & q_o: "All of observational cosmology is the search for two numbers: H_o and q_o" [3]. We notice from equation (6) that a negative sign appears on the right side. The reason for this was that there was such confidence in the expansion of the universe slowing down that a negative sign was assigned to make the value of q positive. For historical reasons, this has remained even today.

The second revolution in cosmology came around 1979 with the advent of the idea of inflation [4]. Inflation occurs during the period $10^{-36} - 10^{-34}/10^{-33}s$ after the big-bang - see figure 2 [2].

We note that inflation solves the horizon, flatness and monopole problems, but does not get rid of the big-bang singularity. Around that time, the big bang model plus inflation was believed to describe the universe to a first approximation.

In 1998, we saw the birth of the third revolution, i.e., the discovery of the accelerated expansion of the universe. One should note that this is usually interpreted as dark enery, but that there are some alternative explanations. The accelerated expansion began around 6

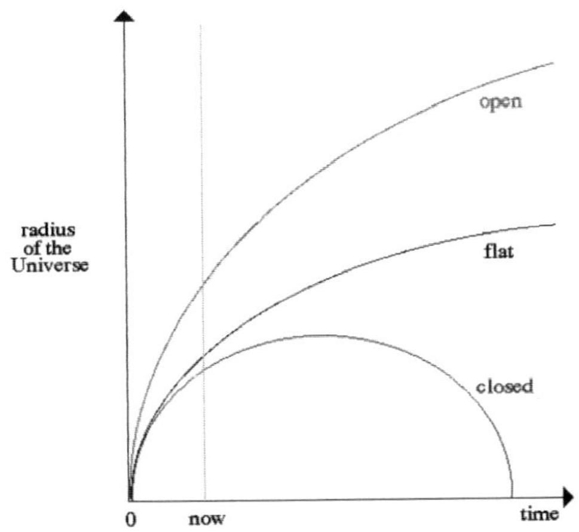

Figure 1. a against t before DE and inflation.

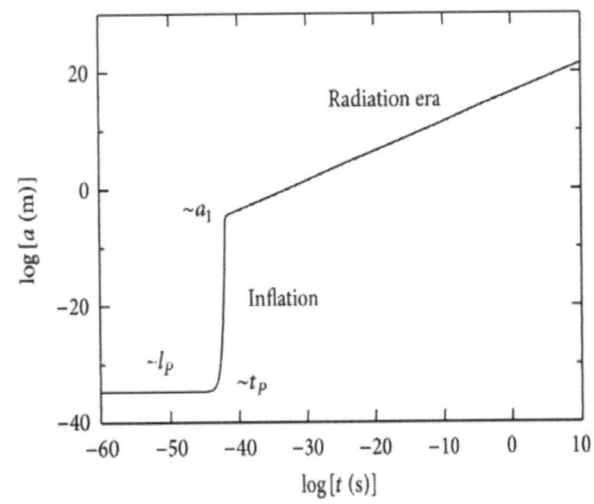

Figure 2. a against t showing early inflation.

billion yrs ago. From the Friedmann (3) and Raychaudhuri-type (4) equations, we can derive a conservation type eq:

$$\dot{\rho} + 3H(\rho + p) = 0. \tag{11}$$

We assume a barotropic eq of state:

$$p = \omega\rho, \qquad \omega = const. \tag{12}$$

For $k = \Lambda = 0$, we can solve equations (3), (4) and (11) to get:

$$a \propto t^{2/(3(1+\omega))}, \qquad \omega \neq -1. \tag{13}$$

For pressure-free matter ($\omega = 0$), $a \propto t^{2/3}$, and for for radiation, ($\omega = 1/3$), $a \propto t^{1/2}$. For $k = 0, \Lambda \neq 0$ or $\rho = const$, $\omega = -1$, we get:

$$a \propto e^{Ht}, \qquad \rho = const, \qquad \omega = -1. \tag{14}$$

This exponential type of expansion occurs during early inflation and the late-time acceleration eras.

From the definition of the deceleration parameter (6), we find that it can be written in a unified way for all three cases as:

$$q = \frac{1}{2}(1 + 3\omega). \tag{15}$$

The pressure can be written in terms of q as:

$$p = H^2(2q - 1), \tag{16}$$

and the age of the universe as:

$$t_o = \frac{1}{H_o(1 + q_o)}. \tag{17}$$

3. Classification in terms of q

Comsological models may be classified according to q in various ways. Firstly, in terms of H as:

- $H > 0,\ q > 0$;
- $H > 0,\ q < 0$;
- $H > 0,\ q = 0$.

For $p = 0$, the relation between q and the spatial geometry can be written as:

- $q > 1/2,\ k = +1$: closed spherical surface
- $q > 1/2,\ k = 0$: open flat space
- $q > 1/2,\ k = -1$: open hyperbolic space

The age of the universe can be written in terms of q and the Einstein-de Sitter age, $t^* = (2/3)H^{-1}$, as:

- $q > 1/2,\ \text{age} < t^*$;
- $q = 1/2,\ \text{age} = t^*$;
- $q < 1/2,\ \text{age} > t^*$.

The energy conditions can be expressed in terms of q as:

- NEC $(\rho + p \geq 0),\ q \geq 1$;
- SEC $(\rho + p \geq 0, \rho + 3p \geq 0),\ q \geq 0$;
- DEC $(\rho \geq 0, \rho \geq |p|),\ q \leq 2$.

A similar analysis may be carried out for multi-fluids also. Now, radiation and matter can only produce deceleration. For $k \neq 0$, the DP (15) for all 3 cases can be written in a unified way as:

$$q = \frac{1}{2}\left(1 + \frac{k}{a^2H^2}\right)(1 + 3\omega). \tag{18}$$

Another way of solving equations (3) and (4) without assuming an equation of state (12) is to make some assumption on q, e.g.,

$$q = const. \tag{19}$$

This yields the solutions (13) and (14) as special cases. Many authors have assumed condition (19) to find solutions in general relativity as well as alternative theories, especially before the discovery of dark energy.

4. Cosmography

Cosmography is concerned with the kinematics of cosmology, i.e., relations independent of the theory. The Hubble parameter H and the deceleration parameter q are the simplest cosmograhic parameters. The procedure in cosmography is to expand the scale factor $a(t)$ (or some other parameter) in Taylor series about the present time t_o:

$$a(t) = a(t_o) + \dot{a}(t_o)(t - t_o) + \frac{1}{2}\ddot{a}(t_o)(t - t_o)^2 + \cdots \tag{20}$$

which can also be written as:

$$\frac{a(t)}{a(t_o)} = 1 + H_o(t - t_o) + \frac{q_o}{2}{H_o}^2(t - t_o)^2 + \cdots \tag{21}$$

from which it can be seen that H and q arise naturally in the expansion.

4.1. Constant deceleration parameter

We assume the cosmological principle, i.e., homogeneity and isotropy. Then expand some measurable quantity in a Taylor series around the present time. Let the deceleration parameter q be expanded about the present time (or redshift, scale factor, Hubble parameter, etc). The simplest is, of course, $q = const.$ This was essentially done by Berman [5] who proposed a law of variation of the Hubble parameter:

$$H = Da^{-n}, \qquad D,\ n = consts. \tag{22}$$

This yields

$$a = (nDt + c)^{1/n}. \tag{23}$$

The deceleration parameter q in this case is

$$q = n - 1, \tag{24}$$

which is constant. So, we may equivalently regard equation (24) as a law of constant deceleration parameter. We observe that $n > 1 \implies$ deceleration, whereas $n < 1 \implies$ acceleration. The idea was to choose values of q consistent with observations. This analysis was carried out before the discovery of dark energy, and we note that a constant q cannot yield a transition from deceleration to acceleration (no "signature flip" - not to be confused with a signature change in the metric). Hence, a variable q was considered which yields the required transition.

4.2. Linearly varying deceleration parameter

After a constant q, the next simplest is a linearly varying q:

$$q(x) = q_o + q_1 x. \tag{25}$$

Some parametrisations that have been considered are:

$$q = q_o + q_1 z; \tag{26}$$

$$q = q_o + q_1 a; \tag{27}$$

$$q = q_o + q_1 t. \tag{28}$$

Akarsu and Dereli [6], and Akarsu et al [7] have given a very discussion of these particular cases and shown that they are consistent with observations. We note that linearity in one parameter $\nRightarrow$ linearity in another.

4.3. Nonlinear deceleration parameter

In this subsection, we focus on some interesting parametrisations of q. Let us firstly consider [8]:

$$H(a) = \alpha(1 + a^{-n}), \tag{29}$$

where $\alpha > 0$ and $n > 1$. For q, this yields:

$$q(z) = \frac{(n-1)(1+z)^n - 1}{1 + (1+z)^n}. \tag{30}$$

or

$$q(t) = -1 + \frac{n}{e^{n\alpha t}}, \quad n, \alpha = const. \tag{31}$$

This is plotted in figure 3 for some suitable values of the parameter n. From the figure, we

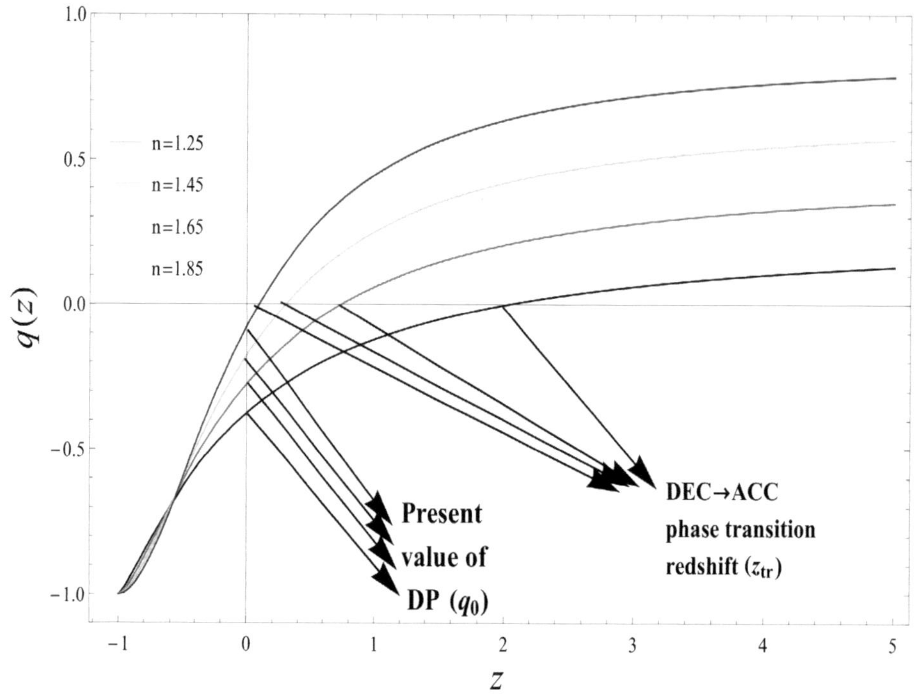

Figure 3. Plot of deceleration parameter *vs* redshift z for different n

see that the deceleration parameter q is positive during early times, implying deceleration, and negative at late times, implying acceleration. There is a transition at a redshift of around $z = 1$. One can get model which fits in with observations in $f(R,T)$ theory.

Secondly, let us consider the deceleration parameter q with a double flip [9]:

$$q = -1 + \frac{\alpha t}{1+t^2}. \tag{32}$$

The matter source in this case is a self interacting scalar field. This allows for a double signature flip, viz., to account for early inflation, and then from deceleration to late-time acceleration. See figure 4. In this case, we notice that the graphs of q above the blue curve cut the t axis at two points. The first point corresponds to the transition form early inflation to the radiation-dominated era, and the second to the transition from the matter era to the currently accelerated era.

4.4. Other forms of the deceleration parameter

Let us illustrate some other interesting forms of q that have been considered. In general, one can write a polynomial in q as a function of the time t:

$$q = q_o + q_1 t + q_2 t^2 + ..., \tag{33}$$

We studied a quadratic form [10], and a cubic form is under preparation. One may also expand q in a series in the redshift z as [11, 12]:

$$q = q_o + (-q_o - 2q_o^2 + j_o)z + \frac{1}{2}(2q_o + 8q_o^2 + 8q_o^3 - 7q_o j_o - 4j_o - s_o)z^2 + O(z^3). \tag{34}$$

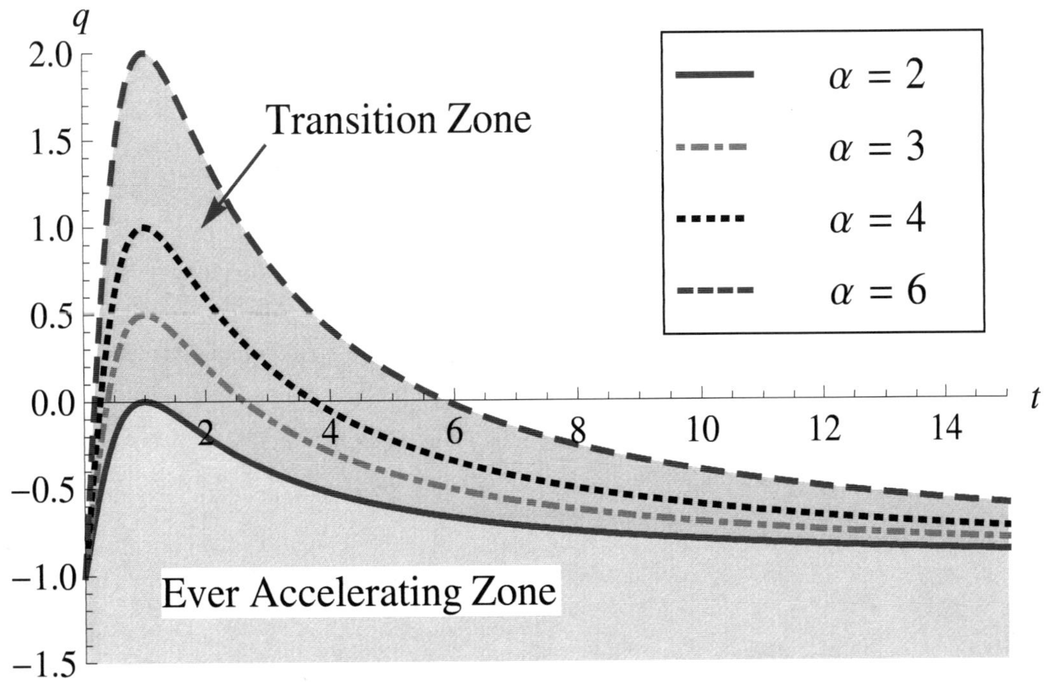

Figure 4. Plot of q against t

Finally, there has been some interest in bouncing solutions recently, i.e., cosmological solutions that do not have a big-bang, but rather a bounce with an infinite series of expansions and contractions. An example of such a model is given by the following periodic form of q:

$$q = m\ cos(nt) - 1. \tag{35}$$

This implies a bouncing solution [13].

5. Conclusion

The basic characteristics of cosmological evolution are expressed by the Hubble parameter H and the deceleration parameter q; H describes the linear part and q relates to the nonlinear part. The standard ΛCDM model is most widely accepted and fits the observations, but it has some problems. Hence, it is worthwhile to examine other possibilities. Many reconstructions have been done with q. Choosing suitable forms of q may explain all stages of evolution of the universe. In this work, we have examined this approach, and looked at the various possibilities for q. It is possible to account for the transition from deceleration to acceleration. Also, nonlinear forms have the potential to explain the initial inflation as well as late-time acceleration. One can construct models that are consistent with observations. Even choices of the scale factor a and Hubble parameter H can be re-expressed as functions of q. We are not sure what lies in store for us in cosmology in the future.

Acknowledgments

This work is based on the research supported wholly/in part by the National Research Foundation of South Africa (Grant Numbers: 118511). The author is grateful for interesting discussions with Binaya Bishi, Ritika Nagpal, Jainendra Singh, Vijay Singh, Rishi Tiwari and Bhupendra Shukla.

References

[1] Hubble E 1929 *Proc. Natl. Acad. Sci. U. S. A.* **15** 168
[2] Gilson, David. The Ever Expanding Universe In Modern Cosmology http://www.laidback.org/ daveg/academic/expandinguniverse/index2.html
[3] Sandage A R 1970 Phys. Today **23** 34
[4] Guth A H 1981 *Phys. Rev. D* **23** 347–356
[5] Berman M S 1983 *Il Nuovo Cim.* **74** 182-186
[6] Akarsu O and Dereli T *Int. J. Theor. Phys.* **51** 612
[7] Akarsu 0, Dereli T, Kumar S and Xu L. *Eur. Phys. J Plus* **129** 22
[8] Nagpal R, Pacif S K J, Singh J K, Bamba K and Beesham A 2018 *Eur. Phys. J. C* **78** 946
[9] Singh V and Beesham A *Int. J. Geom. Meth. Mod. Phys.* **15** 1850145
[10] Tiwari R K, Beesham A and Shukla B K 2021 *Afr. Math.* **32** 983
[11] Bolotin Yu L, Cherkaskiy V A, Lemets O A, Yerokhin D A, and Zazunov L G 2015 Cosmology in terms of the deceleration parameter. Part 1 *gr-qc/1502.00811v1*
[12] Bolotin Yu L, Cherkaskiy V A, Lemets O A, Yerokhin D A, and Zazunov L G 2015 Cosmology in terms of the deceleration parameter. Part 2 *Preprint gr-qc/1506.08918v1*
[13] Hulke N, Singh G P, Bishi B K, and Beesham, A Bianchi-I Dark energy Cosmological Models with generalised Chaplygin gas and time periodic varying deceleration parameter (*Preprint*)

Relic gravitational waves in cosmological models based on the modified gravity theories

I V Fomin[1,2], **S V Chervon** [1,2,3]

[1] Bauman Moscow State Technical University, Moscow, Russia
[2] Ulyanovsk State Pedagogical University, Ulyanovsk, Russia
[3] Kazan Federal University, Kazan, Russia

E-mail: ingvor@inbox.ru, chervon.sergey@gmail.com

Abstract. We consider cosmological models based on the generalized scalar-tensor gravity, which correspond to the observational constraints on the parameters of cosmological perturbations for any model's parameters. The estimates of the energy density of relic gravitational waves for such a cosmological models were made. The possibility of direct detection of such a gravitational waves using modern and prospective methods was discussed as well.

1. Introduction

To describe the physical processes occurring in the early universe, theoretical models of cosmological inflation based on the evolution of a scalar field and Einstein gravity theory [1, 2, 3, 4] or its different modifications [5, 6, 7, 8, 9] are usually used.

In accordance with the theory of cosmological perturbations, during the stage of inflation, quantum fluctuations of the scalar field induce perturbations of the space-time metric. In the context of inflationary cosmology the observed anisotropy and polarization of the CMB [10] can be explained by the action of two types of perturbations, namely, scalar and tensor perturbations (relic gravitational waves). The third type of perturbations, namely vector perturbations, quickly decay in the process of accelerated expansion of the early universe [2]. The observational constraints on the values of the cosmological perturbations parameters can be used for verification of different inflationary models based on GR and its modifications as well [2, 3, 11, 12, 13, 14].

For the case of inflationary models based on Einstein gravity the observational constraints on the values of the cosmological perturbations are satisfied for limited class of the physical potentials of a scalar field that restricts the possible relevant cosmological models [2, 3, 4] However, in the case of modifications of general relativity, one can consider cosmological inflationary models with arbitrary parameters that correspond to the observational constraints (see, for example, in [15, 16]). This approach makes it possible to obtain model-independent estimates of the characteristics of relic gravitational waves, the direct observation of which is an urgent task in the context of the confirmation of application of the inflationary paradigm to the description of physical processes in the early universe [17].

In this paper, we consider the models of cosmological inflation based on the generalized scalar-tensor gravity (STG) theory and the quadratic relationship between the Hubble parameter and coupling function that define the type of STG [18, 19, 20] on the basis of a such approach.

Within the framework of the proposed approach, the relationship between the parameters of cosmological perturbations and the energy scale of inflation which doesn't depend on the type of potential of a scalar field, the type of STG or the type of cosmological dynamics was defined.

We also consider various possibilities of realizing transitions between the end of the inflationary stage and the stage of radiation domination, and estimate the parameters of relic gravitational waves for the proposed scenarios of the evolution of the early universe.

2. The dynamic equations for inflationary models based on STG

The inflationary models based on the generalized scalar-tensor theory are defined by the action [21, 22]

$$S_{GST}=\int d^4x\sqrt{-g}\Big[\frac{1}{2}F(\phi)R-\frac{\omega(\phi)}{2}g^{\mu\nu}\partial_\mu\phi\partial_\nu\phi-V(\phi)\Big]+S_m, \tag{1}$$

in the system of units $8\pi G=m_p^{-2}=c=1$, where g a determinant of the space-time metric $g_{\mu\nu}$, ϕ is a scalar field with the potential $V=V(\phi)$, $F(\phi)$ and $\omega(\phi)$ are the coupling and kinetic functions that determine the type of STG, R is the Ricci scalar curvature of the space-time. The case of vacuum inflationary solutions implies the absence of matter $S_m=0$.

The background dynamic equations in a spatially flat four-dimensional Friedmann-Robertson-Walker (FRW) space-time

$$ds^2=-dt^2+a^2(t)\left(dx^2+dy^2+dz^2\right), \tag{2}$$

corresponding to the action (1) in chosen system of units are

$$3FH^2+3H\dot{F}-\frac{\omega}{2}\dot{\phi}^2-V(\phi)=0, \tag{3}$$

$$3FH^2+2H\dot{F}+2F\dot{H}+\ddot{F}+\frac{\omega}{2}\dot{\phi}^2-V(\phi)=0, \tag{4}$$

$$\omega\ddot{\phi}+3\omega H\dot{\phi}+\frac{1}{2}\dot{\phi}^2\omega'+V'-6H^2F'-3\dot{H}F'=0. \tag{5}$$

where a dot represents a derivative with respect to the cosmic time t, $H\equiv\dot{a}/a$ denotes the Hubble parameter and $F'\equiv\partial F/\partial\phi$.

Also, we note, that the scalar field equation (5) can be derived from the equations (3)–(4). For this reason, equations (3)–(4) completely describe the cosmological dynamics, and can be represented as follow

$$V(\phi)=3FH^2+\frac{5}{2}H\dot{F}+F\dot{H}+\frac{1}{2}\ddot{F}, \tag{6}$$

$$\omega(\phi)\dot{\phi}^2=H\dot{F}-2F\dot{H}-\ddot{F}. \tag{7}$$

There are a lot of cosmological models were considered on the basis of the scalar-tensor gravity theories (see, for example, in [21, 22]). However, we are interested in the inflationary models, corresponding to observational constraints on the parameters of cosmological perturbations without restrictions on the type of the potential $V(\phi)$, the parameters of a scalar-tensor gravity theory $\{F(\phi),\omega(\phi)\}$ and the type of inflationary dynamics $H(t)$.

For this aim we consider the models with quadratic connection between the Hubble parameter and coupling function [18, 19, 20]

$$H=\lambda\sqrt{F}. \tag{8}$$

Under condition (8) dynamic equations (6)–(7) are reduced to the following ones

$$\tilde{F}(\phi) = \lambda^2 F(\phi) = H^2, \tag{9}$$
$$\tilde{V}(\phi) = \lambda^2 V(\phi) = 3H^4 + 6H^2\dot{H} + \dot{H}^2 + H\ddot{H}, \tag{10}$$
$$\tilde{X}(\phi,\dot{\phi}) = \lambda^2 X(\phi,\dot{\phi}) = -\frac{\lambda^2}{2}\omega(\phi)\dot{\phi}^2 = \dot{H}^2 + H\ddot{H}, \tag{11}$$

where λ^2 is the constant parameter which defines the energy scale of inflation.

Based on the definitions of slow-roll parameters with corresponding slow-roll conditions

$$\epsilon = -\frac{\dot{H}}{H^2} \ll 1, \qquad \delta = -\frac{\ddot{H}}{2H\dot{H}} \ll 1, \tag{12}$$

from equations (10)–(11), one can obtain the following expressions

$$\frac{\tilde{V}}{\tilde{F}^2} = 3 - 6\epsilon + 2\epsilon\delta + \epsilon^2, \tag{13}$$
$$\frac{\tilde{X}}{\tilde{F}^2} = 2\epsilon\delta + \epsilon^2, \tag{14}$$

that connect the type of the scalar-tensor gravity, characteristics of a scalar field and the type of cosmological dynamics.

As one can see, the kinetic energy of a scalar field X is represented in the terms of a second order regarding its potential V, therefore the inflationary slow-roll conditions implying $X \ll V$ are satisfied for these models when $\epsilon \ll 1$ and $\delta \ll 1$.

3. The parameters of cosmological perturbations

The parameters of cosmological perturbations in inflationary models based on action (1) with quadratic connection $H = \lambda\sqrt{F}$ were considered in [18, 19, 20].

The expressions of the parameters of cosmological perturbations on the crossing of Hubble radius ($k = aH$) were obtained in these papers in the following form

$$\mathcal{P}_S = A_S = \frac{\lambda^2}{16\pi^2\epsilon(\epsilon-\delta)}, \tag{15}$$
$$\mathcal{P}_T = A_T = \frac{2\lambda^2}{\pi^2}, \tag{16}$$
$$r = \frac{\mathcal{P}_T}{\mathcal{P}_S} = 32\epsilon(\epsilon-\delta), \tag{17}$$
$$n_S = 1 - 4\epsilon + 2\delta + 2\epsilon\delta + (1-\epsilon)\left(\frac{\epsilon\delta - \xi}{\epsilon - \delta}\right), \tag{18}$$
$$n_T = 0, \tag{19}$$

where k is a wave number, $\{A_S, A_T\}$ is the values of the power spectra of scalar and tensor perturbations $\{\mathcal{P}_S, \mathcal{P}_T\}$ on the crossing of Hubble radius, $\{n_S, n_T\}$ are the spectral indices of scalar and tensor perturbations and

$$\xi = \epsilon\delta - \frac{1}{H}\dot{\delta} \ll 1, \tag{20}$$

is the third slow-roll parameter.

In general case, the constraints on the values of the parameters of cosmological perturbations following from observation of anisotropy of CMB by PLANCK [10]

$$A_S = 2.1 \times 10^{-9}, \tag{21}$$
$$n_S = 0.9663 \pm 0.0041, \tag{22}$$
$$r < 0.065, \tag{23}$$

restrict the possible inflationary model's parameters.

From expressions (15), (17) and (21) one has

$$r = \frac{2\lambda^2}{\pi^2 A_S} = 9.5 \times 10^8 \times \frac{\lambda^2}{\pi^2}, \tag{24}$$

therefore, in these inflationary models tensor-to-scalar ratio r depends on the value of the parameter λ^2 only.

Thus, from constraint (23) one has following restriction

$$\lambda^2 < 6.7 \times 10^{-10}, \tag{25}$$

on the energy scale of such a inflationary models.

Taking into account the quasi-de Sitter character of the expansion of the early universe $H \simeq \lambda$, the energy scale of inflation can be estimated from (10) in terms of the Planck mass $m_p^{-2} = 8\pi G = \left(2.4 \times 10^{18}\, GeV\right)^{-2}$ as follows

$$H < 2.6 \times 10^{-5} m_p, \qquad V \simeq 3\lambda^2 m_p^4, \qquad V^{1/4} < 1.6 \times 10^{16}\, GeV, \tag{26}$$

corresponding to estimates given for standard inflationary models based on GR in order of magnitude [2, 23, 24].

Also, under slow-roll conditions $\epsilon \ll 1$ and $\delta \ll 1$ from (18) one has

$$n_S - 1 \simeq -4\epsilon + 2\delta + \left(\frac{\epsilon\delta - \xi}{\epsilon - \delta}\right). \tag{27}$$

Since, the deviations from Harrison-Zel'dovich spectrum (22) are

$$|n_S - 1| \simeq 3 \times 10^{-2}, \tag{28}$$

for the following values of the slow-roll parameters on the crossing of the Hubble radius

$$0 < \epsilon \lesssim 10^{-2}, \qquad -2 \times 10^{-1} \lesssim \delta \lesssim 10^{-2}, \qquad -9 \times 10^{-2} \lesssim \xi \lesssim 10^{-4}, \tag{29}$$

inflationary models based on STG with connection $H = \lambda\sqrt{F}$ and arbitrary type of accelerated expansion of the early universe can satisfy the observational constraint (22) when conditions (29) are fulfilled.

4. The spectrum of relic gravitational waves

As an additional tool for verifying cosmological models under consideration, we will consider the possibility of direct detection of relic gravitational waves through existing and promising methods [25, 26, 27, 28].

It should also be noted that relic gravitational waves have not been observed at the moment, and solving the problem of their direct detection is one of the most important problems of modern observational cosmology and gravitational-wave researching [25, 26, 27, 28].

The energy density of relic gravitational waves at the present time can be obtained from expression [25]

$$\Omega_{GW}(f) = \frac{1}{\rho_c}\frac{d\rho_{GW}}{d\log f} \simeq \frac{\mathcal{P}_T(f)}{24z_{eq}}, \tag{30}$$

where ρ_c is the critical density, f is the frequency of relic gravitational waves and $z_{eq} \simeq 3400$ is the redshift at the matter-radiation equivalence.

The power-spectrum of tensor perturbations can be defined as [26, 27]

$$\mathcal{P}_T(f) \simeq A_T\left(\frac{f}{f_*}\right)^{\alpha(f)+n_T(f)}, \qquad \alpha(f) = 2\left(\frac{3w(f)-1}{3w(f)+1}\right), \tag{31}$$

where f_* is the characteristic frequency corresponding to the pivot scale k_* and $w = w(f)$ is the value of the state parameter when transition from the end of inflation to the BBN epoch occurs. The dependence $n_T = n_T(w)$ in explicit form was considered in [26].

For the models under consideration the spectral index of tensor perturbations $n_T = 0$ and for the case of standard transition between the end of inflation and the BBN epoch implying $w = 1/3$ from (16) and (31) we obtain following energy density of relic gravitational waves

$$\Omega_{GW} \simeq \frac{\lambda^2}{12\pi^2 z_{eq}} < 1.7\times 10^{-15}, \tag{32}$$

which is defined by the energy scale of inflation λ^2 only, and it doesn't depend on the frequency, i.e. in this case one has exactly scale-invariant spectrum of relic gravitational waves.

Comparing this estimate of the energy density of the relic gravitational waves with the sensitivity of LIGO ($\Omega_{GW} \simeq 10^{-9}$) and LISA ($\Omega_{GW} \simeq 10^{-13}$) (for 10^2 Hz and 10^{-3} Hz respectively) [28], one can conclude that the relic gravitational waves predicted in such a models cannot be directly detected by these experiments.

Nevertheless, taking into account the additional stage of predominance of kinetic energy (kination) on the transition between the end of inflation and the BBN epoch for $n_T = 0$ one has $w \simeq 1/2$ [26], and from (30)–(32) we obtain the following expression

$$\Omega_{GW}(f) \simeq \frac{\lambda^2}{12\pi^2 z_{eq}}\left(\frac{f}{f_*}\right)^{2/5}, \tag{33}$$

where $f_* \simeq 10^{-11}$ Hz since the gravitational waves with frequencies below this value reentered the Hubble horizon after BBN [26].

Table 1. The energy density of relic gravitational waves Ω_{GW} at the modern era for different frequencies f.

f, Hz	10^{-3}	10^2	5×10^6	10^8	10^{10}
Ω_{GW}	2.7×10^{-12}	2.7×10^{-10}	2.1×10^{-8}	6.8×10^{-8}	4.3×10^{-7}

The values of the energy density of relic gravitational waves at the modern era for different frequencies are presented in Table 1.

As one can see, the relic gravitational waves from inflation based on scalar-tensor gravity with connection $H = \lambda\sqrt{F}$ can be in principle detected by LISA experiment. Since the energy density in this case is maximal in for high-frequency gravitational waves, the methods of their detection in the high-frequency range [28, 29, 30, 31, 32] can also be considered as relevant ones for verification of the proposed cosmological models on the basis of direct detection of relic gravitational waves.

5. Conclusion

In this paper, we considered the models of cosmological inflation based on generalized scalar-tensor theories of gravity, which satisfy observational constraints on the values of the parameters of cosmological perturbations for arbitrary model's parameters and restricted by the energy scale of inflation only. The proposed approach allows one to give a model-independent estimate of the energy density of relic gravitational waves in the present era of the universe's evolution. It was shown that the verification of the proposed models can be carried out on the basis of the possibility of direct detection of relic gravitational waves in both the low-frequency and high-frequency ranges of their spectrum.

Acknowledgments

I.V.F. and S.V.C. were supported by the Russian Foundation for Basic Research grant No. 20-02-00280 A. I.V.F. was partly supported by the Russian Foundation for Basic Research grant No. 19-29-11015 mk. S.V.C. is grateful for the support of the Program of Competitive Growth of Kazan Federal University.

References

[1] Linde A 1990 *Particle physics and inflationary cosmology* (Boca Raton: CRC Press)
[2] Baumann D and McAllister L 2014 *Inflation and String Theory* (Cambridge: Cambridge University Press)
[3] Chervon S, Fomin I, Yurov V and Yurov A 2019 *Scalar field cosmology* (Singapore: World Scientific)
[4] Martin J, Ringeval C, Vennin V 2014 *Phys. Dark Univ.* **5-6** 75
[5] Nojiri S and Odintsov S 2008 *Phys. Lett.* B **659** 821
[6] Clifton T, Ferreira P, Padilla A and Skordis C 2012 *Phys. Rept.* **513** 1
[7] Nojiri S, Odintsov S and Oikonomou V 2017 *Phys. Rept.* **692** 1
[8] Ishak M 2019 *Living Rev. Rel.* **1** 1
[9] Fomin I and Chervon S 2019 *Phys. Rev.* D **100** 023511
[10] Ade P et al [PLANCK Collaboration] 2016 *Astron. Astrophys.* **594** A13
[11] Melchiorri A, Sazhin M, Shulga V and Vittorio N 1999 *Astrophys. J.* **518** 562
[12] Kopeikin S, Ramirez J, Mashhoon B and Sazhin M 2001 *Phys. Lett. A* **292**, 173
[13] Hwang J and Noh H 2005 *Phys. Rev.* D **71** 063536
[14] Chervon S and Fomin I 2008 *Grav. Cosmol.* **14** 163
[15] Fomin I and Chervon S 2020 *J. Phys. Conf. Ser.* **1557** 012020
[16] Fomin I 2020 *Eur. Phys. J.* C **80** 1145
[17] Baumann D et al [CMBPol Study Team] 2009 *AIP Conf. Proc.* **1141** 10
[18] Fomin I and Chervon S 2018 *Eur. Phys. J.* C **78** 918
[19] Fomin I, Chervon S and Tsyganov A 2020 *Eur. Phys. J.* C **80** 350
[20] Fomin I and Chervon S 2020 *Universe* **6** 199
[21] Fujii Y and Maeda K 2007 *The Scalar-Tensor Theory of Gravitation* (Cambridge: Cambridge University Press)
[22] Faraoni V 2004 *Cosmology in Scalar-Tensor Gravity* (New York: Springer)
[23] Liddle A 1994 *Phys. Rev.* D **49** 739
[24] Guo Z, Schwarz D and Zhang Y 2011 *Phys. Rev.* D **83** 083522
[25] Cabass G, Pagano L, Salvati L, Gerbino M, Giusarma E and Melchiorri A 2016 *Phys. Rev.* D **93** 063508
[26] Boyle L and Buonanno A 2008 *Phys. Rev.* D **78**, 043531
[27] Giovannini M 2020 *Prog. Part. Nucl. Phys.* **112** 103774
[28] Aggarwal N, Aguiar O, Bauswein A, Cella G, Clesse S, Cruise A, Domcke V, Figueroa D, Geraci A and Goryachev M et al [arXiv:2011.12414 [gr-qc]].
[29] Gladyshev V and Morozov A 1993 *Tech. Phys. Lett.* **19** 7 449
[30] Morozov A and Gladyshev V 2002 *Russ. Phys. J.* **45** 113
[31] Esakov A et al. 2015 (in Russian) *Herald of the Bauman Moscow State Tech. Univ. Nat. Sci.* **1** 26
[32] Fomin I and Morozov A 2017 *J. Phys. Conf. Ser.* **798** 012088

Competing processes in generation of the third optical harmonic in air under femtosecond infrared repetitively pulsed excitation

V O Gladyshev[1], E A Sharandin[1], A V Skrabatun[1,2] and P P Nikolaev[1]

[1] Bauman Moscow State Technical University, 2-nd Baumanskaya str. 5, Moscow, 105005, Russia

[2] P.N. Lebedev Physical Institute of Russian Academy of Sciences, Leninskii Pr. 53, Moscow, 119991, Russia

E-mail: gladyshev@bmstu.ru; shar@bmstu.ru; skrabatunav@bmstu.ru; ppn@bmstu.ru

Abstract. Parametric interaction of electromagnetic and gravitational waves with the radiation generation at the third harmonic wavelength is one of the ways to detect gravitational interaction in a material medium. To implement the effect in question, superstrong fields must be used, but in this case competing nonlinear processes arise, leading to the generation of the third harmonic as a result of laser radiation filamentation. This paper investigates the characteristics of the radiation recorded for femtosecond (250 fs) laser pulses with a wavelength of λ = 1032 nm focused in air. The threshold pump power made it possible to observe the formation of a filament with concomitant generation of narrow-band radiation at the focus of the lens at the third harmonic wavelength λ = 344 nm. The research presents spectral and spatial dependences of ultraviolet radiation (λ = 344 nm) at pumping power of infrared radiation (λ = 1032 nm) of 500 mW. Energy dependences of the third harmonic generation efficiency in the power range from 150 to 1750 mW are obtained.

1. Introduction

In was theoretically substantiated that in superstrong electromagnetic fields there occurs a parametric interaction of gravitational and electromagnetic waves [1]. As a result of collinear interactions of these waves, a peak is expected at a combined frequency, which may correspond to the third harmonic (TG) of the laser source. To observe TG radiation, superstrong fields are normally used, with the intensity exceeding 10 TW/cm^2, which causes a large number of nonlinear physical processes. When such radiation is focused, laser filamentation, self-phase modulation, stimulated Raman scattering, generation of higher odd harmonics and, in particular, generation of TG are observed. The most common medium for TG excitation is air [2-5]. A specific feature of TG generation from near-infrared radiation is that the exciting laser radiation can be converted into the ultraviolet wavelength range. This process of TG generation competes with the process of parametric interaction between electromagnetic and gravitational waves, which is also accompanied by TG generation. Further investigation of the parametric interaction process involving of a gravitational wave requires detailed study and identification of all competing processes, primarily, the process of TG generation as a result of filamentation of laser radiation.

2. Experimental technique, results and discussion

To excite nonlinear optical processes in air, an ytterbium femtosecond laser with a central radiation wavelength of 1032 nm was used. The laser generated pulses with a duration of ~ 250 fs with a pulse-repetition rate of 1 kHz. In this case, the average output power ranged from 150 to 1750 mW

Figure 1 shows a schematic of an experimental setup for performing spectral measurements of TG radiation depending on pump intensity. The radiation of a femtosecond laser (1) was focused in air by a lens (2) with a focal length of 85 mm. The maximum peak intensity of exciting laser radiation at the lens focus (2) reached up to 10 TW/cm^2. In the beam waist, TG radiation with 344 nm wavelength was generated along with other nonlinear processes. Radiation passing through a collimating lens (3) was directed at a quartz prism (4), which separated the pump radiation and the TG radiation. When measuring the TG radiation profile (or measuring the energy), a system of light filters was installed behind the lens (3), which attenuated the pumping radiation. The laser beam profilometer with a spatial resolution of ~ 4.5 μm was used to record the transverse distribution of the radiation beam. For far-field beam profiling, an additional focusing lens was installed in front of the profilometer. After that the radiation was directed at a spectrometer (5) connected to a computer (6). The quartz prism and lenses were made of KU quartz glass. The spectrometer provided spectral registration in a wide spectral bandwidth (190–1100 nm) with a resolution of ~ 0.3 nm.

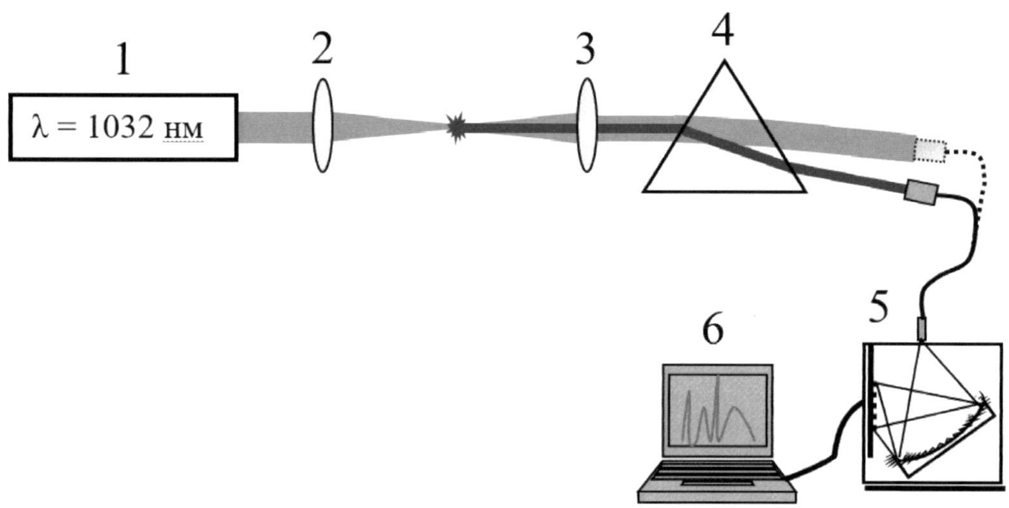

Figure 1. Schematic of an experimental setup for spectral measurements of TG radiation: 1 - radiation source; 2 - focusing lens; 3 - collimating lens; 4 - quartz prism; 5 - spectrometer; 6 – computer.

The radiation spectral characteristics were measured at an average pump power of ~ 500 mW. Figure 2 shows the radiation spectrum of a femtosecond laser after conversion into TG radiation. The plot demonstrates a sharp decrease in the radiation intensity in the 1032 nm wavelength region, which corresponds to the spectral maximum of the femtosecond laser radiation. In this case, the observed intensity dip results from the conversion of exciting radiation into TG radiation with a wavelength of 344 nm (Figure 3). The TG bandwidth is ~ 2 nm.

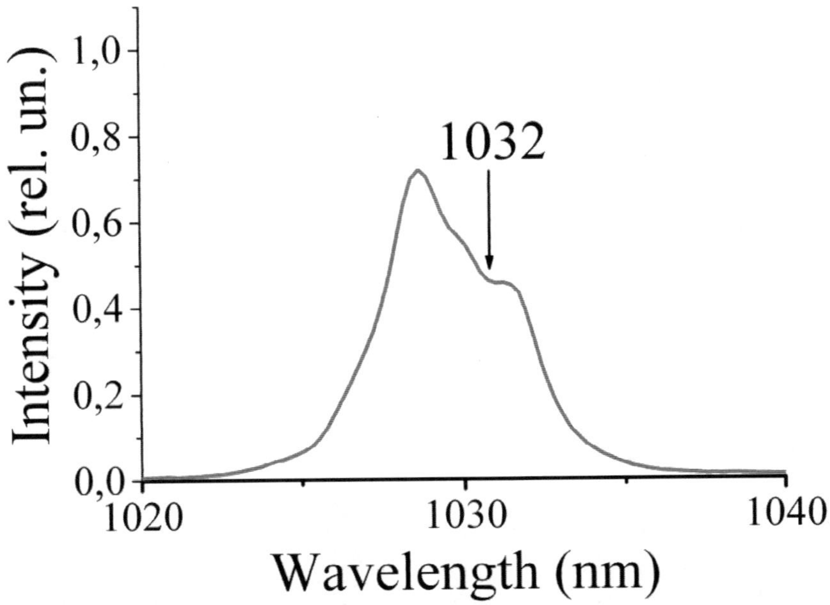

Figure 2. Femtosecond laser radiation spectrum after conversion into TG radiation.

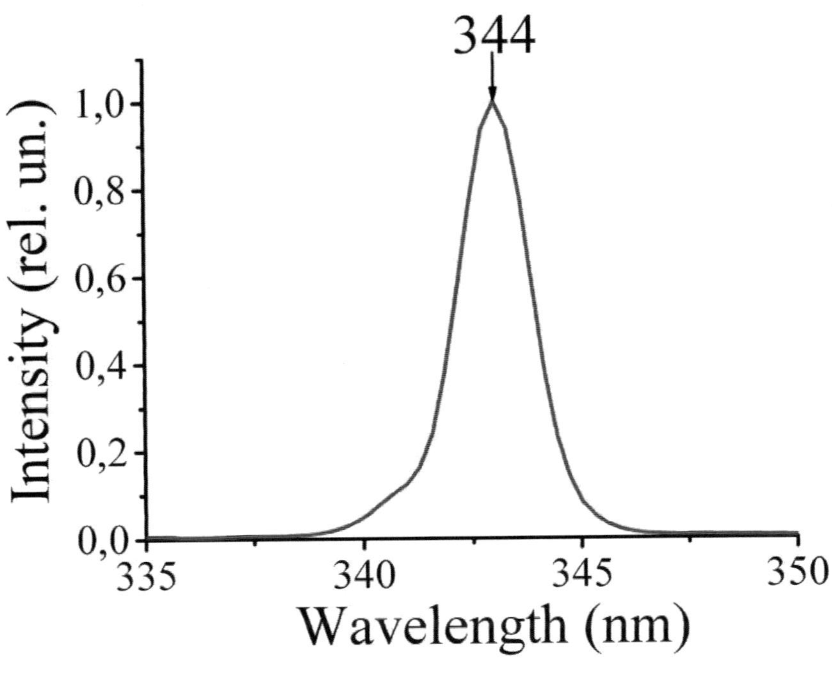

Figure 3. TG radiation spectrum.

The spatial characteristics of the TG radiation were measured at average pump power of ~ 500 mW. Fig. 4 shows the transverse near-field profile of the TG radiation beam. The near-field beam size is 4.5 × 3.6 mm. The beam had an elliptical symmetry caused by a slight dealignment of the focusing and collimating lenses. The far-field beam profile has a symmetric Gaussian profile with a radiation divergence of 0.11 mrad, which corresponds to the diffraction quality of the beam.

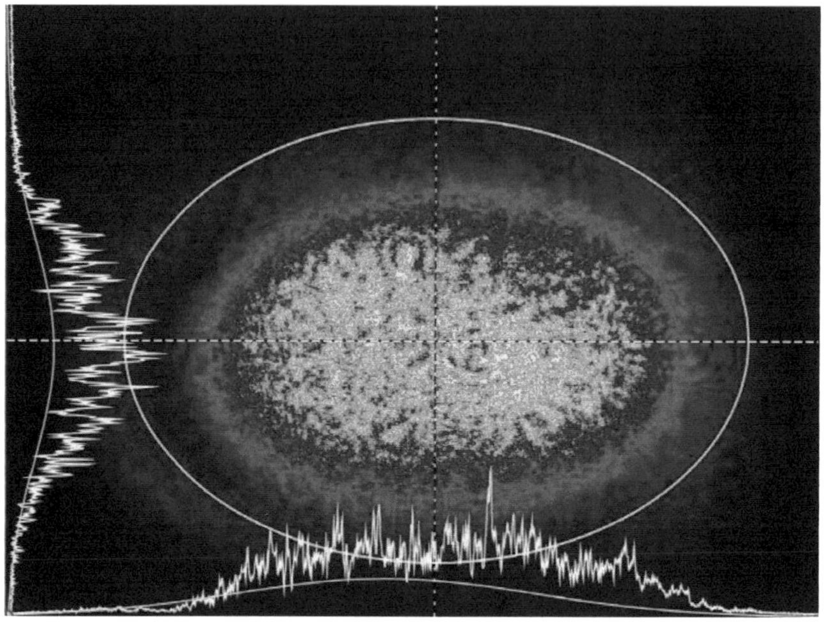

Figure 4. Transverse near-field profile of a TG beam.

Figure 5 shows the measured average TG power and the conversion efficiency of the exciting radiation in the TG as a function of average pump power. In the investigated average pump radiation power range (from 150 to 1750 mW), the obtained dependence has a linear form. The average pump power was limited by the emerging stimulated Raman scattering in lenses and prisms and amounted to ~ 1750 mW. The maximum average TG power is 1.52 mW (Figure 5, left axis) with conversion efficiency in the TG of ~ 0.085% (Figure 5, right axis).

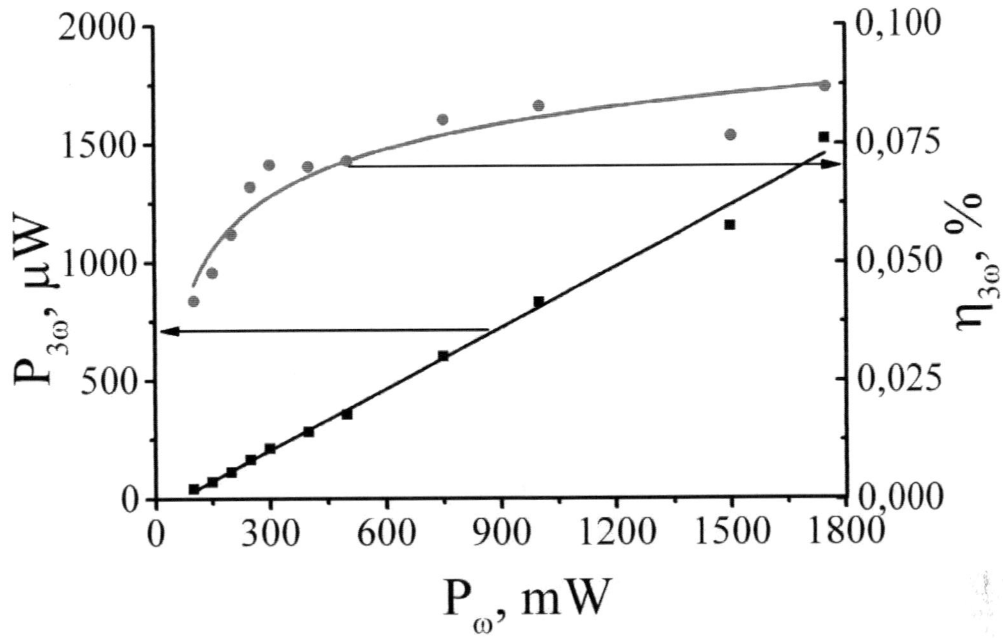

Figure 5. Average TG power ($P_{3\omega}$) and the TG conversion efficiency ($\eta_{3\omega}$) as a function of average pump power.

3. Conclusion

Thus, we present the investigation results for TG generation by focusing femtosecond laser radiation with a wavelength of 1032 nm and duration of ~ 250 fs in air. This paper researches one of the main competing nonlinear processes in a medium. Taking the full characteristics of this into account will allow in the future to observe TG as a result of the interaction between gravitational and electromagnetic waves. To counteract the processes of TG generation due to filamentation of laser radiation, the following methods can be proposed: increasing the transverse dimensions of the beam to reduce the radiation intensity and ensuring a high degree of vacuum.

Acknowledgement

This research was supported by the Russian Science Foundation (grant no. 19-12-00242).

References

[1] Pustovoit V I and Chernozatonskii L A 1981 *JETP Lett.* **34** 229

[2] Backus S, Peatross J, Zeek Z, Rundquist A, Taft G, Murnane M M and Kapteyn H C 1996 *Opt. Lett.* **21** 665

[3] Fedotov A B, Koroteev N I, Loy M M T, Xiao X and Zheltikov A M 1997 *Opt. Commun.* **133** 587

[4] Peatross J, Backus S, Zhou J, Murnane M M and Kapteyn H C 1998 *JOSA B* **15** 186

[5] Aközbek N, Iwasaki A, Becker A, Scalora M, Chin S L and Bowden C M 2002 *Phys. Rev. Lett.* **89** 143901

Some properties of energies in nonsynchronous reference frames in cosmology

Andrey A Grib[1,2] **and Yuri V Pavlov**[3]

[1] Theoretical Physics and Astronomy Department, The Herzen University, 48 Moika, St. Petersburg, 191186, Russia
[2] A. Friedmann Laboratory for Theoretical Physics, St. Petersburg, Russia
[3] Institute of Problems in Mechanical Engineering of Russian Academy of Sciences, 61 Bolshoy, V.O., St. Petersburg, 199178, Russia

E-mail: `andrei_grib@mail.ru`, `yuri.pavlov@mail.ru`

Abstract. It is shown that use of non synchronous reference frame in cosmology can lead to unusual behavior of the energy of particles: it can have not only positive but also negative and zero energies. A special example of the model with the scale factor $a(t) \sim t$ is considered.

1. Introduction

Physics of rotating black holes leads to use of the Boyer-Lindquist [1] coordinates in which the metrical tensor has nondiagonal terms. The consequence of this form of the tensor is existence of negative and zero energies of particles in the region close to the horizon of the black hole called ergosphere [2]. This leads to a well known Penrose effect [3, 4] in decays and collisions of particles. But nondiagonal term in metrical tensor occurs also if one considers the expanding Universe in nonsynchronous reference frames. Differently from the rotating Kerr's black holes this nondiagonal term corresponds to a product not of time and the angle, but of time and the radial coordinate. In our papers [5, 6] it was shown that similarly to the black holes one can have particles with negative and zero energies in this case also. In this paper we continue to study this property of the energy of particles. After studying the general case of of the nonstationary metric of the cosmological model we specially investigate the case with the scale factor $a(t) \sim t$. One of the well known examples of such metric is the Milne's metric [7], i.e. Minkowski space-time in Milne's coordinates.

2. Negative and zero energies in nonsynchronous reference frame in cosmology

The metric of homogeneous isotropic cosmological model can be written as

$$ds^2 = c^2 dt^2 - a^2(t) \left(d\chi^2 + f^2(\chi) d\Omega^2 \right), \tag{1}$$

where c is velocity of light, $d\Omega^2 = d\theta^2 + \sin^2\theta\, d\varphi^2$, the coordinate χ is changing from 0 to π in the closed model with $f(\chi) = \sin\chi$ and from 0 to $+\infty$ for quasi-Euclidean model with $f(\chi) = \chi$, and $f(\chi) = \sinh\chi$ for the open model. The radial distance between points $\chi = 0$ and χ in metric (1) is $D = a(t)\chi$. If t is fixed then in closed model the maximal value of D is $D_{\max} = \pi a(t)$. In open and flat models D is non limited.

Take the new coordinates t, D, θ, φ (see also [8, 9]). Then

$$dD = HD\,dt + a\,d\chi, \qquad d\chi = \frac{1}{a}\,(dD - H\,Ddt) \tag{2}$$

where $H = \dot{a}(t)/a(t)$ is a Hubble parameter, dot above the symbol means the derivative with respect to time t. In new coordinates the interval (1) becomes

$$ds^2 = \left(1 - \left(\frac{HD}{c}\right)^2\right) c^2 dt^2 + 2HD\,dtdD - dD^2 - a^2 f^2\left(\frac{D}{a}\right)\,d\Omega^2. \tag{3}$$

Let $D_s = c/|H(t)|$. The value of D_s defines the boundary of the region out of which the physical bodies cannot be at rest similar to the value of the static limit in the case of rotating black hole. Really, from the condition $ds^2 \geq 0$ one comes to conclusion that no physical object can be at rest in this coordinate system for $D > D_s$. If the universe is expanding then $\dot{a} > 0$ and for $D > D_s$ for the observer using coordinates t, D, θ, φ all bodies move in the direction of growing D (for them $dD > 0$), in case of contraction of the universe one has $dD < 0$.

Note that the value of the static limit D_s in general case is changing in time. The exclusion is the case of constant Hubble parameter H, i.e. scale factor $a \sim e^{Ht}$. In this case the static limit coincides with the event horizon defined by the formula

$$D_H = a(t)c \int\limits_t^\infty \frac{dt}{a(t)}. \tag{4}$$

For $f(\chi) = \chi$ the metric (1) with such scale factor corresponds to de Sitter space-time.

Similar to the ergosphere of rotating black hole in the region out the static limit one can have particles with non-positive energies. Let us show this.

As it was shown in our paper [6] the energy of the particle with mass m for the observer in the reference frame with metric g_{ik} is

$$E = mc^2 g_{0k} \frac{dx^k}{ds}. \tag{5}$$

Note that the energy of the particle in metrics (1), (3) is not conserved because metrics depends on time.

In case with the interval (3) the energy is

$$E = E' \left(1 - \frac{D^2}{D_s^2} + \frac{HD}{c^2}\frac{dD}{dt}\right), \tag{6}$$

where $E' = mc^2 dt/d\tau$ is the energy in comoving coordinate system, τ is the proper time of moving particle. For the analysis of possible values of particle energies one can express its value through the radial velocity in comoving coordinates

$$v_\chi = a\frac{d\chi}{dt}. \tag{7}$$

Condition $ds^2 \geq 0$ leads to

$$\left|a\frac{d\chi}{dt}\right| \leq c, \tag{8}$$

i.e. $-c \leq v_\chi \leq c$. Using notification

$$\operatorname{sgn} x = \begin{cases} 1, & x > 0, \\ 0, & x = 0, \\ -1, & x < 0, \end{cases}$$

write (6) in the form

$$E = E' \left(1 + \frac{HD}{c^2}\frac{a\, d\chi}{dt}\right) = E' \left(1 + \operatorname{sgn}(\dot a)\, \frac{D}{D_s}\frac{v_\chi}{c}\right). \tag{9}$$

So negative energies for particles in the reference frame t, D, θ, φ occur if

$$E < 0 \quad \Rightarrow \quad D > D_s, \quad c \geq |v_\chi| > c\frac{D_s}{D}, \quad \operatorname{sgn}(v_\chi) = -\operatorname{sgn}(\dot a), \tag{10}$$

and one has movement in direction of decreasing of the coordinate χ in case of expanding Universe and in direction of increasing radial coordinate χ in case of contracting cosmological model. To have the negative energy close to the static limit the particle must be close to the the velocity of light in comoving coordinates. Far from the static limit $D \gg D_s$ one can have negative energies for small velocities.

From (9) one can see that zero energies are possible for $D \geq D_s$:

$$E = 0 \quad \Rightarrow \quad D \geq D_s, \quad \frac{d\chi}{dt} = -\frac{c^2}{D\dot a}. \tag{11}$$

3. Negative and zero energies in nonsynchronous reference frame in the model with $a(t) = \sigma ct$

Consider the case of expanding cosmological model with a scale factor $a(t) = \sigma ct$. This model is interesting because if $f(\chi) = \sinh \chi$ and $\sigma = 1$ it describes flat space-time in curvilinear coordinates (Milne Universe).

Using conformal time η, so that $cdt = ad\eta$ one obtains instead of metric (1) the form

$$ds^2 = a^2(\eta) \left(d\eta^2 - d\chi^2 - f^2(\chi) d\Omega^2\right). \tag{12}$$

In this model

$$a = \sigma ct = \sigma ct_0 e^{\sigma(\eta - \eta_0)}, \qquad \eta = \eta_0 + \frac{1}{\sigma}\log\frac{t}{t_0}, \qquad t = t_0 e^{\sigma(\eta-\eta_0)}, \tag{13}$$

where η_0 is arbitrary constant fixing point of the origin of conformal time $t(\eta_0) = t_0$.

Einstein's equations

$$R_{ik} - \frac{1}{2} R g_{ik} + \Lambda g_{ik} = -8\pi \frac{G}{c^4} T_{ik}, \tag{14}$$

where Λ is cosmological constant, G is the gravitational constant, T_{ik} is energy-momentum tensor of background matter, in metric (12) take form

$$T_0^0 = \frac{3c^2}{8\pi G}\left(\frac{\dot a^2 + Kc^2}{a^2} - \frac{c^2}{3}\Lambda\right), \tag{15}$$

$$T_\alpha^\beta = \delta_\alpha^\beta \frac{c^2}{4\pi G}\left(\frac{\ddot a}{a} + \frac{\dot a^2 + Kc^2}{2a^2} - \frac{c^2}{2}\Lambda\right), \tag{16}$$

where $K = 1, 0, -1$ for closed, quasi-Euclidean and open model respectively. So for the scale factor $a(t) = \sigma ct$ the components of the energy-momentum tensor of background matter are

$$T_i^k = \mathrm{diag}\,(\varepsilon, -p, -p, -p), \tag{17}$$

and energy ε and pressure p satisfy the relations

$$p - \frac{c^4}{8\pi G}\Lambda = -\frac{1}{3}\left(\varepsilon + \frac{c^4}{8\pi G}\Lambda\right) = \frac{-c^2}{8\pi G}\frac{(\sigma^2 + K)}{\sigma^2 t^2}. \tag{18}$$

Note that ε and p are the energy density and pressure of background matter in coordinate frame t, D (see [10]). The scalar curvature for a scale factor $a(t) = \sigma ct$ is

$$R = \frac{6}{a^2}\left[\frac{a\ddot{a}}{c^2} + \left(\frac{\dot{a}^2}{c^2} + K\right)\right] = \frac{6}{c^2 t^2}\left(1 + \frac{K}{\sigma^2}\right), \tag{19}$$

and it is either zero for $K = -1$, $\sigma = 1$ or decreases in time. Existence of cosmological constant in the model does not lead to accelerated expansion ($\ddot{a} = 0$) and is compensated by a background matter with equation of state (18).

The static limit in coordinates t, D in this model is equal to $D_s = ct$. The region out of the static limit at the fixed moment t is defined by the inequality $\chi > 1/\sigma$. Particle energy in the reference frame t, D, θ, φ is

$$E = E'\left(1 + \sigma^2 t\chi\frac{d\chi}{dt}\right) = E'\left(1 + \sigma\chi\frac{d\chi}{d\eta}\right). \tag{20}$$

Particle with conserved zero energy during some time must move (noninertially and not on the geodesic line!) according to the law

$$E = 0 \quad \Rightarrow \quad \chi = \sqrt{\chi_0^2 - \frac{2}{\sigma^2}\log\frac{t}{t_0}} = \sqrt{\chi_0^2 - \frac{2}{\sigma}(\eta - \eta_0)}, \qquad \chi_0 \geq \frac{1}{\sigma}. \tag{21}$$

Possible interval of such movement is

$$t \in \left[t_0,\ t_0 \exp\frac{\sigma^2\chi_0^2 - 1}{2}\right], \qquad \eta \in \left[\eta_0,\ \eta_0 + \frac{\sigma}{2}\left(\chi_0^2 - \frac{1}{\sigma^2}\right)\right]. \tag{22}$$

If $\sigma = 1$ than in new coordinates

$$T = t\cosh\chi, \qquad r = ct\sinh\chi, \qquad cT > r > 0 \tag{23}$$

the interval (1) with our scale factor $a(t) = ct$ and $f(\chi) = \sinh\chi$ has the form of the interval in Minkowski space-time [11]

$$ds^2 = c^2 dT^2 - dr^2 - r^2\, d\Omega^2. \tag{24}$$

The arising space-time (Milne universe) with coordinates $t \geq 0$, $\chi \geq 0$ is projected on the future cone in coordinates cT, r.

Let us look what is the region in new coordinates which corresponds to the region out of the static limit in coordinates t, D and what are features of particle movement in flat coordinates if in t, D coordinates frame they have zero and negative energies. Note that in new coordinates $cT/r = \coth\chi$. So the region out of the static limit at the fixed moment t, i.e. the region $\chi \geq 1$ is limited in coordinates cT, r by rays $cT = \coth 1$ and $cT = 1$. This region is shown on Fig. 1

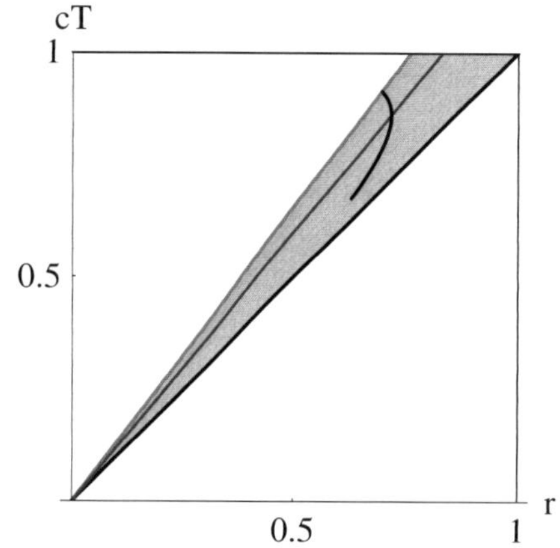

Figure 1. Region out of the static limit for $a = ct$ in coordinates T, r.

by pink color. The curve in this region shows the trajectory of the particle with zero energy (21) for $\chi_0 = 1.65$, $t_0 = 0.25$.

For velocities one obtains

$$\frac{dr}{dT} = \frac{\frac{\partial r}{\partial t} + \frac{\partial r}{\partial \chi}\frac{d\chi}{dt}}{\frac{\partial T}{\partial t} + \frac{\partial T}{\partial \chi}\frac{d\chi}{dt}} = c\frac{\tanh\chi + t\frac{d\chi}{dt}}{1 + t\tanh\chi\frac{d\chi}{dt}}, \quad t\frac{d\chi}{dt} = \frac{\frac{dr}{dT} - c\tanh\chi}{c - \frac{dr}{dT}\tanh\chi}. \tag{25}$$

Particle energy will be negative in the frame t, D, θ, φ if $\chi > 1$ and $t\chi\frac{d\chi}{dt} < -1$ (see (20)). From (25) one can see that this is possible as for positive, as for negative and zero value of the velocity dr/dT in coordinates r, T of the Minkowski space-time. Especially if $r =$ const and $\chi \geq \chi_1$, where $\chi_1 \approx 1.1997$ is the positive root of equation

$$\chi \tanh\chi = 1,$$

then the energy (20) is equal to zero for a particle at rest in coordinates r, T. The region $\chi \geq \chi_1$ where particles at rest in coordinates r, T have negative energy is located on Fig. 1 below the blue line in a pink sector.

4. Conclusion

Our analysis shows that in nonsynchronous reference frame in cosmology due to existence of nondiagonal term in metrical tensor one can find region out of the static limit where particles with negative and zero energies exist like it occurs in the ergosphere of the rotating black hole. However differently from the case of Kerr metric here due to nonstationary form of the metric in cosmology particle energies are not conserved. In special case of the scale factor $a = \sigma ct$ we analysed the trajectories of particles which have for some interval negative and zero energies.

References

[1] Boyer R H and Lindquist R W 1967 Maximal analytic extension of the Kerr metric *J. Math. Phys.* **8** 265–281
[2] Misner C W, Thorne K S and Wheeler J A 1973 *Gravitation* (San Francisco: Freeman)
[3] Penrose R 1969 Gravitational collapse: the role of General relativity *Rivista Nuovo Cimento, Num. Spec.* **I** 252–276
[4] Penrose R and Floyd R M 1971 Extraction of rotational energy from a black hole *Nature Phys. Sci.* **229** 177–179
[5] Grib A A and Pavlov Yu V 2020 Particle properties outside of the static limit in cosmology *Int. J. Mod. Phys. A* **35** 2040044

[6] Grib A A and Pavlov Yu V 2020 Particles with negative energies in nonrelativistic and relativistic cases *Symmetry* **12** 528
[7] Milne E A 1935 *Relativity, Gravitation and World-Structure* (Oxford: Clarendon Press)
[8] Ellis G F R and Rothman T 1993 Lost horizons *Am. J. Phys.* **61** 883–893
[9] Grib A A 1995 *Early Expanding Universe and Elementary Particles* (St. Petersburg: Friedmann Lab. Publ.)
[10] Grib A A and Pavlov Yu V 2021 Some effects of different coordinate systems in cosmology *Eur. Phys. J. Plus* **136** 318
[11] Landau L D and Lifshitz E M 1983 *The Classical Theory of Fields* (Oxford: Pergamon Press)

Newman-Janis Ansatz for rotating wormholes

A C Gutiérrez-Piñeres[1], **N H Beltrán**[2] **and C S López-Monsalvo**[3]

[1] Escuela de Física, Universidad Industrial de Santander, Santander, A. A. 678 Bucaramanga 680002, Colombia
[2]Instituto de Ciencias Nucleares, Universidad Autónoma de México, AP 70543, México, DF 04510, México
[3]Conacyt-Universidad Autónoma Metropolitana Azcapotzalco Avenida San Pablo Xalpa 180, Azcapotzalco, Reynosa Tamaulipas, 02200 Ciudad de México, México

[1]E-mail: acgutier@uis.edu.co
[2] E-mail: nicolas.hernandez7@correo.uis.edu.co
[3]E-mail: cslopezmo@conacyt.mx

Abstract. A central problem in General Relativity is obtaining a solution to describe the source's interior counterpart for Kerr black hole. Besides, determining a method to match the interior and exterior solutions through a surface free of predefined coordinates remains an open problem. In this work, we present the ansatz formulated by the Newman-Janis to generate solutions to the Einstein field equation inspired by the mention problems. We present a collection of independent classes of exact interior solutions of the Einstein equation describing rotating fluids with anisotropic pressures. Furthermore, we will elaborate on some obtained solutions by alluding to rotating wormholes.

1. Introduction

In 1963 Roy Kerr [1] discovered their solution to the Einstein field equations describing a spinning black hole. One year later, Newman and Janis [2] demonstrated another form to derive this solution by making a complex transformation to the Schwarzschild solution. Since then, there is an increasing interest in generating new solutions from a prescribed seed using the original Newman and Janis ideas.

Nowadays, the discussion about how the Newman-Janis Ansatz (NJA) mathematical nature remains open to research. As far as we known, maybe the Drake and Szekeres work [3] and the set by Azreg [4] are currently the most telling about the mathematical view and the applicability of the NJA. In 2016, one of us and H. Quevedo [5] asked themselves if conformal symmetry prevails after using this formalism. The answer was no for a conformastatic seed. This paper shall continue this investigation and explore the possibility of finding new solutions to the Einstein equations describing self-rotating fluids.

Since the paper by Ellis in 1973 [6] and Morris and Thorne in 1988 [7] set up an aesthetical and conceptual renaissance of the notion of wormholes, abundant literature has grown up proposing new examples of non-traversable wormholes. Three among the many problems concerning the original idea remains open to debate: First, how to match the interior solution of the Einstein field equations, whose metric is given by a wormhole, to an exterior solution. Second, to provide wormholes formalism with arbitrary symmetries. Moreover, finally, to construct rotating wormholes with appropriate physical behaviour. For a discussion on matching wormholes to their exterior space-time, see [8], whereas for discussion on rotating wormholes, see [9,10,11].

In section 2 of this paper, we present the ansatz formulated by Newman-Janis to generate solutions to the Einstein field equations for rotating space-times. We employ the differential geometry in the fashion that lets transparent the idea according to which the formalism abandons the transformation between metric spaces in favour of mappings between tetrads.

Section 3 presupposes that the rotating solution obtained using NJA describes a stationary interior space-time corresponding to an anisotropic fluid without heat. We implement this assumption in the Einstein field equations and achieve five independent classes of interior solutions.

Finally, in section 4, we present a short description of the mathematical aspects necessary to generate the rotating counterpart of static wormholes by employing the NJA. Next, we construct the Kerr-Newman wormholes using the Reissner-Nordström wormhole as a seed solution to illustrate the discussed ideas. We conclude this work with some remarks and open proposals for further investigations.

2. The Newman-Janis Ansatz

Let (t,r,θ,ϕ) be the coordinates corresponding to point p in a manifold $\mathcal{M}$, and $(\partial_t,\partial_r,\partial_\theta,\partial_\phi)$ the basis of tangent vectors of $\mathcal{T}_p(\mathcal{M})$. Thus, we have a general metric tensor $\boldsymbol{\mathcal{G}_0} \in \mathcal{T}_p^*(\mathcal{M})$ given in the manner by

$$\boldsymbol{\mathcal{G}_0} = G(r)\, dt\otimes dt - \frac{1}{F(r)}\, dr\otimes dr - H(r)(d\theta\otimes d\theta + sin^2\theta\, d\phi\otimes d\phi). \quad (1)$$

Below, we outline the procedure employed for the first time in 1964 by Newman and Janis [2] to derivative the Kerr metric by performing a complex coordinate transformation on the Schwarzschild solution. We sketch it in modern fashion consisting of fourth steps and employ a static space-time sufficiently general to generate a master metric describing an axially rotating space-time.

2.1 First step

We perform the mapping from spherical coordinates to outgoing Eddington-Filkenstein coordinates. In what follows, we shall use this metric tensor as a seed solution to implement the Newman Janis Ansatz (NJA). To this end, we first perform the transformation

$$\boldsymbol{\Pi}: (t,r,\theta,\phi) \longrightarrow (u,r,\theta,\phi),$$

which maps spherical coordinates to outgoing Eddington-Finkelstein coordinates through the relation

$$u = t - \int \frac{dr}{\sqrt{FG}}.$$

Consistently, the mapping $\boldsymbol{\Pi}$ induces the pullback $\boldsymbol{\Pi}^*$ which carries the metric $\boldsymbol{\mathcal{G}_S}$ into the metric $\boldsymbol{\mathcal{G}_0}$, i.e., $\boldsymbol{\Pi}^*\boldsymbol{\mathcal{G}_S} = \boldsymbol{\mathcal{G}_0}$. Then, the metric (1) can be getting through the action of the map $\boldsymbol{\Pi}^*$ on the metric $\boldsymbol{\mathcal{G}_S} \in \mathcal{T}^*_{\boldsymbol{\Pi}(p)}(\mathcal{M})$ given by

$$\boldsymbol{\mathcal{G}_S} = \boldsymbol{E^t}\otimes\boldsymbol{E^t} - \boldsymbol{E^r}\otimes\boldsymbol{E^r} - \boldsymbol{E^\theta}\otimes\boldsymbol{E^\theta} - \boldsymbol{E^\phi}\otimes\boldsymbol{E^\phi} \quad (2)$$

where the basis vector $\boldsymbol{E^i}$ are given by the orthonormal tetrad

$$\boldsymbol{E^t} \equiv \sqrt{G}\, du + \frac{dr}{\sqrt{F}},$$
$$\boldsymbol{E^r} \equiv \frac{dr}{\sqrt{F}},$$
$$\boldsymbol{E^\theta} \equiv \sqrt{H} d\theta,$$
$$\boldsymbol{E^\phi} \equiv sin\theta\, \sqrt{H}\, d\phi.$$

2.2 Second step

We can turn any real vector space V into a complex vector space $V^{\mathbb{C}}$ by forming the set $V\times V$ of all pairs $(\boldsymbol{E_i},\boldsymbol{E_j})$ with $\boldsymbol{E_i},\boldsymbol{E_j} \in V$ and then writing $(\boldsymbol{E_i},\boldsymbol{E_j})$ as $\boldsymbol{E_i} + i\boldsymbol{E_j}$. With this definition $V^{\mathbb{C}}$ becomes complex vector space (For more details, see [12]). Hence, we can construct a set of vectors forming a basis of $\{\mathbf{L},\mathbf{N},\mathbf{M},\mathbf{W}\}$ in $\mathcal{T}_p^{\mathbb{C}}(\mathcal{M})$ defined by

$$\begin{aligned}\mathbf{L} &\equiv \frac{1}{\sqrt{F}}\left(\boldsymbol{E_t}+\boldsymbol{E_r}\right)=\partial_r,\\ \mathbf{N} &\equiv \frac{\sqrt{F}}{2}\left(\boldsymbol{E_t}-\boldsymbol{E_r}\right)=\frac{\sqrt{F}}{\sqrt{G}}\,\partial_u-\frac{F}{2}\,\partial_r,\\ \mathbf{M} &\equiv \frac{\sqrt{2}}{2}\left(\boldsymbol{E_\theta}+i\boldsymbol{E_\phi}\right)=\frac{\sqrt{2}}{2\sqrt{H}}\,\partial_\theta+\frac{\sqrt{2}\,i}{2sin\theta\,\sqrt{H}}\,\partial_\phi,\\ \mathbf{W} &\equiv \frac{\sqrt{2}}{2}\left(\boldsymbol{E_\theta}-i\boldsymbol{E_\phi}\right)=\frac{\sqrt{2}}{2\sqrt{H}}\,\partial_\theta-\frac{\sqrt{2}\,i}{2sin\theta\,\sqrt{H}}\,\partial_\phi.\end{aligned}$$

This basis constitutes a complex null tetrad, i.e., consists of two real null vectors $\mathbf{L}$, $\mathbf{N}$ and two complex conjugate null vectors $\mathbf{M}$, $\mathbf{W}$ thus, the scalar products of tetrad vectors satisfy: $\mathbf{L}\cdot\mathbf{L}=\mathbf{M}\cdot\mathbf{M}=$ $=\mathbf{N}\cdot\mathbf{N}=\mathbf{L}\cdot\mathbf{M}=\mathbf{M}\cdot\mathbf{N}=0$ and $\mathbf{L}\cdot\mathbf{N}=-\mathbf{M}\cdot\mathbf{W}=1$.

2.3 Third step (a)

Next, we introduce the "rotated" Eddington-Finkelstein coordinates by performing the mapping

$$\boldsymbol{\Pi_R}:(u,r,\theta,\phi)\longrightarrow(u_R,r_R,\theta_R,\phi_R)$$

through the transformation

$$\begin{aligned}u_R &= u-ia\,cos\theta,\\ r_R &= r+ia\,cos\theta,\\ \theta_R &= \theta,\\ \phi_R &= \phi.\end{aligned}\tag{3}$$

The map $\boldsymbol{\Pi_R}$ naturally induces the push-forward $(\boldsymbol{\Pi_R})_*$,

$$(\boldsymbol{\Pi_R})_*:\mathcal{T}_p^{\mathbb{C}}(\mathcal{M})\longrightarrow\mathcal{T}_{\boldsymbol{\Pi_R}(p)}^{\mathbb{C}}(\mathcal{N}).$$

Hence, it is very easy to verify that the basis vector $\xi\equiv\{\mathbf{L},\mathbf{N},\mathbf{M},\mathbf{W}\}\in\mathcal{T}_p^{\mathbb{C}}(\mathcal{M})$ is mapped by $\boldsymbol{\Pi_R}$ into the basis $\xi_R\equiv\{\mathbf{L_R},\mathbf{N_R},\mathbf{M_R},\mathbf{W_R}\}\in\mathcal{T}_{\Pi_R(p)}^{\mathbb{C}}(\mathcal{N})$ and then

$$\begin{aligned}\mathbf{L_R} &= \partial_{r_R},\\ \mathbf{N_R} &= \frac{\sqrt{B}}{\sqrt{A}}\,\partial_{u_R}-\frac{B}{2}\,\partial_{r_R},\\ \mathbf{M_R} &= \frac{\sqrt{2}}{2\sqrt{\Psi}}\left[ia\,sin\theta_R\left(\partial_{u_R}-\partial_{r_R}\right)+\partial_{\theta_R}+\frac{i}{sin\theta_R}\,\partial_{\phi_R}\right],\\ \mathbf{W_R} &= \frac{\sqrt{2}}{2\sqrt{\Psi}}\left[ia\,sin\theta_R\left(\partial_{r_R}-\partial_{u_R}\right)+\partial_{\theta_R}-\frac{i}{sin\theta_R}\,\partial_{\phi_R}\right],\end{aligned}\tag{4}$$

where $A=A(r,\theta), B=B(r,\theta), \Psi=\Psi(r,\theta)$ are functions to be known.

2.4 Third step (b)

As it is evident from (4), ξ_R is a complex null consisting of two real null vectors $\mathbf{L_R}$, $\mathbf{N_R}$ and two complex conjugate vectors $\mathbf{M_R}$, $\mathbf{W_R}$, hence, the scalar products of tetrad vectors satisfy: $\mathbf{L_R}\cdot\mathbf{L_R}=\mathbf{M_R}\cdot\mathbf{M_R}=$ $=\mathbf{N_R}\cdot\mathbf{N_R}=\mathbf{L_R}\cdot\mathbf{M_R}=\mathbf{M_R}\cdot\mathbf{N_R}=0$ and $\mathbf{L_R}\cdot\mathbf{N_R}=-\mathbf{M_R}\cdot\mathbf{W_R}=1$.

Moreover,

$$\eta_R^{ij}=\begin{pmatrix}0&1&0&0\\1&0&0&0\\0&0&0&-1\\0&0&-1&0\end{pmatrix}$$

are the components of a specific metric tensor $\boldsymbol{\mathcal{G}_R}$ respect to the complex null tetrad, i.e.,

$$(\boldsymbol{\mathcal{G}_R})_{ij}=(\mathbf{L_R})_i(\mathbf{N_R})_j+(\mathbf{N_R})_i(\mathbf{L_R})_j-(\mathbf{M_R})_i(\mathbf{W_R})_j-(\mathbf{W_R})_i(\mathbf{M_R})_j.\tag{5}$$

Hence, the metric tensor $\boldsymbol{\mathcal{G}_R}$ can be expressed in terms of the real basis ∂_R as:

$$\mathcal{G}_{\boldsymbol{R}} = A\, du_R \otimes du_R + 2\frac{\sqrt{A}}{\sqrt{B}}\, du_R \otimes dr_R + 2a sin^2\theta_R \left(\frac{\sqrt{A}}{\sqrt{B}} - A\right) du_R \otimes d\phi_R -$$
$$-2a sin^2\theta_R \frac{\sqrt{A}}{\sqrt{B}}\, dr_R \otimes d\phi_R - \Psi\, d\theta_R \otimes d\theta_R + \frac{sin^2\theta_R}{B} \left[a^2 sin^2\theta_R\left(AB - 2\sqrt{AB}\right) - B\Psi\right] d\phi_R \otimes d\phi_R. \quad (6)$$

2.5 Fourth step

We introduce the Boyer Lindquist coordinates by implementing a map

$$\boldsymbol{\Pi}_{\mathbf{BL}}: (u_R, r_R, \theta_R, \phi_R) \longrightarrow (T, R, \Theta, \Phi)$$

from the coordinates $(u_R, r_R, \theta_R, \phi_R)$ into the (T, R, Θ, Φ) through the following transformation:

$$u_R = T - \frac{a^2\sqrt{G} + \sqrt{F}H}{(a^2 + FH)\sqrt{G}} R, \qquad \phi_R = \Phi + \frac{a}{a^2 + FH} R.$$

Thus, the metric (6) admits the form

$$\mathcal{G} = \frac{(a^2 cos^2\Theta + FH)G\Psi}{\left(a^2 cos^2\Theta\sqrt{G} + \sqrt{F}H\right)^2}\, dT \otimes dT - \frac{\Psi}{FH + a^2}\, dR \otimes dR - \Psi\, d\Theta \otimes d\Theta - \Psi\, sin^2\Theta \left[1 + \right.$$
$$\left. + \frac{a^2 sin^2\Theta\sqrt{G}\left(2\sqrt{F}H - FH\sqrt{G} + a^2 cos^2\Theta\sqrt{G}\right)}{\left(a^2 cos^2\Theta\sqrt{G} + \sqrt{F}H\right)^2}\right] d\Phi \otimes d\Phi - \frac{2\left(F\sqrt{G} - F\right) a sin^2\Theta\Psi H\sqrt{G}}{\left(a^2 cos^2\Theta\sqrt{G} + F\sqrt{H}\right)^2}\, dT. \quad (7)$$

This result shows that, by implementing the former successive step, the Newman-Janis Ansatz generates a stationary axially symmetric space-time from static spherically symmetric.

3. Rotating anisotropic fluids from the NJA

To interpret the metric tensor (7), we presuppose that it describes a stationary interior space-time corresponding to an anisotropic fluid without heat. This kind of fluids admits a representation in terms of a tensor's components in the following form,

$$(\mathbf{T})^{ab} = \mu\, (\mathbf{V})^a(\mathbf{V})^b + P_R\, (\mathbf{e_R})^a(\mathbf{e_R})^b + P_\Theta\, (\mathbf{e_\Theta})^a(\mathbf{e_\Theta})^b + P_\Phi\, (\mathbf{e_\Phi})^a(\mathbf{e_\Phi})^b, \quad (8)$$

provided that the space-time endowed with some Lorentzian pseudo-Riemannian metric with signature (+2) [1]

$$(\mathcal{G})^{ab} = -\,(\mathbf{V})^a(\mathbf{V})^b + (\mathbf{e_R})^a(\mathbf{e_R})^b + (\mathbf{e_\Theta})^a(\mathbf{e_\Theta})^b + (\mathbf{e_\Phi})^a(\mathbf{e_\Phi})^b, \quad (9)$$

where $\mu(R,\Theta)$ is the energy density of the fluid and $P_R(R,\Theta), P_\Theta(R,\Theta)$ and $P_\Phi(R,\Theta)$ are the pressure along the corresponding directions. The word-lines of the fluids are integral curves of the velocity vector $\mathbf{V}$, which satisfies the condition $\mathcal{G}(\mathbf{V},\mathbf{V}) = -1$. Here, $(\mathbf{V})^a, (\mathbf{e_R})^a, (\mathbf{e_\Theta})^a$ and $(\mathbf{e_\Phi})^a$ are the components of the vectors of the orthonormal tetrad given by

$$\mathbf{V} = -\frac{1}{\sqrt{\Psi(FH + a^2)}}\left[\left(\frac{\sqrt{F}H}{\sqrt{G}} + a^2\right)\partial_T + a\,\partial_\Phi\right],$$
$$\mathbf{e_R} = \frac{\sqrt{\Psi(FH + a^2)}}{\Psi}\,\partial_R, \quad (10)$$
$$\mathbf{e_\Theta} = \frac{1}{\sqrt{\Psi}}\,\partial_\Theta,$$
$$\mathbf{e_\Phi} = -\frac{1}{\sqrt{\Psi}\, sin\Theta}\left(a\, sin^2\Theta\,\partial_T + \partial_\Phi\right).$$

Thus, the Einstein field equations $\mathbf{G} = 8\pi\mathbf{T}$ with the metric tensor (7) and the energy-momentum tensor given by (8) yields the following system of independent equations for the energy density and the anisotropic pressures

[1] We developed most of the following calculations using a computational routine in Maple 2020. To preserve the consistency with the Differential Geometry package, we changed the signature (-2) to (2) in this section (and the rest

$$\mu = \frac{1}{32\pi\rho^4\Psi^3}\{3\rho^4(\Psi_{,\Theta}^2 + \Delta\Psi_{,R}^2) - 4\rho^4\Psi(\Psi_{,\Theta\Theta} + \Delta\Psi_{,RR}) - \rho^2\Psi_{,R}^2(\rho^2\Delta_{,R} - a^2 sin^2\Theta\, K_{,R}) + 2a^2 sin^2\Theta\, \rho^2\Psi^2 K_{,RR} - 2\cot\Theta\, \rho^2\Psi_{,\Theta}^2(K + a^2) + \Psi^2[3a^2 sin^2\Theta\, K_{,R}^2 - 4(\rho^2 + 4a^2 cos^2\Theta)(K + a^2) + 12cos^2\Theta\, \Delta^2]\}, \tag{11}$$

$$P_R = \frac{1}{32\pi\rho^4\Psi^3}\{4\rho^4\Psi\Psi_{,\Theta\Theta} + 3\rho^4(\Psi_{,R}^2 - \Psi_{,\Theta}^2) + \rho^2\Psi_{,R}^2(\rho^2\Delta_{,R} - 2\Delta K_{,R}) + \rho^2\Psi_{,\Theta}^2[a^2 sin2\Theta + 2\cot\Theta\,(K + a^2)] - 4K(K + 2a^2) + \Psi^2[a^2 sin^2\Theta\, K_{,R}^2 + 4a^2 cos^2\Theta\,(6(\rho^2 + a^2 sin^2\Theta) - (\Delta + a^2 cos^2\Theta))]\}, \tag{12}$$

$$P_\Theta = \frac{1}{32\pi\rho^4\Psi^3}\{4\rho^2\Delta\Psi(\rho^2\Psi_{,RR} - \Psi K_{,RR}) + 2\rho^4\Psi^2(F_{,RR}H + FH_{,RR}) + 2\rho^2[\cot\Theta\,\Psi_{,\Theta}^2(K + a^2(1 + sin^2\Theta)) + \Psi_{,R}^2(\rho^2\Delta_{,R} - \Delta K_{,R})] + 3\rho^4(\Psi_{,\Theta}^2 - \Delta\Psi_{,R}^2) + \Psi^2[K_{,R}^2(8\Delta - a^2 sin^2\Theta) - 6\rho^2 K_{,R}\Delta_{,R} + 4(\rho^4 F_{,R}H_{,R} + a^2 cos^2\Theta\,\Delta)]\}, \tag{13}$$

$$P_\Phi = \frac{1}{32\pi\rho^4\Psi^3}\{4\rho^2\Delta\Psi(\rho^2\Psi_{,RR} - \Psi K_{,RR}) + 2\rho^4\Psi^2(F_{,RR}H + FH_{,RR}) - 3\rho^4(\Psi_{,\Theta}^2 - \Delta\Psi_{,R}^2) + 2\rho^2\Delta(\cot\Theta\,\Psi_{,\Theta}^2 - \Psi^2 K_{,RR}) + \Psi^2[3K_{,R}^2(2\Delta - a^2 sin^2\Theta) - 6\rho^2\Delta_{,R}K_{,R} + 4\rho^4 F_{,R}H_{,R} + 4\Delta(2a^2 cos^2\Theta - K)]\}. \tag{14}$$

where "," denotes derivative and we used the simple notation $\rho^2 \equiv K + a^2 cos^2\Theta, K \equiv H\sqrt{F/G}$ and $\Delta \equiv FH + a^2$. In addition to the above system, the Einstein equations give us the following two non-trivial independent equations:

$$(3\Psi_{,R}\Psi_{,\Theta} - 2\Psi_{,R\Theta})\rho^4 + 3a^2 sin2\Theta\, K_{,R}\,\Psi^2 = 0, \tag{15}$$

$$[(\Psi K_{,R})_{,R} + 2\Psi_{,\Theta}\cot\Theta](K + a^2 cos^2\Theta)^2 - \Psi(2K - a^2 cos^2\Theta\, K_{,R}^2) = 0, \tag{16}$$

which we must solve to determine the previous dynamic quantities fully characterizing the fluid. Hence, a large class of solutions corresponding to rotating anisotropic fluids arose after getting solutions to the system of equations (15)-(16). As we know, there at least five classes of solutions of the above equation system. They are:

$$\text{Class 1: } K = c_1, \Psi = \frac{a^2 cos^2\Theta + c_1}{cos\Theta},$$

$$\text{Class 2: } K = R^2, \Psi = \frac{8c_1^2(a^2 cos^2\Theta + R^2)}{(\sqrt{2}\,c_3 cos\Theta\, R - c_2)^2},$$

$$\text{Class 3: } K = (R - c_1)^2, \Psi = -\frac{c_2^2[a^2 cos^2\Theta + (R - c_1)^2]}{(R - c_1)^2 cos^2\Theta},$$

$$\text{Class 4: } K = R^2 + c_1 R + c_2, \Psi = a^2 cos^2\Theta + R^2 + c_1 R + c_2,$$

$$\text{Class 5: } K = -\frac{[\exp(c_2 + R) + 2]^2}{2\exp(c_2 + R)}, \Psi = \frac{2a^2 cos^2\Theta \exp(c_2 + R) - [\exp(c_2 + R) + 2]^2}{cos^2\Theta\,[\exp(c_2 + R) + 2]^2},$$

c_1, c_2 and c_3 are arbitrary constants, and $K^2 = H^2 F/G$ is a relation between the coefficients of the seed static metric. As we can see, each class defines a kind of specific rotating fluid. In particular, if we introduce $c_1 = c_2$ in the Class 4, we get a solution of the form $\Psi = a^2 cos^2\Theta + R^2$, and $K = R^2$ form. Additionally, if we choose $K \equiv \sqrt{F/G}H$ determining the Schwarzschild solution (i.e., $H = R^2$ and $F = G = 1 - \alpha/R$, with α an arbitrary constant), we achieve a rotating fluid without heat flux with all pressures equal to zero. That case corresponds to Kerr's solution. We will generate rotating wormholes form statics wormholes via the NJA by imposing some geometric and physics conditions on K and Ψ functions; it will be the objective of the following section.

4. NJA for rotating wormholes

4.1 A class of static wormholes fixed by the Newman-Janis Ansatz

To construct static wormholes form solutions of the Einstein field equations for an interior region of the space-time, we first bring the metric (1) to the form:

$$\mathcal{G}_0 = -e^{2\varphi(R)}\, dT\otimes dT + \frac{dR\otimes dR}{1 - \frac{b(R)}{R}} + R^2\,(d\Theta\otimes d\Theta + sin^2\Theta\, d\Phi\otimes d\Phi). \tag{17}$$

Hence, we introduced the usual spherical coordinates (R, Θ, Φ) and performed the identification

$$G(R) = e^{2\varphi(R)}, \qquad F(R) = 1 - \frac{b(R)}{R}, \qquad H(R) = R^2. \tag{18}$$

The functions φ and b determine the gravitational redshift and the spatial shape of the wormhole [7]. Naturally, the match of wormhole to an exterior space-time must occur on a determined value R_s to the radial coordinate R. To guarantee that the Birkhoff theorem is satisfied in the wormhole exterior is mandatory to impose the conditions:

(i) $b(R) = b(R_s) = constant \equiv b_0,\ R > R_s,$

(ii) $\varphi(R) = \frac{1}{2}\ln\left(1 - \frac{b_0}{R}\right),\ R > R_s.$

Additionally, to ensure that we get asymptotically flat solutions, we demand that:

(i) $\lim_{R\to\infty}\left[\frac{b(R)}{R}\right] \longrightarrow 0,$

(ii) $\lim_{R\to\infty} \varphi \longrightarrow 0.$

To embed, in the three-dimensional Euclidean space, a two-dimensional surface with the same geometry we set up $\left(\Theta = \frac{\pi}{2}, T = constant\right)$ in the metric (17)

$$\mathcal{S} = \frac{1}{1-b(R)/R}\, \mathrm{dR}\otimes\mathrm{dR} + R^2\, d\Phi\otimes d\Phi. \tag{19}$$

Then, the surface of rotation

$$\chi(\mathrm{R}, \Phi) = \left(\mathrm{R}\cos\Phi, \mathrm{R}\sin\Phi, \mathrm{Z(R)}\right) \tag{20}$$

determines the spatial geometry of the wormhole spacetime through the single function $Z(R)$, which satisfy the equation

$$\frac{dZ(R)}{dR} = \pm\left(\frac{R}{b(R)} - 1\right)^{\frac{1}{2}}. \tag{21}$$

If we assume that the metric (17) determines the existence of a static wormhole, their rotating counterpart generated by the NJA presents a restriction given by the relation between the metric functions provided by $K^2 = (F/G)\, R^4$. Hence, the unique solutions to the Einstein field equations describing static wormholes suitable to generate rotating wormholes by using NJA satisfy the relation:

$$e^{2\varphi(R)} = \frac{R^4}{K^2}\left[1 - \frac{b(R)}{R}\right] \tag{22}$$

with K determined by an arbitrary static solution of the Einstein field equations, consistent with the five Classes in the former section.

4.2 Space-time for rotating wormholes

The procedure employed here to construct rotating wormholes starts by bringing the stationary axially symmetric space-time previously generated via the NJA to the following form:

$$\mathcal{G} = -e^{2\upsilon(R,\Theta)}\, dT\otimes dT + e^{2\mu_2(R,\Theta)}\, dR\otimes dR + e^{2\mu_3(R,\Theta)}\, d\Theta\otimes d\Theta + e^{2\lambda(R,\Theta)}\, (d\Phi - \omega\, \mathrm{dT})\otimes(d\Phi - -\omega\, \mathrm{dT}), \tag{23}$$

where

$$e^{2\upsilon} = \frac{\Delta\Psi}{\rho^4}\frac{[a^2 + K - a^2 sin^2\Theta]^2}{(a^2+K)^2 - a^2 sin^2\Theta\,\Delta}, \qquad e^{2\mu_2} = \frac{\Psi}{\Delta}, \qquad e^{2\mu_3} = \Psi,$$

$$e^{2\lambda} = sin^2\Theta\, \frac{\Psi}{\rho^4}\left[(a^2+K)^2 - a^2 sin^2\Theta\,\Delta\right], \qquad \omega = \frac{a\,(a^2 + K - \Delta)}{(a^2+K)^2 - a^2 sin^2\Theta\,\Delta},$$

$$K^2 = \frac{F}{G}H^2, \qquad \rho^2 = K + a^2 cos^2\Theta, \qquad \Delta = FH + a^2,$$

where, as we known, each couple (K, Ψ) determine a new rotating solution of the Einstein field equations provided a static solution is known. Next, we suppose that there exists a function $B = B(R, \Theta)$ such that [9,10,11]:

$$e^{2\mu_2} = \mathcal{G}_{RR} = \frac{1}{1 - B/R}, \qquad \lim_{R\to\infty}\frac{B}{R} = 0, \qquad \lim_{R\to\infty}\mathcal{G}_{TT} = 1.$$

Additionally, to avoid conic singularities we assume that, on the axis of rotation ($\Theta = 0$ or $\Theta = \pi$), the following relation it is always possible:

Journal of Physics: Conference Series 2081 (2021) 012005 doi:10.1088/1742-6596/2081/1/012005

$$\mathcal{G}_{\Theta\Theta} = \frac{\mathcal{G}_{\Phi\Phi}}{sin^2\Theta}$$

and, that the near the throat:

$$R\,\partial_R B < B.$$

Furthermore, to ensure the applicability of the NJA, the choices of the metric functions corresponding to the static space-time and the corresponding to the rotating space-time are restricted by the following constrain:

$$F = 1 - \frac{b(R)}{R}, \qquad e^{2\varphi(R)} = \frac{R^4}{K^2}\left[1 - \frac{b(R)}{R}\right], \tag{24}$$

where, as is known, exist a unique correspondence between each choice of K and Ψ determined by the classes of rotating solutions discussed above. Then, finally, the weakest condition to be satisfied by the rotating wormhole is given by the following expression

$$\frac{B(R,\Theta)}{R} = 1 - \frac{R^2}{\Psi}\left(1 - \frac{b(R)}{R} + \frac{a^2}{R}\right), \tag{25}$$

provided the $b(R)$ function defines a static wormhole and $B(R,\Theta)$ satisfies all the conditions previously discussed.

4.3 Example of Rotating wormholes from the NJA

As a straightforward example, let us consider the static solution

$$b(R) = R_S - \frac{q^2}{R}, \tag{26}$$

where R_S and q are real constants. This solution determines the Reissner-Nordström wormhole, the spatial geometry of the wormhole spacetime through the single function $Z(R)$ satisfying

$$dZ(R) = \pm\left(\frac{R}{R_S - q^2/R} - 1\right)^{\frac{1}{2}} \tag{27}$$

and energy-momentum tensor given by [7]

$$\sigma(R) = p(R) = \tau(R) = \frac{q^2}{8\pi R^4}. \tag{28}$$

Accordingly, if we choose, for example K and Ψ as given by the Class 2 of rotating solutions obtained by the NJA, then we achieve a Kerr-Newman wormhole:

$$\frac{B(R,\Theta)}{R} = 1 - \frac{R^2}{R^2 + a^2cos^2\Theta}\left(1 - \frac{q^2}{R} + \frac{a^2}{R}\right).$$

The wormhole rotates with angular velocity given by

$$\omega = \frac{q^2}{(a^2 + R^2)^2 - a^2sin^2\Theta\,(a^2 + R^2 + q^2 - R_S R)},$$

and their corresponding energy-momentum tensor given by the dynamical quantities:

$$\sigma = \tau = p = \frac{q^2}{8\pi\,(a^2cos^2\Theta + R^2)^2}.$$

In other words, the rotating counterpart of the Reissner-Nordström wormhole obtained by the NJA corresponds to the Kerr-Newman wormhole.

5. Conclude and remark

This paper presented the ansatz formulated by Newman-Janis to generate solutions to the Einstein field equations for rotating space-times. We performed the algorithm's structure based on modern differential geometry to keep open the discussion about their nature. Actuality, we discuss the algorithm step setting successive maps on the tangent and cotangent spaces, and compile some independent classes of solutions of the Einstein field equations for rotating space-times, all obtained from prescribed static and spherically space-times.

We have elaborated on the possibility of constructing static wormholes and their rotating counterpart by choosing any of the classes of solutions generated by the Newman-Janis Ansatz. These classes restrict the nature of both static and rotating wormholes. To illustrate the discussed ideas, we generated the

Kerr-Newman wormhole using the Reissner-Nordström wormhole as a seed solution. It will be fructiferous to produce many examples of rotating wormholes operating with the ideas presented in this work. The problem of match wormholes to their exterior space-time remains open. One emergent here is studying the wormholes' asymptotically flat characteristics and using a C^3 junction technique [13,14]. We will explore that in subsequent work.

Acknowledgments
A.C.G-P is grateful for the invitation to participate in XXII International Meeting Physical Interpretations of Relativity Theory – 2021 (Moscow, Russia). This work was partially supported by Programa Capital Semilla para Investigación, Proyecto 2490, VIE-UIS.

References

[1] Kerr R P 1963 *Physical review letters* **11** 237
[2] Newman E T and Janis A 1995 *Journal of Mathematical Physics* **6** 915
[3] Drake S and Szekeres P 2000 *General relativity and Gravitation* **32** 445
[4] Azreg-Ainou M 2014 *The European Physical Journal C* **74** 2865
[5] Gutiérrez-Piñeres A C and Quevedo H 2016 *General Relativity and Gravitation* **48** 146
[6] Ellis H G 1973 *Journal of Mathematical Physics* **14** 104
[7] Morris M S, Thorne K S and Yurtsever U 1988 *Physical Review Letters* **61** 1446
[8] Lemos J P, Lobo F S and de Oliveira S Q 2003 *Physical Review D* **68** 064004
[9] Teo E 1998 *Physical Review D* **58** 024014
[10] Azreg-Ainou M 2016 *The European Physical Journal C* **76** 3835
[11] Azreg-Ainou M 2016 *The European Physical Journal C* **76** 3836
[12] De Felice and Clarke C J S 1992 *Relativity on curved Manifolds* (Cambridge University Press)
[13] Gutiérrez-Piñeres A C and Quevedo H 2019 *Classical and Quantum Gravity* **36** 135003
[14] Gutiérrez-Piñeres A C and Quevedo H 2021 *Preprint* arXiv:2106.08679

Non-Commutative Worlds and Relativity

Louis H Kauffman

Department of Mathematics, Statistics and Computer Science,
University of Illinois at Chicago, 851 South Morgan Street, Chicago, IL, 60607-7045

E-mail: loukau@gmail.com

Abstract. This paper shows how aspects of gauge theory, Hamiltonian mechanics and quantum mechanics arise naturally in the mathematics of a non-commutative framework for calculus and differential geometry.

1. Introduction to Non-Commutative Worlds

Calculus was formulated in a commutative framework by Newton and his successors. We are all familiar with the limit definitions for derivatives and we take for granted that classical mechanics is formulated in this framework. The advent of quantum mechanics brought with it formulations of physical theory that are deeply related to non-commutativity. Heisenberg invented a calculus of quantum quantities that did not commute with one another and obeyed specific identities such as the famous

$$QP - PQ = \hbar i.$$

Schrödinger gave a formulation of quantum mechanics using standard partial differential equations and then discovered that the operators $Q = x$ and $P = -i\hbar\partial/\partial x$ obeyed exactly Heisenberg's relations and gave the translation between his work and the Heisenberg viewpoint. Dirac discovered a key to quantization via the replacement of the Poisson bracket of Hamiltonian mechanics with the commutators of quantum operators. The curvatures in differential geometry and general relativity were seen, through the work of Weyl and others to correspond to measurements in differences in parallel translation and correspondingly, in the commutators of covariant derivatives. This emphasis on parallel translation led to the generalizations of differential geometry fundamental to gauge theory. Gauge theory began, with the work of Hermann Weyl [20] as a generalization of differential geometry where lengths as well as angles were dependent upon the choice of paths. Weyl saw how to incorporate electromagnetism into general relativity along these lines. Later the ideas of Weyl were adopted in the context of quantum mechanics and became the basis for the understanding of nuclear forces. General relativity and gauge theory have deep relationships beyond Weyl via loop quantum gravity and the Ashtekar reformulation of general relativity in terms of gauge theory [1]. Throughout all this development, the underlying calculus remained in the commutative and continuum realms.

In this paper we shall begin by formulating calculus in non-commutative domains. Our constructions are motivated by discrete calculus. Discrete calculus is naturally embedded in a non-commutative context, and there can be adjusted so that it satisfies the derivative of a product in the form of the Leibniz rule $D(FG) = D(F)G + FD(G)$. We explain this point in Section 6, an Appencix on discrete calculus. We then show how the approach via non-commutativity leads to direct connections wtih gauge theory and make a beginning at relativistic developments related to Weyl [20] and to our previous work [15, 16]. It is our intent to go beyond the present paper to relationships with loop quantum gravity.

If we take commutators $[A, B] = AB - BA$ in an abstract algebra and define $DA = [A, J]$ for a fixed element J, then D acts like a derivative in the sense that $D(AB) = D(A)B + AD(B)$ (the Leibniz rule). As soon as we have calculus in such a framework, concepts of geometry are immediately available. For example, if we have two derivatives $\nabla_J A = [A, J]$ and $\nabla_K A = [A, K]$, then we can consider the commutator of these derivatives

$$[\nabla_J, \nabla_K]A = \nabla_J \nabla_K A - \nabla_K \nabla_J A = [[J, K], A].$$

(The verification of this last inequality is an exercise for the reader.) We can regard $R_{JK} = [J, K]$ as the *curvature* associated with ∇_J and ∇_K. Note that the commutator of the derivations ∇_J and ∇_K. is represented by $R_{JK} = [J, K]$ so that when the representatives J and K for two derivations commute, then the derivations themselves commute and the curvature vanishes. The non-commutation of derivations corresponds to curvature in geometry. We shall see that the emergence of curvature in this context is the formal analog of the curvature of a gauge connection.

Aspects of gauge theory, Hamiltonian mechanics and quantum mechanics arise naturally in the mathematics of a non-commutative framework for calculus and differential geometry. This paper consists in 6 sections including the introduction. Section 2 outlines the general properties of calculus in a non-commutative domain, where derivatives are all represented by commutators. This includes a special element H such that the total time derivative $\dot{F} = [F, H]$ for any F in the non-commutative domain. We show how the formal analog of Hamilton's equations arises naturally in a flat coordinate system, and we show how Schrödinger's equation arises from the time derivative made appropriately complex. Section 3 explores the consequences of defining dynamics in the form

$$\dot{X}_i = dX_i/dt = \mathcal{G}_i$$

where $\{\mathcal{G}_1, \cdots, \mathcal{G}_d\}$ is a collection of elements of the non-commutative domain $\mathcal{N}$. Write $\mathcal{G}_i$ relative to flat coordinates via $\mathcal{G}_i = P_i - A_i$. This is a definition of A_i with $\partial_i F = \partial F/\partial X_i = [F, P_i]$. The formalism of gauge theory appears naturally via the curvature of ∇_i with $\nabla_i(F) = [F, P_i - A_i]$ that is given by the formula

$$R_{ij} = \partial_i A_j - \partial_j A_i + [A_i, A_j].$$

With $\dot{X}_i = \mathcal{G}_i = P_i - A_i$, we let $g_{ij} = [X_i, \dot{X}_j]$ and show that this is a natural choice for a generalized metric. In particular, we show that for a quadratic Hamiltonian with these metric coefficients, the formula $g_{ij} = [X_i, \dot{X}_j]$ is a consequence. Furthermore, we show that for any F,

$$\dot{F} = \frac{1}{2}(\dot{X}_i \partial_i(F) + \partial_i(F)\dot{X}_i).$$

This means that with this choice of Hamiltonian the non-commutative world satisfies a correct analog to standard time derivative formula in commutative contexts. Higher order constraints of this sort are considered in [15]. Here we work at the level of this first constraint. We show how a covariant version of the Levi-Civita connection arises naturally in this commutator calculus. This connection satisfies the formula

$$\Gamma_{kij} + \Gamma_{ikj} = \nabla_j g_{ik} = \partial_j g_{ik} + [g_{ik}, A_j].$$

and so is exactly a generalization of the connection defined by Hermann Weyl in his original gauge theory [20]. In the non-commutative world $\mathcal{N}$ the metric indeed has a wider variability than the classical metric and its angular holonomy. Weyl's idea was to work with such a wider variability of the metric. The present formalism provides a new context for Weyl's original idea. The rest of Section 3 discusses the possibility of using this Levi-Connection to formulate a corresponding covariant derivative, Einstein tensor and general relativity. A direct formulation gives a curvature tensor without the usual symmetries (due to the presence of the covariant derivatives ∇_i) and so the corresponding Einstein tensor will not have vanishing divergence. The same issue arises in the original Weyl theory. Here it becomes a problem for further exploration. Section 4 recapitulates our constructs of the Levi-Civita connection in an index-free fashion. Section 5 is an appendix on the structure of the Einstein tensor and how the Bianchi identity can be seen from the Jacobi idenity in a non-commutative world. Section 6 explains how discrete calculus embeds in non-commutative calculus.
Remark. In our papers [15, 16] we have examined relationships of non-commutative worlds with general relativity in relation to higher order constraints. The results of this paper will be compared and combined with that previous work. Here we have not mentioned the relationships with the Feynman-Dyson derivation of electromagnetism from commutator calculus [2, 4, 17, 18] that were the initial impetus for this work [6]. It was that impetus that led to our work on non-commutative worlds including the following references [7, 8, 9, 10, 11, 12, 13, 14, 15, 16]. In this paper we return to this subject in a new way.

2. Calculus in Non-Commutative Worlds

Our constructions are performed in a Lie algebra $\mathcal{N}$. One may take $\mathcal{N}$ to be a specific matrix Lie algebra, an abstract Lie algebra or as an associative algebra closed under the operation of commutation. That is, if A and B are elements of $\mathcal{N}$, then $[A, B] = AB - BA$ is also an element of $\mathcal{N}$. If $\mathcal{N}$ is taken to be an abstract Lie algebra, then it is convenient to use the universal enveloping algebra so that the Lie product can be expressed as a commutator.

On $\mathcal{N}$, a variant of calculus is built by defining derivations as commutators (or more generally as Lie products). For a fixed N in $\mathcal{N}$ one defines

$$\nabla_N : \mathcal{N} \longrightarrow \mathcal{N}$$

by the formula

$$\nabla_N F = [F, N] = FN - NF.$$

∇_N is a derivation satisfying the Leibniz rule.

$$\nabla_N(FG) = \nabla_N(F)G + F\nabla_N(G).$$

In $\mathcal{N}$ there are as many derivations as there are elements of the algebra, and these derivations behave quite wildly with respect to one another. If one takes. as in the introduction to the present paper, the concept of *curvature* as the non-commutation of derivations, then $\mathcal{N}$ is a highly curved world indeed. Within $\mathcal{N}$ one can build a tame world of derivations that mimics the behaviour of flat coordinates in Euclidean space. In order to have flat coordinates, we need that the derivations for those coordinate directions commute with one another. This, in turn, is implied by the commuting of the representatives for those derivatons. That is, suppose that X and Y are coordinates and that P_X and P_Y represent derivatives in thes directions so that one writes

$$\partial_X F = [F, P_X]$$

and

$$\partial_Y F = [F, P_Y].$$

Then

$$\partial_X \partial_Y = \partial_Y \partial_X$$

when $[P_X, P_Y] = 0$, as we have seen in the introduction. A flat coordinate system corresponds to a collection of commutator equations.
We take a collection of special elements $X_1, X_2, \cdots, X_d$ to represent coordinates. With $d = 3$ these can be the familiar pattern of three dimensional spatial coordinates. With $d = 4$ we can take X_4 to represent time, if this is desired. There is no a priori restriction on the value of d. As *flat coordinates* the X_i satisfy the commutator equations below with the P_j chosen to represent differentiation with respect to X_j.:

$$[X_i, X_j] = 0$$
$$[P_i, P_j] = 0$$
$$[X_i, P_j] = \delta_{ij}.$$

Derivatives are represented by commutators.

$$\partial_i F = \partial F/\partial X_i = [F, P_i],$$

$$\hat{\partial}_i F = \partial F/\partial P_i = [X_i, F].$$

The time derivative is represented by commutation with a special element H of the algebra:

$$dF/dt = [F, H].$$

The element H corresponds to the Hamiltonian in classical physics or the Hamiltonian operator in quantum physics. In the abstract world we are constructing, it is neither of these, but can be compared and represented by the classical or quantum Hamiltonians. For quantum mechanics, we take $i\bar{}hdA/dt\!= [A, H]$.

These non-commutative coordinates are the simplest flat set of coordinates for description of temporal phenomena in a non-commutative world. Note that *Hamilton's equations are a consequence of these definitions.* The very short proof of this fact is given below.
Hamilton's Equations.

$$dP_i/dt = [P_i, H] = -[H, P_i] = -\partial H/\partial X_i$$

$$dX_i/dt = [X_i, H] = \partial H/\partial P_i.$$

These are exactly Hamilton's equations of motion. The pattern of Hamilton's equations is built into the system.

3. Dynamics, Gauge Theory and the Weyl Theory.

One can take the general dynamical equation in the form

$$\dot{X}_i = dX_i/dt = \mathcal{G}_i$$

where $\{\mathcal{G}_1, \cdots, \mathcal{G}_d\}$ is a collection of elements of $\mathcal{N}$. Write $\mathcal{G}_i$ relative to the flat coordinates via $\mathcal{G}_i = P_i - A_i$. This is a definition of A_i and $\partial_i F = \partial F/\partial X_i = [F, P_i]$. The formalism of gauge theory appears naturally. In particular, defining derivations corresponding to the $\mathcal{G}_i$ by the formulas

$$\nabla_i(F) = [F, \mathcal{G}_i],$$

then one has the curvatures (as commutators of derivations)

$$[\nabla_i, \nabla_j]F = [[\mathcal{G}_i, \mathcal{G}_j], F]$$

(by the formula in the introduction to this paper)

$$= [[P_i - A_i, P_j - A_j], F]$$

$$= [[P_i, P_j] - [P_i, A_j] - [A_i, P_j] + [A_i, A_j], F]$$
$$= [\partial_i A_j - \partial_j A_i + [A_i, A_j], F].$$

Thus the curvature is given by the formula

$$R_{ij} = \partial_i A_j - \partial_j A_i + [A_i, A_j].$$

We see that our curvature formula is the well-known formula for the curvature of a gauge connection when A_i is interpreted or represeented as such a connection. We shall now see that aspects of geometry arise naturally in this context, including the Levi-Civita connection (which is seen as a consequence of the Jacobi identity in an appropriate non-commutative world).

With $\dot{X}_i = P_i - A_i$, the commutator $[X_i, \dot{X}_j]$ takes the form

$$g_{ij} = [X_i, \dot{X}_j] = [X_i, P_j - A_j] = [X_i, P_j] - [X_i, A_j] = \delta_{ij} - [X_i, A_j].$$

Thus we see that the "gauge field" A_j provides the deviation from the Kronecker delta in this commutator. We have $[\dot{X}_i, \dot{X}_j] = R_{ij}$, so that these commutators represent the curvature.

Before proceeding further it is necessary to explain why $g_{ij} = [X_i, \dot{X}_j]$ is a correct analogue to the metric in the classical physical situation. In the classical picture we have $ds^2 = g_{ij} dx_i dx_j$ representing the metric. Hence $ds^2/dt^2 = g_{ij}(dx_i/dt)(dx_j/dt) = g_{ij} p_i p_j$ and so (with m = 1) the Hamiltonian is $H = \frac{1}{2} g_{ij} p_i p_j$. By convention we sum over repeated indices.

For the commutator $[X_i, \dot{X}_j] = g_{ij}$, this equation is a consequence of the right choice of Hamiltonian. In a given non-commutative world, we choose an H in the algebra to represent the total time derivative so that $\dot{F} = [F, H]$ for any F. Suppose we have elements g_{ij} such that

$$[g_{ij}, X_k] = 0,$$

$$[g_{ij}, P_k] = 0$$

and

$$g_{ij} = g_{ji}.$$

We choose

$$H = (g_{ij} P_i P_j + P_i P_j g_{ij})/4 = \frac{1}{2} g_{ij} P_i P_j.$$

This is the non-commutative analog of the classical $H = \frac{1}{2} g_{ij} p_i p_j$. We now show that this choice of Hamiltonian implies that $[X_i, \dot{X}_j] = g_{ij}$.

Lemma. Let g_{ij} be given such that $[g_{ij}, X_k] = 0$ and $g_{ij} = g_{ji}$. Define

$$H = \frac{1}{2} g_{ij} P_i P_j$$

(where we sum over the repeated indices) and

$$\dot{F} = [F, H].$$

Then

$$[X_i, \dot{X}_j] = g_{ij}.$$

Proof. Consider

$$[X_k, g_{ij}P_iP_j] = g_{ij}[X_k, P_iP_j]$$
$$= g_{ij}([X_k, P_i]P_j + P_i[X_k, P_j])$$
$$= g_{ij}(\delta_{ki}P_j + P_i\delta_{kj}) = g_{kj}P_j + g_{ik}P_i$$
$$= 2g_{kj}P_j.$$

Thus we have shown that

$$\dot{X}_k = g_{kj}P_j.$$

Then

$$[X_r, \dot{X}_k] = [X_r, g_{kj}P_j] = g_{kj}[X_r, P_j] = g_{kj}\delta_{rj} = g_{kr} = g_{rk}.$$

This completes the proof.

We can further remark that

Lemma. With the same hypotheses as the previous Lemma and with F any element of the given non-commutative world $\mathcal{N}$, we have the formula

$$\dot{F} = \frac{1}{2}(\dot{X}_i\partial_i(F) + \partial_i(F)\dot{X}_i).$$

Proof. Note from the previous Lemma that $\dot{X}_k = g_{kj}P_j$.

$$\dot{F} = [F, H] = [F, \frac{1}{2}g_{ij}P_iP_j] = \frac{1}{2}g_{ij}[F, P_iP_j]$$

$$= \frac{1}{2}g_{ij}([F, P_i]P_j + P_i[F, P_j])$$
$$= \frac{1}{2}g_{ij}P_i[F, P_j] + [F, P_i]\frac{1}{2}g_{ij}P_j$$
$$= \frac{1}{2}(\dot{X}_i\partial_i(F) + \partial_i(F)\dot{X}_i).$$

This completes the proof.

With this Lemma we see that the quadratic Hamiltonian not only connects the abstract metric coefficients g_{ij} of the non-commutative world with the metric coefficients in classical worlds, but also the basic time derivative formula

$$\dot{F} = \dot{X}_i\partial_i(F)$$

has its correct (symmetrized) non-commutative counterpart. Elsewhere [15] we have said that with the quadratic Hamiltonian, the non-commutative world satisfies the *First Constraint.*

One can consider the consequences of the commutator $[X_i, \dot{X}_j] = g_{ij}$, deriving that

$$\ddot{X}_r = G_r + F_{rs}\dot{X}^s + \Gamma_{rst}\dot{X}^s\dot{X}^t,$$

where G_r is the analogue of a scalar field, F_{rs} is the analogue of a gauge field and Γ_{rst} is the Levi-Civita connection associated with g_{ij}. This decompositon of the acceleration is uniquely determined by the given framework. See [14, 12] for the details of this result. Here we discuss the Levi-Civita connection.

A stream of consequences follows by differentiating both sides of the equation

$$g_{ij} = [X_i, \dot{X}_j].$$

We show how the form of the Levi-Civita connection appears naturally.
In the following we shall use D as an abbreviation for d/dt.
The Levi-Civita connection

$$\Gamma_{kij} = (1/2)(\nabla_i g_{jk} + \nabla_j g_{ik} - \nabla_k g_{ij})$$

associated with the g_{ij} comes up almost at once from the differentiation process described above. To see how this happens, view the following calculation where

$$\hat{\partial}_i \hat{\partial}_j F = [X_i, [X_j, F]].$$

We apply the operator $\hat{\partial}_i \hat{\partial}_j$ to the second time derivative of X_k.
Lemma 3.1 Let $\Gamma_{kij} = (1/2)(\nabla_i g_{jk} + \nabla_j g_{ik} - \nabla_k g_{ij})$ where $\nabla_i(F) = [F, \dot{X}_i]$, the covariant derivative generated by $\dot{X}_i = P_i - A_i$. Then

$$\Gamma_{kij} = (1/2)\hat{\partial}_i \hat{\partial}_j \ddot{X}_k.$$

Proof. Note that by the Leibniz rule

$$D([A,B]) = [A, \dot{B}] = [\dot{A}, B] + [A, \dot{B}],$$

we have

$$\dot{g_{jk}} = [\dot{X}_j, \dot{X}_k] + [X_j, \ddot{X}_k].$$

Therefore

$$\hat{\partial}_i \hat{\partial}_j \ddot{X}_k = [X_i, [X_j, \ddot{X}_k]]$$

$$= [X_i, \dot{g_{jk}} - [\dot{X}_j, \dot{X}_k]]$$

$$= [X_i, \dot{g_{jk}}] - [X_i, [\dot{X}_j, \dot{X}_k]]$$

(Now use the Jacobi identity $[A,[B,C]] + [C,[A,B]] + [B,[C,A]] = 0$.)

$$= [X_i, \dot{g_{jk}}] + [\dot{X}_k, [X_i, \dot{X}_j]] + [\dot{X}_j, [\dot{X}_k, X_i]]$$

$$= -[\dot{X}_i, g_{jk}] + [\dot{X}_k, [X_i, \dot{X}_j]] + [\dot{X}_j, [\dot{X}_k, X_i]]$$

$$= \nabla_i g_{jk} - \nabla_k g_{ij} + \nabla_j g_{ik}$$

$$= 2\Gamma_{kij}.$$

This completes the proof.
Remark. The upshot of this derivation is that it confirms our interpretation of

$$g_{ij} = [X_i, \dot{X}_j] = [X_i, P_j] - [X_i, A_j] = \delta_{ij} - \partial A_j / \partial P_i$$

as an abstract form of metric (in the absence of any actual notion of distance in the non-commutative world). *This calls for a re-evaluation and reconstruction of differential geometry based on non-commutativity and the Jacobi identity.* This is differential geometry where the fundamental concept is no longer parallel translation, but rather a non-commutative version of a physical trajectory.

Note that given

$$\Gamma_{kij} = (1/2)(\nabla_i g_{jk} + \nabla_j g_{ik} - \nabla_k g_{ij}),$$

we have

$$\Gamma_{ikj} = (1/2)(\nabla_k g_{ji} + \nabla_j g_{ki} - \nabla_i g_{kj}).$$

Hence

$$\Gamma_{kij} + \Gamma_{ikj} = \nabla_j g_{ik} = \partial_j g_{ik} + [g_{ik}, A_j].$$

In other words, we see that the Levi-Civita connection that we have derived differs from the classical Levi-Civita connection via the replacement of the partial derivatives ∂_j with the covariant derivatives $\nabla_j = \partial_j + A_j$ where $A_j(F) = [F, A_j]$. Since A_j is our analog of a gauge field, this formalism can be compared with the corresponding Levi-Civita connection in Weyl's theory combining aspects of general relativity with electromagnetism, and we see that this is exactly how Weyl did modify the Levi-Civita connection. See [20] Chapter 35, page 290 and [19] p. 85. By following the non-commutative world formalism we have reproduced the core of Weyl's original approach to gauge theory in a new context.

We have already seen how electromagnetism and gauge theories arise in this setting and indeed the motion of a particle as described by the equation [14, 12]

$$\ddot{X}_r = G_r + F_{rs}\dot{X}^s + \Gamma_{rst}\dot{X}^s\dot{X}^t$$

is also the analogue of a particle moving in the Weyl theory where the new Levi-Civita connection controls its acceleration. Note also that our metric coefficients g_{ij} are subject to the curvature associated with the gauge A_i and so undergo holonomy just as did Weyl's metric. It is useful to make these comparisons because the intent of the original Weyl theory was to extend general relativity to include electromagnetism in terms the holonomy associated with the metric. This full extension was not developed due to criticisms of the program and also because the gauge idea related directly to quantum theory in a highly fruitful way where the metric holonomy was replaced by holonomy for the phase of the wave function. Our present formalism suggests a new investigation of these connections that is beyonnd the scope of the present paper.

Nevertheless, note that classical general relativity begins with the standard Levi-Civita connection, and so defining curvature via its interpretation as a parallel displacement. To this end, recall the formalism of parallel translation. The infinitesimal parallel translate of A is denoted by $A' = A + \delta A$ where

$$\delta A^k = -\Gamma^k_{ij} A^i dX^j$$

where here we are writing in the usual language of vectors and differentials with the Einstein summation convention for repeated indices. We assume that the Christoffel symbols satisfy the symmetry condition $\Gamma^k_{ij} = \Gamma^k_{ji}$. The inner product is given by the formula

$$< A, B >= g_{ij} A^i B^j$$

Note that here the bare symbols denote vectors whose coordinates may be indicated by indices. The requirement that this inner product be invariant under parallel displacement is the requirement that $\delta(g_{ij}A^iA^j) = 0$. Calculating, one finds

$$\delta(g_{ij}A^iA^j) = (\partial_k g_{ij})A^iA^j dX^k + g_{ij}\delta(A^i)A^j + g_{ij}A^i\delta(A^j)$$

$$= (\partial_k g_{ij})A^i A^j dX^k - g_{ij}\Gamma^i_{rs}A^r dX^s A^j - g_{ij}A^i\Gamma^j_{rs}A^r dX^s$$

$$= (\partial_k g_{ij})A^i A^j dX^k - g_{ij}\Gamma^i_{rs}A^r A^j dX^s - g_{ij}\Gamma^j_{rs}A^i A^r dX^s$$

$$= (\partial_k g_{ij})A^i A^j dX^k - g_{sj}\Gamma^s_{ik}A^i A^j dX^k - g_{is}\Gamma^s_{jk}A^i A^j dX^k$$

Hence

$$(\partial_k g_{ij}) = g_{sj}\Gamma^s_{ik} + g_{is}\Gamma^s_{jk}.$$

From this it follows that

$$\Gamma_{ijk} = g_{is}\Gamma^s_{jk} = (1/2)(\partial_k g_{ij} - \partial_i g_{jk} + \partial_j g_{ik}).$$

Certainly these notions of variation can be imported into our abstract context. The question remains how to interpret the new connection that arises when we replace ∂_i with $\nabla_i = \partial_i + A_i$. This yields our new Levi-Civita connection (here denoted by the old symbol).

$$\Gamma_{ijk} = g_{is}\Gamma^s_{jk} = (1/2)(\nabla_k g_{ij} - \nabla_i g_{jk} + \nabla_j g_{ik}).$$

One could forge forward and define a new covariant derivative by the formula

$$\hat{\nabla}_i X^j = \partial_i X^j + \Gamma^j_{ki} X^k$$

and then attempt to head toward a generalization of general relativity via its curvature tensors. The question is how the curvature of this connection interfaces with the gauge potentials that gave rise to the metric in the first place. The theme of this investigation has the flavor of gravity theories with a qauge theoretic background. The difficulty is that due to the nature of the new covariant derivatives there is not known (to this author) a suitable analogue for the Einstein tensor. The new Riemann tensor will no longer have the many symmetries of the classical Riemann tensor and a simple copy of the original Einstein tensor will not have vanishing divergence.

All this said, the above becomes a significant program for further investigation. It would appear that this program was implicit in Weyl's original work as well, and examination of his papers and the book "Space Time Matter" [20] suggests that he did not carry out a full unification of his gauge theory with general relativity. Our approach suggests a new start on this problem.

4. Recapitulation - Curvature, Jacobi Identity and the Levi-Civita Connection

In this section, we go back to basics and examine the context of calculus defined via commutators. We recapitulate what we have accomplished so far, and set the stage for the next level of structure. We shall use a partially index-free notation. In this notation, we avoid nested subscripts by using different variable names and then using these names as subscripts to refer to the relevant variables. Thus we write X and Y instead of X_i and X_j, and we write g_{XY} instead of g_{ij}. It is assumed that the derivation DX has the form $DX = [X, J]$ for some J.

The bracket $[A, B]$ is not assumed to be a commutator. It is assumed to satisfy the Jacobi identity, bilinearity in each variable, and the Leibniz rule for all functions of the form $\delta_K(A) = [A, K]$. That is we assume that

$$\delta_K(AB) = \delta_K(A)B + A\delta_K(B).$$

Recall that in classical differential geometry one has the notion of a covariant dervative, defined by taking a difference quotient using parallel translation via a connection. Covariant derivatives in different directions do not necessarily commute. The commutator of covariant derivatives gives rise to the curvature tensor in the form

$$[\nabla_i, \nabla_j]X^k = R^k_{lij}X^l.$$

If derivatives do not commute then we regard their commutator as expressing a curvature. In our non-commutative context this means that curvature arises *prior* to any notion of covariant derivatives since *even the basic derivatives do not commute.*

We shall consider derivatives in the form

$$\nabla_X(A) = [A, \Lambda_X].$$

Examine the following computation:

$$\begin{aligned}\nabla_X\nabla_Y F = [[F, \Lambda_Y], \Lambda_X] &= -[[\Lambda_X, F], \Lambda_Y] - [[\Lambda_Y, \Lambda_X], F] \\ &= [[F, \Lambda_X], \Lambda_Y] + [[\Lambda_X, \Lambda_Y], F] \\ &= \nabla_Y\nabla_X F + [[\Lambda_X, \Lambda_Y], F].\end{aligned}$$

Thus

$$[\nabla_X, \nabla_Y]F = R_{XY}F$$

where

$$R_{XY}F = [[\Lambda_X, \Lambda_Y], F].$$

We can regard R_{XY} as a curvature operator.

The analog in this context of flat space is abstract quantum mechanics! That is, we assume position variables (operators) $X,\ Y,\ \cdots$ and momentum variables (operators) $P_X,\ P_Y,\ \cdots$ satisfying the equations below.

$$[X, Y] = 0$$

$$[P_X, P_Y] = 0$$

$$[X, P_Y] = \delta_{XY}$$

where δ_{XY} is equal to one if X equals Y and is zero otherwise. We define

$$\partial_X F = [F, P_X]$$

and

$$\partial_{P_X} F = [X, F].$$

In the context of the above commutation relations, note that these derivatives behave correctly in that

$$\partial_X(Y) = \delta_{XY}$$

and

$$\partial_{P_X}(P_Y) = \delta_{XY}$$

$$\partial_{P_X}(Y) = 0 = \partial_X(P_Y)$$

with the last equations valid even if $X = Y$. Note also that iterated partial derivatives such as $\partial_X\partial_Y$ commute. Hence the curvature R_{XY} is equal to zero. We shall regard these position and momentum operators and the corresponding partial derivatives as an abstract algebraic substitute for flat space.

With this reference point of (algebraic, quantum) flat space we can define

$$\hat{P_X} = P_X - A_X$$

for an arbitrary algebra-valued function of the variable X. In indices this would read

$$\hat{P_i} = P_i - A_i,$$

and with respect to this deformed momentum we have the covariant derivative

$$\nabla_X F = [F, \hat{P_Y}] = [F, P_Y + A_Y] = \partial_Y F + [F, A_Y].$$

The curvature for this covariant derivative is given by the formula

$$R_{XY}F = [\nabla_X, \nabla_Y]F = [[\lambda_X, \lambda_Y], F]$$

where $\lambda_X = P_X - A_X$. Hence

$$R_{XY} = [P_X - A_X, P_Y - A_Y] = -[P_X, A_Y] - [A_X, P_Y] + [A_X, A_Y]$$

$$= \partial_X A_Y - \partial_Y A_X + [A_X, A_Y].$$

With indices this reads

$$R_{ij} = \partial_i A_j - \partial_j A_i + [A_i, A_j].$$

and the reader will note that this has the abstract form of the curvature of a Yang-Mills gauge field, and specifically the form of the electromagnetic field when the potentials A_i and A_j commute with one another.

Continuing with this example, we compute

$$[X, \hat{P_Y}] = [X, P_Y - A_Y] = \delta_{XY} - [X, A_Y].$$

Let

$$g_{XY} = \delta_{XY} - [X, A_Y]$$

so that

$$[X, \hat{P_Y}] = g_{XY}.$$

We will shortly consider the form of this general case, but first it is useful to restrict to the case where $[X, A_Y] = 0$ so that $g_{XY} = \delta_{XY}$ (for the space coordinates). In order to enter this domain, we set

$$\dot{X} = DX = \hat{P_X} = P_X - A_X.$$

We then have

$$[X_i, X_j] = 0$$

$$[X_i, \dot{X_j}] = \delta_{ij}$$

and

$$R_{ij} = [\dot{X_i}, \dot{X_j}] = \partial_i A_j - \partial_j A_i + [A_i, A_j].$$

Generalizing, we wish to examine the structure of the following special axioms for a bracket.

$$[X, DY] = g_{XY}$$

$$[X, Y] = 0$$

$$[Z, g_{XY}] = 0$$

$$[g_{XY}, g_{ZW}] = 0$$

Note that

$$Dg_{YZ} = D[Y, DZ] = [DY, DZ] + [Y, D^2Z].$$

and that $D[X, g_{XY}] = 0$ implies that

$$[g_{XY}, DZ] = [Z, Dg_{XY}].$$

Define two types of derivations as follows

$$\nabla_X(F) = [F, DX]$$

and

$$\nabla_{DX}(F) = [X, F].$$

These are dual with respect to g_{XY} and will act like partials with respect to these variables in the special case when g_{XY} is a Kronecker delta, δ_{XY}. If the form g_{XY} is invertible, then we can rewrite these derivations by contracting the inverse of g to obtain standard formal partials.

$$\nabla_{DX}\nabla_{DY}D^2Z = [X, [Y, D^2Z]]$$

$$= [X, Dg_{YZ} - [DY, DZ]] = [X, Dg_{YZ}] - [X, [DY, DZ]]$$

$$= [g_{YZ}, DX] - [X, [DY, DZ]]$$

$$= \nabla_X(g_{YZ}) - [X, [DY, DZ]].$$

Now use the Jacobi identity on the second term and obtain

$$\nabla_{DX}\nabla_{DY}D^2Z = \nabla_X(g_{YZ}) + [DZ, [X, DY]] + [DY, [DZ, X]]$$

$$= \nabla_X(g_{YZ}) - \nabla_Z(g_{XY}) + \nabla_Y(g_{XZ}).$$

This is the formal Levi-Civita connection.

At this stage we face once again the mystery of the appearance of the Levi-Civita connection. We have seen in this section that it is quite natural for curvature in the form of the non-commutativity of derivations to appear at the outset in a non-commutative formalism. We have also see that this curvature and connection can be understood as a measurement of the deviation of the theory from the flat commutation relations of ordinary quantum mechanics. Electromagnetism and Yang-Mills theory can be seen as the theory of the curvature introduced by such a deviation. On the other hand, from the point of view of metric differential geometry, the Levi-Civita connection is the unique connection that preserves the inner product defined by the metric under the parallel translation defined by the connection.

5. Appendix – Einstein's Equations and the Bianchi Identity

The purpose of this section is to show how the Bianchi identity (see below for its definition) appears in the context of non-commutative worlds. The Bianchi identity is a crucial mathematical ingredient in general relativity. We shall begin with a quick review of the mathematical structure of general relativity (see for example [3]) and then turn to the context of non-commutative worlds.

The basic tensor in Einstein's theory of general relativity is

$$G^{ab} = R^{ab} - \frac{1}{2} R g^{ab}$$

where R^{ab} is the Ricci tensor and R the scalar curvature. The Ricci tensor and the scalar curvature are both obtained by contraction from the Riemann curvature tensor R^a_{bcd} with $R_{ab} = R^c_{abc}, R^{ab} = g^{ai}g^{bj}R_{ij}$, and $R = g^{ij}R_{ij}$. Because the Einstein tensor G^{ab} has vanishing divergence, it is a prime candidate to be proportional to the energy momentum tensor $T^{\mu\nu}$. The Einstein field equations are

$$R^{\mu\nu} - \frac{1}{2} R g^{\mu\nu} = \kappa T^{\mu\nu}.$$

The reader may wish to recall that the Riemann tensor is obtained from the commutator of a covariant derivative ∇_k, associated with the Levi-Civita connection $\Gamma^i_{jk} = (\Gamma_k)^i_j$ (built from the space-time metric g_{ij}). One has

$$\lambda_{a:b} = \nabla_b \lambda_a = \partial_b \lambda_a - \Gamma^d_{ab} \lambda_d$$

or

$$\lambda_{:b} = \nabla_b \lambda = \partial_b \lambda - \Gamma_b \lambda$$

for a vector field λ. With

$$R_{ij} = [\nabla_i, \nabla_j] = \partial_j \Gamma_i - \partial_i \Gamma_j + [\Gamma_i, \Gamma_j],$$

one has

$$R^a_{bcd} = (R_{cd})^a_b.$$

(Here R_{cd} is *not* the Ricci tensor. It is the Riemann tensor with two internal indices hidden from sight.)

One has explicitly that [3]

$$R_{\mu\nu\rho\sigma} = \frac{1}{2}(g_{\mu\sigma,\nu\rho} - g_{\nu\sigma,\mu\rho} - g_{\mu\rho,\nu\sigma} + g_{\nu\rho,\mu\sigma}) + \Gamma_{\beta\mu\sigma}\Gamma^{\beta}_{\nu\rho} + \Gamma_{\beta\mu\rho}\Gamma^{\beta}_{\nu\rho}.$$

Many symmetries of the Riemann tensor follow from this formula. If the derivatives in this formula are replaced by covariant derivatives, the symmetries do not all survive and the story we are about to tell about the divergence of the Einstein tensor will have to be modified. That is a project for future work.

One way to understand the mathematical source of the Einstein tensor, and the vanishing of its divergence, is to see it as a contraction of the Bianchi identity for the Riemann tensor. The Bianchi identity states

$$R^a_{bcd:e} + R^a_{bde:c} + R^a_{bec:d} = 0$$

where the index after the colon indicates the covariant derivative. Note also that this can be written in the form

$$(R_{cd:e})^a_b + (R_{de:c})^a_b + (R_{ec:d})^a_b = 0.$$

Journal of Physics: Conference Series 2081 (2021) 012006 doi:10.1088/1742-6596/2081/1/012006

The Bianchi identity is a consequence of local properties of the Levi-Civita connection and consequent symmetries of the Riemann tensor. One relevant symmetry of the Riemann tensor is the equation $R^a_{bcd} = -R^a_{bdc}$.

We will not give a classical derivation of the Bianchi identity here, but it is instructive to see how its contraction leads to the Einstein tensor. To this end, note that we can contract the Bianchi identity to

$$R^a_{bca:e} + R^a_{bae:c} + R^a_{bec:a} = 0$$

which, in the light of the above definition of the Ricci tensor and the symmetries of the Riemann tensor is the same as

$$R_{bc:e} - R_{be:c} + R^a_{bec:a} = 0.$$

Contract this tensor equation once more to obtain

$$R_{bc:b} - R_{bb:c} + R^a_{bbc:a} = 0,$$

and raise indices

$$R^b_{c:b} - R_{:c} + R^{ab}_{bc:a} = 0.$$

Further symmetry gives

$$R^{ab}_{bc:a} = R^{ba}_{cb:a} = R^a_{c:a} = R^b_{c:b}.$$

Hence we have

$$2R^b_{c:b} - R_{:c} = 0,$$

which is equivalent to the equation

$$(R^b_c - \frac{1}{2}R\delta^b_c)_{:b} = G^b_{c:b} = 0.$$

From this we conclude that $G^{bc}_{:b} = 0$. The Einstein tensor has appeared on the stage with vanishing divergence, courtesy of the Bianchi identity!

Bianchi Identity and Jacobi Identity. Now lets turn to the context of non-commutative worlds. We have infinitely many possible convariant derivatives, all of the form

$$F_{:a} = \nabla_a F = [F, N_a]$$

for some N_a elements in the non-commutative world. Choose any such covariant derivative. Then, as in the introduction to this paper, we have the curvature

$$R_{ij} = [N_i, N_j]$$

that represents the commutator of the covariant derivative with itself in the sense that $[\nabla_i, \nabla_j]F = [[N_i, N_j], F]$. Note that R_{ij} is not a Ricci tensor, but rather the indication of the external structure of the curvature without any particular choice of linear representation (as is given in the classical case as described above). We then have the Jacobi identity

$$[[N_a, N_b], N_c] + [[N_c, N_a], N_b] + [[N_b, N_c], N_a] = 0.$$

Writing the Jacobi identity in terms of curvature and covariant differention we have

$$R_{ab:c} + R_{ca:b} + R_{bc:a}.$$

Thus in a non-commutative world, every covariant derivative satisfies its own Bianchi identity. This gives an impetus to study general relativity in non-commutative worlds by looking for covariant derivatives that satisfy the symmetries of the Riemann tensor and link with a metric in an appropriate way. We have only begun this aspect of the investigation. The point of this section has been to show the intimate relationship between the Bianchi idenity and the Jacobi identity that is revealed in the context of non-commutative worlds.

6. Appendix - Discrete Calculus Reformulated with Commutators

There are many motivations for replacing derivatives by commutators. If $f(x)$ denotes (say) a function of a real variable x, and $\tilde{f}(x) = f(x+h)$ for a fixed increment h, define the *discrete derivative* Df by the formula $Df = (\tilde{f} - f)/h$, and find that the Leibniz rule is not satisfied. One has the basic formula for the discrete derivative of a product:

$$D(fg) = D(f)g + \tilde{f}D(g).$$

Correct this deviation from the Leibniz rule by introducing a new non-commutative operator J with the property that

$$fJ = J\tilde{f}.$$

Define a new discrete derivative in an extended non-commutative algebra by the formula

$$\nabla(f) = JD(f).$$

It follows at once that

$$\nabla(fg) = JD(f)g + J\tilde{f}D(g) = JD(f)g + fJD(g) = \nabla(f)g + f\nabla(g).$$

Note that

$$\nabla(f) = (J\tilde{f} - Jf)/h = (fJ - Jf)/h = [f, J/h].$$

In the extended algebra, discrete derivatives are represented by commutators, and satisfy the Leibniz rule. One can regard discrete calculus as a subset of non-commutative calculus based on commutators.

Discrete Measurement. Consider a time series $\{X, X', X'', \cdots\}$ with commuting scalar values. Let

$$\dot{X} = \nabla X = JDX = J(X' - X)/\tau$$

where τ is an elementary time step (If X denotes a times series value at time t, then X' denotes the value of the series at time $t + \tau$.). The shift operator J is defined by the equation $XJ = JX'$ where this refers to any point in the time series so that $X^{(n)}J = JX^{(n+1)}$ for any non-negative integer n. Moving J across a variable from left to right, corresponds to one tick of the clock. This discrete, non-commutative time derivative satisfies the Leibniz rule.

This derivative ∇ also fits a significant pattern of discrete observation. Consider the act of observing X at a given time and the act of observing (or obtaining) DX at a given time. Since X and X' are ingredients in computing $(X' - X)/\tau$, the numerical value associated with DX, it is necessary to let the clock tick once, Thus, if one first observe X and then obtains DX, the result is different (for the X measurement) if one first obtains DX, and then observes X. In the second case, one finds the value X' instead of the value X, due to the tick of the clock.

(i) Let $\dot{X}X$ denote the sequence: observe X, then obtain $\dot{X}$.
(ii) Let $X\dot{X}$ denote the sequence: obtain $\dot{X}$, then observe X.

The commutator $[X, \dot{X}]$ expresses the difference between these two orders of discrete measurement. In the simplest case, where the elements of the time series are commuting scalars, one has

$$[X, \dot{X}] = X\dot{X} - \dot{X}X = J(X' - X)^2/\tau.$$

Thus one can interpret the equation

$$[X, \dot{X}] = Jk$$

(k a constant scalar) as

$$(X' - X)^2/\tau = k.$$

This means that the process is a walk with spatial step

$$\Delta = \pm\sqrt{k\tau}$$

where k is a constant. In other words, one has the equation

$$k = \Delta^2/\tau.$$

This is the diffusion constant for a Brownian walk. A walk with spatial step size Δ and time step τ will satisfy the commutator equation above exactly when the square of the spatial step divided by the time step remains constant. This shows that the diffusion constant of a Brownian process is a structural property of that process, independent of considerations of probability and continuum limits.

References

[1] Baez J and Muniain J P 1994 *Gauge Fields, Knots and Gravity* (Singapore: World Scientific)
[2] Dyson F J 1990 *Am. J. Phys.* **58** 209
[3] Dirac P A M 1975 *General Theory of Relativity* (New York: John Wiley and Sons)
[4] Hughes R J 1992 *Am. J. Phys.* **60** 301
[5] Kauffman L H 1991 *Knots and Physics* (World Scientific Pub.)
[6] Kauffman L H and Noyes H Pierre 1996 *Proc. of the Royal Soc. Lond. A* **452** 81
[7] Kauffman L H and Noyes H Pierre 1996 *Physics Letters A* **218** 139
[8] Kauffman L H 1996 *Twistor Newsletter* **41**
[9] Kauffman L H 1998 *Physica D* **120** 125
[10] Kauffman L H 1998 *Intl. J. Gen. Syst.* **27** 241
[11] Kauffman L H 1999 A non-commutative approach to discrete physics *Aspects II - Proceedings of ANPA 20* 215
[12] Kauffman L H 2000 Differential geometry in non-commutative worlds *Quantum Gravity* pp 61-76
[13] Kauffman L H 2003 Non-commutative calculus and discrete physics *Boundaries- Scientific Aspects of ANPA 24* pp 73-128
[14] Kauffman L H 2004] *New Journal of Physics* **6** 2
[15] Kauffman Louis H 2018 *Entropy* **7** 25
[16] Deakin A M and Kauffman L H 2018 Cosmological theories of the extra terms *Chiasmus - Proceedings of ANPA* **37, 38** 107-202
[17] Montesinos M and Perez-Lorenzana A 1999 Minimal coupling and Feynman's proof *arXiv:quant-phy/9810088*
[18] Tanimura Shogo 1992 *Annals of Physics* **220** 229
[19] O'Raifeartaigh L 1997 *The Dawning of Gauge Theory* (Princeton, N. J.: Princeton University Press)
[20] Weyl H 1952 *Space - Time - Matter* (New York: Dover Publications)

The field-theoretical methods in Lovelock gravity

A N Petrov

Sternberg Astronomical Institute, M.V. Lomonosov Moscow State University, Universitetskii pr. 13, Moscow 119992, Russia

E-mail: alex.petrov55@gmail.com

Abstract. The field-theoretical methods are used to construct conserved currents and related superpotentials for perturbations on arbitrary backgrounds in the Lovelock gravity. The perturbations are considered as a dynamic field configuration propagating in a given spacetime. The field-theoretical formalism is exact (without approximations) and equivalent to the original metric theory. As Lagrangian based formalism, it allows us to apply the Noether theorem. As a result, we construct conserved currents and superpotentials, where we use arbitrary displacement vectors, not only the Killing ones or other special vectors. The developed formalism is checked in calculating mass of the Schwarzschild-anti-de Sitter (AdS) black hole. The new formalism is adopted to the case of a so-called pure Lovelock gravity, where in the Lagrangian only a one polynomial in Riemannian tensor presents. We construct conserved charges and currents for static and dynamic black holes of the Vaidya type with AdS, dS and flat asymptotics. New properties of the solutions under consideration have been found. The more results are discussed. The first section in your paper

1. Introduction

1.1. Motivation

The Lovelock gravity [1] becomes very popular in the last years. The researchers develop its theoretical foundations, find various solutions and analyze them, see, for example [2,3] and references therein. Many methods have been suggested to construct conserved quantities in the Lovelock theory. However, by our opinion, these approaches do not cover all the arising problems. In particular, construction of conserved quantities for perturbations is restricted, as a rule, to maximally symmetric backgrounds, not arbitrary ones; one chooses usually Killing vectors as vectors of displacements, not other vectors.

1.2. Goals

Keeping in mind the above, we plan to close the mentioned gap, at least, particularly. The theoretical goal of the present paper is to develop a so-called field-theoretical formalism (as a more universal one) in metric theories [4-7] in the framework of the Lovelock gravity. We check it, calculating the mass of the Schwarzschild-anti-de Sitter (AdS) black hole. The other goal is to adopt the new general formalism to the case of a so-called pure Lovelock gravity, see, for example [8] and references therein. Then, using powerful possibilities of the formalism, we construct conserved quantities for Vaidya-type black holes (static and dynamic) in the pure Lovelock gravity, solutions for which have been suggested in [9,10]. We plan to construct both global and quasi-local charges (energy expressions) presented both on the backgrounds of asymptotic solutions and backgrounds of related static black holes. Besides, our formalism is valid to calculate energy densities and densities of energy fluxes measured by freely falling observers on the backgrounds of static black holes. The presentation is based on our papers [11,12] and one of the goals here is to unite their results and attract to them the attention of readers.

2. The field-theoretical formalism in an arbitrary metric theory

Let us recall the history of development of the field-theoretical methods. Many solutions in general relativity are considered when perturbations are propagating on various fixed backgrounds. Then equations are reformulated in a perturbed form. A part linear in metric perturbations is placed at the left hand side, whereas all the other terms are transferred to the right hand side and interpreted as an effective energy-momentum. For example, choosing as a background Minkowski space with the metric $\eta^{\mu\nu}$, defining the metric perturbations as $h^{\mu\nu} = g^{\mu\nu} - \eta^{\mu\nu}$, one has:

$$G^L_{\mu\nu}\left(h\right) = \kappa t^{eff}_{\mu\nu}. \tag{1}$$

Here and below, Greek indices mean spacetime components; Latin indices mean space components. Because the divergence of the left hand side of (1) disappears identically, $\partial_\nu G^{\mu\nu}_L\left(h\right) \equiv 0$, one obtains the conservation law for the effective energy-momentum $\partial_\nu t^{\mu\nu}_{eff} = 0$.

In 1950-th and 1960-th the above picture has been provided by iterations and approximations. In the famous work [13], Deser has united the previous results in the exact and closed form:

$$G^L_{\mu\nu}\left(h\right) = \kappa t^{tot}_{\mu\nu}\left(h, \Gamma, \phi^A\right). \tag{2}$$

It is obtained in the framework of the Lagrangian based method by varying the related action with respect to $h^{\mu\nu}$; the total energy-momentum is obtained by varying with respect to the background metric; it was used the first order formalism with variables $h^{\mu\nu}$ and $\Gamma^\alpha_{\beta\gamma}$; ϕ^A is a generalized matter variable. However, Deser's formalism has been presented for the Minkowski background only. We [4] have generalized his approach in general relativity for the case of arbitrary curved backgrounds including a background matter. Later we have developed this approach for the case of many modified metric theories and examined various solutions of these theories [5-7].

Because the field-theoretical methods are quite universal, we outline them below on the example of an arbitrary covariant metric theory in D dimensions. A unique requirement is that its field equations are not more than of second order in derivatives, like in the Lovelock gravity. Let the Lagrangian of such a theory be

$$L = -\frac{1}{2\kappa} L_g\left(g_{\mu\nu}\right) + L_m\left(g_{\mu\nu,} \Phi^A\right), \tag{3}$$

with metric $g_{\mu\nu,}$ and generalized matter Φ^A variables. Here and below, $\kappa = 2\Omega_{D-2} G_D$, where G_D is D-dimensional Newton's constant, Ω_{D-2} is a square of $(D-2)$-dimensional sphere with unit radius. The system (3) is considered as a perturbed one with respect to the background system:

$$\bar{L} = -\frac{1}{2\kappa} \bar{L}_g + \bar{L}_m, \tag{4}$$

where $\bar{L} = L(\bar{g}_{\mu\nu}, \bar{\Phi}^A)$, $\bar{L}_g = L_g(\bar{g}_{\mu\nu})$, $\bar{L}_m = L_m(\bar{g}_{\mu\nu}, \bar{\Phi}^A)$; $\bar{g}_{\mu\nu}$ and $\bar{\Phi}^A$ present a known (fixed) solution of the theory. Next we define perturbations:

$$\begin{aligned} \kappa_{\mu\nu} &= g_{\mu\nu} - \bar{g}_{\mu\nu}, \\ \phi^A &= \Phi^A - \bar{\Phi}^A \end{aligned} \tag{5}$$

and interpret them as a configuration of dynamic variables. Generalizing the Deser prescription [13], we define the Lagrangian for $\kappa_{\mu\nu}$ and ϕ^A in the form [5]:

$$L^{dyn} = L\left(\overline{g} + \kappa, \overline{\Phi} + \phi\right) - \kappa_{\mu\nu}\frac{\delta \overline{L}}{\delta \overline{g}_{\mu\nu}} + \phi^A \frac{\delta \overline{L}}{\delta \overline{\Phi}^A} - \overline{L}\,. \tag{6}$$

Varying this Lagrangian with respect to $\kappa_{\mu\nu}$ one obtains the field equations in the form:

$$G^L_{\mu\nu} + F^L_{\mu\nu} = \kappa\left(t^g_{\mu\nu} + t^m_{\mu\nu}\right) = \kappa t_{\mu\nu}\,, \tag{7}$$

where the left hand side is linear in $\kappa_{\mu\nu}$ and ϕ^A. The right hand side of (7) is the metric energy-momentum tensor density defined for the system (6): $t_{\mu\nu} = 2\delta L^{dyn}/\delta \overline{g}^{\mu\nu}$. In the case of a vacuum background, $\overline{\Phi}^A = 0$, one has $F^L_{\mu\nu} = 0$ and $\overline{\nabla}^\nu G^L_{\mu\nu} \equiv 0$. Then, one has again the conservation law for the energy-momentum: $\overline{\nabla}^\nu t_{\mu\nu} = 0$. Here and below $\overline{\nabla}_\nu$ means a covariant derivative in the background spacetime. In the case of a non-vacuum background one has $F^L_{\mu\nu} \neq 0$ and $\overline{\nabla}^\nu (G^L_{\mu\nu} + F^L_{\mu\nu}) \neq 0$. As a result, there is no a conservation for the energy-momentum in general. The problem is resolved only when one constructs a conserved current including $t_{\mu\nu}$ together with terms, which interact with a complicated background.

To obtain conserved currents one can apply the Noether procedure directly to the Lagrangian (6). However, in this case one meets a very complicated calculations. There is another more economical way. Let us consider an auxiliary Lagrangian:

$$L_1 = -\frac{1}{2\kappa}\kappa_{\mu\nu}\frac{\delta \overline{L}_g}{\delta \overline{g}_{\mu\nu}} \tag{8}$$

that is linear in $\kappa_{\mu\nu}$ and is a part of (6). Because L_1 is a scalar density one can use the diffeomorphism invariance and apply the Noether theorem to (8). As a result, one obtains the identities:

$$\overline{\nabla}_\mu i^\mu_1 \equiv \partial_\mu i^\mu_1 \equiv 0 \tag{9}$$

for the current i^μ_1 (vector density) that is expressed through a divergence of a superpotential $i^{\mu\nu}_1$ (antisymmetrical tensor density) as

$$i^\mu_1 \equiv \overline{\nabla}_\nu i^{\mu\nu}_1 \equiv \partial_\nu i^{\mu\nu}_1\,. \tag{10}$$

However, (9) and (10) are the identities only. To transform them to physically sensible conservation laws we use the field equations (7). In the result one obtains:

$$\overline{\nabla}_\mu I^\mu = \partial_\mu I^\mu = 0\,, \tag{11}$$

$$I^\mu = \overline{\nabla}_\nu I^{\mu\nu} = \partial_\nu I^{\mu\nu}\,. \tag{12}$$

Thus, we have a differential conservation law (11) for the current I^μ.

It is important to consider its structure $I^\mu = \tau^{\mu\nu}\xi_\nu - z^\mu_1(\xi)$. The generalized energy-momentum has the form:

$$\tau_{\mu\nu} = t_{\mu\nu} - \frac{1}{\kappa}F^L_{\mu\nu} + \frac{1}{2\kappa}\frac{\delta \overline{L}_g}{\delta \overline{g}_{\alpha\beta}}\left(\kappa_{\alpha\beta}\overline{g}_{\mu\nu} - \kappa_{\alpha\mu}\overline{g}_{\beta\nu} - \kappa_{\alpha\nu}\overline{g}_{\beta\mu}\right). \tag{13}$$

Above we have remarked that $t_{\mu\nu}$ itself is not conserved on complicated backgrounds with a background matter. Formula (13) shows this explicitly. Indeed, only with interacting terms in (13) we have a conservation law for the current in (11). The term $z^\mu_1(\xi)$ disappears if the displacement vector ξ^μ is a Killing vector of the background spacetime. The structure of the superpotential is important as well:

$$I^{\mu\nu} = \tfrac{3}{4}\left(2\xi^\sigma \overline{\nabla}_\lambda \omega_\sigma{}^{[\mu|\nu]\lambda} - \omega_\sigma{}^{[\mu|\nu]\lambda}\overline{\nabla}_\lambda \xi^\sigma\right). \tag{14}$$

The key quantity here that is valid for an arbitrary background is defined as $\omega^{\rho\lambda|\mu\nu} = \partial L_1 / \partial \bar{g}_{\rho\lambda|\mu\nu}$ with the symmetry properties $\omega^{\rho\lambda|\mu\nu} = \omega^{\lambda\rho|\mu\nu} = \omega^{\rho\lambda|\nu\mu} = \omega^{\mu\nu|\rho\lambda}$.

Differential conservation laws give a possibility to define integral conserved quantities. Thus, the integration of (11) leads to constructing a quantity

$$P(\xi) = \int_\Sigma d^{D-1}x I^0(\xi) \tag{15}$$

defined on a spacelike section $\Sigma := x^0 = const$; interpretation of $P(\xi)$ is determined by a choice of ξ^μ. The conservation law (12) allows us to represent (15) in the form of a surface integral:

$$P(\xi) = \oint_{\partial\Sigma} ds_i I^{0i}(\xi), \tag{16}$$

where $\partial\Sigma$ is a boundary of Σ, and ds_i is a coordinate volume element at $\partial\Sigma$.

3. Conserved quantities for perturbations in Lovelock gravity

3.1. The Lovelock formulae in general

All the formulae from the above section can be used for concrete calculations in the Lovelock gravity. We consider the Lagrangian (3), where the pure gravitational part presents the Lovelock Lagrangian in D dimensions [1]:

$$L_\ell = \sqrt{-g} \sum_{p=0}^{m} \frac{\alpha_p}{2^p} \delta^{[\mu_1\mu_2\ldots\mu_{2p}]}_{[\nu_1\nu_2\ldots\nu_{2p}]} R^{\nu_1\nu_2}_{\mu_1\mu_2} \ldots R^{\mu_{2p-1}\mu_{2p}}_{\nu_{2p-1}\nu_{2p}} . \tag{17}$$

Then the auxiliary Lagrangian (8) acquires the explicit form:

$$L_{1\ell} = -\frac{\sqrt{-g}}{2\kappa} \kappa^\rho_\sigma \sum_{p=0}^{m} \frac{\alpha_p}{2^{p+1}} \delta^{[\sigma\mu_1\mu_2\ldots\mu_{2p}]}_{[\rho\nu_1\nu_2\ldots\nu_{2p}]} \bar{R}^{\nu_1\nu_2}_{\mu_1\mu_2} \ldots \bar{R}^{\mu_{2p-1}\mu_{2p}}_{\nu_{2p-1}\nu_{2p}} . \tag{18}$$

The key quantity, $\omega^{\rho\lambda|\mu\nu}$, with the use of which one can construct both superpotential (14) and current (12), is:

$$\omega^{\rho\lambda|\mu\nu} = -\frac{\sqrt{-g}}{2\kappa} \kappa^\alpha_\beta \sum_{p=1}^{m} \frac{p\alpha_p}{2^{p+1}} \delta^{[\beta\pi\sigma\mu_3\mu_4\ldots\mu_{2p}]}_{[\alpha\varphi\psi\nu 3\nu_4\ldots\nu_{2p}]} \bar{R}^{\nu_1\nu_2}_{\mu_1\mu_2} \ldots \bar{R}^{\mu_{2p-1}\mu_{2p}}_{\nu_{2p-1}\nu_{2p}} \bar{g}^{\varphi\tau} \bar{g}^{\psi\kappa} D^{\rho\lambda\mu\nu}_{\pi\sigma\tau\kappa} ; \tag{19}$$

$$D^{\rho\lambda\mu\nu}_{\pi\sigma\tau\kappa} \equiv \tfrac{1}{2}\left(\delta^\rho_\pi \delta^\lambda_\kappa + \delta^\rho_\kappa \delta^\lambda_\pi\right)\left(\delta^\mu_\sigma \delta^\nu_\tau + \delta^\mu_\tau \delta^\nu_\sigma\right).$$

The main advantage of our approach with respect to others is that it is quite universal. By this, we have a possibility, first, to study perturbations on arbitrary backgrounds, including a background matter, second, to use arbitrary displacement vectors.

3.2. Perturbations on AdS background

To check the formalism one has to apply it to calculate conserved quantities for known models. One of them is the Schwarzschild-AdS black hole, and we calculate a global mass for it. From the start we adopt our formulae for perturbations propagating on AdS backgrounds. Thus, let us derive vacuum background equations in the Lovelock gravity:

$$\frac{\delta \bar{L}_g}{\delta \bar{g}_{\rho\sigma}} = \sqrt{-\bar{g}}\, \bar{g}^{\pi\rho} \sum_{p=0}^{m} \frac{\alpha_p}{2^{p+1}} \delta^{[\sigma\mu_1\mu_2\ldots\mu_{2p}]}_{[\pi\nu_1\nu_2\ldots\nu_{2p}]} \bar{R}^{\nu_1\nu_2}_{\mu_1\mu_2} \ldots \bar{R}^{\mu_{2p-1}\mu_{2p}}_{\nu_{2p-1}\nu_{2p}} = 0 . \tag{20}$$

The AdS space has to satisfy these equations with the Riemannian tensor $\bar{R}^{\rho\lambda}_{\mu\nu} = -\ell^{-2}_{eff} \delta^{[\rho\lambda]}_{[\mu\nu]}$, where the effective AdS radius ℓ_{eff} is to be the solution to the equation:

$$V(x)\big|_{x=\ell^{-2}_{eff}} = \sum_{p=0}^{m} \frac{(D-3)!}{(D-2p-1)!} \alpha_p (-1)^{p-1} \left(\ell^{-2}_{eff}\right)^p = 0 . \tag{21}$$

The AdS metric in a more popular form is

$$d\bar{s}^2 = -\bar{f}(r)dt^2 + \bar{f}^{-1}(r)dt^2 + r^2 \sum_{a,b}^{D-2} q_{ab} dx^a dx^b, \tag{22}$$

where $\bar{f}(r) = 1 + r^2 \ell_{eff}^{-2}$ and q_{ab} is the metric at the $(D-2)$ -dimensional unit sphere. Then the quantity (19) for the metric perturbations (5) on the background (22) becomes

$$\omega^{\rho\lambda|\mu\nu} = \omega_h^{\rho\lambda|\mu\nu} V'(x)\big|_{x=\ell_{eff}^{-2}} = \omega_h^{\rho\lambda|\mu\nu} \left[\sum_{p=1}^{m} \frac{(D-3)!}{(D-2p-1)!} \alpha_p \left(\ell_{eff}^{-2}\right)^{p-1} \right], \tag{23}$$

where $\omega_h^{\rho\lambda|\mu\nu}$ is the quantity for the Einstein theory in D dimensions:

$$\omega_h^{\rho\lambda|\mu\nu} = -\frac{\sqrt{-\bar{g}}}{4\kappa} \left[\bar{g}^{\mu\nu}\kappa^{\rho\lambda} + \bar{g}^{\rho\lambda}\kappa^{\mu\nu} - \bar{g}^{\rho(\mu}\kappa^{\nu)\lambda} - \bar{g}^{\lambda(\mu}\kappa^{\nu)\rho} - \kappa(\bar{g}^{\mu\nu}\bar{g}^{\rho\lambda} - \bar{g}^{\rho(\mu}\bar{g}^{\nu)\lambda} \right] . \tag{24}$$

Finally, superpotential on the background (22) acquires a concrete form:

$$I^{\mu\nu}(\xi) = \frac{\sqrt{-\bar{g}}}{\kappa} V'(\ell_{eff}^{-2}) \left[\xi_\rho \bar{\nabla}^{[\mu}\kappa^{\nu]\rho} - \xi^{[\mu}\bar{\nabla}_\rho \kappa^{\nu]\rho} + \xi^{[\mu}\bar{\nabla}^{\nu]}\kappa + \kappa^{\rho[\mu}\bar{\nabla}^{\nu]}\xi_\rho + \tfrac{1}{2}\kappa\bar{\nabla}^{[\mu}\xi^{\nu]} \right] . \tag{25}$$

3.3. *Mass of the Schwarzschild-AdS black hole*

Now, we will check the formula (25) to calculate the mass of the Schwarzschild-AdS black hole. After substituting its metric in the form:

$$ds^2 = -f(r)dt^2 + f^{-1}(r)dt^2 + r^2 \sum_{a,b}^{D-2} q_{ab} dx^a dx^b \tag{26}$$

into the vacuum equations (20) and integrating their *rr*-component one obtains

$$\sum_{p=0}^{m} \frac{\alpha_p}{(D-2p-1)!} \left(\frac{1-f}{r^2} \right)^p = \frac{m_0}{(D-3)!\, r^{D-1}}, \tag{27}$$

where m_0 is a constant of integration (mass parameter). In general case it is impossible to derive f, However, we assume that there is a black hole solution. Then one can define asymptotics of f. In this case we have a possibility to define the asymptotics of perturbations:

$$\Delta f = f - \bar{f} \sim \frac{1}{V'(\ell_{eff}^{-2})} \frac{m_0}{r^{D-3}}; \qquad \kappa_{00} \sim -\Delta f, \quad \kappa_{11} \sim -\Delta f / \bar{f}^2 . \tag{28}$$

To calculate the mass we choose a displacement vector as a timelike Killing vector $\xi^\mu = (-1, 0, ..., 0)$, take perturbations (28) and the background metric (22). Then we substitute the superpotential (25) into (16) and obtain:

$$M = \lim_{r\to\infty} \oint_{\partial\Sigma} dx^{D-2} I^{01} = \frac{D-2}{2\kappa} m_0 \oint_{\partial\Sigma} dx^{D-2} \sqrt{\det q_{ij}} = \frac{D-2}{4G_D} m_0 . \tag{29}$$

It is the standard result that has been obtained in many researches, see, for example, [14-16].

4. The pure Lovelock gravity and conserved quantities for the Vaidya type black holes

The Lovelock theory of the general type (17) has $p+1$ connection constants. In this case, a physical interpretation of them becomes unclear. Therefore, in many researches authors reduce the number of these constants by various ways. One of the very popular ways is to preserve α_0 (it can be $\alpha_0 = 0$ or $\alpha_0 \neq 0$) and the unique $\alpha_p \neq 0$ from other constants in (17). Such a theory is called a *pure* Lovelock gravity. The key expression (19), which totally determines the superpotential (14), is simplified correspondently to this assumption, namely, one has the term with α_0 and the other term with $\alpha_p \neq 0$ without summing Σ. In our calculations to simplify expressions we use the redefinition of the connection constants: $\alpha_0(D-1)^{-1}(D-2)^{-1} \equiv -\ell^{-2}$ and $\alpha_p(D-3)![(D-2p-1)!]^{-1} \equiv \alpha^{2p-2}$.

4.1. Generalized Vaidya solutions

Below, applying the field-theoretical formalism, we calculate conserved quantities both for static [9] and for dynamic black holes of the Vaidya type constructed in [10]. These solutions are presented in the general form:

$$ds^2 = -f(r,v)dv^2 + 2dvdr + r^2 q_{ij} dx^i dx^j, \tag{30}$$

where the null coordinate $v = x^0$, and $r = x^1$. In this subsection we outline main properties of these solutions. It is known that for the classic Vaidya solution one considers the energy-momentum tensor of the null fluid: $T_{\alpha\beta} = \mu n_\alpha n_\beta$, where n_α are ingoing unit null vectors. The authors of [10] have generalized this energy-momentum in the pure Lovelock gravity to

$$T_{\alpha\beta} = \mu n_\alpha n_\beta + P_r\left(n_\alpha l_\beta + l_\alpha n_\beta\right) + P_\theta r^2 q_{\alpha\beta}. \tag{31}$$

Here, l^α are outgoing unit null vectors; $P_r = -C(v) r^{-(D-1)(1-\sigma)}$ is a radial pressure of the null fluid with undetermined up to now function $C(v)$; $P_\theta = -\sigma P_r$ is a tangential pressure of the null fluid with a constant σ. The energy density is defined as $\mu = \kappa^{-1}\left(\dot{m}(v) + \dot{C}(v)\Theta(r)\right) r^{-(D-2)}$. The dominant energy conditions lead to the next restrictions: $\mu \geq 0$, $C(v) \leq 0$ and $-1 \leq \sigma \leq 0$. More details for outline of $\Theta(r)$ are as follows: first, $\sigma = -1/(D-2) \rightarrow \Theta(r) = \ln(r)$, second, $-1 \leq \sigma \leq 0$ when $\sigma \neq -1/(D-2) \rightarrow \Theta(r) = r^{(D-2)\sigma+1}/[(D-2)\sigma+1]$. The authors of [10] have found the general solutions in the pure Lovelock gravity with the matter presented by the energy-momentum (31). Among these solutions we chose only the solutions which present black holes and we do not consider naked singularities and other exotic objects. These black hole solutions 1) with $1/\ell^2 < 0$ have AdS asymptotics; 2) with $1/\ell^2 > 0$ have dS asymptotics and 3) with $1/\ell^2 = 0$ have a flat asymptotics. In below subsections we derive concrete formulae for aforementioned solutions.

4.2. General formulae

Here, before considering the aforementioned variants of black holes we give the general formulae, which are valid for all the cases. The background is chosen in the Eddington-Finkelstein coordinates in the form:

$$ds^2 = -\bar{f}(r)dv^2 + 2dvdr + r^2 q_{ij} dx^i dx^j. \tag{32}$$

Non-zero components of metric perturbations in the general form are $\kappa_{00} = -f + \bar{f} = -\Delta f$ and $\kappa_0^1 = \kappa^{11} = \kappa_{00}$. The displacement vector we chose in the form preserving the spherical symmetry, but without other restrictions: $\xi^\alpha = (\xi^0(v,r), \xi^1(v,r), 0, \ldots, 0)$. Then, keeping in mind all the above, we found out that the superpotential (25) has the unique non-zero component:

$$I^{01}(\xi) = -I^{10}(\xi) = (D-2)\frac{p\sqrt{\det q_{ij}}}{2\kappa} r^{D-3}\left[\frac{\alpha^2}{r^2}\left(1-\bar{f}\right)\right]^{p-1} \Delta f \xi_1. \tag{33}$$

Then the components of the current (12) are only: $I^0(\xi) = \partial_1 I^{01}(\xi)$ and $I^1(\xi) = \partial_0 I^{10}(\xi)$.

For the background (32) it is convenient to consider sections $\Sigma := v = const$. If one chooses a timelike Killing vector $\bar{\xi}^{\alpha} = (-1,0,...,0)$ the charge (16) with (33) presents the quasi-local energy when $r = r_0$ or global energy when $r \to \infty$:

$$P(\bar{\xi}) = \oint_{\partial\Sigma} dx^{D-2} I^{01}(\bar{\xi}), \tag{34}$$

where $\partial\Sigma$ is the boundary of Σ and is an intersection of Σ with the surface $r = const$; the quantity (34) is considered at each of the moment v. The flux of the energy (34) through $\partial\Sigma$ is defined by

$$\dot{P}(\bar{\xi}) = -\oint_{\partial\Sigma} dx^{D-2} I^{1}(\bar{\xi}) \tag{35}$$

that can be obtained by differentiating (34) with respect to v. The minus sign in (35) is chosen because the direction of coordinate r is opposite to a contraction of black hole.

The field-theoretical formalism allows us to use arbitrary curved backgrounds. For studying the black holes with the use of (34) and (35) we can consider the background metric (32), first, as a metric of asymptotic maximally symmetric spacetime with respect to both dynamic and static black holes. Second, the background metric (32) can be considered as a metric of static black holes (not maximally symmetric) with respect to the metric of dynamic black holes. The other advantage of the field-theoretical methods is that one can use arbitrary displacement vectors. Together with the Killing vectors $\bar{\xi}$ we use proper vectors of freely falling observers $\tilde{\xi}$ on the background of static black holes. As perturbations we consider solutions for related dynamic black holes and construct for them components $I^{0}(\tilde{\xi})$ (energy density) and $I^{1}(\tilde{\xi})$ (density of energy flux) measured by the observers. Below we have a possibility only briefly to outline these results, for a more detail see [12].

4.3. *Black holes with AdS asymptotics. The case* $1/\ell^2 < 0$

For black holes with AdS asymptotics it is natural to choose the AdS space as a background:

$$\bar{f}(r) = 1 + \frac{r^2}{\alpha^{2-2/p}} \frac{1}{|\ell|^2}. \tag{36}$$

The static black hole is describe by

$$f(r) = 1 + \frac{r^2}{\alpha^{2-2/p}} \left(\frac{1}{|\ell|^2} - \frac{m_0}{r^{D-1}} \right)^{1/p}, \tag{37}$$

whereas the dynamic of Viadya type black hole is presented by

$$f(r,v) = 1 + \frac{r^2}{\alpha^{2-2/p}} \left(\frac{1}{|\ell|^2} - \frac{m_0 + m(v)}{r^{D-1}} - \frac{C(v)\Theta(r)}{r^{D-1}} \right)^{1/p}. \tag{38}$$

It was calculated the global mass for (37) on the background (36) that exactly coincides with the standard result (29). The mass of the dynamic black hole (38) on the background (36) is $P(\bar{\xi})|_{r\to\infty} = (D-2)\left(m_0 + m(v)\right)/4G_D$. Besides, our formalism allows us to calculate the mass of (38) on the background (37), it is $P(\bar{\xi})|_{r\to\infty} = (D-2)m(v)/4G_D$. First, the calculation of global mass in the AdS case restricts the parameters of the null fluid. The requirement of the finiteness leaves only: $-1 \le \sigma < -1/(D-2) \to \Theta(r) = r^{(D-2)\sigma+1}/[(D-2)\sigma+1]$. Second, the pressure does not contribute to the global expressions. Third, the global energy of the dynamic black holes turns out additive as calculated on the backgrounds (36) and (37).

Analogously to the global mass it was calculated the quasi-local mass at the finite radius $r = r_0$ in the three aforementioned cases. This gives a possibility for the next conclusions. First, all the expressions depend on the radial pressure, although one has to preserve the previous restrictions for the null fluid:

$-1 \le \sigma < -1/(D-2) \to \Theta(r) = r^{(D-2)\sigma+1}/[(D-2)\sigma+1]$. Second, there is no additivity discussed above. Third, the quasi-local energy expressions calculated outside of the horizons allows us to conclude that the density of the total (energy perturbations and null matter) energy is negative. This result generalizes the textbook [17] conclusions in general relativity on the negative energy density of the potential gravitational field.

4.4. Black holes with dS asymptotics. The case $1/\ell^2 > 0$

For black holes with dS asymptotics it is natural to choose the dS space as a background:

$$\bar{f}(r) = 1 - \frac{r^2}{\alpha^{2-2/p}} \frac{1}{\ell^2}. \tag{39}$$

In this case one cannot consider the asymptotics at $r \to \infty$ because there is a cosmological horizon. For the static solution

$$f(r) = 1 - \frac{r^2}{\alpha^{2-2/p}} \left(\frac{1}{\ell^2} + \frac{m_0}{r^{D-1}} \right)^{1/p} \tag{40}$$

we consider only the situation when the parameters allow the existence of the cosmological horizon and the black hole event horizon. We do not consider other variants leading to exotic objects. Concerning the dynamic solution

$$f(r,v) = 1 - \frac{r^2}{\alpha^{2-2/p}} \left(\frac{1}{\ell^2} + \frac{m_0 + m(v)}{r^{D-1}} + \frac{C(v)\Theta(r)}{r^{D-1}} \right)^{1/p}, \tag{41}$$

we restrict the time interval of falling the null matter to preserve the relation of parameters necessary for the existence of the mentioned two horizons.

Again both the solutions (40) and (41) can be considered on the background of (39). Besides, the dynamic black hole (41) can be considered on the background of the related static black hole (40). By the impossibility to provide the asymptotics at $r \to \infty$ we cannot construct global conserved quantities. As a result we cannot introduce additional restrictions for parameters of null matter. Thus, we construct only quasi-local conserved energy at finite $r = r_0$ in all the three cases. We have found the next results. First, the same as in the AdS case, all the expressions depend on the radial pressure. Second, there is no additivity for the quasi-local charges. Third, the quasi-local energy expressions calculated outside of the event horizons allows us to conclude that the density of the total (metric perturbations and null matter) energy is *positive*. It is unusual result. However, it is not surprising. Indeed, analogous exotic conclusions frequently appear in literature when black holes with dS asymptotics are considered, see, for example, [18,19].

4.5. Black holes with flat asymptotics. The case $1/\ell^2 = 0$

For the black holes with a flat asymptotics one has

$$\bar{f}(r) = 1. \tag{42}$$

The related static black holes are defined by

$$f(r) = 1 - \frac{r^2}{\alpha^{2-2/p}} \left(\frac{m_0}{r^{D-1}} \right)^{1/p}, \tag{43}$$

whereas for the dynamic black hole one has

$$f(r,v) = 1 - \frac{r^2}{\alpha^{2-2/p}} \left(\frac{m_0 + m(v)}{r^{D-1}} + \frac{C(v)\Theta(r)}{r^{D-1}} \right)^{1/p}. \tag{44}$$

It is impossible to calculate the charges on the background (42). The reason is that the superpotential (33) vanishes for (42) when $p \neq 1$. However, a special trick, see [12], allows us to calculate the global mass only, not a quasi-local mass. Thus for the static solution (43) we have again the standard result (29), and for the dynamic solution (44) one has $P(\bar{\xi})\,|_{r\to\infty} = (D-2)\big(m_0 + m(v)\big)\big/4G_D$ at each of the moment v. At the same time, we calculate without difficulties the global and quasi-local mass for (44) on the background (44). Finally, we find that there is no the additivity of both global and quasi-local energies. Considering the quasi-local energy the density of the total (metric perturbations and null matter) energy is negative. But it is determined *only* by the pressure of the null fluid, when the pressure is absent one has a zero total energy density. In this case the energy density of the gravitational field and null matter are compensated exactly.

5. Concluding remarks

In the present format, there is no a possibility to include the results giving fluxes of energy of various kinds noted in (35). It is impossible to give the very complicated expressions for energy density and density of energy flux measured by freely falling observers on the geometry of static black holes. For a detail see [12].

References

[1] Lovelock D 1971 *J. Math. Phys.* **12** 498
[2] Crisostomo J, Troncoso R and Zanelli J 2000 *Phys. Rev.* D **62** 084013
[3] Aros R and Estrada M 2019 *Eur.Phys. J.* C **79** 259
[4] Grishchuk L P, Petrov A N and Popova A D 1984 *Commun. Math. Phys.* **94** 379
[5] Popova A D and Petrov A N 1988 *Int. J. Mod. Phys. A* **3** 2651
[6] Petrov A N, Kopeikin S M, Lompay R R and Tekin B 2017 *Metric Theories of Gravity: Perturbations and Conservation Laws* (Germany: de Gruyter)
[7] Petrov A N and Pitts J B 2019 *Space, Time and Fundamental Interactions* **4** 66
[8] Dadhich N 2016 *Euro. Phys. J. C* **76** 104
[9] Cai R-G and Ohta N 2006 Phys. Rev. D **74** 064001
[10] Cai R-G, Cao L-M, Hu Y-P and Kim S P 2008 *Phys. Rev. D* **78** 124012
[11] Petrov A 2019 *Class. Quantum Grav.* **36** 235021
[12] Petrov A 2021 *Class. Quantum Grav.* **38** 155017
[13] Deser S 1970 *Gen. Relativ. Gravit.* **1** 9
[14] Allemandi G, Francaviglia M and Raiteri M 2003 *Class. Quantum Grav.* **20** 5103
[15] Okuyama N and Koga J-I 2005 *Phys. Rev. D* **71** 084009
[16] Deruelle N, Katz J and Ogushi S 2004 *Class. Quantum Grav.* **21** 1971
[17] Misner C W, Thorne K S and Wheeler J A 1973 *Gravitation* (San Francisco: W H Freeman and Company)
[18] Balasubramanian V, de Boer J and Minic D 2002 *Phys. Rev. D* **65** 123508
[19] Cai R-G, Myung Y S and Zhang Y-Z 2002 *Phys. Rev. D* **65** 084019

Acknowledgments

The author has been supported by the Interdisciplinary Scientific and Educational School of Moscow University “Fundamental and Applied Space Research”.

Estimating the Chirp Mass of Eccentric Inspiraling Binary Systems from Time-Frequency Representations of their Gravitational Radiation

Innocenzo M Pinto

University of Naples *Federico II*, INFN, LVK Collaboration
Department of Electrical Engineering and Information Technologies (DIETI), v. Claudio 21, 80126 Napoli, Italy

E-mail: innocenzo.pinto@ligo.org

Abstract. Using the simplest yet meaningful Peters-Mathews model describing the orbital damping of a compact binary system under the emission of gravitatonal radiation, we show that the chirp-mass of an eccentric inspiraling binary, and its (Keplerian) orbital eccentricity at some reference time, can be estimated from the time-frequency skeleton of its gravitational wave signal. The estimation algorithm is nicely simple, and is robust against the non-ideal (non Gaussian, non stationary) features of detector noise.

1. Introduction

Several mechanisms exist whereby compact binary systems with large eccentricities may form [1], including close-encounters and other many-body interactions in dense stellar environments [2]-[5], and the Lidov-Kozai mechanism [6],[7]. These may result in the formation of systems with both relatively high orbital frequencies, within the observational frequency window of present-day Earth-based detectors, and large eccentricities, that wouldn't otherwise form as an effect of radiation damping alone [8].

Circular-orbit template banks are known since long to be inefficient for matched-filter detection of gravitational waves from eccentric binaries [9]-[11], due to accumulating phase-mismatch [12] resulting in large SNR loss and unacceptable false-dismissal rate; and/or eventually in selection effects, and misestimation of the source parameters [13].

Efforts to construct gravitational wave template families with eccentricity, in the time-domain and/or in the frequency domain, are ongoing following different approaches, including post-Newtonian expansions, effective-one-body models, numerical solutions, etc. - see e.g. [14]-[23]. As of today, however, no ready-to-use template family is available covering the full eccentricity range, and incorporating spin effects as well - which is needed in view of the observed degeneracy[1] between eccentricity and spin [25].

Model independent methods, based on excess-power detection in time-frequency (TF) representations [26],[27], proved effective in reconstructing injected waveforms obtained from hybrid (analytic/numeric) modeling of highly eccentric sources [28], [29]. Re-examination of

[1] Degeneracies can be circumvented, at the expense of some loss in physical insight, by orthogonalization and dimensional reduction of the parameter basis [24].

the data gathered during the 1st and 2nd observing run of the advanced LIGO and Virgo detectors, using a well tested pipeline belonging to this class [30], didn't find any signal, though, beyond those already confidently detected as originating from circular binaries [31]. Among the gravitational wave events detected by LIGO/Virgo in the O3 science-run, on the other hand, at least one (GW190521) has been indicated as possibly originating from a highly-eccentric source [32].

The possible use of TF representations for parameter estimation, other than detection of GWs from eccentric binaries was suggested in [33]-[35], and certainly deserves further investigation. It is worth noting that while the orbital phase (needed for template-based data analysis) is strongly affected by higher-order PN corrections, the orbital frequency (the main observable of TF representation based analysis) is affected to a much lesser extent.

This communication adds to the TF based approach, and proposes a simplest method for estimating the source chirp mass, and its (Keplerian) eccentricity at some reference time, using the time-frequency skeleton of its gravitational waves. For the sake of simplicity and readability, we adopt the seminal Peters-Mathews model, that makes the whole derivation straightforward (Sect. 2 and Appendix), allows to track explicitly the intensity evolution of the GW spectral lines, while they sweep the detector observational frequency-window (Sect. 3), and suggests almost naturally the proposed algorithm (Sect. 4).

2. Simplest Orbital Damping Model

The first orbital evolution model of a binary systems made of compact bodies under the emission of gravitational radiation was formulated in 1963 by J. Mathews (Caltech) and P.C. Peters (his student, at the time) [36], [37]. The relevance of the the Peters-Mathews (PM) model is threefold: historical (it supported the first indirect evidence of gravitational wave emission from continued observation of the orbital evolution of the binary pulsar PSR1913+16 [38], [39]); conceptual (all higher-order post-Newtonian models are rooted in it), and, not least, pedagogical, in view of its full mathematical solvability, and the high physical readability of the ensuing solution, describing a *universal* evolutionary scenario.

The PM analysis tracks the binary evolution starting from known initial values of the orbital frequency ω_{orb} and Keplerian orbital eccentricity e, and can be boiled down to the following equations [40] (see Appendix for details) :

$$T_c[e(0)=0] = \frac{5}{256}\omega_{orb}^{-8/3}(0)\left(\frac{c^3}{GM_\odot}\right)^{5/3}\overline{\mathcal{M}}^{-5/3} \tag{1}$$

yielding the lifetime (fiducial time to coalescence) of a circular-orbit ($e(0)=0$) binary, where $\omega_{orb}(0)$ is the orbital frequency at $t=0$, $\overline{\mathcal{M}}$ is the binary chirp-mass in solar masses, and all other symbols have the usual meaning;

$$\frac{T_c[e(0)]}{T_c(0)} = [1-e^2(0)]^4\left[1+\frac{121}{304}e^2(0)\right]^{-3480/2299}\Omega[e^2(0)] \tag{2}$$

yielding the lifetime reduction due to a non-zero initial eccentricity $e(0)\in]0,1[$, where

$$\Omega(x) = {}_2F_1\left[\frac{24}{19},\frac{3}{2},-\frac{1181}{2299},\frac{43}{19};x,-\frac{121}{304}x\right], \tag{3}$$

${}_2F_1(\cdot)$ being Appell hypergeometric function of the first kind[2];

$$\frac{\tau}{T_c[e(0)]} = 1-\left[\frac{e^2(\tau)}{e^2(0)}\right]^{24/19}\frac{\Omega[e^2(\tau)]}{\Omega[e^2(0)]} \tag{4}$$

[2] The Appell function ${}_2F_1(\cdot)$ is available, e.g., in the standard WOLFRAM MATHEMATICA® package.

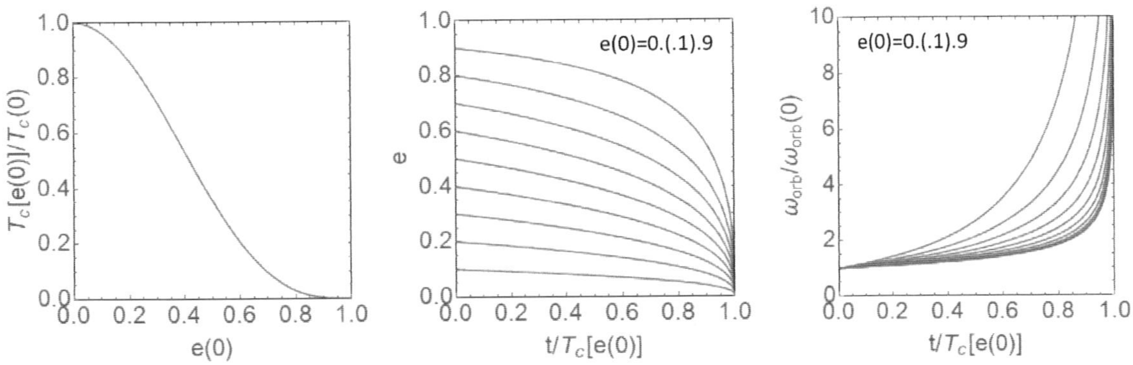

Figure 1. The PM model in a nutshell: left to right: the ratio $T_c[e(0)]/T_c(0)$ vs. $e(0)$ from eq. (2); the Keplerian eccentricity vs $t/T_c[e(0)]$, from eq. (4); the scaled orbital frequency $\omega_{orb}/\omega_{orb}(0)$ vs $t/T_c[e(0)]$, from eq. (5), for different values of $e(0)$. These curves, together with eq. (1) provide a *universal* description of binary evolution under gravitational emission.

relating (upon inversion[3]) orbital eccentricity to scaled time $t/T_c[e(0)]$, and

$$\frac{\omega_{orb}(t)}{\omega_{orb}(0)} = \left[\frac{e(t)}{e(0)}\right]^{-18/19} \left[\frac{1-e^2(t)}{1-e^2(0)}\right]^{3/2} \left[\frac{1+\frac{121}{304}e^2(t)}{1+\frac{121}{304}e^2(0)}\right]^{-1305/2299} \tag{5}$$

relating $\omega_{orb}(t)$ and $e(t)$ throughout the binary lifetime[4]. Equations (1) to (5) provide a simple, general and readable description of binary evolution under gravitational radiation, illustrated in Figure 1.

3. Gravitational Waves from Eccentric Binaries

The steady-state gravitational wave luminosity spectrum of an elliptical-orbit binary was also first derived in [36], and further elaborated in [42]-[44]. The luminosity of the n-th harmonic of the orbital frequency (sum over both polarizations) can be accordingly written

$$\mathcal{L}_n = \Pi\, \overline{\mathcal{M}}^{10/3} \left(\frac{\omega_{orb}(t)}{\omega_{orb}(0)}\right)^{10/3} g_{tot}(n,e), \tag{6}$$

where $\Pi = (32/5)G^{7/3}c^{-5}\left(M_{\odot}\omega_{orb}(0)\right)^{10/3}$ (erg/sec), and

$$g_{tot}(n,e) = \frac{n^4}{32}\left\{\left[J_{n-2}(ne) - J_{n+2}(ne) - 2e\left[J_{n-1}(ne) - J_{n+1}(ne)\right] + \frac{2}{n}J_n(ne)\right]^2 + \right.$$
$$\left. + (1-e^2)\left[J_{n-2}(ne) - 2J_n(ne) + J_{n+2}(ne)\right]^2 + \frac{4}{3n^2}J_n^2(ne)\right\}. \tag{7}$$

[3] The function $\Omega(x)$ is strictly monotonic in $[0,1[$, and numerical inversion of (4) is straightforward to any desired accuracy.

[4] Note that in the asymptotic circular-orbit limit, $e \sim e(0) \sim 0$, $\Omega(x) \sim 1$, $T_c[e(0)] \sim T_c(0)$, $1 - t/T_c(0) \sim [e(\tau)/e(0)]^{48/19}$, and, from eq. (5),

$$\omega_{orb}(t) \sim \omega_{orb}(0)\left(1 - \frac{t}{T_c(0)}\right)^{-3/8}.$$

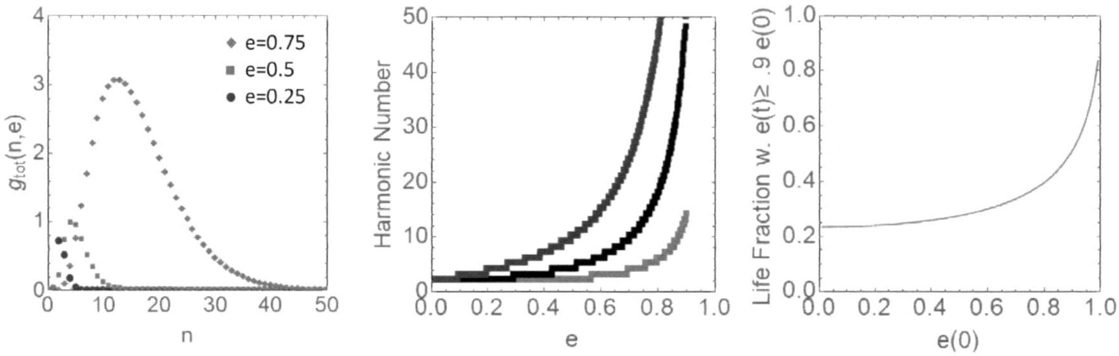

Figure 2. Left to right: steady state gravitational wave luminosity spectrum $g_{tot}(n, e)$ for three different values of e, from eq. (7); harmonic-order of the strongest line in the steady-state spectrum (black) and of the lowest (red) and highest (blue) frequency lines such that $\mathcal{L}_n \geq 0.1\mathcal{L}_{max}$, vs. orbital eccentricity; fraction of the binary lifetime where $e(t) \geq 0.9e(0)$ as a function of $e(0)$.

$J_n(\cdot)$ being the Bessel function of the first kind[5].

The function $g_{tot}(n, e)$ is shown in Figure 2 for different values of e.

It is seen that as Keplerian eccentricity increases, the order of the strongest harmonic in the spectrum and the spectral width also increase.

A non-trivial consequence of eq.s (2) and (4) is that while the binary lifetime is a decreasing function of $e(0)$, the *fraction* of the binary lifetime over which $e(t) \approx e(0)$ is an increasing function of $e(0)$, as shown in Figure 2.

It is possible to combine eq. (7) with eqs. (4) and (5) allowing to express the orbital eccentricity and frequency as functions of (scaled) time $t/T_c[e(0)]$, to see how the spectral lines and their luminosities evolve during the binary lifetime[6].

Doing so, we should keep in mind that GW data from interferometric detectors are *pass-banded*, in order to get rid of the technical noises [45] occurring below and above some typical frequencies, denoted here as f_{in} and f_{out}. Hence, the n-th spectral line will become visible (provided its luminosity exceeds the noise-floor of the detector) only in the time interval where $f_{in} \leq f_n(t) \leq f_{out}$.

In general, different harmonics will be dominant in different time-intervals, until the $n = 2$ line will take the lead, and higher harmonics will eventually vanish as $t/T_c[e(0)] \to 1$, because of asymptotic orbit circularization, for any $e(0) \in]0, 1[$. The higher harmonics, however will exceed, or remain comparable to, the 2nd one throughout a fraction of the binary lifetime that increases with $e(0)$. To illustrate the above, we consider two sources[7] with $e(0) = 0.75$ and $e(0) = 0.25$, respectively, $f_{orb}(0) = 1\text{Hz}$ for both, and assume $f_{in} = 20\text{Hz}$, $f_{out} = 150\text{Hz}$. In Figure 3 all harmonics for which $\mathcal{L}_n \geq 0.01\mathcal{L}_{max}$ at $t = 0$ are included, as they sweep through the spectral window (f_{in}, f_{out}). It is seen that a bunch of higher harmonics of the more eccentric source falls in the detector spectral window already at $t = 0$; the situation is different for the less eccentric source, whose spectral lines enter the observational window much later.

[5] Efficient computational tools for evaluating the Bessel functions in eq. (7) were first discussed in [44].

[6] For simulation purposes, the evolving transverse-traceless components of the emitted gravitational waves from [37],[44] can be consistently computed using the orbital phase evolution equations derived in [12].

[7] By suitable choice of their chirp masses, these two sources can be designed using eqs. (1) and (2) to have the same lifetime. Their luminosities will be different, though, according to eq. (6).

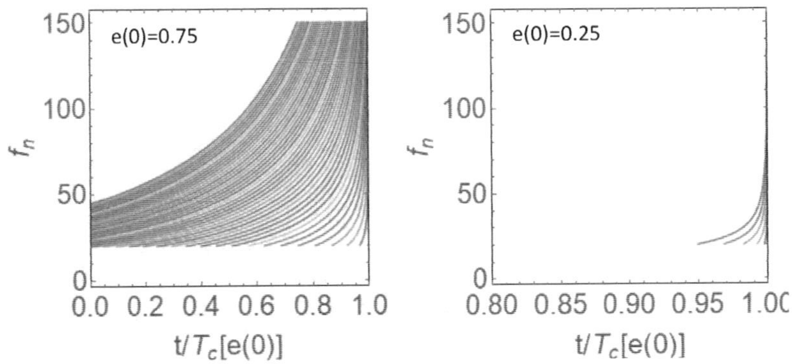

Figure 3. Frequency evolution of gravitational wave lines (harmonics of the orbital frequency) for which $\mathcal{L}_n \geq 0.01\mathcal{L}_{max}$ at $t = 0$, as they sweep the detector observational frequency window (here between 20Hz and 150Hz). Left: source with $e(0) = 0.75$; right: source with $e(0) = 0.25$; $f_{orb}(0) = 1$Hz for both sources.

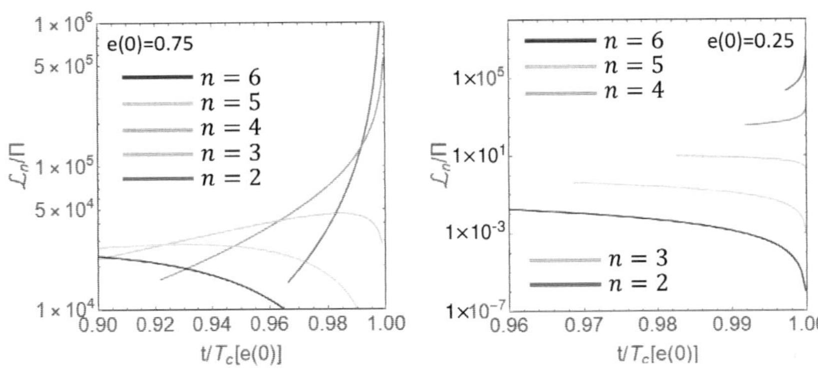

Figure 4. Evolution of the luminosity of the five strongest gravitational wave lines (harmonics of the orbital frequency) in the lifetime fraction where they sweep the detector observational freuency window (here from 20Hz to 150Hz). Left: source with $e(0) = 0.75$; right: source with $e(0) = 0.25$; $f_{orb}(0) = 1$Hz for both sources.

In Figure 4 the luminosity of the five most intense spectral lines in the late evolutionary phase is shown, again for both cases. It is seen that for the source with $e(0) = 0.25$, all harmonics with $n > 2$ are fainter by a factor $\sim 10^3$ compared to the 2nd one, throughout the interval where they sweep through the detector spectral window, between $f_{in} = 20$Hz and $f_{out} = 150$Hz; for the source with $e(0) = 0.75$, one or more of the higher harmonics exceed, or are comparable to, the asymptotically dominant 2nd one over extended fractions of the lifetime.

We may thus expect a fairly richer time-frequency scenario, compared to the simplest circular-orbit case, when observing binaries with relatively large eccentricities.

3.1. *Time-Frequency Representations*

Time-frequency (TF) representations aim at displaying the energy content of nonstationary signals as a function of time *and* frequency. Many such representations have been proposed, each with its own pros and cons [46], [47]. See [48] for a comparative discussion in the context of GW data analysis.

For multi-line *chirping* GW signals, consisting of harmonically-related quasi-monochromatic signals whose frequency and amplitude change adiabatically in time, TF representations should be rated according to their ability to produce uniformly sharp, faithful and clean (devoid of, e.g., nonlinear artifacts) estimates of the time-dependent (instantaneous) frequencies of the various harmonic components throughout the time-frequency plane - such ability being ultimately limited by Gabor-Heisenberg inequality.

The instantaneous frequency lines (henceforth IFLs) form the so called *skeleton* or *ridge* of the TF representation. In practice, *discrete* TF representations are used, where at each (discrete) time, the *hottest* (or a few hottest) pixel is selected to extract the skeleton. For an eccentric binary, the skeleton will accordingly consists of different tracks spanning different time intervals, reflecting the locally dominant harmonics.

The tracks representing such harmonics can be combined into a *single* line representing the (estimated) IFL of the source orbital frequency.

In the next section we shall assign, for simplicity, a unit level to all TF pixels on the orbital IFL, and a zero level value to all other pixels[8].

4. A Simple Chirp Mass Estimator for Eccentric Sources

The PM model suggests a simple algorithm for estimating, given the orbital IFL, the chirp mass of an eccentric binary together with its (Keplerian) orbital eccentricity at some reference time[9]. The first equation in (A.1), integrated between t_1 and t_2 and combined with (A.5) can be formally solved for the normalized chirp mass $\overline{\mathcal{M}} = \mathcal{M}/M_{\odot}$, yielding

$$\overline{\mathcal{M}} = \left(\frac{c^3}{GM_{\odot}}\right)\left\{\frac{5}{256}\frac{\omega_{orb}^{-8/3}(t_1) - \omega_{orb}^{-8/3}(t_2)}{\displaystyle\int_{t_1}^{t_2} f[e(t)]dt}\right\}^{3/5}. \tag{8}$$

From (8) we may obtain as many values of the chirp mass $\overline{\mathcal{M}}$, as the number of available TF pixel-pairs [10] $\{\{t_1, \omega_{orb}(t_1)\}, \{t_2, \omega_{orb}(t_2)\}\}$ from the given orbital IFL. In the absence of noise and numerical (roundoff) errors, one would obtain from (8) the *same* value of $\overline{\mathcal{M}}$ from *all* TF pairs . In practice, the computed values of $\overline{\mathcal{M}}$ will be different, and can be binned to form a histogram, $\mathcal{D}(\overline{\mathcal{M}})$, to obtain an estimate of the source chirp-mass [49].

The computation of (8) is straightforward for circular orbits[11] where the integral in the denominator is simply $(t_2 - t_1)$. In the general case of elliptical orbits, in order to compute the integral in (8) we first need to set a time origin $t = 0$ at some (arbitrary) time-sample in the available IFL, and retrieve the corresponding $\omega_{orb}(0)$. Then, for any *trial value* of $e(0) \in [0, 1[$, we can derive an instantaneous eccentricity line (IEL) from the available IFL and the above determined $\omega_{orb}(0)$, by numerical inversion of (5), and use it to compute the integral in the denominator of (8). We may accordingly obtain a *family* of histograms corresponding to a suitable set of trial values of $e(0)$.

The results of such procedure are illustrated by examples in Table 1, and suggest the following simple algorithm for estimating *both* the chirp-mass and the Keplerian eccentricity at the reference time $t = 0$:

[8] Extension to the more general case where each TF pixel (and hence each pixel pair) has a non-binary level is straightforward, and amounts to adding the (non-unitary) levels of the pairs to the histogram bin counts.

[9] A possible choice for this latter is e.g. the time when some (visible) spectral line enters the detector spectral window.

[10] In practice, only pairs for which $|t_2 - t_1| \gg \delta_t$ and $|\omega_{orb}(t_2) - \omega_{orb}(t_2)| \gg \delta_\omega$ should be used, δ_t, δ_ω being the time and frequency resolution of the available TF representation.

[11] Note that retrieving the chirp mass from the IFL via eq. (8) does *not* require differentiation of the IFL, at variance of the conceptually similar method proposed in [34].

Table 1. Retrieving the chirp mass from eccentric inspirals, using eq. (8). Two illustrative examples are considered: left: $\{e(0) = 0.5, \overline{\mathcal{M}} = 2.8\}$, and right: $\{e(0) = 0.1, \overline{\mathcal{M}} = 30\}$. We use noise-free one-bit coded orbital IFLs, and a typical number of time-frequency TF pixel pairs $\sim 5 \cdot 10^3$. The quantities $\widehat{\overline{\mathcal{M}}}$ and $\widehat{\sigma}$ are the median and std. deviation of the chirp mass histograms computed from eq. (8) for various values of $e(0)$. For both examples $t = 0$ was chosen as the time when the 2nd harmonic of the orbital frequency enters the detector observational frequency window at $f_{in} = 20$Hz. The trial value e_k that most closely matches the true (unknown) $e(0)$ is signaled by minima in both the spread ($\widehat{\sigma}$) and the L^2 error between the actual IFL and the one reconstructed from the trial eccentricity and the corresponding estimated chirp mass.

e_k	$\widehat{\overline{\mathcal{M}}}$	$\widehat{\sigma}$	L^2 error
0.1	4.94	0.78	1.81
0.15	4.84	0.72	1.68
0.2	4.69	0.65	1.46
0.25	4.49	0.55	1.07
0.3	4.25	0.45	0.56
0.35	3.95	0.33	1.42
0.4	3.61	0.22	2.32
0.45	3.22	0.11	1.90
0.5	**2.80**	**0.02**	**0.45**
0.55	2.37	0.05	0.52
0.6	1.94	0.10	0.85
0.65	1.52	0.13	1.04
0.7	1.14	0.14	1.16
0.75	0.80	0.12	1.24
0.8	0.51	0.09	1.31

e_k	$\widehat{\overline{\mathcal{M}}}$	$\widehat{\sigma}$	L^2 error
0	30.65	0.23	0.06
0.025	30.61	0.21	0.05
0.05	30.49	0.17	0.04
0.075	30.29	0.11	0.02
0.1	**30.00**	**0.02**	**0.00**
0.125	29.63	0.10	0.02
0.15	29.17	0.23	0.04
0.175	28.64	0.38	0.06
0.2	28.02	0.55	0.08
0.225	27.31	0.73	0.11
0.25	26.53	0.92	0.17
0.275	25.67	1.11	0.24
0.3	24.73	1.30	0.31
0.325	23.72	1.48	0.38
0.35	22.64	1.66	0.44

(i) construct the histograms $\mathcal{D}_k(\overline{\mathcal{M}})$ corresponding to a set of possible values $\{e_k | k = 1, 2, \ldots, N\}$ of $e(0)$ in $[0, 1[$, and let $\widehat{\overline{\mathcal{M}}}_k$ the corresponding chirp-mass estimate (e.g., the median), and Δ_k the L^2 error between the available orbital IFL and the *reconstructed* IFL obtained by letting $\{e(0), \overline{\mathcal{M}}\} = \{e_k, \widehat{\overline{\mathcal{M}}}_k\}$;

(ii) find k for which $\mathcal{D}_k(\overline{\mathcal{M}})/\Delta_k$ has the highest peak, and take $\{e_k, \widehat{\overline{\mathcal{M}}}_k\}$ as your estimate of the source parameters.

The different heights of the histogram peaks resolve the estimation ambiguity originating from the degeneracy between chirp mass and eccentricity pointed out in [51] and further discussed in [52],[53]. An example of the histogram family obtained using simulated data added to a stretch of (public domain) LIGO noise is shown in Figure 5.

4.1. Robustness

A quantitative assessment of the impact of noise in the data in terms of bias and spread of the obtained estimates is in progress, based on massive numerical simulations.

Additive locally stationary/Gaussian noise in the data manifests itself in the 1-bit coded TF skeleton as frequency *jitter*, whereby $\omega_{orb}(t_n)$ may be slightly different from its true/expected value.

More interesting is the case where strong transients of untraced environmental/instrumental

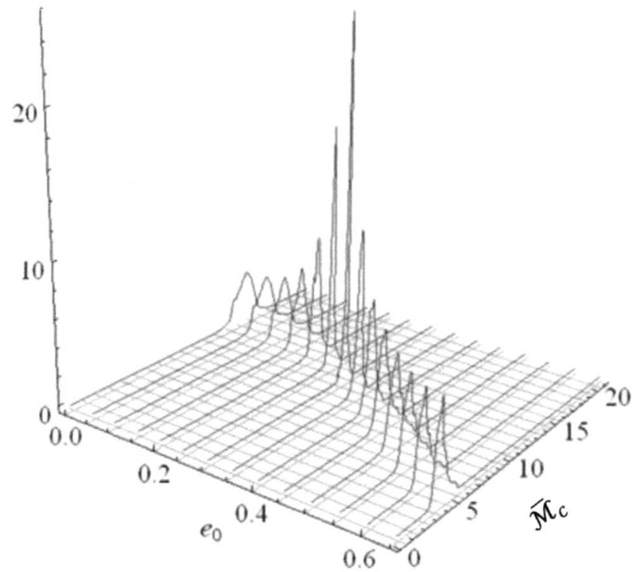

Figure 5. The weighted histogram $\mathcal{D}_k(\overline{\mathcal{M}})/\Delta_k$ for an eccentric inspiral with $e(0) = 0.3$ and $\overline{\mathcal{M}} = 12.16$ (courtesy P. Addesso [50]). The waveform was generated using [12] and [44] and added to a stretch of (public domain) LIGO background noise for an $SNR = 25dB$. The TF representation used to produce the orbital IFL was a sparsified smoothed Wigner-Ville [48].

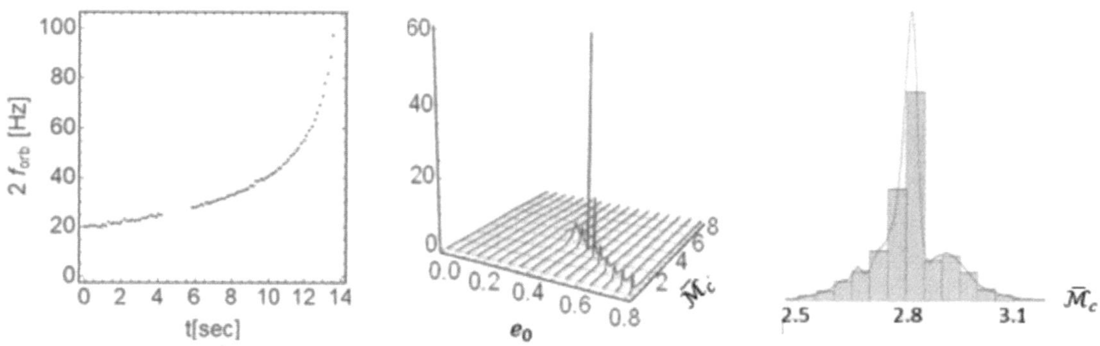

Figure 6. Left: noisy one-bit encoded IFL with excision. Source parameters: $e(0) = 0.5, \overline{\mathcal{M}} = 2.8$. The shown jitter corresponds roughly to an (energy) SNR ~ 20 in the simulated data; center: the corresponding family of smoothed histograms $\mathcal{D}_k/\Delta_k$; right: a close-up of the smoothed and raw $e_k = 0.5$ histogram; a tiny second lobe is visible.

origin (nicknamed *glitches*) are present. These should be excised from the data, yielding a *lacunary* IFL. Preliminary results indicate that the poposed algorithm is nicely robust against both kinds of noises above. Qualitatively, the main effect of IFL jitter is to *broaden the peaks* of the $\overline{\mathcal{D}}_k$ histograms; the main effect of lacunarity is to produce moderate *multi-modality* in the histograms, as illustrated in Figure 6.

5. Conclusions

The PM binary evolution model may be deemed as exceedingly simple to describe a realistic binary system, where e.g., higher order post-Newtonian corrections, spin and precession effects, may be important. However, it provides a neat conceptual framework that may be used as a guideline to build more sophisticated, yet readable and predictive, investigations.

In addition, it suggests a simple and nicely informative algorithm for estimating the source chirp mass and the Keplerian eccentricity at some reference time from the TF skeleton of the gravitational wave data, useful to restrict a full fledged template-based analysis, whose computational burden would be otherwise unaffordable, to a suitably reduced subset of the whole source parameter space.

Acknowledgements

The author acknowledges several useful discussions with V. Pierro, P. Addesso, V. Matta, A. Fusco, E. Mejuto-Villa and G.C. Cella. Special thanks go to the PIRT-21 Organizers and the whole Conference Staff.
This work is dedicated to the memory of Theo Demma.

Appendix

The Peters-Mathews equations ruling two compact (point) masses in a Keplerian orbit are:

$$\begin{cases} \langle \frac{da}{dt} \rangle = -\frac{64}{5}\frac{G^3}{c^5}\frac{m_1 m_2(m_1+m_2)}{a^3(1-e^2)^{7/2}}\left(1+\frac{73}{24}e^2+\frac{37}{96}e^4\right) \\ \langle \frac{de}{dt} \rangle = -\frac{304}{15}\frac{G^3}{c^5}\frac{m_1 m_2(m_1+m_2)}{a^4(1-e^2)^{5/2}}\left(1+\frac{121}{304}e^2\right)e \end{cases} \tag{A.1}$$

where a and e are the (Keplerian) orbital semimajor axis and eccentricity, $m_{1,2}$ the companion masses, and $\langle\cdot\rangle$ denotes averaging over an orbit. Equations (A.1) are obtained from conservation of energy and angular momentum, after using the quadrupole gravitational radiation formula. They are valid to lowest meaningful order in the usual v/c and r_g/r parameters, r_g being the binary gravitational radius, and under the assumption that the orbital parameters change adiabatically, viz., $T(d/dt)\log(T) \ll 1 \wedge T(d/dt)\log(e) \ll 1$. Using Kepler law relating the semimajor axis to the orbital frequency,

$$a(t) = \omega_{orb}^{-2/3}[G(m_1+m_2)]^{1/3} \tag{A.2}$$

equations (A.1) can be written

$$\begin{cases} \frac{d}{dt}\left[\frac{\omega_{orb}(t)}{\omega_{orb}(0)}\right]^{-8/3} = -\frac{f[e(t)]}{T_c(0)} \\ \frac{de}{dt} = -\left[\frac{\omega_{orb}(t)}{\omega_{orb}(0)}\right]^{8/3}\frac{19}{48}\frac{g[e(t)]}{T_c(0)} \end{cases} \tag{A.3}$$

where

$$f(e) = (1-e^2)^{-7/2}\left(1+\frac{73}{24}e^2+\frac{37}{96}e^4\right),\ g(e) = (1-e^2)^{-5/2}\left(1+\frac{121}{304}e^2\right)e \tag{A.4}$$

and

$$T_c(0) = \frac{5}{256}\omega_{orb}^{-8/3}(0)\left(\frac{c^3}{GM_\odot}\right)^{5/3}\overline{\mathcal{M}}^{-5/3} \tag{A.5}$$

$\overline{\mathcal{M}}$ being the binary chirp mass in units of $M_\odot$, and all other symbols having the usual meaning. If $e(0) = 0$, it is seen that $e(t) = 0, \forall t$, and the first equation in (A.3) yields

$$\omega_{orb}(t) = \omega_{orb}(0) \left(1 - \frac{t}{T_c(0)}\right)^{-3/8} \tag{A.6}$$

whereby $T_c(0)$ is understood as the fiducial binary *lifetime* for a circular orbit, i.e., the time until the orbit radius vanishes ($a = 0$). We shall denote as $T_c[e(0)]$ the binary lifetime for the general case of non-zero initial eccentricity. If $e(0) \neq 0$ the two equations in (A.3) can be combined to obtain

$$\frac{d}{de} \log \left[\frac{\omega_{orb}(t)}{\omega_{orb}(0)}\right]^{-8/3} = \frac{48}{19} \frac{f[e(t)]}{g[e(t)]} \tag{A.7}$$

which is readily integrated, yielding the following relationship between $e(t)$ and $\omega_{orb}(t)$, valid throughout the binary lifetime.

$$\frac{\omega_{orb}(t)}{\omega_{orb}(0)} = \left[\frac{e(t)}{e(0)}\right]^{-18/19} \left[\frac{1 - e^2(t)}{1 - e^2(0)}\right]^{3/2} \left[\frac{1 + \frac{121}{304} e^2(t)}{1 + \frac{121}{304} e^2(0)}\right]^{-1305/2299} \tag{A.8}$$

This can used in (A.3) to obtain

$$\frac{de}{dt} = -\frac{19}{48} \left[\frac{e(t)}{e(0)}\right]^{-48/19} \left[\frac{1 - e^2(t)}{1 - e^2(0)}\right]^4 \left[\frac{1 + \frac{121}{304} e^2(t)}{1 + \frac{121}{304} e^2(0)}\right]^{-3480/2299} \frac{g[e(t)]}{T_c(0)}. \tag{A.9}$$

This latter can be integrated between $t = \tau$ and the coalescency time $t = T_c[e(0)]$, where the eccentricity vanishes,

$$\frac{T_c[e(0)] - \tau}{T_c(0)} = \frac{48}{19} [e(0)]^{-48/19} [1 - e^2(0)]^4 \left[1 + \frac{121}{304} e^2(0)\right]^{-3480/2299} \cdot \\ \cdot \int_0^{e(\tau)} e^{29/19} [1 - e^2]^{-3/2} \left[1 + \frac{121}{304} e^2\right]^{1181/2299} de. \tag{A.10}$$

Upon letting $e^2 = x e^2(\tau)$, the integral on the r.h.s. of (A.10) becomes

$$\frac{[e(\tau)]^{48/19}}{2} \int_0^1 x^{5/19} [1 - x e^2(\tau)]^{-3/2} \left[1 + \frac{121}{304} x e^2(\tau)\right]^{1181/2299} dx, \tag{A.11}$$

that is expressible in terms of P.E. Appell's generalized hypergeometric function of the first kind ([41], eq. 3.211). Hence:

$$\frac{T_c[e(0)] - \tau}{T_c(0)} = \left[\frac{e(\tau)}{e(0)}\right]^{48/19} [1 - e^2(0)]^4 \left[1 + \frac{121}{304} e^2(0)\right]^{-3480/2299} \Omega[e^2(\tau)]. \tag{A.12}$$

where

$$\Omega(x) = {}_2F_1 \left[\frac{24}{19}, \frac{3}{2}, -\frac{1181}{2299}, \frac{43}{19}; x, -\frac{121}{304} x\right]. \tag{A.13}$$

For $\tau = 0$ this becomes

$$\frac{T_c[e(0)]}{T_c(0)} = [1 - e^2(0)]^4 \left[1 + \frac{121}{304} e^2(0)\right]^{-3480/2299} \Omega[e^2(0)] \tag{A.14}$$

whereby (A.12) can be further written,

$$\frac{\tau}{T_c[e(0)]} = 1 - \left(\frac{e^2(\tau)}{e^2(0)}\right)^{24/19} \frac{\Omega[e^2(\tau)]}{\Omega[e^2(0)]}. \qquad \text{(A.15)}$$

Equations (A.5), (A.8), (A.14) and (A.15) are the general exact solution of PM equations, first obtained in [40].

As stressed in Sect. 1, this solution is both computationally cheap and numerically accurate.

References

[1] Tagawa H *et al* 2021 *Astrophys. J. Lett.* **907** L20
[2] Gondan L *et al* 2018 *Astrophys J.* **860** 5
[3] Samsing J 2018 *Phys. Rev.***D97** 103014
[4] Gold *et al Phys. Rev.* **D 86** 121501(R)
[5] Vick M and Lai D 2019 *Phys. Rev.* **D100** 063001
[6] Naoz S 2016 *Ann. Rev. Astron. Astroph.* **54** 441
[7] Liu B *et al* 2019 *Astrophys. J.* **881** 41
[8] Kowalska I *et al* 2010 *Astronomy & Astrophys.* **527** A70
[9] Martel E K and Poisson E, 1999 *Phys. Rev.* **D60** 124008
[10] Brown D A and Zimmermann P J 2010 *Phys. Rev.* **D81** 024007
[11] Favata M *et al* 2014 *Phys. Rev. Lett.* **112** 101101
[12] Pierro V and Pinto I M 1996 *Nuovo Cimento* **B111** 1517
[13] Ramos Buades A *et al* (2020) *Phys. Rev.* **D101** 083015
[14] Damour T *et al* 2004 *Phys. Rev.* **D70** 064028
[15] Moore B *et al* 2016 *Phys. Rev.* **D 93** 124061
[16] Tanay S *et al* 2016 *Phys. Rev.* **D93** 064031
[17] Huerta E *et al* 2018 *Phys. Rev.* **D97** 024031
[18] Csizmadia *et al* 2012 *Class. Quantum Grav.* **29** 245002
[19] East W *et al* 2013 *Phys. Rev.* **D87** 043004
[20] Hinderer T and Babak S 2017 *Phys. Rev.* **D96** 104048
[21] Cao Z and Han W B 2017 *Phys. Rev.* **D96**
[22] Liu X *et al* 2020 *Phys. Rev.* **D101** 044049
[23] Chiaramello D and Nagar A 2020 *Phys. Rev.* **D101** 101501(R)
[24] Barta D. and Vasúth M 2018 *Phys. Rev.* **D97** 124011
[25] Romero-Shaw I. *et al* 2020 *Astrophys J. Lett.* **903** L5
[26] Klimenko S *et al* 2008 *Class. Quantum Grav.* **25** 114029
[27] Cornish N J *et al* 2021 *Phys. Rev.* **D103** 044006
[28] Tai K-S *et al* 2014 *Phys. Rev.* **D90** 103001
[29] Dalya G *et al* 2021 *Class. Quantum Grav.* f38 065002
[30] Klimenko S *et al* 2016 *Phys. Rev.* **D93** 042004
[31] Abbott B P *et al* 2019 *Astrophys. J.* **883** 149
[32] Gayathry V *et al* 2020 *preprint* ArXiV:2009:05461
[33] Coughlin M *et al* 2015 *Phys. Rev.* **D91** 063004
[34] Tiwari V *et al* 2016 *Class. Quantum Grav.* **33** 01LT01
[35] Tiwari V *et al* 2016 *Phys. Rev.* **D93** 043007
[36] Peters P C and Mathews J 1963 *Phys. Rev.* **131** 435
[37] Peters P C 1964 *Phys. Rev.* **136** B1224
[38] Taylor J H and Weisberg J M 1982 *The Astrophysical Journal* **253** 908
[39] Weisberg J M and Huang Y 2016 *The Astrophysical Journal* **829** 55
[40] Pierro V and Pinto I M 1996 *Nuovo Cimento* **B111** 631
[41] Gradshteyn I S and Ryzhik I M 1980 *Tables of Integrals Series and Products* (New York, NY : Academic Press)
[42] Zel'dovich Ya B and Novikov I N 1967 *Relativistic Astrophysics*, vol. 1, sect. 1.13 (Chicago, IL : Chicago Univ. Press)
[43] Moreno-Garrido C *et al* 1994 *Mon. Not. Roy. Astron. Soc.* **266**16; ibid. 1995 **274** 115
[44] Pierro V *et al* 2001 *Mon. Not. Roy. Astron. Soc.* **325** 358; ibid. 2002 **334** 855
[45] Martynov D V *et al* 2016 *Phys. Rev* **D93** 112004; ibid. 2018 **D97** 059901
[46] Boashash B B 2016 *Time-Frequency Signal Analysis and Processing*, (New York, NY : Academic Press)

[47] Flandrin P 1998 *Time-Frequency/Time-Scale Analysis*, (New York, NY : Academic Press)
[48] Addesso P *et al* 2016 *preprint* arXiv:1605.03496
[49] Mejuto-Villa E 2019 *PhD Thesis* (Benevento: University of Sannio)
[50] Addesso P *et al* 2021 LIGO Document G2100118
[51] Lenon A K *et al* 2020 *Mon. Not. Roy. Astron. Soc.* **497** 1966
[52] Bose N and Pai A 2021 *preprint* arXiv:2107.14736
[53] Favata M *et al* 2021 *preprint* arXiv:2108.05861

High frequency gravitational waves: generation, detection

V Pustovoit, V Gladyshev, V Kauts, A Morozov, P Nikolaev, I Fomin, E Sharandin, A Kayutenko

Physics Department, Bauman Moscow State Technical University, 2-nd Baumanskaya street 5, Moscow, 105005, Russia

E-mail: ingvor@inbox.ru

Abstract. We consider the different approaches to the generation and detection of the high-frequency gravitational waves. The estimations of the amplitude of gravitational waves generated by these methods were made. The characteristics of the high-frequency gravitational waves induced by perturbations of matter and the electromagnetic field are compared. The possibility of the direct detection of gravitational waves on the basis of the different methods was considered as well.

1. Introduction

At present, various astrophysical processes [1] and cosmological perturbations in the early universe [1,2] are considered as natural sources of gravitational-wave radiation. The low-frequencies gravitational waves arising from black holes and neutron stars mergers with amplitude $h_0 \sim 10^{-20} - 10^{-21}$ were detected in the LIGO and Virgo experiments [3, 4] based on the interference method proposed in [5]. The relic gravitational waves of cosmological genesis waves have a wide frequency range for different models of the early universe, which are restricted by BBN bound and energy scale of inflation as well [6]. Various types of inflationary models of the early universe predict a significant increase in the amplitude of the high-frequency relic gravitational waves $f_g \sim 10^9$ Hz to the values of the order of $h_0 \sim 10^{-30} - 10^{-27}$ [7-9]. However, at the moment, relic gravitational waves have not been detected.

Also, various laboratory sources of gravitational waves are considered, based both on the interaction of electromagnetic radiation and matter [10, 11], and for the electromagnetic field itself as a source of gravitational waves [12, 13, 14].

Thus, gravitational-wave fluctuations of the space-time metric can be induced both by perturbations of the medium and the electromagnetic field, and the predominance of the first or second factor is associated with the specifics of the process of generating gravitational waves.

According to estimates given in [10,11], the interaction of a short pulse of high-power laser radiation and matter makes it possible to generate gravitational waves with an amplitude $h_0 \sim 10^{-40}$ which is much lower than the sensitivity limit of modern detectors.

Nevertheless, in [12,13,14] it was shown that taking into account the processes of conversion of intense laser radiation into a gravitational wave in nonlinear dielectric media makes it possible to consider these methods of generating and recording high-frequency gravitational waves as relevant ones.

In this paper we give brief review of the methods for generating the high-frequency gravitational waves by perturbations of the medium and the electromagnetic field $T_{\mu\nu} = T^{M}_{\mu\nu} + T^{EM}_{\mu\nu}$ in two regimes, namely for $T^{M}_{\mu\nu} \gg T^{EM}_{\mu\nu}$ and $T^{M}_{\mu\nu} \ll T^{EM}_{\mu\nu}$. The possibility of detecting high-frequency gravitational waves generated on the basis of these methods is also discussed.

2. The equations of a gravitational field

Artificial sources of gravitational waves imply gravitational radiation of small amplitude, and thus, weak gravitational waves are considered as small perturbations $|h_{\mu\nu}| \ll 1$ of Minkowski space-time $\eta_{\mu\nu}$, and the metric of perturbed space-time is defined as $g_{\mu\nu} = \eta_{\mu\nu} + h_{\mu\nu}$.

In this case, the equations of gravitational field can be written as follow [1]

$$\Box \bar{h}_{\mu\nu} = -\frac{16\pi G}{c^4} T_{\mu\nu}, \tag{1}$$

where $T_{\mu\nu}$ is the energy-momentum tensor of matter, G is the gravitational constant, c is the speed of light in vacuum, $\Box$ is the d'Alembert operator, and tensor $\bar{h}_{\mu\nu}$ satisfies the following gauge conditions

$$\partial^{\mu} \bar{h}_{\mu\nu} = 0,\ \bar{h}_{\mu\nu} = h_{\mu\nu} - \frac{1}{2}\eta_{\mu\nu} h. \tag{2}$$

The solution of equation (1) can be written as follows [1]

$$h_{ij}(t, \hat{x}) = \frac{4G}{rc^4} \Lambda_{ij,kl}(\hat{n}) \int d^4x' T_{kl}\left(t - \frac{r}{c} + \frac{\hat{x}' \cdot \hat{n}}{c}, \hat{x}'\right), \tag{3}$$

where $r = |\hat{x} - \hat{x}'|$ and tensor $\Lambda_{ij,kl}$ is defined as

$$\Lambda_{ij,kl}(\hat{n}) = P_{ik}P_{jl} - \frac{1}{2}P_{ij}P_{kl}, \quad P_{ij}(\hat{n}) = \delta_{ij} - n_i n_j, \tag{4}$$

where vector n_i determines the direction of propagation of the gravitational wave.

For the case when the gravitational waves are induced by the perturbations of the medium with the small velocities $v \ll c$ of the particles oscillations, the quadrupole formalism implying approximate solutions of the gravitational field equations is usually used.

The leading non-zero (quadrupole) term in multipole expansion of expression (3) can be written as follow [1]

$$h_{ij}(t, \hat{x}) = \frac{2G}{rc^4} \Lambda_{ij,kl}(\hat{n}) \left[\frac{d^2}{dt^2} Q^{kl}\left(t - \frac{r}{c}\right)\right], \tag{5}$$

where quadrupole moment is defined as

$$Q^{ij} = \frac{1}{c^2} \int d^3x T^{00} \left(x^i x^j - \frac{1}{3} r^2 \delta^{ij}\right). \tag{6}$$

Thus, to determine the characteristics of gravitational waves induced by perturbations of matter, it is sufficient to know the quadrupole moment of the system. The different types of the sources of the high-frequency gravitational waves were considered earlier on the basis of the quadrupole formalism (see, for example, in [1, 10, 11] and in many other works as well).

On the other hand, equations (1) can be represented as

$$\Box h_{\mu\nu} = 2\kappa \left(T_{\mu\nu} - \frac{1}{2}\eta_{\mu\nu} T_{\rho}{}^{\rho}\right), \tag{7}$$

were $\kappa = \left(8\pi G / c^4\right)$ and $T_{\rho}{}^{\rho} = T_{\rho\sigma}\eta^{\rho\sigma} = T$ is the trace of the energy-momentum tensor.

Also, we note, that gauge conditions (2) are correct when the energy-momentum conservation

$$\partial^{\mu} T_{\mu\nu} = 0, \tag{8}$$

is satisfied.

This representation of the gravitational field equations is useful for the case when the electromagnetic field is considered as a source of gravitational waves, since for electromagnetic field one has $T = 0$, and equations (7) are reduced to the following form

$$\Box h_{\mu\nu} = 2\kappa T_{\mu\nu}, \tag{9}$$

implying the analysis of the characteristic of gravitational waves based on the exact solutions of these equations.

The wave solutions of the gravitational field equations outside the source in empty space $T_{\mu\nu} = 0$ can be noted as the plane gravitational waves propagating with the speed of light [1]

$$h_{\mu\nu} = A_{\mu\nu} \cos\left(k_{\rho} x^{\rho}\right), \tag{10}$$

where $A_{\mu\nu}$ is the amplitude of gravitational waves which is orthogonal to the wave vector $A_{\mu\nu} k^{\rho} = 0$, and $k^0 = \omega/c$, where ω is the frequency of gravitational waves.

The characteristic of the gravitational waves in empty space are defined by the type of the energy-momentum tensor of material source.

3. Perturbations of a medium as the source of gravitational waves

Irradiation of matter with the high-power laser radiation is considered as a possible mechanism for generating high-frequency gravitational waves, which implies an increase in the characteristics of gravitational wave radiation with an increase in the power of modern lasers [10]. In this case, one can neglect the gravitational field of electromagnetic waves and consider the reaction of the medium to external influence as the only source of gravitational waves.

Shock waves arising from the irradiation of matter with short pulses of high-power laser radiation were considered as sources of high-frequency gravitational waves in [10].

The amplitude of flat gravitational waves induced by the shock waves was estimated on the basis of equations (5)-(6) as follows

$$h_0 = \frac{7}{2} \frac{G P_L \tau}{r c^4} \sqrt{\frac{\rho_c}{\rho_0}}, \tag{11}$$

where P_L is the laser power and τ is the laser pulse duration, ρ_0 is the density of the medium and

$$\rho_c = \frac{\epsilon_0 m_e m_i}{Z e^2} \frac{(2\pi c)^2}{\lambda_L^2}, \tag{12}$$

where ϵ_0 is permittivity of vacuum, m_e is the rest mass of the electron, m_i is the mass of the ion, e is the charge of electron, λ_L is the wavelength of the laser, Z is the atomic number.

The maximal value of the amplitude of gravitational waves on the distance $r = 10\,\mathrm{m}$ from such a source for realistic parameters of experiment was estimated as $h_0 \sim 10^{-40}$ [10].

The other approach to generation of the high-frequency gravitational waves by the perturbations of the medium based on a model of an ionic crystal interacting with an electromagnetic wave was considered in [11].

The amplitude of flat gravitational waves induced by this method was estimated on the basis of equations (5)-(6) as

$$h_0 = \frac{G}{r c^4} \gamma_{\omega} l^2 \omega_T^2 m_i N,, \tag{13}$$

where $\gamma_\omega \equiv \dfrac{\varepsilon_\infty - \varepsilon_0}{2\pi}$ is the dispersion factor, l is the electromagnetic field-induced ion movement, m_i is the mass of ion, N is the number of ions affected by electromagnetic radiation, ω_T is the natural frequency of the elastic transverse oscillations of the crystal.

For realistic values of the parameters in expression (13) the maximal amplitude of gravitational waves on the distance $r = 10\,\text{m}$ from source was estimated as $h_0 \sim 10^{-40}$ [11], which is coincide with result in previous approach.

4. Electromagnetic field as the source of gravitational waves

The other approach to analyze the generating high frequency gravitational waves is considering the electromagnetic field itself as the source of gravitational waves in the electromagnetic resonators. In the context of this approach one can neglect the gravitational waves induced by the action of electromagnetic field on the walls of the resonator.

In the papers [15, 16] the electromagnetic waves in the Fabry-Perot resonator are considered as the source of coupled gravitational waves inside resonator and free gravitational waves in empty space outside resonator.

After obtaining the inner exact solutions of equation (9) corresponding to gravitational waves coupled with electromagnetic ones, one can reconstruct the characteristics of the gravitational waves in empty space on the basis of the continuity of the energy density flux

$$ct^{01} = -\frac{c}{4\kappa}\left(\partial_0 h_{\alpha\beta}\partial_1 h^{\alpha\beta}\right), \tag{14}$$

after reflection of electromagnetic waves by the walls of resonator.

The amplitude and frequency of the gravitational waves for this case can be noted as follow [16]

$$h_0 = \frac{GE_0^2}{4\omega^2 c^2} = \frac{G}{c^4}\left(\frac{E_0 L}{2\pi q}\right)^2, \tag{15}$$

$$\omega_g = 2\omega = \frac{2\pi qc}{L}, \tag{16}$$

where E_0 is the electric field strength (or magnetic field $H_0 = E_0$) of the electromagnetic wave, L is the length of the Fabry-Perot resonator and $\omega = q(\pi c/L)$ is the frequency of electromagnetic wave, where $q = 1,2,3,...$ correspond to the number of the longitudinal optical mode in the resonator.

For the optical range $\omega \sim 10^{14} - 10^{15}\,sec^{-1}$ of electromagnetic waves with realistic electric field strength E_0 and realistic Fabry-Perot resonator length L, the amplitude of gravitational waves induced by such a way was estimated as $h_0 \sim 10^{-44} - 10^{-46}$ [16].

After placing the Fabry-Perot resonator in a constant magnetic field H_{00} transverse to the direction of propagation of electromagnetic waves, it is possible to significantly increase the amplitude of the induced gravitational waves due to the process of conversion of electromagnetic waves into gravitational ones [17-19].

The amplitude of the gravitational waves induced by this approach is [20, 21]

$$h_0 = \frac{2G}{c^3\omega} E_0 H_{00} L, \tag{17}$$

and its maximal realistic value is estimated as $h_0 \sim 10^{-32} - 10^{-30}$ [20].

Thus, the amplitude of gravitational waves induced by an electromagnetic field is ten orders of magnitude higher than in the case of the amplitude of gravitational waves induced by the matter perturbations, which was considered earlier in [22].

5. The detection of the high-frequency gravitational waves

There are different methods for detecting the high-frequency gravitational waves (see, for example, in [23-27]). As the promising method for detection of the high-frequency gravitational waves in optical range of frequencies one can use the electromagnetic response of gravitational waves passing through the magnetic field or inverse Gertsenshtein effect [18, 19].

This approach is based on the detection of photon flux arising as the result of graviton to photon conversion when gravitational waves passing through the magnetic field. The minimal detectable amplitude of the high-frequency gravitational waves by this method can be defined as follows [27]

$$h_0^{min} \simeq \left(\frac{4N_{exp}}{AH_{00}^2 L^2 \varepsilon_\gamma(\omega)\Delta\omega} \right)^{1/2}, \tag{18}$$

where $\omega = 2\pi f = \omega_g$ is the main frequency of the photons, N_{exp} is the total detected number of photons per second in the bandwidth $\Delta\omega$, A is, H_{00} is the magnetic field amplitude, L is the distance extension of the magnetic field and $\varepsilon_\gamma(\omega)$ is the quantum efficiency of the detector.

The minimal detectable amplitude of the high-frequency gravitational waves with frequencies $f_g = (2.7-14)\times 10^{14}$ Hz is estimated as $h_0^{min} \simeq 6\times 10^{-26}$ [27].

Also, for detection of the gravitational waves with frequencies $f_g < f_{opt}$ one can use the other methods, for example, the gravitational-optical resonance in a multi-beam interferometers [28-30], where condition

$$f_g = \frac{n}{2}\left(\frac{c}{L}\right), \quad n = 1,2,3,\ldots, \tag{19}$$

is fulfilled.

For the case $n = 1$ the relative variation of the power of the laser radiation in the Fabri-Perrot interferometer in the field of a gravitational wave is [30]

$$\frac{\delta W(t)}{W} = \frac{QL}{\lambda_e} h(t), \tag{20}$$

where Q is the quality factor of interferometer, λ_e is the laser wavelength and W is the laser power at the input to the Fabry-Perot interferometer.

For the following parameters $Q = 10^6$, $\lambda_e = 1.064\times 10^{-6}$ m, $L \sim 1$ m, $f_g \sim 10^8$ Hz and sensitivity $(\delta W/W)_{min} \sim 10^{-12}$ one has $h_0^{min} \sim 10^{-24}$. We also note that averaging the spectral density of variations in the power of laser radiation for the case of long-term measurements can improve the sensitivity by two orders of magnitude [30].

Thus, at the moment, the discrepancy between the maximum amplitude of artificial high-frequency gravitational waves and the sensitivity of the detectors is at least several orders of magnitude.

6. Conclusion

In this paper, we considered the methods for generating the high-frequencies gravitational waves by perturbations of medium and the electromagnetic field. The maximum amplitude of gravitational waves obtained by the first method was estimated as $h_0 \sim 10^{-40}$, while the generation of high-frequency gravitational waves by means of electromagnetic field gives a much higher value $h_0 \sim 10^{-32} - 10^{-30}$.

It is necessary to note, that modern detection methods do not allow direct observation of artificial high-frequency gravitational waves in contrast to low-frequency gravitational waves from astrophysical sources [3, 4]. Thus, the development of methods for generating and detecting artificial high-frequency gravitational waves requires fundamentally new approaches.

Nevertheless, the further development of the existing methods for detecting high-frequency gravitational waves implies the possibility of direct observation of relic gravitational waves of cosmological genesis [7-9], which can be considered as a promising direction for further research in the field of gravitational wave astronomy.

References

[1] Maggiore M 2007 *Gravitational waves. Volume 1: theory and experiments* (Oxford University Press) p 576
[2] Chervon S, Fomin I, Yurov V and Yurov A 2019 *Scalar field cosmology* (Singapore: World Scientific)
[3] Abbott B *et al* 2016 *Phys. Rev. Lett.* **116** 061102
[4] Abbott B *et al* 2017 *Astrophys. J.* **848** L13
[5] Gertsenshtein M and Pustovoit V 1962 *Sov. Phys. JETP* **16** 433
[6] Cabass G, Pagano L, Salvati L, Gerbino M, Giusarma E and Melchiorri A 2016 *Phys. Rev. D* **93** 063508
[7] Boyle L and Buonanno A 2008 *Phys. Rev. D* **78** 043531
[8] Giovannini M 2020 *Prog. Part. Nucl. Phys.* **112** 103774
[9] Ito A, Soda J 2016 *JCAP* **04** 35
[10] Kadlecová G, Klimo O, Weber S and Korn G 2017 *Eur. Phys. J. D* **71** 89
[11] Fomin I *et al* 2020 *J. Phys.: Conf. Ser.* **1705** 012004
[12] Gorelik V S *et al* 2018 *J. Phys.: Conf. Ser.* **1051** 012001
[13] Pustovoit V *et al* 2019 *J. Phys.: Conf. Ser.* **1348** 012008
[14] Pustovoit V *et al* 2020 *J. Phys.: Conf. Ser.* **1557** 012034
[15] Morozov A and Pustovoit V 2020 *Herald of the Bauman Moscow State Tech. Univ. Nat. Sci.* **1** 46
[16] Morozov A, Pustovoit V and Fomin I 2021 *Gravit. Cosmol.* **27** 24
[17] Gertsenshtein M 1962 *Sov. Phys. JETP Lett.* **14** 84
[18] Zeldovich Ya 1973 *Sov. Phys. JETP* **65** 1311
[19] Zeldovich Ya and Novikov I 1983 *Structure and Evolution of the Universe* (Chicago: University of Chicago Press)
[20] Kolosnitsyn N and Rudenko V 2015 *Phys. Scripta* **90** 074059
[21] Morozov A, Pustovoit V and Fomin I 2020 *Eur. Phys. J. Plus* **135** 950
[22] Grishchuk L 1976 *Phys. Lett. A* **56** 255
[23] Cruise A and Ingley R 2006 *Class. Quant. Grav.* **23** 6185
[24] Nishizawa A, Kawamura S, Akutsu T and Arai K 2008 *Phys. Rev. D 77* 022002
[25] Armano M *et al* 2016 *Phys. Rev. Lett.* **116** 231101
[26] Li F *et al* 2009 *Phys. Rev. D* **80** 064013
[27] Aggarwal N *et al* (*Preprint* arXiv:2011.12414 [gr-qc])
[27] Ejlli A, Ejlli D, Cruise A, Pisano G and Grote H *2019 Eur. Phys. J. C* **79** 1032
[28] Esakov A, Morozov A, Tabalin S and Fomin I 2015 *Herald of the Bauman Moscow State Tech. Univ., Nat. Sci.* **1** 26
[29] Fomin I, Morozov A 2017 *J. Phys. Conf. Ser.* **798** 012088
[30] Golyak I, Esakov A, Morozov A, Nazolin A, Tabalin S, Fomin I 2021 *Radio Engineering* **02** 13

Acknowledgments

V Pustovoit, V Gladyshev, V Kauts, A Morozov, P Nikolaev, I Fomin, E Sharandin were supported by the Russian Science Foundation under Grant No. 19-12-00242.

An approach to Grand Unification

Peter Rowlands

Department of Physics, Oliver Lodge Laboratory, University of Liverpool, Oxford St, Liverpool, L69 7ZX, UK

E-mail: p.rowlands@liverpool.ac.uk

Abstract. Around the time of the completion of the Standard Model of particle physics in the 1970s, schemes were put forward for unifying the three gauge interactions (electric, strong and weak) using the renormalization equations at an energy approaching the Planck mass. Though these looked promising, the exact unification never materialised, and doubts have been raised about whether this Grand Unification can be achieved. It may be possible, however, to create Grand Unification at the Planck mass if we start with a radical examination of the nature of the colour model of quarks.

1. Introduction

The Standard Model is the most successful theory ever devised. It was completed about 1973 and very little has happened since to change it in any significant way. If we want to calculate the results of an experiment in a particle collider we have an equation which can be solved iteratively to take it to the order of magnitude desired. We know how all the fundamental fields work and we know all the interacting particles (Table 1). We know how everything happens. We just don't know *why*. We have four interactions, gravity and the three gauge interactions (strong, weak and electric), which have seemingly different force laws or gauge symmetries, and we don't know why they are different and how they all connect.

Table 1. The Standard Model particles.

	Quarks		*Gauge Bosons*	*Higgs Boson*
u	c	t	g	H
d	s	b	γ	
	Leptons			
e	μ	τ	Z	
ν_e	ν_μ	ν_τ	W	

Two things that happened in the 1960s and 1970s seemed to give us an inkling of a way forward. The first was the Glashow-Weinberg-Salam theory 'combining' the weak and electric interactions (1967). The second was the possibility of a 'Grand Unification' of the three gauge forces at a high enough energy regime [1]. In the electroweak theory the electric and weak interactions are combined mathematically in an $SU(2) \times U(1)$ gauge group, which combines the original weak isospin fields W_1, W_2, and W_3, and the weak hypercharge field B, into the familiar weak W^+, W^-, Z^0 and the massless γ (photon) for the electric component. The mixing is specifically between W_3 and B, which creates the new fields Z^0 and γ. It is quantified by the weak mixing angle θ_W, whose sine squared fixes the ratio of the charge squared for the weak and electric interactions, and whose cosine defines the mass ratio of the W and Z bosons. Current measurements suggest that $\sin^2\theta_W = 0.231$ at energies close to the values of the W and Z masses (80 to 91.2 GeV). The non-unit value means that we should be careful about implying that the electric and weak forces have become a single force, as some authors have been inclined to do. We have shown that they are connected, but they have different sources, and don't have the same coupling constant. Connected does not mean interchangeable, and weak and electric sources are not interchangeable.

The second potential development, grand unification, became possible because, though charges are conserved as units, their value in terms of energy of interaction varies with the strength of the field. This is because a particle such as an electron will polarize the vacuum around it to produce virtual electron-positron pairs, which then produce further pairs, and so on. All the gauge forces act in this way, but, fortunately, the potentially infinite series of interactions can be shown to yield a finite sum, using the process we call renormalization. The parameter that we use is called the 'fine structure constant' for the interaction. This is the ratio of the charge squared to the square of the Planck charge ($\hbar c$). So the electric fine structure constant is given by

$$\alpha = e^2 / \hbar c = 1/137.0359 \ldots$$

(Here, and elsewhere, it will be assumed that e^2 automatically includes the factor $1/4\pi\varepsilon_0$ introduced with SI units.)

The charge responsible for each interaction is given by its own fine structure constant α_n. At any given energy m it has a value determined by its renormalizaton, according to the gauge symmetry which governs the interaction. The first order equations for the $SU(2)$ weak and $SU(3)$ strong interactions are as follows:

$$\frac{1}{\alpha_2(\mu)} = \frac{1}{\alpha_G} - \frac{5}{6\pi} ln \frac{M_X^2}{\mu^2}$$

$$\frac{1}{\alpha_3(\mu)} = \frac{1}{\alpha_G} - \frac{7}{4\pi} ln \frac{M_X^2}{\mu^2}.$$

Here M_X is the GU energy scale, α_G is the fine structure constant at this energy and μ is the energy scale of measurement. These two fine structure constants decrease with increasing energy, but the remaining one, for the electric interaction, will increase.

$$\frac{1}{\alpha(\mu)} = \frac{1}{\alpha_G} + \frac{5}{3\pi} ln \frac{M_X^2}{\mu^2}.$$

So, there is a possibility that they could converge, and this was put forward by Georgi and Glashow in 1974. One of the problems is that the weak coupling cannot be measured directly in the same way as the strong and electric couplings. It can, however, be determined through calculation of $\sin^2\theta_W$, which can be found on the assumption that the interactions can be accommodated within a single grand unified gauge group. The simplest group accommodating each of $SU(3)$, $SU(2)$ and $U(1)$ is $SU(5)$.

2. Grand Unification

A Grand Unified Theory (GUT), to unite electric, weak and strong interactions, based on the $SU(5)$ group, was first proposed by Georgi and Glashow in 1974, following on from the GWS $SU(2) \times U(1)$ unification of the electric and weak interactions, governed by the weak mixing angle parameter $\sin^2\theta_W$, which is effectively the ratio (α / α_2) between the weak and electric couplings (α_2 and α). Georgi and Glashow showed that, in any GU scheme determined by a single GU gauge group, not necessarily $SU(5)$, $\sin^2\theta_W$ would be given by the ratio of the sum of all the squared units of weak isospin ($t_3 = \pm \frac{1}{2}$) for the fermions of the SM to the sum of all their squared units of electric charge (Q).

$$sin^2\theta_W = \frac{Tr\left(t_3^2\right)}{Tr\left(Q^2\right)}$$

Taking the weak components with only left-handed contributions to weak isospin, for the first generation of quarks and leptons, that is, for 3 colours of u, 3 colours of d, and the leptons e and ν, we obtain:

$$Tr\left(t_3^2\right) = \frac{1}{4} \times 8 = 2.$$

For, the fractional electric charges of the Gell-Mann-Zweig quark model (1964), with both LH and RH contributions in the first generation, we obtain

$$Tr\left(Q^2\right) = 2 \times \left(\frac{4}{9} \times 3 + \frac{1}{9} \times 3 + 1 + 0\right) = \frac{16}{3},$$

from which

$$\sin^2\theta_W = 0.375.$$

The SU (5) theory raised many hopes because of the seeming 'near convergence' at an energy of about 10^{15} GeV, which was only a few orders of magnitude below the magical 'Planck mass', the key value for quantum gravity at 1.22×10^{19} GeV, and suggesting a route to a four-force unification.

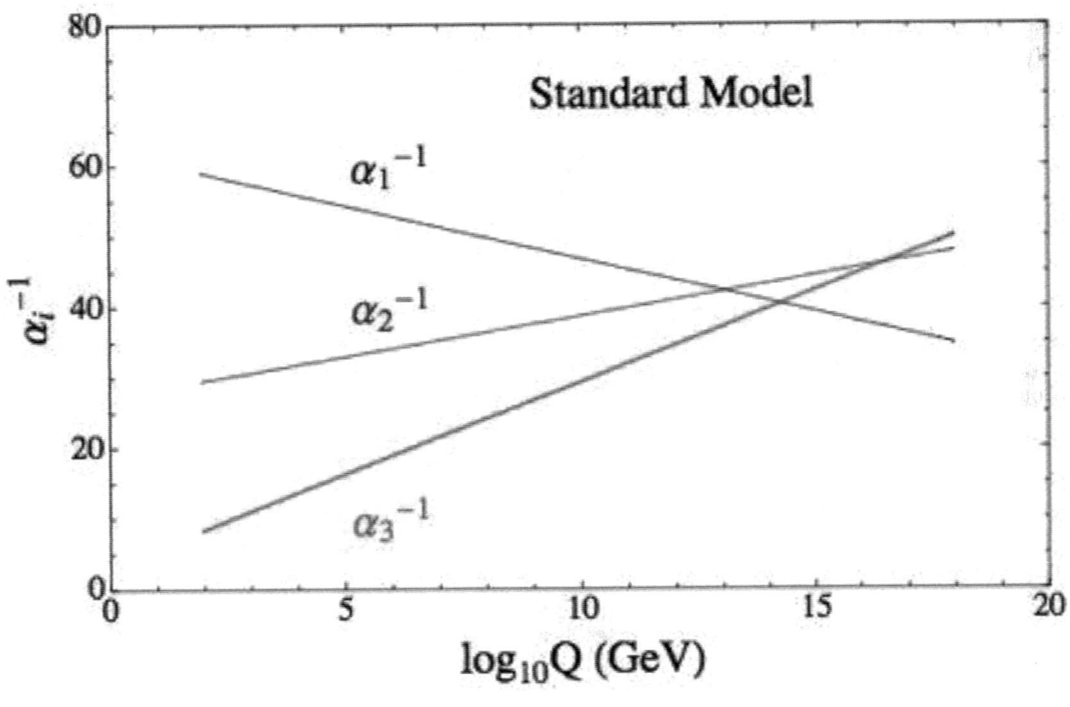

Figure 1. The SU(5) Grand Unification (from [2]).

Many decades later, these hopes have not been met. First, the hoped for convergence never materialised. Second, the SU (5) predicted a new range of X and Y bosons, carrying all three gauge interactions, which seemed to require a decay of the proton, which again has not materialised. Finally, the value for $\sin^2\theta_W$, calculated from the theory at 0.375, was nothing like the experimental value of 0.231, leading to a claim that the parameter would be 'renormalized' at higher energies to ~0.2. Another attempt to save the situation was to invoke supersymmetry [3]. If exactly the right amount of the right kind of supersymmetry was included, then the values of the a terms would be reduced, leading to an exact unification at ~10^{16} GeV. It was even claimed – and still is – that this was *evidence* for supersymmetry – a claim beyond all reasonable interpretations of the scientific method. The fact that no such evidence has been found suggests that we should abandon this idea as well. Ethan Siegel, an influential and incisive commentator on many aspects of modern physics, suggests that GU 'may be a dead end for physics' [4].

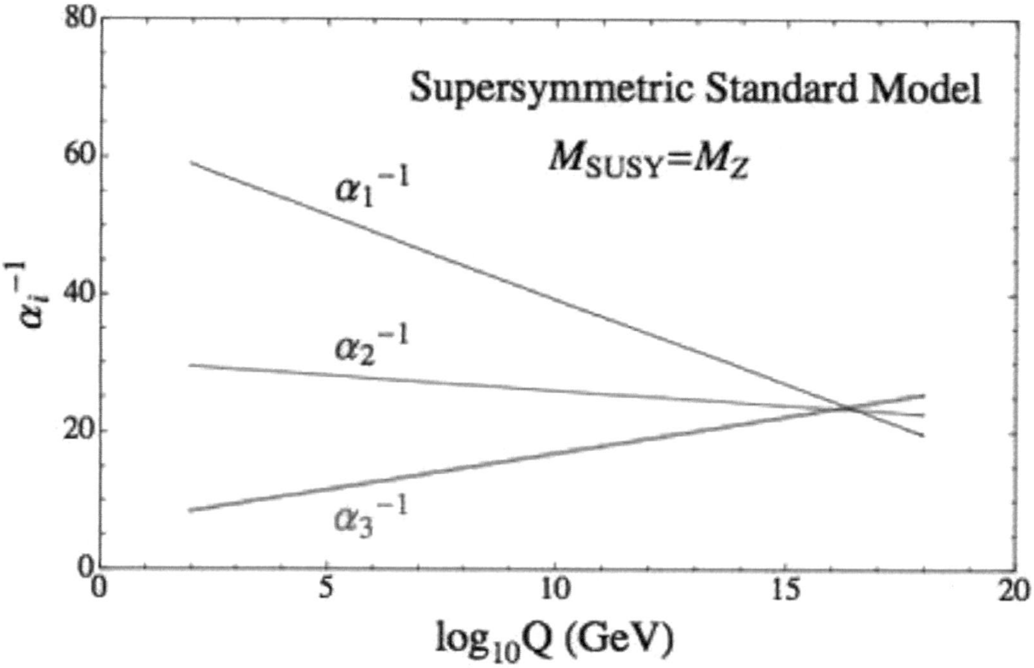

Figure 2. Grand Unification including supersymmetry (from [2]).

3. Two versions of quark theory

Does this mean that we should abandon all hope of GU? When I first started seriously investigating the subject in the 1990s, partly with my colleague John Cullerne, I put down the equations for renormalization in the three gauge theories, and could not reconcile them [5-7]. If I put in the value of 10^{15} or 10^{16} GeV for the GU mass I got nonsense for the three individual equations. They simply didn't work at all. Nor did $\sin^2\theta_W$ 'renormalize' in the way suggested. Instead of decreasing with energy it *increased*, becoming 0.6 at GU, not 0.2! It looked like the supposed convergence was merely a result of compensating errors. But, even more significant, was the fact that the graphs did not represent the convergence of the three gauge interactions at all, for the electric α was replaced by an 'electroweak' α_1.

Instead of using

$$\frac{1}{\alpha(\mu)} = \frac{1}{\alpha_G} + \frac{5}{3\pi} ln \frac{M_X^2}{\mu^2}$$

we now have

$$\frac{1}{\alpha_1(\mu)} = \frac{1}{\alpha_G} + \frac{1}{6\pi} ln \frac{M_X^2}{\mu^2},$$

where

$$\frac{5}{3\alpha_1(\mu)} + \frac{1}{\alpha_2} = \frac{1}{\alpha}$$

is introduced from the electroweak theory. It doesn't seem to make sense to unify electroweak, weak and strong rather than electric, weak and strong. But we can see why this was done. If we use a rather than a_1, we get nothing resembling convergence at all.

But I also came into the problem with a more flexible attitude to the nature of quarks than has been usual since about 1980, and this may lead us to a remarkable solution of the problem. have no doubt at all that experimental QED studies using deep inelastic scattering tell us that up quarks behave as though they have +2/3 of the value of the fundamental unit of charge (that of the electron), while down quarks behave as though they have –1/3. This is true at all energies and there is no evidence of any transition at higher energies. So the quark charge quantum numbers would be as follows:

Table 2. Quark charges: Gell-Mann Zweig (GMZ).

	Blue	Green	Red
up	$2e / 3$	$2e / 3$	$2e / 3$
	$B / 3$	$B / 3$	$B / 3$
down	$-e / 3$	$-e / 3$	$-e / 3$
	$B / 3$	$B / 3$	$B / 3$

Here, B = baryon number, effectively strong charge.

Experiment shows that QED correctly uses fractional charges. But we don't know, in principle, if these are *fundamental* or if they stem from a broken symmetry. Experimental measurements of fractional charges may be a result of *competing* symmetries, and we will show later how this could be possible. Certainly, there are problems with fractional charges. Do we, for example, have to redefine the fundamental charge unit as e / 3, instead of e? If so, does this mean that the electron is not really a fundamental particle? How does it manage to incorporate 3 repelling units in a point-like structure, and how does an 'elementary' up quark incorporate 2?

But the original and now standard Gell-Mann-Zweig version was not the only quark model proposed. As O. W. Greenberg, one of the major players, wrote in 2015: 'In 1965 Moo-Young Han and Yoichiro Nambu introduced a model in which an *SU*(3) color group is explicit. They avoided fractional electric charges by introducing nine quarks with integer charges …' [8, 9] (The origin of quark color. *Physics Today*, 1 January 2015, p. 33) He continued, 'but they paid the price that color was not an exact symmetry. (Moreover, the notion of integer quarks ended up conflicting with later experimental evidence.)' (Both these claims are wrong.) The Han-Nambu model, the first to introduce colour as a gauge symmetry, can be summarised as follows:

Table 3. Quark charges: Han-Nambu (HN).

	Blue	Green	Red
up	e	e	0
	B	0	0
down	0	0	$-e$
	B	B	B

We may note here that B (strong charge) is on only one quark at a time.

Han and Nambu may originally have thought that colour singlets or separated quarks might emerge at higher energies, but it seems that the colour symmetry is never broken. The symmetry is exact because the strong interaction is perfectly gauge invariant and has massless bosons. Greenberg's claim that in their model 'color was not an exact symmetry' is utterly incorrect. There is no problem with colour because the symmetry is exact, and it is entirely due to the strong force or 'colour charge', and has nothing to do with what other charges are present. There is also no problem with experiment because QED and QCD (the theory of the strong interaction) are totally independent.

QED phenomenology clearly doesn't decide the question of which basic structure of charges to use in other areas such as GU, or the *gauge relations between the interactions*. Our fundamental methodology suggests that we go for the best fit with the symmetry, and, if that leads to modifications in our usual practice, see if this has any interesting consequences. In fact, as we shall see, there is one of exceptional interest.

A result of immense significance came with the remarkable discovery, by Laughlin in 1983 [10], of the fractional quantum Hall effect, where electrons assumed effective charges of $e / 3$, $e / 5$ and many other fractional values, by forming pseudobosonic combinations with odd numbers of magnetic flux lines and so became shared out between them. As long as the total number of 'fermions' (i.e. electrons and flux lines) is even, then collective bosonic states can form. For example, 3 electrons can combine with 5 flux lines to produce effective charges on each flux line of $3\,e / 5$. This discovery could have made a difference to the understanding of the quark model, but it didn't.

The Han-Nambu quarks and the fractional quantum Hall electrons form a remarkable comparison. In each case the QED phenomenology is defined by a totally external agent. In the first case a totally gauge invariant *strong* interaction acts independently of any electric charges. This is the interaction which makes the composite baryon (e.g. neutron or proton) by combining quarks into a collective structure, which then determines its QED. In the second case a gauge invariant *weak* interaction (fermion / boson converter) acts independently of any electric charges. The collective state of electrons and flux lines governed by this weak interaction then determines its QED.

4. The effect of the HN model on GU

How would the HN model affect the quantities observed in gauge interactions? First of all, it would make no difference to the squared weak components with only left-handed contributions to weak isospin, which, for the first generation of quarks and leptons, would still be:

$$Tr\left(t_3^2\right) = \frac{1}{4} \times 8 = 2\,.$$

But for the electric charge, the summation would be different. For, fractional charges, with both LH and RH contributions in the first generation, we would now obtain

$$Tr\,(Q^2) = 2 \times (1 + 1 + 0 + 0 + 0 + 1 + 1 + 0) = 8$$

and

$$sin^2\theta_W = \frac{Tr\left(t_3^2\right)}{Tr\left(Q^2\right)} = 0.25\,.$$

While, for Steven Weinberg, 0.375 is in 'gross disagreement' with the experimental value for $\sin^2\theta_W$ of 0.231 at $M_Z = 91.2$ GeV (a sure sign that there is something intrinsically wrong) [11], 0.25 is relatively close to this value, and the true value may well be 0.25 at the vacuum expectation value of the Higgs field (246 GeV), and close to this at M_W or M_Z. There are also small second order corrections, and the effect of the direct production of W and Z bosons, producing a distinct dip in the measured value [12], to be taken into account. 0.25 is also the value that would be obtained purely from the leptonic contribution, and it would be somewhat strange if the value for a purely electroweak parameter should be different in the quark and lepton sectors. The HN model, unlike GMZ, makes quarks leptonlike, allowing a real unification of these sectors.

Note that this is not dependent on the GU gauge group used. So the Georgi-Glashow group $SU(5)$ is not the only one for which it is valid. In fact, Georgi and Glashow's use of the simple group $SU(5)$ was preceded by Pati and Salam's semisimple $SU(4)_L \times SU(2)_L \times SU(2)_R$, in which the $SU(4)_L$ requires an extension of the colour symmetry of $SU(3)$ from 3 to 4 'colours', representing another route to combining of quarks and leptons [13]. Significantly $SO(10)$, which has been favoured by many, incorporates both $SU(5)$ and $SU(4)_L \times SU(2)_L \times SU(2)_R$ as components. In this sense, the Pati-Salam theory, though not directly based on a simple group, is based on a larger simple group. In this context, the Georgi-Glashow argument for $\sin^2\theta_W$ would still exist for Pati-Salam (which provides an alternative argument for 0.25), but the real unification would be through $SO(10)$, which could similarly be applied to $SU(5)$.

There are two significant consequences of the calculation of an independent and credible value for $\sin^2\theta_W = \alpha / \alpha_2 = 0.25$ at the electroweak scale. The first is that we can now produce a much simpler calculation for M_X without making assumptions about the group structure, by avoiding the problematic running coupling constant equation for $1 / \alpha_1$, using only the more secure equations for $1 / \alpha_2$ and $1 / \alpha_3$. In addition, the hypercharge numbers for the $U(1)$ electromagnetic running coupling equation will now be no longer identical to those for a quark model based purely on QED phenomenology. Looking at the original three graphs for the fine structure constants, we see that those for α_2 and α_3 cross at a distinctively higher value of energy m than when combined with that for α_1. Because α_2 depends on $\sin^2\theta_W$, when this is reduced from 0.375 to 0.25, the crossing point is pushed to an even higher energy. We can use this to find M_X or the GU energy scale, and α_G, which is the fine structure constant at this energy. Let us look again at the two key equations:

$$\frac{1}{\alpha_2(\mu)} = \frac{1}{\alpha_G} - \frac{5}{6\pi} ln \frac{M_X^2}{\mu^2},$$

$$\frac{1}{\alpha_3(\mu)} = \frac{1}{\alpha_G} - \frac{7}{4\pi} ln \frac{M_X^2}{\mu^2}.$$

We have good experimental values of

$$\alpha_3(M_Z^2) = 0.118 \text{ (or } 0.12)$$

and

$$\alpha(M_Z^2) = 1 / 128$$

at the electroweak energy $\mu = M_Z = 91.2$ GeV. We also have a calculated $\sin^2\theta_W = 0.25$ or close, which allows us to convert $\alpha M_Z^2)$ to $\alpha_2(M_Z^2)$. If we solve the two equations, we obtain a value for the GU energy scale ($M_X = 2.8 \times 10^{19}$ GeV) which is extraordinarily close to the Planck value (1.22×10^{19} GeV), and may well be exactly so, as purely first-order calculations overestimate the value of M_X by more than 50 % [12]. Assuming that M_X *is* the Planck mass, we obtain

$$\alpha_G \text{ (the GU value for all interactions)} = 1 / 52.4$$

and

$$\alpha_2(M_Z^2) = 1 / 31.5,$$

which is exactly the kind of value we would expect for the weak coupling with $\sin^2\theta_W = 0.25$ close to M_Z.

5. Including the electric force

We have shown that the weak and strong coupling constants suggests GU at the Planck mass. Can we include the electric force? The original GU of 1974 used a combined electroweak parameter which made assumptions about group structure, and relied on a particular value for the squared 'Clebsch-Gordan coefficient' of the group, $C^2 = 1 / \sin^2\theta_W - 1 = 5 / 3$, that had no experimental or theoretical justification. However, the fermionic contribution to QED vacuum polarization is, for *fundamental* fractional charges,

$$\frac{4}{3}\times\frac{1}{2}\times\left(\frac{1}{36}\times3+\frac{1}{36}\times3+\frac{1}{9}\times3+\frac{4}{9}\times3+\frac{1}{4}\times1+\frac{1}{4}\times1+1\right)\frac{n_g}{4\pi}=\frac{5}{3\pi},$$

where $n_g = 3$ is the number of fermion generations, and the terms in the bracket represent, respectively, the squared average charge in the isospin quark doublet, the squared charges of the quarks, the squared average charge of the isospin lepton doublet, and the squared charges of the leptons, all for both left- and right-handed states. Modifying this for fundamental integral charges, we obtain:

$$\frac{4}{3}\times\frac{1}{2}\times\left(\frac{1}{4}\times3+\frac{1}{4}\times3+1+1+0+0+0+\frac{1}{4}\times1+\frac{1}{4}\times1+1\right)\frac{n_g}{4\pi}=\frac{3}{\pi}.$$

This result corresponds to a change in the squared Clebsch-Gordan coefficient from $C^2 = 5 / 3$ to $C^2 = 3$, when $\sin^2\theta_W = 1 / (1 + C^2)$ changes from 0.375 to 0.25. With the new values we have obtained for the hypercharge numbers, the running coupling of the pure electromagnetic interaction, will be:

$$\frac{1}{\alpha(\mu)}=\frac{1}{\alpha_G}+\frac{3}{\pi}ln\frac{M_X^2}{\mu^2}.$$

Using just the well-established equations for $1 / \alpha_2$ and $1 / \alpha_3$, and $\sin^2\theta_W = \alpha / \alpha_2$, we obtain

$$sin^2\theta_W(\mu)=\alpha(\mu)\left(\frac{1}{\alpha_3(\mu)}+\frac{11}{6\pi}ln\frac{M_x}{\mu}\right).$$

We can *now* make direct use of the equation we have derived for $1 / \alpha$, with the new hypercharge numbers and GU at the Planck mass, to obtain $1 / \alpha(M_Z^2) = 128$, which is, of course, exactly the value obtained experimentally at energies corresponding to $\mu = M_Z$. Conversely, using $1 / \alpha(M_Z^2) = 128$ gives $M_X = 1.42 \times 10^{19}$ GeV. This appears to be a striking confirmation of the assumptions made in the first calculation, leading to M_X, as coincidental agreements are most unlikely for equations involving logarithmic terms, and it is also potentially very significant, for it would now appear that the unification which occurs at M_X might well involve a direct numerical equalization of the strengths of the three, or even four, physical force manifestations, without reference to the exact unification structure.

We can now summarise the Grand Unification theory in four equations:

$$\frac{1}{\alpha(\mu)}=\frac{1}{\alpha_G}+\frac{3}{\pi}ln\frac{M_X^2}{\mu^2}$$

$$\frac{1}{\alpha_2(\mu)}=\frac{1}{\alpha_G}-\frac{5}{6\pi}ln\frac{M_X^2}{\mu^2}$$

$$\frac{1}{\alpha_3(\mu)}=\frac{1}{\alpha_G}-\frac{7}{4\pi}ln\frac{M_X^2}{\mu^2}$$

$$sin^2\theta_W(\mu)=\alpha(\mu)\left(\frac{1}{\alpha_3(\mu)}+\frac{11}{6\pi}ln\frac{M_x}{\mu}\right).$$

The results are represented in graphical form in Figure 3 for the coupling constants and in Figure 4 for $\sin^2\theta_W$. In this representation, the electroweak parameter $\sin^2\theta_W$ notably becomes 1 at Grand Unification, the only true unification of electric and weak.

At GU, we may suppose, all four forces are reduced to scalar phases, with $U(1)$ symmetry and purely Coulombic interaction, all distinguishing aspects of the weak and strong interactions having diminished to zero. One of the most significant aspects of the calculation is that it leads to completely testable predictions, as the values of the three coupling constants can be calculated for any energy with relative precision from the known values of α_G and M_X. In particular, the value of α changes rapidly in a way that can be determined at energies now available to us experimentally. At 14 TeV, for instance, it would have the value of 1 / 118, compared to 1 / 125 from the minimal *SU*(5) theory of Georgi and Glashow. This is a very definite prediction which can be tested with current experimental facilities.

The GU calculation reduces the number of free parameters. The GU energy at the Planck value is purely a consequence of fundamental constants – it's a pure number. It suggests a unification with gravity, something like gravity-gauge theory correspondence, with a quantum aspect. $\sin^2\theta_W$ at the electroweak scale is a result of pure numerical calculation and it rises to 1 at GU. Here, we see that, at the electroweak scale, the parameter is a measure of the *lack of unification* of the two forces, as true unification requires the value 1. There is no electroweak term at GU. At GU, the logarithmic terms zero in the renormalization equations and only the coupling constant, equivalent to the Coulomb component, remains.

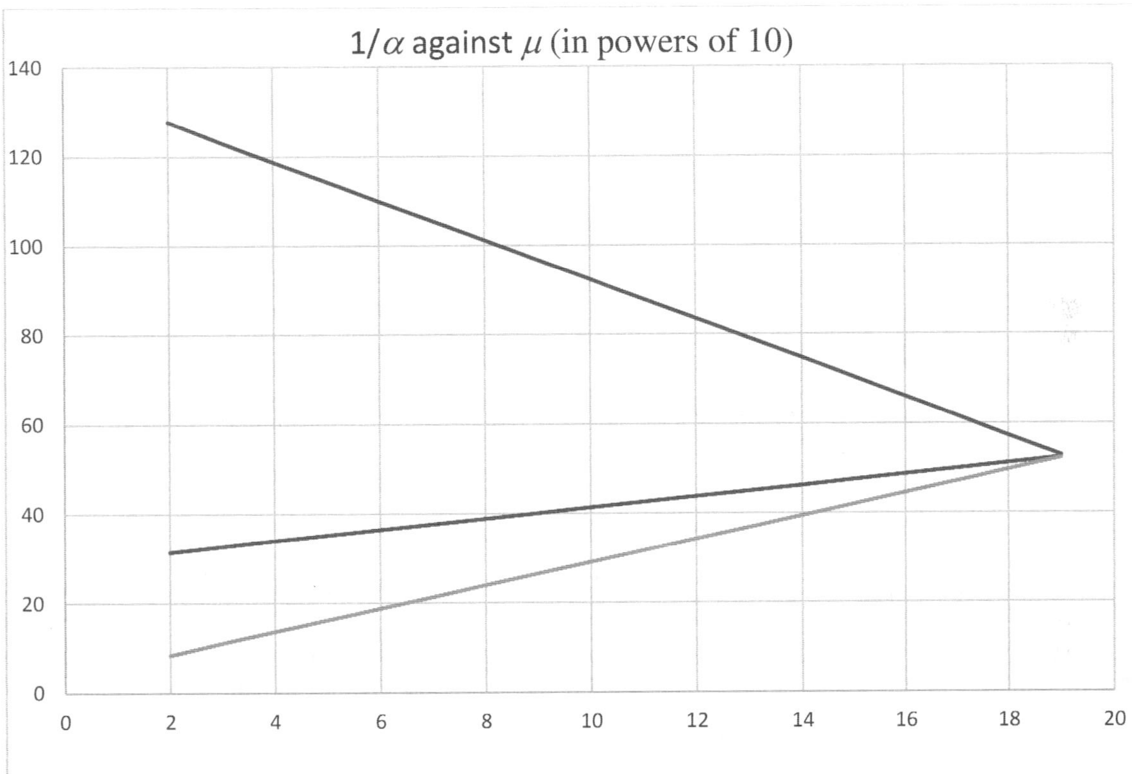

Figure 3. The coupling constants as a function of energy of interaction.

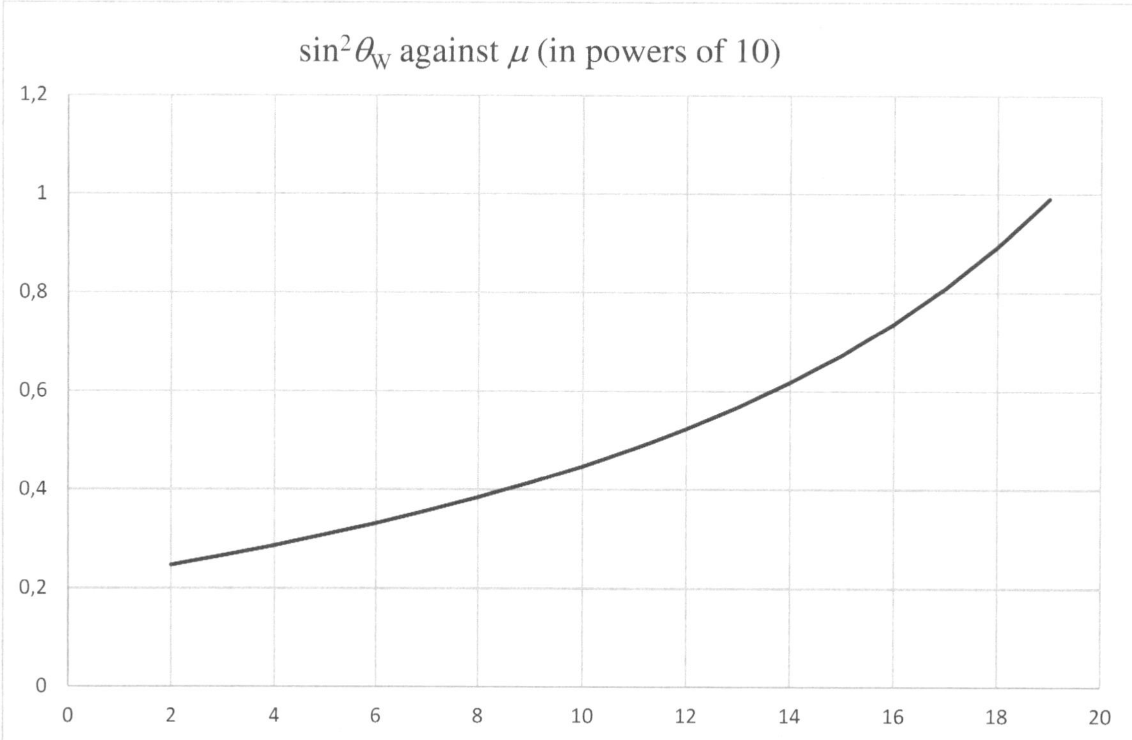

Figure 4. The electroweak parameter as a function of energy of interaction.

6. Conclusion

Grand unification was one of those powerful ideas, like supersymmetry, which seemed destined never quite to live up to its promise. I think the problem stemmed from the lack of understanding of the differences between reality seen as the result of observation (1) and reality seen as underlying structure (2). Broken symmetries, in my view, often stem from clashes between these modes of explanation because they develop from hidden symmetries (reality 2) which are not directly concerned with the observations being made (reality 1). This is what I believe has happened with quarks, where the underlying group structure needed for GU operates on different rules from the QED of observation.

We can overcome this problem by returning to a model that was proposed almost as soon as the quark theory was created and which has never been disproved, but simply fallen out of fashion because more immediate needs have taken precedence. As soon as we do, we obtain remarkable results: a credible value for the electroweak mixing parameter, GU at the Planck mass; pure Coulombic forces at this energy; a connection with gravity, and a meaningful connection between quarks and leptons. And we have predictions that can be tested with realisable experiments.

References

[1] Georgi H and Glashow S 1974 Unity Of All Elementary Particle Forces *Phys. Rev. Lett.* **32** 438
[2] Langacker P 2012 Grand unification scholarpedia.org, doi:10.4249/scholarpedia.11419
[3] Amaldi U de Boer W and Fürstenau H 1991 Comparison of grand unified theories with electroweak and strong coupling constants measured at LEP *Phys. Lett.* **260** 447
[4] Siegel E 2016 Grand Unification May Be A Dead End For Physics *Forbes Magazine* 7
[5] Rowlands P and Cullerne J P 2001 The connection between the Han-Nambu quark theory, the Dirac equation and fundamental symmetries *Nuclear Physics* A **684** 713
[6] Rowlands P 2007 *Zero to Infinity: The Foundations of Physics* (Singapore and Hackensack, N.J: World Scientific)

[7] Rowlands P 2014 *The Foundations of Physical Law* (Singapore and Hackensack: N.J: World Scientific)
[8] Han M Y and Nambu Y 1965 Three-Triplet Model with Double SU(3) Symmetry *Phys. Rev.* **139** 1006
[9] Greenberg O W 2015 The origin of quark color *Physics Today* p 33
[10] Laughlin R B 1983 Anomalous quantum Hall effect: an incompressible field with fractionally charged excitations *Phys. Rev. Lett.* **50** 1395
[11] Weinberg S 1996 *The Quantum Theory of Fields* (Cambridge University Press) pp. 327-32
[12] Kounnas C Masiero A Nanopoulos D V and Olive A 1984 *Grand Unification with and without Supersymmetry and Cosmological Implication* (World Scientific)
[13] Pati J C and Salam A 1974 Lepton number as the fourth color.*Phys. Rev.* **D10** 275–289

Optimization of the Euro-Asian network of gravitational detectors for detecting the radiation of collapsing objects

V Rudenko[1,2,3],S Andrusenko[3], D Krichevskiy[1,3] and G Manucharyan[1,3]

[1] Sternberg Astronomical Institute, Lomonosov Moscow State University, Universitetskii Prospect 13, 119234 Moscow, Russia
[2] Faculty of Physics, Lomonosov Moscow State University, 119991 Moscow, Russia
[3] Department of Physics, Bauman Moscow State Technical University, 2-nd Baumanskaya Street 5, Building 1, 105005 Moscow, Russia

E-mail: valentin.rudenko@gmail.com

Abstract. A Euro-Asian network of four gravitational-wave (GW) interferometers is considered, taking into account the plan to create such a detector in Novosibirsk. The efficiency of the network is assessed by typical numerical criteria, which also depend on the characteristics of the received signal. In this work, we calculate the optimal orientation of the Novosibirsk detector for the problem of detecting GW radiation accompanying the collapse of the progenitor star with an initial angular momentum. The specificity of the scenario is the presence of the so-called. bar stage deformation, for which the shape of the emitted GW signal is known.

1. Introduction

In September 2015, the first direct registration of a gravitational-wave burst from the merger of a relativistic binary, whose components were evaluated as black holes (BH), took place. The detection of this event was carried out using LIGO detectors [1]. This was followed by the registration of a big number of events. A qualitative step was the registration by three detectors (including a similar interferometer VIRGO in Europe) GW170814 burst from the merger of BH binary ($M = 30M_{\odot}$) from the distance of 540 Mps [2], which allowed to reduce the localization zone of the source on the celestial sphere by an order of magnitude, up to $\sim 60\ deg^2$. A gravitational wave (GW) signal from neutron stars (NS) merger was registered, coinciding with GRB170817A gamma burst (with 1.7 s delay) [3]. One of the most recent significant steps is the registration of gravitational waves from the neutron star – black hole binaries coalescence [4]. All these facts allow to claim confidently real occurrence of a new gravitational-wave channel of astrophysical information and heuristic value of multi-messenger astronomy, i.e. strategy of parallel observation of transients on detectors of different physical nature.

However, there is still no observation of gravitational waves from supernova as well as significant coincidence between LIGO/VIGRO data and neutrino detectors like NOvA [5] and IceCube [6] experiments. Gravitational waves from core-collapse supernovae can serve as an important source of information about the processes occurring during a given event [7].

In this situation, the problem of creating an optimal terrestrial network of gravitational antennas both in terms of their location, orientation and interconnection (methods of processing common signals against the background of noises) is of current interest [8]. In this context, we continue the discussion of a European-Asian network (EAN) which consists of four antennas in the northern hemisphere: VIRGO in Italy, KAGRA in Japan, LIGO-India in India and the planned new additional antenna in Novosibirsk. In order to assess the scientific feasibility and efficiency of such a network the calculation of its main characteristics was performed in the approach developed in the articles [9, 10]. In this paper we consider a gravitational signal from core-collapse supernovae and define the optimal orientation of Novosibirsk detector while the analysis for the case of relativistic binary coalescence was carried out in the previous paper [11]. In order to select the optimal detector orientation angle in Novosibirsk, we perform numerical calculation of criteria [9]. On this basis, a generalized integral efficiency criterion is formed, maximizing which by the angle of orientation of the detector, we find the most effective angle of Novosibirsk detector.

Table 1 shows the coordinates of the detectors in question. The detector orientation angle γ is defined as the angle between the southward direction at the detector location and the bisector of the angle formed by its arms, measured counterclockwise. Figure 1 shows geographic location of all detectors.

Table 1. Detector data; all angles given in degrees.

Detector	Latitude l	Longitude L	Orientation γ
VIRGO	43.6	-10.5	206.5
KAGRA	36.4	-137.3	163.3
LIGO India	19.6	-77.0	254.0
Novosibirsk	55.0	-82.9	to be defined

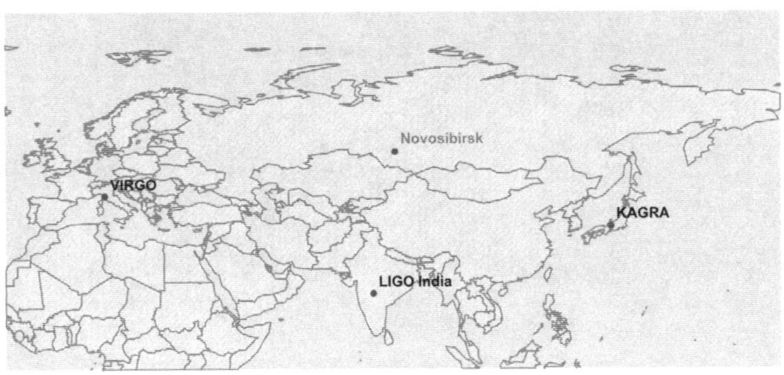

Figure 1. Geographic location of all detectors.

2. Criteria of a network

To estimate efficiency of a network of ground based detectors it's necessary to construct power patterns of individual components and the whole network. Let's result the basic information necessary for construction of the pattern.

In the long wavelength approximation (the GW wavelength is much larger than the interferometer arm length L) the detector response can be evaluated as

$$h(t) = \frac{\delta L}{L} = F_+(\theta, \varphi, \psi) h_+(t) + F_\times(\theta, \varphi, \psi) h_\times(t), \tag{1}$$

where $F_+(\theta, \varphi, \psi)$, $F_\times(\theta, \varphi, \psi)$ are the antenna pattern functions for the two polarizations, which are functions of the polar angle θ and the azimuth angle φ of the spherical coordinate system (XY is the detector plane) and the polarization angle of the GW ψ.

Antenna pattern functions have the following form (in the coordinate frame, which basis vectors, coincide with the direction of the detector arms):

$$F_+ = \frac{1}{2}(1 + \cos^2\theta) \cos 2\varphi \cos 2\psi - \cos\theta \sin 2\varphi \sin 2\psi, \tag{2}$$

$$F_\times = \frac{1}{2}(1 + \cos^2\theta) \cos 2\varphi \sin 2\psi + \cos\theta \sin 2\varphi \cos 2\psi. \tag{3}$$

In [12] it's shown that for a network of N detectors network antenna power pattern P^N:

$$P^N = \sum_{k=1}^{4} (F_{+,k}^2 + F_{\times,k}^2). \tag{4}$$

Figure 2 shows EAN network power pattern.

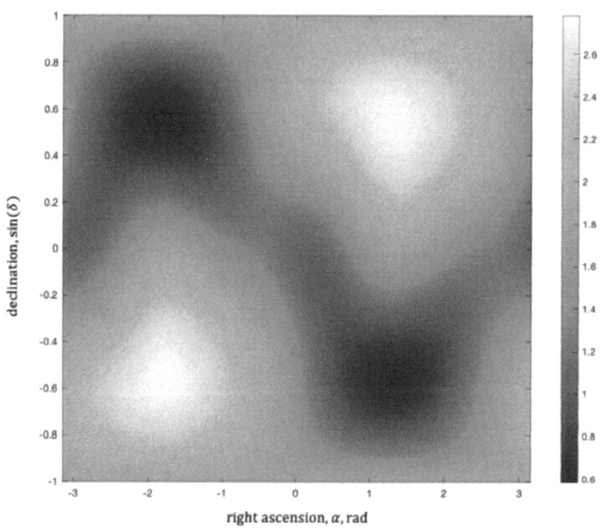

Figure 2. EAN network power pattern.

In order to choose the optimal detector angle in Novosibirsk, we use 3 independent criteria presented in the works [9, 10]. These three conditions form an integral criterion, which is to be maximized by changing the orientation of the Novosibirsk detector, to find the most effective angle.

2.1. Polarization criterion I

Criterion I characterizes ability of the network to assess the polarization of the received GW. Following [9] we define + and × integral functions for a network of four detectors:

$$F^N = \frac{1}{2}\sqrt{F_1^2 + F_2^2 + F_3^2 + F_4^2}, \tag{5}$$

where N - stands for a network function and $F_1 ... F_4$ all either correspond to the + or × polarization. Obviously F^N depend on the polarization angle of ψ.

Calculation of I is carried out in dominant polarization frame (DPF) [13]. In DPF for each point on the celestial sphere $(\alpha;\delta)$ (in equatorial coordinate system point is defined by right ascension $\alpha \in [-\pi;\pi]$ and declination $\delta \in [-\frac{\pi}{2};\frac{\pi}{2}]$) a polarization angle that maximizes the network factor F_+^N and minimizes $F_\times^N$ is chosen. Consequently, for this direction $(\alpha;\delta)$ the condition $F_+^N \geq F_\times^N$ is valid. The condition of approximate equality of factors $F_\times^N$ and F_+^N has to be kept, i.e. $\frac{F_\times^N}{F_+^N} \approx 1$. This means that the gravity detector network will be sensitive to both gravity wave polarizations. It follows that a minimum difference of $|F_\times^N - F_+^N|$ should be sought for all $(\alpha;\delta)$. This leads to the quantitative formulation of the polarization criterion I [9]:

$$I = \left(\frac{1}{4\pi}\oiint \left|F_+^N(\alpha;\delta) - F_\times^N(\alpha;\delta)\right|^2 d\Omega\right)^{-1/2}, \tag{6}$$

where averaging of $|F_\times^N - F_+^N|$ over celestial sphere takes place ($d\Omega$ - solid angle). The bigger is I the smaller is the averaged difference $|F_\times^N - F_+^N|$.

2.2. Localization criterion D

Criterion D characterizes the ability of a network to define angular position of a source. In astrometry the problem of a source localization on celestial sphere of a radiation source is solved by a method of triangulation. Triangulation is based on the difference in time between the registration of signals by network detectors. The further apart the detectors are, the greater the time delay is. To maximize the source location accuracy on the celestial sphere, the telescopes should be placed as far apart from each other as possible. According to [9], for a network of four detectors D is calculated as the area of the triangle formed by the three detectors in the network, which has the largest area among all possible combinations. If O - the center of the Earth, A, B, C - points where the detectors are located, the area of the corresponding triangle:

$$S_{ABC} = 1/2|[\overrightarrow{AC}, \overrightarrow{AB}]| = 1/2|[\overrightarrow{OC} - \overrightarrow{OA}; \overrightarrow{OB} - \overrightarrow{OA}]|. \tag{7}$$

2.3. Parameters reconstruction criterion R

Criterion R characterizes the possibility of reconstruction the parameters of the signal of a known analytical form. According to the Maximum likelihood estimation in the additive Gaussian noise background model, the parameters of the received signal are evaluated by the Rao-Cramer bound. The best possible estimates are obtained using the Fisher information matrix $\Gamma_{\alpha\beta}$ [14] in accordance with the formula

$$\Gamma_{\alpha\beta} = Re\left\{4\int_{f_{min}}^{f_{max}} \frac{\overline{\partial_\alpha \tilde{h}(f)}\partial_\beta \tilde{h}(f)}{S_n(f)} df\right\}, \tag{8}$$

where $\tilde{h}(f)$ is the Fourier image of response of the detector, the line above the Fourier image of response represents the complex conjugate, and $S_n(f)$ is the spectral noise density of a single detector. In this paper we assume for simplicity that all detectors have the same noise properties presented in [15].

Rao-Cramer bound determines the best possible accuracy of parameter P estimation [14]:

$$\delta P^2 = \left(\Gamma_N^{-1}\right)_{PP}, \tag{9}$$

where $\Gamma_N = \sum_{i=1}^{N} \Gamma_i$, i.e. the Fisher information matrix for detector network, is the sum of the corresponding detector matrices constituting the network. The inverse value of the celestial-averaged relative error is a numerical expression of criterion R:

$$R = \left(\frac{1}{4\pi} \oiint \left(\frac{\delta P}{P}\right)^2 d\Omega\right)^{-1/2} = \left\langle\frac{\delta P}{P}\right\rangle^{-1}, \tag{10}$$

Maximization of criterion R leads to the minimum relative error averaged over the celestial sphere in the estimation of the parameter.

2.4. Integral criterion C

Integral criterion C is used to compare different configurations of a network:

$$C = \sqrt{(\frac{I}{I_{max}})^2 + (\frac{D}{D_{max}})^2 + (\frac{R}{R_{max}})^2} \tag{11}$$

Criteria I, D and R together define a three-dimensional space that can be used to define a point which describes a particular configuration. Maximization of C by orientation angle of the detector in Novosibirsk gives the optimal orientation angle and leads to values $I(\gamma_{Nsk}^{max}) = I_{max}, D(\gamma_{Nsk}^{max}) = D_{max}$ and $R(\gamma_{Nsk}^{max}) = R_{max}$ (from the definition it's clear $C_{max} = 1$).

3. Source

In this paper we considered EAN efficiency of in registration of gravitational waves from core-collapse supernovae. During the core-collapse there exist many mechanisms of gravitational waves radiation on different stages of the process [16]. As Fisher matrix approach (8) requires analytical form of the signal we have considered gravitational waves from long-lived rotational instabilities of a proto-neutron star. If the key result of this instabilities is bar deformation that the radiation can be simulated by radiation from a rotating cylinder (axis of rotation is a bisector of the cylinder axis) with a Gaussian exponent which is introduced phenomenologically to take into account finiteness of the signal:

$$h(t) = \sqrt{F_+^2 + F_\times^2}\frac{GM\omega^2L^2}{3c^4r}(1 - 3\varepsilon^2)e^{-\frac{4(t-t_0)^2}{t_0^2}} cos(2\omega t + \phi), \tag{12}$$

$$\phi = \pi + \tan^{-1}\frac{F_+}{F_\times}, \tag{13}$$

where G is Newtonian constant of gravitation, c - speed of light, M - mass of the source, L - length of the cylinder, R - radius of the cylinder, ω - source angular frequency, t_0 - characteristic signal length and parameter $\varepsilon = \frac{R}{L}$ characterizes the degree of deformation and is used for calculation of criterion R (10). Gravitational radiation carries away energy from the system so ω decreases over time. However for typical values of source parameters [16] $\omega = 2\pi \times 700$ rad/s, $L = 20$ km, $R = 5$ km, $M = M_\odot$ and $t_0 = 1$ s calculation of gravitational waves luminosity via Einstein's formula [17]:

$$L_{GW} = \frac{G}{5c^5}\left\langle \sum_{j,k=1}^{3} \dddot{Q}_{jk}(t - \frac{r}{c})\dddot{Q}_{jk}(t - \frac{r}{c})\right\rangle, \tag{14}$$

where Q_{jk} - reduced quadrupole moment and dot denotes time derivative, leads to $\frac{\Delta\omega}{\omega} \sim 10^{-7}$, so we consider ω to be constant.

4. Numerical results and conclusions

In the case of gravitational collapse the strain $h(t)$ is expected to be several orders of magnitude less than for binary coalescence ($h \sim 10^{-22} - 10^{-20}$ at 10 kpc) so we can't assume isotropic distribution of sources over the celestial sphere. Instead we integrate over the Milky Way disk (neglecting the fact that the sensitivity may be sufficient for detection of the signal from several nearby galaxies). Due to rotation of the Earth antenna pattern functions F_+ and $F_\times$ depend on time (and consequently $P = P(t)$, $\phi = \phi(t)$) thus we average functions of time over 24 hours which also makes analytical calculation of Fourier image $\tilde{h}(f)$ in (8) possible (see Figure 3 as an example).

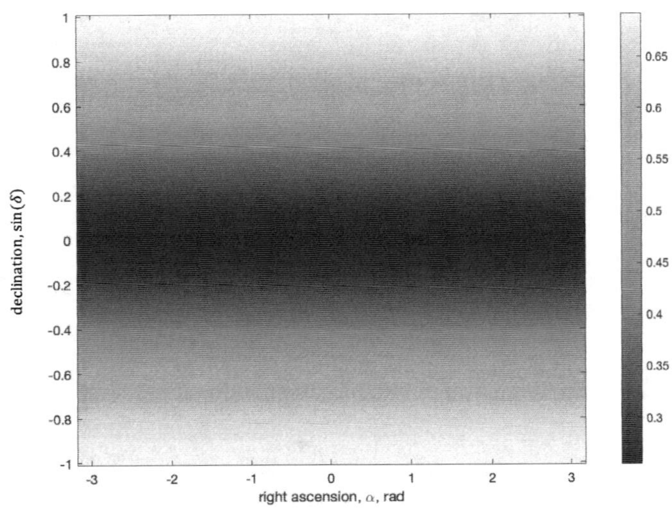

Figure 3. Averaged antenna power pattern for detector in Novosibirsk.

The calculation results are shown in Figure 4 and Figure 5. Criterion D obviously does not depend on the orientation angle of the detector in Novosibirsk. I is the most sensitive criterion to changes of orientation angle. The maximum value of integral criterion C is achieved at $\gamma_{Nsk}^{max} = 40°$ (see Figure 6). It's worth noting that the optimal angle differs from the case of binary coalescence - 13° [11]. The final choice depends on the most topical problems at the time of construction of the detector in Novosibirsk. To conclude we notice that the choice of the source is a limitation of this paper, because more significant physics is encrypted in a more complex structure of the signal from the core collapse, but such signals are model-dependent and do not have an analytical form (e.g. [18]), which does not allow using them within the framework of this approach.

Acknowledgments

The authors are grateful to V.I.Pustovoit, A.N.Morozov and V.L.Kauts for their attention to this work and useful discussions.

This work was supported by the grant RFBR 19-29-11010.

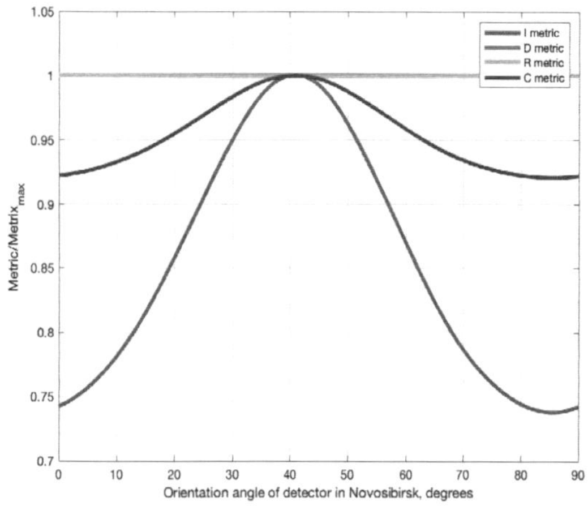

Figure 4. Dependence of all criteria on orientation angle of detector in Novosibirsk.

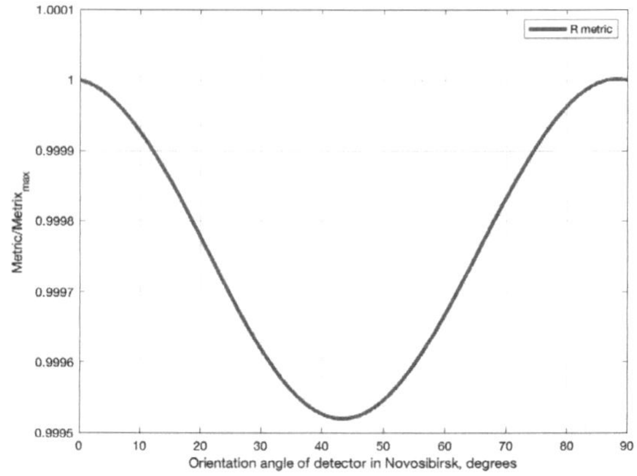

Figure 5. Dependence of R criterion on orientation angle of detector in Novosibirsk.

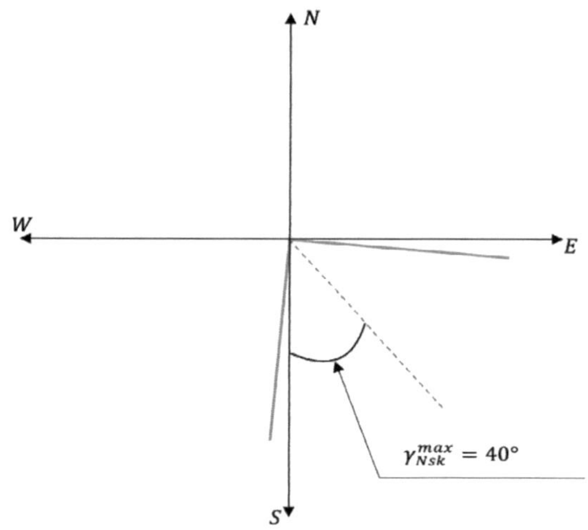

Figure 6. Optimal orientation of the detector in Novosibirsk.

References

[1] Abbott B P *et al.* 2016 *Phys. Rev. Lett.* **116**(6) 061102
[2] Abbott B P *et al.* 2017 *Phys. Rev. Lett.* **119**(14) 141101
[3] Abbott B P *et al.* 2017 *Phys. Rev. Lett.* **119**(16) 161101
[4] Abbott R *et al.* 2021 *ApJL* **915** L5
[5] Acero M A *et al.* (NOvA Collaboration) 2020 *Phys. Rev. D* **101**(11) 112006
[6] Veske D *et al.* 2021 *PoS* **ICRC2021** 950
[7] Fryer C L and New K C B 2011 *Living Rev. Relativ.* **14** 1
[8] Gusev A and Rudenko V 2019 *Moscow Univ. Phys.* **74** 115–123
[9] Raffai P *et al.* 2013 *Class. Quantum Grav.* **30** 155004
[10] Hu Y M *et al.* 2015 *Class. Quantum Grav.* **32** 105010
[11] Rudenko V, Andrusenko S, Krichevskiy D and Manucharyan G 2020 *Universe* **6** 140
[12] Finn L S 2001 *Physical Review D* **63** 102001

[13] Klimenko S *et al.* 2011 *Phys. Rev. D* **83**(10) 102001
[14] Kocsis B, Haiman Z, Menou K and Frei Z 2007 *Phys. Rev. D* **76**(2) 022003
[15] LIGO 2009 lIGO Document T0900288-v3 URL `https://dcc.ligo.org/LIGO-T0900288/public`
[16] Gossan S E *et al.* 2016 *Phys. Rev. D* **93**(4) 042002
[17] Landau L D and Lifshitz E M 1980 *The Classical Theory of Fields: Volume 2* vol 2 (Butterworth-Heinemann)
[18] Dimmelmeier H *et al.* 2008 *Phys. Rev. D* **78**(6) 064056

A new approach to the derivation of the law of universal gravitation from Kepler's laws

P N Antonyuk

Faculty of Mechanics and Mathematics, Lomonosov Moscow State University, Moscow, Russia
Moscow Institute of Physics and Technology, Dolgoprudny, Russia
S. Vavilov Institute for the History of Science and Technology, RAS, Moscow, Russia

E-mail: pavera@bk.ru

Abstract. Everyone knows that the inverse square law follows from Kepler's third law. Let us prove more: the law of universal gravitation follows from Kepler's third law.

1. Kepler's third law

German mathematician and astronomer Johannes Kepler (1571-1630) discovered three empirical laws of planetary motion in the Solar System. These laws are true provided that the masses of the planets are negligible compared to the mass of the Sun. The English mathematician, physicist and astronomer Isaac Newton (1643-1727) established a close connection between Kepler's laws and the law of universal gravitation discovered by him.

Let us show that the law of universal gravitation (discovered in 1666, published in 1687) is a consequence of Kepler's third law (discovered in 1618, published in 1619). Kepler's first and second laws (published in 1609) also follow from Kepler's third law, since these laws are derived from the law of universal gravitation.

Kepler's third law [1]

$$\frac{a^3}{T^2} = C \tag{1}$$

connects for each planet the semi-major axis a of the elliptical orbit with the orbital period T. Constant C is the same for all planets. Modern measurements show that

$$C = 2509.45482 \cdot 10^{16}\ km^3 day^{-2},$$

if the mass m of the planet and the mass M of the Sun are related by the inequality

$$0 < m \ll M\ .$$

The law will also be true for other celestial systems, for example, for the Jupiter-satellites system and the Mars-satellites system. The corresponding values of Kepler's constant will already be different:

$$C = 2.3955 \cdot 10^{16}\ km^3 day^{-2}\ ,$$

$$C = 0.0008098 \cdot 10^{16}\ km^3 day^{-2}\ .$$

We see that the central body of the system (Sun, Jupiter, Mars) significantly affects the value of Kepler's constant.

2. Inverse square law

In the case of a circular orbit, the distance R of the planet from the Sun remains constant, then

$$a = R\ . \tag{2}$$

Dutch mathematician, physicist and astronomer Christiaan Huygens (1629-1695) and Newton independently found a formula for centripetal force, as well as for centrifugal force. For the case of a circular orbit, Huygens and Newton equated the force F of the planet's attraction to the Sun and the centripetal force of the planet:

$$F = m\omega^2 R\ . \tag{3}$$

The planet's circular velocity ω is related to its period T by a simple formula

$$\omega = \frac{2\pi}{T}\ . \tag{4}$$

Four equalities (1), (2), (3), (4) give the formula

$$F = \frac{4\pi^2 Cm}{R^2} \tag{5}$$

known as the inverse square law. Newton argued that Wren, Hooke and Halley also came to this law independently of each other [2].

3. Kepler's constant depends only on the mass of the Sun

Let us formulate a number of statements concerning the Kepler constant.

The constant C does not depend on the parameters of the planet and on the distance R.

The force F of the planet's attraction to the Sun depends only on one parameter of the planet, on the mass m.

The methodological principle of symmetry allows us to assert that the Sun attracts the planet with the same force with which the planet attracts the Sun. Newton's third law of mechanics confirms this statement.

From the same principle it follows that the force F of the Sun's attraction to the planet depends only on one parameter of the Sun, on the mass M.

The constant C depends on only one parameter, the mass of the Sun M.

Considering the above, formula (5) takes the form:

$$F = \frac{4\pi^2 C(M)m}{R^2}\ . \tag{6}$$

4. Functional equation for Kepler's constant

The properties of homogeneity and isotropy of three-dimensional Euclidean space make it possible to assert that the force of attraction will not change if the Sun and the planet are mutually swapped. In other words, formula (6) remains valid if the masses M and m are interchanged:

$$F = \frac{4\pi^2 C(M)m}{R^2} = \frac{4\pi^2 C(m)M}{R^2}\ .$$

For the unknown function $C(\cdot)$, we find the functional equation

$$C(M)m = C(m)M \quad \text{or} \quad \frac{C(M)}{M} = \frac{C(m)}{m}\ .$$

Simple functional equation

$$f(x) = f(y)$$

has a solution $f(x) = const$. Hence,

$$C(x) = kx, k = const.$$

As a result, we get the formula

$$F = \frac{4\pi^2 kMm}{R^2} .$$

The gravitational constant G is equal to the force of attraction of two unit masses located at a unit distance:

$$M = 1, m = 1, R = 1 \Rightarrow G = F .$$

Hence,

$$G = 4\pi^2 k .$$

Finally, we get Newton's law of universal gravitation [2]

$$F = \frac{GMm}{R^2} . \tag{7}$$

5. What is Kepler's constant equal to?

The found formulas make it possible to simultaneously refine Kepler's law for a circular orbit:

$$\frac{R^3}{T^2} = \frac{GM}{4\pi^2} .$$

According to the methodological principle of correspondence, Kepler's law for an elliptical orbit takes the form:

$$\frac{a^3}{T^2} = \frac{GM}{4\pi^2} .$$

A well-known result of Newton for a pair of bodies (Sun and planet),

$$\frac{a^3}{T^2} = \frac{G(M+m)}{4\pi^2} ,$$

goes beyond Kepler's third law, since the right part of the last formula is not a constant value and depends on the planet's mass m. As a result, the natural condition,

$$0 < m \ll M ,$$

necessary for a rigorous formulation of Kepler's three laws is violated.

6. The average distance of a planet to the Sun is equal to the semi-major axis of the planet's orbit

This statement first appears as a hypothesis when we come across formula (2). According to Kepler's first law, the planet moves along an elliptical trajectory. The proof of the statement is related to the central symmetry in the plane of the orbit relative to the central point of the orbit. Two points P (planet) and S (Sun – the first focus) are translated by symmetry into points P' (a point in the orbit) and S' (the second focus), respectively. Points P, S, P', S' form a parallelogram. The averaging of the planet-Sun distances is performed for all pairs of distances PS and $P'S$. The parallelogram gives the averaging of each pair equal to a, from which our statement follows.

7. The uniqueness of the law of universal gravitation

Four variables F, M, m, R and constant G can be connected together by only one formula (7). Let us prove this statement. The so-called pi-theorem and the definition of the gravitational constant in the case of attraction of two equal masses m give the formula

$$F = \frac{Gm^2}{R^2} .$$

The principle of superposition of gravitational forces, formulated by Newton [2], and a simple identity

$$Mm + mM = (M + m)^2 - M^2 - m^2 ,$$

applied to the masses M and m, finally give the formula (7).

The uniqueness of the law means that numerous old attempts to present the law of universal gravitation in a "corrected" form

$$F = \frac{GMm}{R^{2+\varepsilon}} , -1 \ll \varepsilon \ll 1 ,$$

lead to only one conclusion: $\varepsilon = 0$.

The results of the article presented in paragraphs 4, 6 and 7 belong to the author.

References

[1] Keppler I 1619 *Harmonices mundi libri V* (Linz)
[2] Newton I 1687 *Philosophiae naturalis principia mathematica* (Londini)

Pulsar: physical generalization of the galactic time-space

A E Avramenko

Pushchino Radio Astronomy Observatory, Russia

E-mail: avr@itaec.ru

Abstract. The article considers a complex of geometric representations of space-time, based on general dynamic theories of celestial mechanics in close connection with pulsar astrometry as the physical basis of coordinate-time transformations within the solar system and galactic space as a whole. The pulsar time scale is considered as a certain material system with continuous and stable motion, representing a certain measurable parameter – the rotation period *P*, which changes as a function of the independent time variable – its derivative. The physical pulsar scale is a sequence of measured daily increments of the initial radiation period within any duration. According to observations at the BSA radio telescope (Pushchino) of the pulsar B0950+08, the time scale was determined with an initial period of P_0=0.2530653211840410 s on the date MJD_0 = 58971 (02.05.2020; 21h.58m.07s). A measured daily increment △P = 1.4441·10E-11 s corresponds to the measured value of the derivative P = 1.6759949886E-16, which is determined by the observational timing data. Measured △P are defined in the 25th decimal place. Up to 14-15 digits, △ P there is a pulsar time scale with femtosecond resolution. From 15-16 to 25 digits △ P is presumably sequential fixation of discrete states of microparticles during quantum-mechanical interactions of matter and electromagnetic radiation of a pulsar. According to our hypothesis, the diversity of the material world and physical processes occurring in celestial and quantum mechanics are finite and it can be generalized. This implies the inseparable unity of physical laws in four-dimensional space of celestial and quantum mechanics, detectable on pulsar time scales under the same conditions.

1. Introduction

From a geometric point of view, the theory of space and time has historically developed in two directions: as a theory of homogeneous (Galilean) space and as a theory of inhomogeneous (Riemannian) space based on Einstein's theory of gravitation. The uniformity of Galilean space and time is determined by the group of transformations of the four-dimensional interval, the form of which is directly related to the form of the basic laws of physics, namely the law of motion of a free material point and the law of propagation of a light (electromagnetic) wave in free space.

The mathematical theory of gravitation, on the contrary, allows the uncertainty of the time-space metric and, as a consequence, the ambiguity of solutions. Fock [1,3], considering both theories as a whole, proved the existence of the necessary unambiguous solution of the equations of gravitation by imposing additional conditions compatible with them (1955). However, the law of transformation of physical quantities must be determined by the Lorentz transformation by the condition of preserving the form of general equations in the transition from one frame of reference to another. Fock proved that in an isolated mass system, similar to the solar system, there is a class of harmonic coordinate systems that are very close in their properties to the Galilean coordinates, also related to each other by the Lorentz transformation. In particular, the currently used time scales in modern ephemeris astronomy are defined in harmonic coordinates [9].

Historically, the possibility of establishing the basic physical laws of the motion of bodies in celestial mechanics arose much earlier, even at a time when observational astronomy had only direct visual methods. As Abalakin notes in his fundamental publication "O.V. Struve – G.V. Schiaparelli. Correspondence 1859-1904 (St. Petersburg, "NAUKA", 2005)", in the process of development of world astronomical science two directors of the Nikolaev Main Astronomical Observatory in Pulkovo and the Royal Astronomical Observatory of Brera in Milan - academicians O.V Struve (1819-1905) and G. V. Schiaparelli (1835-1910) united close attention to one of the remarkable classes of celestial objects - to binary and multiple star systems.

This attention and interest were combined in them with the art of observation and a deep scientific and theoretical approach, which allowed everyone to skillfully use their own series of observations of the position of stars to accurately derive the parameters of the orbits of the components of multiple stellar systems. Even earlier, Kepler (1571-1630) was the first to formulate the laws of motion of the planets of the solar system in the form of kinematic solutions of the equations of their elliptical orbits. The shape of the ellipse and the position of the planets in orbits directly follows from the physical laws of Newton. This unity of Kepler's kinematic laws and Newton's dynamical laws reveals their exhaustive completeness and sufficiency as applied to celestial mechanics as a whole.

The further development of celestial mechanics, on the general dynamic theories of which ephemeris astronomy is based, is associated with the practical establishment of space-time coordinate systems representing certain approximations to an ideal inertial reference frame. The requirements for the time scale in such a frame of reference were formulated by Abalakin [4] as follows. The time scale in ephemeris astronomy is considered as a certain material system with continuous and stable motion and representing a certain measurable parameter P, which changes with time as a function of the independent variable of time t.

It is precisely such a measurable parameter that satisfies these requirements is the period of rotation and, accordingly, the period of the electromagnetic radiation of a pulsar - a neutron star rotating with a gradual deceleration determined by the constant value of the derivative of the period. The values of the pulsar emission period measured from observations at the BSA radio telescope in Pushchino are used as the basis for the ephemeris coordinate pulsar time scales, which are uniquely related by the four-dimensional Lorentz transformation.

The paper considers the physical features and metric properties of the coordinate pulsar time scales according to the observational data on the timing of the pulsar B0850+08. The complete correspondence of physical pulsar and calculated ephemeris time scales of the solar system, satisfying the conditions of coordinate transformations and axiomatic transfer of coordinate time scales along a fixed origin within any duration, is shown.

By analogy with the solar second, canceled in 1967, a new definition of the pulsar second was given as a time interval consisting of a countable number of periods of radiation of the pulsar B0850+08 at a geodetic point with the coordinates of the BSA radio telescope for a specified date and time. The definition of the pulsar second equally applies both to the calculated ephemeris time and to the physical coordinate time in inertial systems, being a necessary condition for the simultaneity of physical processes occurring in any of them.

2. Parametric approximation of pulsar time scales

The monotonic deceleration of the rotation of the pulsar with a constant derivative of the period is fully consistent with the requirements of the time scale, which were formulated in ephemeris astronomy. The time scale is considered as any material system with continuous and stable motion and representing a certain measurable parameter P, changing as a function of the independent variable of time.

For an objective perception of time in space, the measurement of time in astronomy is based on the time scale, defined as outlined in [4]:

1) some material system with a continuous and stable movement and representing a certain measurable parameter P, changing with time;

2) a theory giving the values of this parameter P in a function of the independent variable t, called as time. The function $P = f(t)$ should be unambiguous and allow for the distinction between its particular values. Then the time it can be expressed in terms of the measured values of the determining parameter P in the form $t = \varphi(P)$. The scale, thus determined makes it possible to order events in the sense of "earlier" or "later" with respect to a certain time t_0, depending on whether there is $t_0 > t$ or $t_0 < t$, if it is the point in time at which the events occurred. Thus, when measuring time, it is necessary to strictly distinguish the point in time (or the epoch of the event) from the time interval (or time scale interval).

Pulsar time scales PT are defined as functions of the rotation period P and its derivatives $\dot{P}$, $\ddot{P}$ in the spatial coordinate reference systems adopted in the solar system: topocentric (TT) and barycentric (TB). Each of them is a digital series of intervals obtained by integrating the observed rotation period $P(t)$ determined for each date within the selected observation interval:

$$PT_{i,j} = \frac{1}{P_0^*}\int_{t_i}^{t_j} P(t)\,dt\,. \tag{1}$$

The period $P(t)$ within the observation interval is determined by the initial value P_0^* and derivatives $\dot{P}$, $\ddot{P}$:

$$P(t) = P_0^* + \left(\dot{P}\cdot t + \frac{1}{2}\ddot{P}\cdot t^2\right); \quad -\infty < t < +\infty\ . \tag{2}$$

Our studies [11] of the nature of the physical process of slowing down the rotation of pulsars based on the results of measuring the derivatives and the generalized braking index showed that its numerical values are in the range $n = -(0{,}9 \pm 0{,}2)$, that is, they are grouped near a negative unit with about near – 0,9 mean values.

So, the analytical model of pulsar time scales determines the unique consistent values P_0^*, $\dot{P}$, $\ddot{P}$, corresponding to the coherence of periodic radiation of pulsar, with the time of arrivals (ToAs) of the observed pulses only. Random variations of ToAs due to imprecise knowledge of physical variables and unmodelled noise of observations, have no significant impact on the modelled intervals of coherent periodic radiation of pulsar, which are identified with a machine accuracy and sub-nanosecond resolution by the consistent rotation parameters of the pulsar.
As a result of dividing the observed intervals, the random variations are not contained detectable rotation components. They can be analyzed additionally to identify with affecting factors, for example, the distortion of an atomic time scale of the radio telescope, for possible physical interpretation. It is assumed also, that the pulsar rotation parameters are independent of any physical factors, which are involved in a parametrical fitting [12].

Since both derivatives determine practically the same deceleration index for all pulsars, and the numerical value of the second derivatives (10^{-31}–10^{-23}) s^{-1} is many orders of magnitude less than the first ones (10^{-16}–10^{-13}) $s \cdot s^{-1}$, then in practical calculations of the pulsar scales in expression (2) we will take into account only the first derivative:

$$P(t) = P_0^* + \dot{P}\cdot t\,; \quad -\infty < t < +\infty\,. \tag{3}$$

Thus, we have two mutually agreed definitions of the time scale: a) traditional – for an integrated scale, in accordance with expression (1) – as a sequence of intervals of a certain duration, counted from a certain beginning within any duration, and b) – for a differential time scale, in accordance with the definition of Abalakin [4] for the ephemeris time scale.

The differential pulsar time scale is expressed as a sequence of measured daily increments ΔP_i of the initial period of the pulsar emission:

$$P_i = P_0^* \pm \Delta P_i \; ; \quad -\infty < i < +\infty . \tag{4}$$

Considering that all daily increments ΔP_i are the same and equal to ΔP, the value of the period P_k for any date can be calculated from the fixed difference k of the initial MJD_0 and the selected MJD_k dates for the observation epoch:

$$P_k = P_0^* \pm k \cdot \Delta P \; ; \quad -\infty < k < +\infty$$

$$k = MJD_0 - MJD_k . \tag{5}$$

Thus, we have determined the physical time scale, the requirements for which in ephemeris astronomy were formulated by Abalakin. The observed parameter P is the rotation period of the pulsar, which monotonically increases as a function of the independent variable of time.

3. Physical generalization of the proper time of an observer on a geoid

The coordinate time TCB in the BRS reference frame and the TCG coordinate time in the GRS reference frame refer to unobservable points of the barycenter and geocenter, respectively, and are expressed by arguments in the mathematical equations of the motion of the Sun and the rotating Earth [4]. Meanwhile, by direct observation with a radio telescope, taking into account the geographic location of the observation point, the observer's own time τ can be measured, which coincides with the coordinate time τ of the corresponding topocentric reference system (TRS), estimated at its origin on the geoid.

The relationship between TAI and TT in the form TT = TAI + 32,184 s, refers to the geocentre, the point at which the averaged result of the readings of atomic clocks distributed over the real Earth's surface is localized. For a given unit of time, the scale TDB differs from the TT scale (TAI) only by relativistic nonlinear and periodic components resulting from the theories of the motion of the Earth, planets and the Moon. The proper time of the observer in any point on the Earth's surface coincides with the coordinate time τ of the corresponding topocentric reference system (TRS), estimated at its beginning. The time is associated with the TT (TAI) relativistic transformation, including the GRS observer speed, its height above the geoid and the quadrupole tidal gravitational potential of the outer masses. This involves visualizing ideal clocks on the Earth's surface, the TAI of which extends within the hypersurface TT = const and is synchronized with the TT.

It should be noted that this assumption does not explicitly take into account the coordinate topocentric time τ, that is, the observer's proper time, depending on its location on the Earth's surface. For each observer, it is different, whereas ideal atomic clocks that reproduce the standard SI second-on-time can not by themselves react to the relativism of topocentric coordinate time and, therefore, cannot perform the synchronization function of topocentric time scales in accordance with the principle of simultaneity of events observed in different coordinate systems [13].

Fomalont (1984) [5], and later Kopejkin (1990) [6], by the VLBI method measured the difference τ–TCG, thereby determining the observer's own physical time in dynamic coordinate systems on the geoid. All values prescribed by general relativity with all the necessary accuracy provided by the theory of general relativity at a fixed point were taken as the initial ones for calculating the difference. As a result, a picosecond measurement accuracy is achieved, which is inherent in the observed object - a pulsar.

Now we will logically continue these measurements by successive movements of the second VLBI radio telescope point by point along any arbitrary closed geodetic line. The set of measured differences τ –TCG upon returning to the starting point, this difference will coincide with what it would have been at this point, as in the absence of movement of the radio telescope.

Thus, direct VLBI observation of pulsed radio emission from a neutron star, excluding the uncertainty of the gravity factor at fixed observation points, removes the uncertainty of the proper time τ during coordinate transformations at any point in space along a closed geodesic line. This confirms the uniformity of the pulsar time scales in the coordinate systems of the observer.

It is known [1], in the general case, that in a space of 4 dimensions, a group of transformations that leaves the expression for the square of the distance between infinitely close points unchanged can contain no more than 10 parameters. This coincides with the number of parameters in the Lorentz transformation for which the space is maximally homogeneous, like the Galileo space. In such a space (a) all points and moments of time are equal, which corresponds to 4 parameters: three initial coordinates and an initial moment of time; (b) all directions are equal, which corresponds to 3 parameters of rotation of the coordinate axes: three angles; and (c) all inertial systems with translational moving relative to each other are equal, which corresponds to 3 parameters: three components of the relative velocity.

Later, Logunov [2] clarified the signs of homogeneity by assuming an infinite set of inertial systems, in each of which space is orthogonal to time, as a result of which the rotation of the coordinate axes does not affect the coordinate time, which is the same at any point of the inertial coordinate system, coinciding with the time at its origin. It is in such a space-time that all physical processes take place, which under the same conditions in these systems are identical; that is why the laws of conservation of energy-momentum and angular momentum are fulfilled, and all natural standards in all inertial reference frames are the same. Therefore, 3 parameters of rotation of the coordinate axes are excluded from the group of transformations in inertial systems.

In addition, of the three components of the relative velocity, taking into account the remoteness of the source of electromagnetic radiation - the pulsar, only one remains - in the direction to the source along which the velocity is measured.

Thus, out of the possible 10 parameters of a 4-dimensional coordinate transformation, in a homogeneous Galilean space with electromagnetic radiation of a galactic source - a pulsar, only 5 physically significant measurable parameters remain: 3 initial coordinates of the radio telescope (Xo, Yo, Zo), the initial period Po of the coordinate time scale and the speed of the pulsar V towards the radio telescope.

Poincaré's physical theory of relativity, based on Lorentz transformations and extended to all inertial coordinate systems, unambiguously corresponds to the laws of motion of the planets of the Kepler-Newton-Galileo solar system and is fully consistent with homogeneous time-space.

All coordinate systems in which the listed homogeneity criteria are fulfilled are inertial. Physical processes occurring in these systems under the same conditions are identical. The same applies to the process of periodic radiation and propagation of the front of the electromagnetic wave of the pulsar.

This numerical experiment on the propagation of pulsar time scales into two points dist ant from the radio telescope, taken as a test case. For coordinate calculations, a specialized ERA software package, developed by the IPA RAS, was used. The estimated error of modern ephemeris time scales during coordinate transformations is on the order of 1 ns within a secular interval (Pitjeva, 2015) [7].

4. Physical pulsar time scales in the space of the solar system

We will consider pulsar scales as any material system with continuous and stable motion and representing a measurable parameter - the period of rotation P. The value of the period P is expressed as a function of the independent variable of time - the derivative of the period, which, in the process of a gradual deceleration of the pulsar's rotation, remains constant for an unlimited duration of the observations.

We took the daily data on the timing of the pulsar B0950+08 in the topocentric (TT) and barycentric (TB) coordinate systems from observations at the BSA radio telescope in Pushchino during 2020. Pulsar B0950+08 in a supernova remnant (age 1.7·107 years) in the vicinity of the Solar system (distance 0.26 kpc) can be attributed to older stars, the lifetime of which together with the Solar system is estimated to be approximately 4.6 ·109 years, as well as the Sun [10].

The parameters of the time scale of the pulsar B0950 + 08, which correspond to formula (5), are as follows:

$$P_0^* = 0{,}2530653211840410 \text{ s at } MJDo = 58971 \ (02.05.2020;\ 21\text{h}.58\text{m}.07\text{s})$$

$$\Delta P = 1{,}4441 \cdot 10^{-11}, \text{ s}$$

$$k = MJD_n - MJDo,$$

hence:

$$P_k = 0{,}2530653211840410 \pm k \cdot 1{,}4441 \cdot 10^{-11}; \quad -\infty < k < +\infty$$

or:

$$P_k = P_0^* \pm 1{,}4441 \cdot 10^{-11}\left(MJDn - MJDo\right).$$

According to the observational data on timing, the translational motion of the pulsar was detected and its velocity was measured in the topocentric (TT) and barycentric (TB) coordinate systems:

$V_{TT}=V_{TB}=0{,}05137\pm 2\cdot 10^{-5}$ m/s.

Since the speed of movement of the pulsar is constant and small, the movement itself was not detected either in observations with a radio telescope, or, even more so, in barycentric residual deviations, which led to a distortion of the derivative $\dot{P}$ = 2.2915E-16 indicated in the catalog. In fact, the true measured value of the derivative, corresponding to the observational timing data, is $\dot{P}$ = 1.6759949886E-16, or, in the format of the catalog for ephemeris calculations, $\dot{P}$ = 1.676E-16.

Thus, all five measurable parameters: coordinates (Xo, Yo, Zo) of the BSA radio telescope (Pushchino), the period of rotation of the Po pulsar at a fixed observation date, and the velocity V of the pulsar in the radial direction completely determine any coordinate-time transformations within the solar system.

In this case, according to the value of the period Po measured on a fixed date and the constant daily increment of the period, axiomatically, only by the difference between the specified and fixed dates, the value of the period is determined for any date in any coordinate system. Thus, in any inertial system, the beginning of the pulsar time scale is set on the selected date, which operates without change within any duration, both in the past and in the future.

Table 1 shows a fragment of the measurement data from daily observations on the BSA radio telescope (Pushchino) of the daily increments of the rotation period of the pulsar B0950+08 in the topocentric (TT) and barycentric (TB) coordinate systems.

The measured daily period increments ΔP show the absolute determinism of the pulsar rotation period with a resolution of up to 25 decimal places with a constant derivative of the period $\dot{P}$ = 1.676E-16. In this case, up to the 14-15th decimal place, corresponding to the femtosecond resolution of the pulsar time scale, the daily increments of the period are strictly constant within the entire observation interval. The remaining decimal places, from 14-15th to 25th, are expressed in a countable number of randomly repeating combinations of them. This testifies, according to our hypothesis, that here we are dealing with the quantum-mechanical interaction of a wave of electromagnetic radiation from a pulsar with a matter field, considered in the physics of nonequilibrium processes associated with the self-organization of dissipative structures [14]. The pulsar time scale thus detects these wave interactions and determines their sequence in time, thereby confirming that the pulsar time scales are the same for measuring any processes occurring in galactic space, at any time scale, including quantum mechanical ones.

According to de Broglie's hypothesis, a microparticle moving in regions whose sizes are comparable to the sizes of atoms has wave properties, and these properties cannot be neglected. In this case, the motion of a particle is associated with a change in the time of its mechanical states, and a continuous change of states corresponds to the motion of a particle along a certain trajectory.

Table 1. Measured daily increments of the rotation period of the pulsar B0950 + 0.

MJD	ΔP_TT, s	ΔP_TB, s	N
59108	0,00000000001444105945935800	0,00000000001444161457087030	1
59109	0,00000000001444111497050930	0,00000000001444178110432400	2
59110	0,00000000001444111497050930	0,00000000001444178110432400	2
59111	0,00000000001444111497050930	0,00000000001444178110432400	2
59112	0,00000000001444111497050930	0,00000000001444183661547530	3
59113	0,00000000001444105945935800	0,00000000001444183661547530	3
59114	0,00000000001444111497050930	0,00000000001444183661547530	3
59115	0,00000000001444111497050930	0,00000000001444189212662650	4
59116	0,00000000001444105945935800	0,00000000001444189212662650	4
59117	0,00000000001444111497050930	0,00000000001444194763777770	5
59118	0,00000000001444111497050930	0,00000000001444189212662650	4
59119	0,00000000001444111497050930	0,00000000001444216968238270	6
59120	0,00000000001444105945935800	0,00000000001444189212662650	4
59121	0,00000000001444111497050930	0,00000000001444200314892900	7
59122	0,00000000001444111497050930	0,00000000001444211417123140	8
59123	0,00000000001444111497050930	0,00000000001444183661547530	3
59124	0,00000000001444111497050930	0,00000000001444205866008020	9
59125	0,00000000001444111497050930	0,00000000001444211417123140	8
59126	0,00000000001444111497050930	0,00000000001444211417123140	8
59127	0,00000000001444105945935800	0,00000000001444216968238270	6
59128	0,00000000001444111497050930	0,00000000001444205866008020	9
	Celestial Mechanics ↑ *Quantum Mechanics*	Celestial Mechanics ↑ *Quantum Mechanics*	

Since the Schrödinger equation is invariant under the Lorentz transformations based on Galileo's principle of relativity, this implies the existence of a number of operators of quantum and the existence of quantum mechanical invariants associated with Galileo transformations, similar to the transformations in celestial mechanics.

Thus, the front of a wave of electromagnetic radiation of a pulsar, in the process of interaction with particles with wave properties, on its scale records the change in time their mechanical state, and according to the observed sequence of alternating discrete states, their movement along a certain trajectory is recorded.

Note on the table that the uncertainty of the particle motion along the trajectory, observed when changing mechanical states on the TB scale, is approximately an order of magnitude greater (14th decimal place) than on the TT scale (15th decimal place). This is due to the uncertainty of the TB scale itself, introduced by its post-observational mathematical fit, while the TT scale is determined more strictly from direct physical measurements on a radio telescope. But on the other hand, on the TB scale, we have the opportunity to detect an order of magnitude more mechanical states of a particle than on the TT scale.

Such an unprecedented resonance stability of microparticles on the global effect of external fields is possible, apparently only on a large scale of space. As we are convinced, this is impossible, for example, in the local space of the measuring device of the atomic second. The practice here is that the maximum approximation of the measured second to the reference one, determined during the transition between two hyperfine levels of the ground state of the cesium 133 atom, is achieved only in a short, within a few hours, interval, but later, on a daily or more scale, these values diverge irreversibly.

5. Pulsar analogue of the ephemeris solar second

Ephemeris time, established back in the 19th century on the basis of the theory of planetary motions and their practical applications, refers to the form of classical dynamic time. The unit and the beginning of the ephemeris time T from 1960 to 1967 were determined by adopting a numerical expression for the geometric mean longitude L of the Sun, established by Newcomb [4]:

L=279°41^' 48^",04+129602768″,13 T+1″,089 T^2,

here time T is counted in Julian centuries, consisting of 36,525 days with a duration of 86400 s. The second was defined as 1/86400 of an average solar day, which corresponds to 1/31556 925.9747 part of a tropical year – as the interval during which all seasons change according to the movement of the Sun:

1s = 1/31556 925,9747 = 3,16888E-06 – by astronomical measurements

1s= 1/36525· 86400 = 3,16881E-06 – calculated value by ephemeris.

The relative discrepancy between the measured and calculated seconds is of the order of 10E-05 within a century. This, for comparison, is 3 orders of magnitude less than the estimated divergence of the orbital motions of the planets of the solar system from the resonant calculated Keplerian orbits due to small interactions of perturbations of planets orbiting the Sun and other factors of gravity.

By analogy with the definition of the solar second as a fraction of a tropical year we introduce a new definition of the unit of time the pulsar as the number of periods of the pulsar radiation at a fixed point in space at a given date and ephemeris time (epoch): a pulsar second is a time interval consisting of 3.951 548 933 378 950 periods of the radiation of the pulsar B0950+08 according to observations at the BSA radio telescope of the Pushchino Radio Astronomical Observatory with geodetic coordinates (Xo, Yo, Zo) and an initial value of the period P = 0.253 065 321 184 041 s, measured on the date 02.05.2020; time 16h 36m 25s GMT + 3.

Both definitions - solar and pulsar seconds, based on the motion of stars, have a certain functional continuity. Their fundamental difference is that the periodic radiation of a pulsar determines not only a unit of time - a second, but also highly stable time scales based on it in the entire physically significant, axiomatically transferred to any coordinate system of four-dimensional time-space of the solar system.

Like the previous definition of the second, canceled in 1967, it is also associated with kinematic characteristics, only now not the Sun, but the neutron star - the pulsar B0950+08.

Both definitions of solar and pulsar seconds are based on the kinematics of stars, the ephemerides of which are reliably established in long-term observations, and they have a certain functional continuity. The fundamental difference is that the periodic electromagnetic radiation of a pulsar determines not only a unit of time - a second, but also highly stable time scales based on it in the entire physically significant axiomatically transferred into any coordinate system of four-dimensional time-space of the solar system.

6. Conclusion

According to the observed electromagnetic periodic radiation of a galactic neutron star - a pulsar, the exhaustive completeness and sufficiency of Kepler's kinematic laws and Newton's dynamic laws as applied to celestial mechanics as a whole is manifested. All physical processes, according to A. Poincaré, take place in a homogeneous time-space in the Galilean coordinates of the inertial system. Physical states of the material world are generalized in 4-dimensional homogeneous time-space and are axiomatically transformed in inertial coordinate systems of celestial mechanics.

The study of the dynamics of celestial bodies and astronomical phenomena is directly related to the measurements of periodic pulsar radiation. For an objective perception of time in space, the measurement of time is based on the pulsar time scales PT, associated with a continuous and measurable rotation period P of galactic neutron stars, changing with time, so that pulsar time scales it can be expressed in the form PT = f (P).

Thus, in any inertial system - topocentric and barycentric, the beginning of the pulsar time scale is established on the selected date, acting without change within any duration, both in the past and in the future.

The front of a wave of electromagnetic radiation from a pulsar interacting with microparticles with wave properties, on its scale, records the change in their mechanical state in time, and according to the observed sequence of alternating discrete states, it detects their movement along a certain trajectory.

Hypothesis. The variety of the material world and its physical states in celestial and quantum mechanics is finite and can be generalized under the same conditions. This implies the inseparable unity of physical laws in four-dimensional space of celestial and quantum mechanics, detectable on pulsar time scales.

References

[1] Fok V A 2014 *Theory of Space, Time and Gravity* (Moscow: Lenand)
[2] Logunov A A 2004 *Henry Poincare and the Relativity Theory* (Moscow: Nauka)
[3] Fok V A 2010 *Theory of Einstein and Physical Relativity* (Moscow: Librokom)
[4] Abalakin V K 1979 *Basics of Ephemeris Astronomy* (Moscow: Nauka)
[5] Fomalont E R *et al* 1984 *Astron. Soc.* **210** 113
[6] Kopejkin S M 1990 *Astron. Zh.* **1**
[7] Pitjeva E V 2012 The IAA RAS fundamental ephemerides of planets and the Moon (EPM): their model, parameters, accuracy *Proc. IAA RAS* **26** 54
[8] Audoin C, Guinot B 1998 *Les fondements de la mesure du temps* (Paris: Masson)
[9] Brumberg V A, Kopejkin S M 1990 *Relativistic Time Scales in the Solar System Celestial Mechanics and Dynamical Astronomy* **48** 23
[10] Pilkington J D H, Hewish A, Bell S J, Cole T W 1968 *Nature* **218** 126
[11] Avramenko A E 2017 *Pulsar: stable rotation, coherent radiation, monotone slowdown* (Lambert Academic Publishing)
[12] Avramenko A E, Losovsky B Y *at al* 2018 *International Journal of Astronomy and Astrophysics* **8** 24
[13] Avramenko A E 2017 *J. Phys.: Conf. Ser.* **1051** 012004
[14] Prigogin I and Stengers I 1984 *Order out Chaos: Man's New Dialogue with Nature* (London: Heinemann)

Acknowledgments

Prof. I.F. Malov for critical discussions on theoretical approximations and physical consistency in observational pulsar astrometry;
Dr. B.Ya.Losovskiy and Dr. V.D. Pugachev for systematizing and combining long-term archives of observational data on pulsar timing at the BSA radio telescope at Pushchino;
Prof. E.V. Pitjeva and Dr. D.A. Pavlov for their active support and kind methodological assistance from the Laboratory of Ephemeris Astronomy of the Institute of Astronomy of the Russian Academy of Sciences in testing the coordinate transformations of pulsar time scales;
Head of the VLBI ground station (Pushchino) as part of the international project RADIOASTRON, Dr. A.V. Kovalenko, for active support and participation in comparative measurements of pulsar and atomic scales in real time.

The structure of the curvature tensor of plane gravitational waves

O V Babourova[1], B N Frolov[2], M S Khetzeva[3] and D V Kushnir[4]

[1] Moscow Automobile and Road Construction State Technical University (MADI), Department 'Physics', Leningradsky pr., 64, Moscow, 125319, Russia
[2,3,4] Moscow Pedagogical State University (MPGU), Institute of Physics, Technology and Information Systems, M. Pirogovskaya ul. 29, Moscow 119992, Russia

E-mail: 1ovbaburova@madi.ru, 2bn.frolov@mpgu.su, 3ms.khetzeva@m.mpgu.edu, 4kushnir@gmail.ru

Abstract. Plane gravitational waves in the Riemann space of General Relativity is considered. The criterion of plane gravitational waves is used based on the analogy between plane gravitational and electromagnetic waves. The Theorem is proved that the action of the Lie derivative on the plane wave curvature 2-form in the direction of the vector generating the invariance group of this wave in the Riemann space is equal to zero. It is justified that the gravitational waves can be used to transmit information in the Riemann space.

1. Introduction

In previous papers, the structure of space-time formed by plane gravitational waves in the Riemann–Cartan space [1] and general affine-metric space [2] was studied. In these works, the criterion of plane gravitational waves was used, based on the analogy between plane gravitational and electromagnetic waves.

By analogy with a plane electromagnetic wave, the metric tensor of space-time must have a group of motions G5, which does not change the isotropic hypersurface describing a plane wave front with constant amplitude.

In this paper, the structure of the curvature tensor is determined for the case of plane waves in the Riemann space of the General Relativity Theory.

2. Plane wave metrics in Riemann space

The generator (infinitesimal operator) of the group G_5 is the vector field $\vec{X}=(a+b'x+c'y)\partial_{\rm v}+b\partial_x+c\partial_y$, where $b(u)$, $c(u)$ are arbitrary functions, and $b'(u)$, $c'(u)$ are their derivatives ($a=const$). Vectors $\vec{e}_\alpha\ (\alpha=\hat{0},\hat{1},\hat{2},\hat{3})$: $\vec{e}_{\hat{0}}=\partial_{\rm v}$, $\vec{e}_{\hat{1}}=\partial_u$, $\vec{e}_{\hat{2}}=\partial_x$, $\vec{e}_{\hat{3}}=\partial_y$ are the basis vectors of a special coordinate system v, u, x, y, where u is the phase of the wave (delayed time), and x and y are the coordinates on the wave surface $(u,{\rm v})=const$. The vector $\vec{e}_{\rm v}$ is covariant constant, isotropic and directed along the wave ray. The vector $\vec{e}_u$ is also isotropic. Vectors $\vec{e}_x$ and $\vec{e}_y$

are both space-like, commute and orthogonal to the vector $\vec{e}_{\mathrm{v}}$. Basic vectors correspond to basic 1-forms θ^{α} $(\alpha = \hat{0},\hat{1},\hat{2},\hat{3})$, $\tilde{\theta}^{\hat{0}} = \mathrm{dv}$, $\tilde{\theta}^{\hat{1}} = \mathrm{du}$, $\tilde{\theta}^{\hat{2}} = \mathrm{d}x$, $\tilde{\theta}^{\hat{3}} = \mathrm{d}y$, d – an operator of external differentiation.

For a plane gravitational wave, the metric of the Riemann space V_4 satisfies the condition $\mathrm{L}_{\vec{X}} g_{\alpha\beta} = 0$, where $\mathrm{L}_{\vec{X}}$ is the Lie derivative in the direction of the vector $\vec{X}$. This condition leads to a metric of Kundt type

$$\hat{g} = 2H(u,x,y)du^2 + 2du d\mathrm{v} - dx^2 - dy^2,$$

where

$$H(u,x,y) = \frac{1}{2}A(u)x^2 + B(u)xy + \frac{1}{2}C(u)y^2,$$

and the functions $A(u)$, $B(u)$, $C(u)$ satisfy the conditions

$$b'' + bA(u) + cB(u) = 0, \qquad c'' + bB(u) + cC(u) = 0.$$

It is convenient to go over to a nonholonomic basis $\vec{e}_a$ $(a = 1,2,3,4)$:

$$\vec{e}_0 = \vec{e}_{\hat{0}},\ \vec{e}_1 = -H(u,x,y)\vec{e}_{\hat{0}} + \vec{e}_{\hat{1}},\ \vec{e}_2 = \vec{e}_{\hat{2}},\ \vec{e}_3 = \vec{e}_{\hat{3}},$$

which corresponds to the following nonholonomic basic 1-forms:

$$\theta^0 = H(u,x,y)\mathrm{d}u + \mathrm{dv},\ \theta^1 = \mathrm{d}u,\ \theta^2 = \mathrm{d}x,\ \theta^3 = \mathrm{d}y.$$

The transition from holonomic basic 1-forms θ^{α} to nonholonomic one is carried out by a linear transformation $\theta^a = h^a{}_{\alpha}\theta^{\alpha}$, where the matrix $h^a{}_{\alpha}$ is equal to

$$h^a{}_{\alpha} = \theta^a(\vec{e}_{\alpha}) = \begin{pmatrix} 1 & H & 0 & 0 \\ 0 & 1 & 0 & 0 \\ 0 & 0 & 1 & 0 \\ 0 & 0 & 0 & 1 \end{pmatrix},\quad h^{\alpha}{}_a = \theta^{\alpha}(\vec{e}_a) = \begin{pmatrix} 1 & -H & 0 & 0 \\ 0 & 1 & 0 & 0 \\ 0 & 0 & 1 & 0 \\ 0 & 0 & 0 & 1 \end{pmatrix}.$$

The components of the metric corresponding to the nonholonomic basis are equal:

$$g_{ab} = \hat{g}(\vec{e}_a, \vec{e}_b) = h^{\alpha}{}_a h^{\beta}{}_b g_{\alpha\beta} = \begin{pmatrix} 0 & 1 & 0 & 0 \\ 1 & 0 & 0 & 0 \\ 0 & 0 & -1 & 0 \\ 0 & 0 & 0 & -1 \end{pmatrix} = g^{ab}.$$

The nonzero components of these metrics are equal, $g_{01} = g_{10} = 1$, $g_{22} = g_{33} = -1$. The components of the metrics in the coordinate space are equal

$$g_{\alpha\beta} = \hat{g}(\vec{e}_{\alpha}, \vec{e}_{\beta}) = h^a{}_{\alpha} h^b{}_{\beta} g_{ab}.$$

3. Plane waves of curvature in Riemann space

The curvature is determined by the tensor-valued curvature 2-form,

$$\mathrm{R}^a{}_b = (2\Gamma^a{}_{b[\nu,\mu]} + 2\Gamma^a{}_{c[\mu}\Gamma^c{}_{|b|\nu]})\theta^{\mu} \wedge \theta^{\nu},$$

where $\Gamma^a{}_{b\mu}$ is the 1-form of the Lorentzian connection, defined in terms of the Christoffel symbols and tetrad coefficients in the form, $\Gamma^a{}_{b\mu} = h^a{}_{\lambda} h^{\nu}{}_b \Gamma^{\lambda}{}_{\nu\mu} - h^a{}_{\nu,\mu} h^{\nu}{}_b$.

The components of the non-zero Lorentzian connection 1-form are,

$$\Gamma^0{}_{2a}\theta^a = \Gamma^2{}_{1a}\theta^a = \mathrm{d}u\,\partial_x H,\quad \Gamma^0{}_{3a}\theta^a = \Gamma^3{}_{1a}\theta^a = \mathrm{d}u\,\partial_y H.$$

The nonzero components of the curvature 2-form are equal,

$$\mathrm{R}^{0}{}_{2\mu\nu}\,\theta^{\mu}\wedge\theta^{\nu} = \mathrm{R}^{2}{}_{1\mu\nu}\,\theta^{\mu}\wedge\theta^{\nu} = -\partial^{2}{}_{xx}H\,\mathrm{d}u\wedge\mathrm{d}x - \partial^{2}{}_{xy}H\,\mathrm{d}u\wedge\mathrm{d}y,$$

$$\mathrm{R}^{0}{}_{3\mu\nu}\,\theta^{\mu}\wedge\theta^{\nu} = \mathrm{R}^{3}{}_{1\mu\nu}\,\theta^{\mu}\wedge\theta^{\nu} = -\partial^{2}{}_{xy}H\,\mathrm{d}u\wedge\mathrm{d}x - \partial^{2}{}_{yy}H\,\mathrm{d}u\wedge\mathrm{d}y\,.$$

Theorem. The action of the Lie derivative on the curvature 2-form $\dot{\mathrm{R}}^{a}{}_{b}$ of a plane gravitational wave in the direction of the vector $\vec{X}$ generating the 5-parameter invariance group G_5 of this wave in the Riemann space is equal to zero, $\mathrm{L}_{\vec{X}}\mathrm{R}^{a}{}_{b}=0$.

Proof. The components of the Lie derivative acting on an arbitrary vector in a nonholonomic basis $\vec{\mathrm{v}}=\mathrm{v}^{a}\vec{e}_{a}$ are equal to

$$\left(\mathrm{L}_{\vec{X}}\vec{\mathrm{v}}\right)^{a} = X^{b}\vec{e}_{b}\mathrm{v}^{a} - \mathrm{v}^{c}\vec{e}_{c}X^{a} + X^{b}\mathrm{v}^{c}\left[\vec{e}_{b},\vec{e}_{c}\right]^{a}.$$

The components of the Lie derivative acting on an arbitrary 1-form $\sigma=\sigma_{a}\theta^{a}$ in a nonholonomic basis are equal to

$$(\mathrm{L}_{\vec{X}}\sigma)_{a} = X^{b}(\vec{e}_{b},\sigma_{a}) + (\vec{e}_{a}X^{b})\sigma_{b} + X^{b}\left[\vec{e}_{a},\vec{e}_{b}\right]^{c}\sigma_{c}\,.$$

Applying these formulas to the curvature 2-form, we find that the components of the Lie derivative acting on it have the form:

$$(\mathrm{L}_{\vec{X}}\mathrm{R})^{a}{}_{b\mu\nu} = X^{d}(\vec{e}_{d}\,\mathrm{R}^{a}{}_{b\mu\nu}) - (\vec{e}_{d}X^{a} + X^{c}\left[\vec{e}_{d},\vec{e}_{c}\right]^{a})\mathrm{R}^{d}{}_{b\mu\nu} +$$
$$+ (\vec{e}_{b}X^{d} + X^{c}\left[\vec{e}_{b},\vec{e}_{c}\right]^{d})\mathrm{R}^{a}{}_{d\mu\nu} + (\vec{e}_{\mu}X^{\sigma})\mathrm{R}^{a}{}_{b\sigma\nu} + (\vec{e}_{\nu}X^{\sigma})\mathrm{R}^{a}{}_{b\mu\sigma}\,.$$

From here, for example, for the values $\mu=\hat{1},\ \nu=\hat{2}$ we find,

$$(\mathrm{L}_{\vec{X}}\mathrm{R})^{0}{}_{2\hat{1}\hat{2}} = (a+b'(u)x+c'(u)y)(\partial_{\mathrm{v}}\mathrm{R}^{0}{}_{2\hat{1}\hat{2}}) + b(\partial_{\mathrm{x}}\mathrm{R}^{0}{}_{2\hat{1}\hat{2}}) + c(\partial_{\mathrm{y}}\mathrm{R}^{0}{}_{2\hat{1}\hat{2}})\,.$$

Similarly, for all nonzero values of the Lie derivative, we obtain,

$$(\mathrm{L}_{\vec{X}}\mathrm{R})^{0}{}_{2\hat{1}\hat{2}} = -b\partial^{3}{}_{xxx}H - c\partial^{3}{}_{xxy}H,\quad (\mathrm{L}_{\vec{X}}\mathrm{R})^{0}{}_{2\hat{1}\hat{3}} = -c\partial^{3}{}_{xyy}H - b\partial^{3}{}_{xxy}H,$$
$$(\mathrm{L}_{\vec{X}}\mathrm{R})^{0}{}_{3\hat{1}\hat{2}} = -c\partial^{3}{}_{xyy}H - b\partial^{3}{}_{xxy}H,\quad (\mathrm{L}_{\vec{X}}\mathrm{R})^{0}{}_{3\hat{1}\hat{3}} = -b\partial^{3}{}_{xyy}H - c\partial^{3}{}_{yyy}H,$$
$$(\mathrm{L}_{\vec{X}}\mathrm{R})^{2}{}_{1\hat{1}\hat{2}} = -b\partial^{3}{}_{xxx}H - c\partial^{3}{}_{xxy}H,\quad (\mathrm{L}_{\vec{X}}\mathrm{R})^{2}{}_{1\hat{1}\hat{3}} = -c\partial^{3}{}_{xyy}H - b\partial^{3}{}_{xxy}H,$$
$$(\mathrm{L}_{\vec{X}}\mathrm{R})^{0}{}_{3\hat{1}\hat{2}} = -c\partial^{3}{}_{xyy}H - b\partial^{3}{}_{xxy}H,\quad (\mathrm{L}_{\vec{X}}\mathrm{R})^{0}{}_{3\hat{1}\hat{3}} = -b\partial^{3}{}_{xyy}H - c\partial^{3}{}_{yyy}H.$$

For a specific value of the function $H(u,x,y)$ of a plane gravitational wave, the values of all third-order derivatives with respect to the variables x and y are equal to zero. As a result, *we obtain the proof of the theorem.*

As a consequence of the expression for the function $H(u,x,y)$, we find that the components of the curvature tensor of the Riemann space depend on all three arbitrary functions that define the metrics,

$$\mathrm{R}^{0}{}_{212} = -A(u),\quad \mathrm{R}^{0}{}_{213} = -B(u),\quad \mathrm{R}^{0}{}_{312} = -B(u),\quad \mathrm{R}^{0}{}_{313} = -C(u),$$
$$\mathrm{R}^{2}{}_{112} = -A(u),\quad \mathrm{R}^{2}{}_{113} = -B(u),\quad \mathrm{R}^{3}{}_{112} = -B(u),\quad \mathrm{R}^{3}{}_{113} = -C(u).$$

In this case, the curvature scalar is equal to zero, and the Ricci tensor is equal to,

$$\mathrm{R}_{11} = A(u) + C(u).$$

The dependence of the curvature tensor on arbitrary functions allows us to assert that the curvature tensor of a plane gravitational wave of the Riemann space in General Relativity can transfer information encoded in the source of the gravitational wave using the arbitrary functions $A(u)$ and $C(u)$.

4. Conclusion

We have considered plane gravitational waves in the Riemann space of General Relativity using the criterion of plane gravitational waves based on the analogy between plane gravitational and electromagnetic waves. This criterion states that the metric tensor of the gravitational wave space-time has a group G_5 of motion, which leaves invariant the isotropic hypersurface of a plane wave front with constant amplitude.

The Theorem is proved that the action of the Lie derivative on the plane wave curvature 2-form in the Riemann space in the direction of the vector generating the invariance group G_5 of this wave is equal to zero.

In the Riemann space of General Relativity, the components of the curvature tensor depend on three arbitrary functions that determine the metrics. But Einstein's equations define only two functions, $A(u)$ and $C(u)$, because the Ricci tensor has the form, $R_{11} = A(u) + C(u)$, and the curvature scalar is equal to zero. The third function, $B(u)$, remains arbitrary. Therefore, the curvature tensor of plane gravitational waves can carry information encoded by these two functions in the wave source. This fact may be significant, if gravitational waves in Riemann space can be used in future to transmit information.

References

[1] Babourova O V, Frolov B N and Scherban' V N 2013 *Gravit. Cosmol.* **19** 144
[2] Babourova O V, Frolov B N, Khetseva M S and Markova N V 2018 *Class. Quantum Grav.* **35** 175011

Decrease of the effective cosmological constant in the Poincare gauge theory of gravity with a scalar field

O V Babourova[1] **and B N Frolov**[2]

[1] Moscow Automobile and Road Construction State Technical University (MADI), Department 'Physics', Leningradsky pr., 64, Moscow, 125319, Russia
[2] Moscow Pedagogical State University (MPGU), Institute of Physics, Technology and Information Systems, M. Pirogovskaya ul. 29, Moscow 119992, Russia

E-mail: ovbaburova@madi.ru, bn.frolov@mpgu.su

Abstract. Cosmological consequences of the Poincare gauge theory of gravity are considered. An effective cosmological constant depending from the Dirac scalar field is introduced. It is proved that at the super-early Universe, the effective cosmological constant decreases exponentially from a huge value at the Big Bang to its extremely small value in the modern era, while the scale factor sharply increases and demonstrates inflationary behavior. This fact solves the well-known "cosmological constant problem" also in the Poincare gauge theory of gravity.

1. Introduction

Earlier in [1-3] it has been shown that in the framework of the Poincare–Weyl gauge theory of gravity there is a solution in which one has the sharp exponential decrease of the effective cosmological constant at small time and the scale-invariant inflation-like exponential increase of the scale factor.

These results represent a development of the authors' ideas since 2012, based on the assumption made by Harrison and Zeldovich about the approximate scale invariance of the early stage of the evolution of the Universe, which underlies the calculation of the initial part of the spectrum of primary fluctuations of matter density in the early Universe (Harrison–Zeldovich Plateau, [4,5]) and have been confirmed by results of the COBE experiment for the study of the anisotropy of the brightness of the background radiation.

We shall show that a similar result can be obtained in the Poincare gauge theory of gravity in the Riemann–Cartan space without nonmetricity. But only if at the same time we admit the existence of a cosmological scalar field, which simulates dark energy.

2. Lagrangian density and field equations

A truncated version of Poincare gauge theory of gravity is considered, the Lagrangian of which contains the squares of the torsion tensor (which thus plays a dynamic role in the theory), but excludes the squares of the curvature tensor:

$$\begin{aligned} \mathrm{L} = 2f_0\Big[&(1/2)\mathcal{R}^a{}_b \wedge \eta_a{}^b + \beta^4\Lambda_0\eta + \rho_1\beta^2\mathcal{T}^a \wedge {*}\mathcal{T}_a + \rho_2\beta^2(\mathcal{T}^a \wedge \theta_b)\wedge{*}(\mathcal{T}^b \wedge \theta_a) \\ &+\rho_3\beta^2(\mathcal{T}^a \wedge \theta_a)\wedge{*}(\mathcal{T}^b \wedge \theta_b) + l\,\mathrm{d}\beta \wedge {*}\mathrm{d}\beta\Big], \end{aligned} \quad (1)$$

where $\theta^a\ (a=0,1,2,3)$ are basis 1-forms, $*$ is the Hodge dualization operation, η is a volume 4-form with components η_{abcd}, and $\eta_{ab}=(1/2!)\,\theta^c\wedge\theta^d\eta_{abcd}$.

According to E. Gliner [6], the cosmological constant Λ in the Einstein equations is interpreted as the energy of the hypothetical vacuum-like medium, which is currently called vacuum energy and also dark energy. The term $\beta^4\Lambda_0$ in (1) describes the effective cosmological constant (interpreted here as dark energy), depending on the Weyl–Dirac scalar field β [7] (Λ_0 is the theory parameter providing the correct rate of inflation).

The derivation of the variational field equations is based on the use of the lemma on commutation of variation and dualization operations of external forms in the Riemann–Cartan space, proved in [8] and then modified taking into account the presence of the scalar Weyl–Dirac field.

As a result of the variation procedure, we have obtained three field equations: Γ-equation, θ-equation, and β-equation:

Γ-equation:

$$\begin{aligned}&f_0\Big[(1/2)\mathcal{T}_c\wedge\eta_a{}^{bc}+\mathrm{d}\ln\beta\wedge\eta_a{}^{b}\\&+2\rho_1\theta^b\wedge*\mathcal{T}_a+2\rho_2\theta^b\wedge\theta_c\wedge*(\mathcal{T}^c\wedge\theta_a)+2\rho_3\theta^b\wedge\theta_a\wedge*(\mathcal{T}^c\wedge\theta_c))\Big]=0\,.\end{aligned}\tag{2}$$

θ-equation:

$$\begin{aligned}&\frac{1}{2}\mathcal{R}^b{}_c\wedge\eta_b{}^c{}_a+\beta^2\Lambda_0\eta_a\\&+\rho_1\Big[2\mathrm{D}*\mathcal{T}_a+\mathcal{T}_b\wedge*(\mathcal{T}^b\wedge\theta_a)+*(*\mathcal{T}^b\wedge\theta_a)\wedge*\mathcal{T}_b+4\mathrm{d}\ln\beta\wedge*\mathcal{T}_a\Big]\\&+\rho_2\Big[2\mathrm{D}(\theta_b\wedge*(\mathcal{T}^b\wedge\theta_a))+2\mathcal{T}^b\wedge*(\theta_b\wedge\mathcal{T}_a)-*(\mathcal{T}^b\wedge\theta_c\wedge\theta_a)(\mathcal{T}^c\wedge\theta_b)\\&-*(*(\mathcal{T}^c\wedge\theta_d)\wedge\theta_a)\wedge*(\mathcal{T}^d\wedge\theta_c)+4\mathrm{d}\ln\beta\wedge\theta_b\wedge*(\mathcal{T}^b\wedge\theta_a)\Big]\\&+\rho_3\Big[2\mathrm{D}(\theta_a\wedge*(\mathcal{T}^b\wedge\theta_b))+2\mathcal{T}_a\wedge*(\mathcal{T}^b\wedge\theta_b)-*(\mathcal{T}^b\wedge\theta_b\wedge\theta_a)(\mathcal{T}^c\wedge\theta_c)\\&-*(*(\mathcal{T}^b\wedge\theta_b)\wedge\theta_a)\wedge*(\mathcal{T}^c\wedge\theta_c)+4\mathrm{d}\ln\beta\wedge\theta_a\wedge*(\mathcal{T}^b\wedge\theta_b)\Big]\\&+l[-\mathrm{d}\ln\beta\wedge*(\mathrm{d}\ln\beta\wedge\theta_a)-*(*\mathrm{d}\ln\beta\wedge\theta_a)\wedge*\mathrm{d}\ln\beta)]=0\,.\end{aligned}\tag{3}$$

β-equation:

$$\begin{aligned}&\mathcal{R}^a{}_b\wedge\eta_a{}^b+4\beta^2\Lambda_0\eta\\&+\rho_1 2\mathcal{T}^a\wedge*\mathcal{T}_a+\rho_2 2(\mathcal{T}^a\wedge\theta_b)\wedge*(\mathcal{T}^b\wedge\theta_a)+\rho_3 2(\mathcal{T}^a\wedge\theta_a)\wedge*(\mathcal{T}^b\wedge\theta_b)\\&+l(-2\mathrm{d}*\mathrm{d}\ln\beta-2\mathrm{d}\ln\beta\wedge\mathrm{d}\ln\beta)=0\,.\end{aligned}\tag{4}$$

3. Cosmological model

The scale invariant super-early stage of the Universe evolution is considered. Using the spatially flat FRW metric with the scale factor $a(t)$,

$$ds^2=dt^2-a^2(t)(dx^2+dy^2+dz^2)\,,\tag{5}$$

let us solve the field equations (2)-(4). In this case the torsion 2-form is completely determined by the 1-form of its trace $\mathcal{T}$, $\mathcal{T}_a=(1/3)\mathcal{T}\wedge\theta_a$.

Then the Γ-equation (2) becomes equivalent to the equations,

$$\mathcal{T} = s(\mathrm{d}\beta/\beta), \quad s=-3, \quad \rho_1 - 2\rho_2 = 0. \tag{6}$$

For the θ -equation (3) we obtain the only non-null expressions

$$3H^2 + 6HU + U^2(l+3) = \Lambda_0\beta^2, \tag{7}$$

$$2H_t + 2U_t + 4HU + 3H^2 - U^2(l-1) = \Lambda_0\beta^2, \tag{8}$$

where we use notations,

$$H = a_t/a, \quad U = \beta_t/\beta, \quad a_t = da/dt, \quad \beta_t = d\beta/dt. \tag{9}$$

For the β -equation (4) we have the only non-null expression

$$3H_t + 6H^2 + (U_t + U^2 + 3HU)(l+3) = 2\Lambda_0\beta^2. \tag{10}$$

After subtracting the equation (7) from the equation (8) we have,

$$H_t + U_t - HU - U^2(l+1) = 0. \tag{11}$$

Then let us subtract from the equation (10) the double equation (8) taking into account the equation (11). As a result we obtain the equation

$$l(U_t + 3HU + 2U^2) = 0, \qquad l \neq 0. \tag{12}$$

When $l \neq 0$, we have three equations (7), (11) and (12) for two unknown functions $a(t)$ and $\beta(t)$, and the system of equations is over-determined.

In order to overcome this result, we hypothesize that the interaction constant l is extremely small. In this case in equations (7) and (11), we may neglect the terms with the coefficient l. Then the equation (7) can be reduced to the equation

$$H + U = (\lambda/3)\beta, \quad \lambda = \sqrt{3\Lambda_0}. \tag{13}$$

The sign "+" has been chosen here.

In this case we make sure that the equation (11) with regard to (12) becomes a consequence of the equation (13), and now the system of equations includes the two equations (12) and (13) for two unknown functions $a(t)$ and $\beta(t)$. Therefore the system of equations is well defined.

The equations (12) and (13) yield the equation,

$$U_t + \lambda\beta U - U^2 = 0, \tag{14}$$

which is equivalent to the equation

$$\beta\beta_{tt} + \lambda\beta_t\beta^2 - 2\beta_t^2 = 0. \tag{15}$$

Using the substitution $\beta_t = \beta^2 z(\beta)$, this equation is integrated with the first integral,

$$U = \beta_t/\beta = -\lambda\beta\ln(C\beta), \tag{16}$$

(C is an integration constant).

The second equation of the system is obtained by substituting (16) into the equation (13),

$$H = a_t/a = \lambda\beta\left(\ln(C\beta) + 1/3\right). \tag{17}$$

We shall use a boundary condition,

$$C\beta \to 1, \text{ when } t \to \infty. \tag{18}$$

The limit $\beta \to 1/C$ occurs exponentially, $\sim e^{\frac{-\lambda t}{C}}$, for large t.

To choose the value of the constant C we require that the value of the effective cosmological constant $\beta^2(t)\Lambda_0$ currently coincides with the modern cosmological constant value Λ,

$$\beta^2(t_U)\Lambda_0 = \Lambda. \tag{19}$$

Here t_U is the Universe lifetime (t_U = 13. 8 billion years), moreover the equality $\beta_0^2\Lambda_0 = 10^{120}\Lambda$ should be fulfilled (see [9-11]), where β_0 is the Weyl–Dirac scalar field value at $t = 0$.

The result of numerical integration of the system of equations (16) and (17) (see [2,3]) with the boundary condition (18) is shown in Fig. 1 and Fig. 2, on which one can see for small and large time values the inflationary behavior of the scale factor $a(t)$, as well as a sharp exponential decrease in the effective cosmological constant $\beta^2(t)\Lambda_0$ from a huge value at the beginning of the Big Bang to an extremely small (but not zero) value in the modern epoch, which coincides with its observed value.

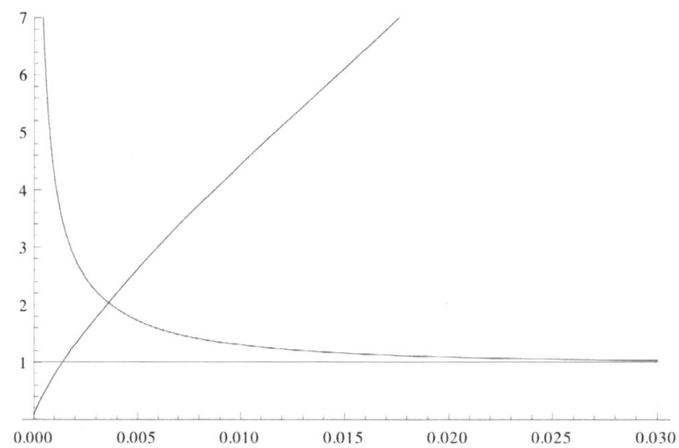

Figure 1. The functions $a(t)$, $\beta(t)$ for small values of time t.

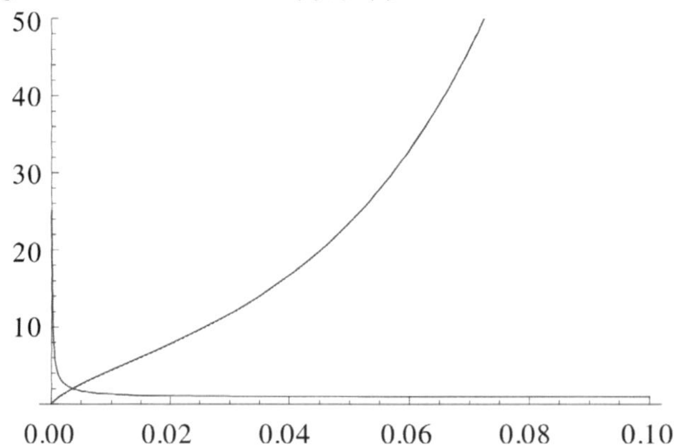

Figure 2. The functions $a(t)$, $\beta(t)$ for large values of time t.

4. Conclusion

We have shown that the sharp exponential decrease of the effective cosmological constant and the inflationary increase of the scale factor at the early stage of the Universe evolution can be realized not only in the Cartan–Weyl space with curvature, torsion and nonmetricity, but also on the basis of the Poincare gauge theory of gravity in the presence of only torsion along with curvature under the following conditions:

1) A cosmological scalar field (the Weyl–Dirac scalar field), which simulates dark energy, exists in Nature.
2) Dark energy has extremely weak self-action.

We follow the idea of E. Gliner about a homogeneous vacuum-like medium generating the gravitational field subsequently named as dark energy. In our theory this medium determines the non-constant effective cosmological "constant", which depends on the Weyl–Dirac scalar field.

We also follow the idea of Harrison and Zeldovich that the Universe at the early stage has the property of approximate scale invariance. As a consequence, we get a scale invariant inflation.

Under the condition that the constant of the self-action of the field β is substantially small (but not equal to zero), a cosmological solution is found, which is similar to that obtained in [1,2]. The peculiarity of this solution consists in an exponentially sharp decrease in the effective cosmological constant (within the time of inflation) with a transition to its present value at a large value of time. At the same time, the scale factor $a(t)$ demonstrates inflationary behavior.

The hypothesis on the sharply decrease of the effective cosmological constant as a consequence of the fields dynamics in the super-early Universe was expressed in the monograph [12] and the earlier papers [13–15]. The main thing of the solution obtained and that found earlier by the authors is that these solutions demonstrate for the effective cosmological constant sharply decrease at small t and approach Λ (but not zero) at large values of t.

It can be assumed that the found solution can serve as a basis for solving the well-known cosmological constant problem also in the Poincare gauge theory of gravity, and not only in the Poincare–Weyl gauge theory of gravity [1,2].

References

[1] Babourova O V, Frolov B N 2020 Harrison–Zeldovich scale invariance and the exponential decrease of the "cosmological constant" in the super-early Universe *Preprint* 2001.05968 gr-qc
[2] Babourova O V, Frolov B N 2020 *Universe* **6** 230
[3] Baburova O V, Frolov B N 2020 *Journal of Physics: Conference Series* **1557** 012011
[4] Sazhin M V 2004 *Physics-Uspekhi.* **47** 187
[5] Rubakov V A 2009 Harrison–Zeldovich spectrum from conformal invariance *Preprint* 0906.3693 [hep-th]
[6] Gliner E B 2002 *Physics-Uspekhi* **45** 213
[7] Dirac P A M 1973 *Proc. Roy. Soc.* A **333** 403
[8] Babourova O V, Frolov B N and Klimova E A 1999 *Class. Quantum Grav.* **16** 14 (*Preprint* gr-qc/9805005).
[9] Weinberg S 1989 *Rev. Mod. Phys.* **61** 1
[10] Gorbunov D S and Rubakov V A 2011 *Introduction to the theory of the Early Universe: Hot Big Bang Theory* (World Scientific) p 488
[11] Peebles P J E and Ratra B 2003 *Rev. Mod. Phys.* **75** 559 (*Preprint* astro-ph/0207347v2)
[12] Babourova O V and Frolov B N 2012 *Mathematical problems of the modern theory of gravity* (Moscow: Moscow Pedagogical State University Press) p 120
[13] Babourova O V, Frolov B N 2011 Dark energy, Dirac's scalar field and the cosmological constant problem *Preprint* 1112.4449 [gr-qc]
[14] Babourova O V, Frolov B N and Lipkin K N 2012 *Gravitation and Cosmology* **18** 225

[15] Babourova O V, Frolov B N 2015 *Physics Research International* 952181

Journal of Physics: Conference Series 2081 (2021) 012016 doi:10.1088/1742-6596/2081/1/012016

Gravitation in Theory of Space-Time Film and Galactic Soliton

Alexander A Chernitskii

[1] Department of Mathematics, St. Petersburg State Chemical Pharmaceutical University, Prof. Popov str. 14, 197022 St. Petersburg, Russia
[2] A. Friedmann Laboratory for Theoretical Physics, The Herzen University, 191186 St. Petersburg, Russia

E-mail: AAChernitskii@mail.ru

Abstract. The scalar field of extremal space-time film is considered as unified fundamental field. Metrical interaction between solitons-particles as gravitational interaction is considered here in approximation of a weak fundamental field. It is shown that the signature of metrics $\{-,+,+,+\}$ in the model formulation provides the observable gravitational attraction to a region with bigger energy density of the fundamental field. The induced gravitational interaction in the space-time film theory is applied to stars in a galaxy. The conception of galaxy soliton of space-time film is introduced. A weak field asymptotic solution for a galaxy soliton is proposed. It is shown that the effective metrics for this solution can provide the observable velocity curves for galaxies and explains their spiral structure. Thus a solution for so-called dark matter problem in the framework of space-time film theory is proposed.

1. Introduction

We consider the scalar field of space-time film as unified fundamental field [1]. This scalar field model is related to well known Born – Infeld nonlinear electrodynamics [2, 3]. The model under consideration is attractive because it has relatively simple and geometrically clear form. It can be considered as a relativistic generalization of the minimal surface or minimal thin film model in three-dimensional space.

The model under consideration provides the necessary effects which are required for a realistic unified filed model.

In particular, there are exact solutions of the model equation which can be considered as photons [4].

Also, in the space-time film theory, distant interactions of two kinds between particles-solitons, namely force and metrical interactions [5], correspond to electromagnetic [6] and gravitational [7] interactions of particles respectively.

Here we apply the induced gravitational interaction in the framework of the space-time film theory [8] to stars in a galaxy. So we try to solve the known so called dark matter problem.

2. Extremal space-time film

The following variational principle and the action which has the world volume form are considered:

$$\delta\mathcal{A} = 0\,, \qquad \mathcal{A} = \int_{\overline{V}} \sqrt{|\mathfrak{M}|}\;(\mathrm{d}x)^4 = \int_{\overline{V}} \pounds\; \mathrm{d}\overline{V}\,, \tag{1a}$$

where $\mathfrak{M} \doteqdot \det(\mathfrak{M}_{\mu\nu})$, $(\mathrm{d}x)^4 \doteqdot \mathrm{d}x^0\mathrm{d}x^1\mathrm{d}x^2\mathrm{d}x^3$, $\overline{V}$ is space-time volume, $\mathrm{d}\overline{V} \doteqdot \sqrt{|\mathfrak{m}|}\;(\mathrm{d}x)^4$ is four-dimensional volume element,

$$\mathfrak{M}_{\mu\nu} = \mathfrak{m}_{\mu\nu} + \chi^2\,\frac{\partial\Phi}{\partial x^\mu}\,\frac{\partial\Phi}{\partial x^\nu}\,, \qquad \pounds \doteqdot \sqrt{\left|1 + \chi^2\,\mathfrak{m}^{\mu\nu}\,\frac{\partial\Phi}{\partial x^\mu}\,\frac{\partial\Phi}{\partial x^\nu}\right|}, \tag{1b}$$

$\mathfrak{m}_{\mu\nu}$ are components of metric tensor for flat four-dimensional space-time, Φ is scalar real field function, χ is dimensional constant. The Greek indices take values $\{0, 1, 2, 3\}$. The tensor $\mathfrak{M}_{\mu\nu}$ used here can be called also as world tensor.

The model (1) can be considered as a relativistic generalization of the appropriate expression for the mathematical model of two-dimensional minimal thin film in the tree-dimensional space of our everyday experience.

We consider two possible signatures of metrics in the expression of Lagrangian (1b): $\{+, -, -, -\}$ and $\{-, +, +, +\}$. But an analysis of the induced gravitational interaction in the framework of this model admits the signature $\{-, +, +, +\}$ only in (1b) for a weak fundamental field approximation. This is shown below.

We have the following symmetrical canonical energy-momentum density tensor of the model in Cartesian coordinates

$$\overline{T}^{\mu\nu} = \frac{1}{4\pi}\left(\frac{\Phi^\mu\,\Phi^\nu}{\pounds} - \frac{\underline{\mathfrak{m}}^{\mu\nu}}{\chi^2}\,\pounds\right)\,, \qquad \Phi^\alpha \doteqdot \underline{\mathfrak{m}}^{\alpha\beta}\,\frac{\partial\Phi}{\partial x^\beta}\,, \tag{2}$$

where $\underline{\mathfrak{m}}^{\mu\nu}$ is Minkowski metrics.

To use finite integral characteristics of solutions in infinite space-time we introduce regularized energy-momentum density tensor with the following formula:

$$T^{\mu\nu} = \overline{T}^{\mu\nu} - \overset{\infty}{T}{}^{\mu\nu}\,, \qquad \overset{\infty}{T}{}^{\mu\nu} = -\frac{1}{4\pi\,\chi^2}\,\underline{\mathfrak{m}}^{\mu\nu}\,, \tag{3}$$

where $\overset{\infty}{T}{}^{\mu\nu}$ is the constant regularizing symmetrical energy-momentum density tensor.

The field equation in Cartesian coordinates can be written in the following form:

$$\overline{T}^{\mu\nu}\,\frac{\partial^2\,\Phi}{\partial x^\mu\,\partial x^\nu} = 0\,, \tag{4a}$$

where $\overline{T}^{\mu\nu}$ is the canonical energy-momentum density tensor (2).

By multiplying (4a) to $(-4\pi\,\chi^2\,\pounds)$, the field equation can be written in the following form without radicals:

$$\left(\underline{\mathfrak{m}}^{\mu\nu}\left(1 + \chi^2\,\underline{\mathfrak{m}}_{\sigma\rho}\,\Phi^\sigma\,\Phi^\rho\right) - \chi^2\,\Phi^\mu\,\Phi^\nu\right)\frac{\partial^2\,\Phi}{\partial x^\mu\,\partial x^\nu} = 0\,. \tag{4b}$$

Equation (4) transforms to ordinary linear wave equation with $\chi = 0$:

$$\underline{\mathfrak{m}}^{\mu\nu}\,\frac{\partial^2\,\Phi}{\partial x^\mu\,\partial x^\nu} = 0\,. \tag{5}$$

The model under consideration have the following notable characteristic equation:

$$\tilde{\mathfrak{m}}^{\mu\nu}\,\frac{\partial\mathcal{S}}{\partial x^{\mu}}\,\frac{\partial\mathcal{S}}{\partial x^{\nu}} = 0\;, \qquad \tilde{\mathfrak{m}}^{\mu\nu} \doteqdot -4\pi\,\chi^{2}\,\bar{\mathcal{T}}^{\mu\nu}\;, \tag{6}$$

where $\tilde{\mathfrak{m}}^{\mu\nu}$ is the effective Riemann metrics which is proportional to the canonical energy-momentum tensor $\bar{\mathcal{T}}^{\mu\nu}$. Equation $\mathcal{S} = 0$ gives a three-dimensional characteristic hypersurface of the field model in four-dimensional space-time.

3. Gravitation as metrical interaction of solitons-particles

We consider the solitons-particles of two kind. Luminal solitons-particles have the speed of light and subluminal solitons-particles have a subluminal speed. For a subluminal soliton, there is an intrinsic coordinate system in which the soliton rests as a whole. A subluminal soliton has a form of standing wave in its intrinsic coordinate system in general case.

There are also such space localized solutions for the linear wave equation (5). We call these solutions weak solitons. For example, we have the following elementary oscillating solution of the linear wave equation in the spherical coordinate system:

$$\Phi = \frac{a}{\underset{.}{r}}\,\sin(\underset{.}{\omega}\,\underset{.}{r})\,\sin(\underset{.}{\omega}\,\underset{.}{x}^{0})\;, \tag{7}$$

where a is an amplitude of the weak soliton, $\underset{.}{\omega}$ is its intrinsic frequency, the the point under symbols denotes the belonging to the intrinsic coordinate system.

For an arbitrary Cartesian coordinate system, the wave vector $\{k_{\mu}\}$ of a subluminal soliton satisfies the following dispersion relation:

$$\left|\underline{\mathfrak{m}}^{\mu\nu}\,k_{\mu}\,k_{\nu}\right| = \underset{.}{\omega}^{2}\;. \tag{8}$$

And the dispersion relation (8) of a test subluminal soliton is modified under the influence of distant solitons-particles:

$$\left|\tilde{\mathfrak{m}}^{\mu\nu}\,k_{\mu}\,k_{\nu}\right| = \underset{.}{\omega}^{2}\;, \tag{9}$$

where $\tilde{\mathfrak{m}}^{\mu\nu}$ is effective metrics (6) with the substitution of the field of distant solitons. Thus here we have $\tilde{\mathfrak{m}}^{\mu\nu} = \tilde{\mathfrak{m}}^{\mu\nu}(x^{\rho})$ and $\bar{k}_{\mu} = \bar{k}_{\mu}(x^{\rho})$.

The modified dispersion relation (9) gives the trajectory equation for the subluminal soliton in the following geodesic line equation form [8]:

$$\frac{d\mathfrak{u}^{\mu}}{d\mathfrak{s}} + \tilde{\Gamma}^{\mu}_{\nu\rho}\,\mathfrak{u}^{\nu}\,\mathfrak{u}^{\rho} = 0\;, \qquad \mathfrak{u}^{\mu} \doteqdot \frac{d\mathfrak{x}^{\mu}}{d\mathfrak{s}}\;, \qquad \tilde{\Gamma}^{\mu}_{\nu\rho} \doteqdot \frac{1}{2}\,\tilde{\mathfrak{m}}^{\mu\delta}\left(\frac{\partial\check{\tilde{\mathfrak{m}}}_{\delta\nu}}{\partial x^{\rho}} + \frac{\partial\check{\tilde{\mathfrak{m}}}_{\delta\rho}}{\partial x^{\nu}} - \frac{\partial\check{\tilde{\mathfrak{m}}}_{\nu\rho}}{\partial x^{\delta}}\right)\;, \tag{10}$$

where $\mathfrak{x}^{j} = \mathfrak{x}^{j}(x^{0})$ is a trajectory of energy center for the subluminal soliton, $\mathfrak{x}^{0} = x^{0}$, $\check{\tilde{\mathfrak{m}}}_{\mu\nu}$ is the inverse tensor to the tensor $\tilde{\mathfrak{m}}^{\mu\nu}$. Here the parameter of movement $\mathfrak{s}$ satisfies the relation

$$d\mathfrak{s}^{2} = \left|\check{\tilde{\mathfrak{m}}}_{\mu\nu}\,d\mathfrak{x}^{\mu}\,d\mathfrak{x}^{\nu}\right|\;. \tag{11}$$

In small-speed approximation for test particle $\mathfrak{v} \ll 1$ the geodesic line equation (10) has the form

$$\frac{d\mathfrak{v}^{i}}{dx^{0}} = -\tilde{\Gamma}^{i}_{00}\;, \qquad \mathfrak{v}^{i} \doteqdot \frac{d\mathfrak{x}^{i}}{dx^{0}}, \tag{12}$$

where $\mathfrak{v}_{i}$ are components of three-dimensional velocity for test soliton-particle. The Latin indices take values $\{1, 2, 3\}$.

In the approximation for a weak distant solitons field, we can represent the effective metrics with the power series form in square of the nonlinearity constant χ^2 with keeping the linear terms only:

$$\tilde{\underline{\mathfrak{m}}}^{\mu\nu} \approx \underline{\mathfrak{m}}^{\mu\nu} - \chi^2 \left(\Phi^\mu \, \Phi^\nu - \frac{1}{2} \underline{\mathfrak{m}}^{\mu\nu} \, \underline{\mathfrak{m}}_{\alpha\beta} \, \Phi^\alpha \, \Phi^\beta \right) , \tag{13a}$$

$$\check{\tilde{\underline{\mathfrak{m}}}}_{\mu\nu} \approx \underline{\mathfrak{m}}_{\mu\nu} + \chi^2 \left(\Phi_\mu \, \Phi_\nu - \frac{1}{2} \underline{\mathfrak{m}}_{\mu\nu} \, \underline{\mathfrak{m}}^{\alpha\beta} \, \Phi_\alpha \, \Phi_\beta \right) . \tag{13b}$$

In this approximation, we have the following expression for $\tilde{\Gamma}^i_{00}$ via the derivatives of fundamental field Φ:

$$\tilde{\Gamma}^i_{00} \approx \underline{\mathfrak{m}}^{ii} \left(\frac{\partial \check{\tilde{\mathfrak{m}}}_{0i}}{\partial x^0} - \frac{1}{2} \frac{\partial \check{\tilde{\mathfrak{m}}}_{00}}{\partial x^i} \right) \approx \underline{\mathfrak{m}}^{ii} \, 4\pi \, \chi^2 \left(\frac{\partial \bar{\mathcal{P}}^i}{\partial x^0} - \frac{1}{2} \frac{\partial \bar{\mathcal{E}}}{\partial x^i} \right) \qquad (\not\Sigma) , \tag{14a}$$

$$\bar{\mathcal{E}} \doteqdot \frac{1}{8\pi} \left((\Phi_0)^2 + (\Phi_1)^2 + (\Phi_2)^2 + (\Phi_3)^2 \right) , \quad \bar{\mathcal{P}}^i \doteqdot \frac{1}{4\pi} \, \Phi_0 \Phi_i , \tag{14b}$$

where $\bar{\mathcal{E}}$ and $\bar{\mathcal{P}}$ are energy and momentum densities for the linearized model (5).

As it can be seen from (12) and (14), we have the attraction of test particle-soliton to a region with bigger energy density for the case $\underline{\mathfrak{m}}^{ii} = 1$. Thus to have the observable gravitational attraction in this model we must choose the signature of metric $\{-,+,+,+\}$ in assumption formula (1) for the weak fundamental field approximation.

4. About solving the dark matter problem

So called the dark matter problem has appeared from the discrepancy of available theoretical and observable velocity curves for galaxies. The velocity curves for galaxies are the dependence of rotational velocity for stars of a galaxy from the distance to its center.

In the case of circular orbit and gravitational potential φ, which is proportional to inverse radius, we obtain the dependence of the orbital velocity for a star from the distance to a massive galaxy center:

$$\varphi = -G \, \frac{M}{r} , \tag{15a}$$

$$\frac{v^2}{r} = G \, \frac{M}{r^2} \qquad \Longrightarrow \qquad v = \sqrt{G \, \frac{M}{r}} \, . \tag{15b}$$

But the observable velocities of distant stars in a galaxy has a very small dependence from the distance to its center.

If we assume that each visible star in a galaxy has the potential of type (15a), we can calculate the summary gravitational potential of galaxy. But in this case we also do not obtain the observable rotation curves. For a solution of the problem we can assume an existence of a dark matter or we can abandon the galaxy potential which is obtained by ordinary summation of star potentials of type (15a).

The conception of induced gravitation in the framework of the space-time film theory gives a radical solution of this problem.

All observable galaxies has a disk form, usually with helicoidal structure. Thus they have a certain axis symmetry. In the point of view of the space-time film theory we must have an appropriate galaxy solution or galaxy soliton or soliton-galaxy. Far from the region of space with high energy density, we can assume that the fundamental field Φ satisfies the linearized equation of the model or ordinary linear wave equation (5). It is natural to consider in this connection its solutions with the appropriate galaxy symmetry.

As the possible reductive asymptotics of the galactic soliton, let us consider the following solution of the linear wave equation (5) in cylindrical coordinate system $\{\rho, \varphi, z\}$:

$$\Phi = \Re\left[\left(J_m\left(\sqrt{\lambda^2+\omega^2}\,\rho\right) + \mathrm{i}\,Y_m\left(\sqrt{\lambda^2+\omega^2}\,\rho\right)\right)\mathrm{e}^{-\lambda|z|+\mathrm{i}\left(m\,\varphi-\omega\,x^0\right)}\right]. \tag{16}$$

where J_m and Y_m are Bessel functions of the first and second kinds, $J_m + \mathrm{i}\,Y_m$ is the Hankel function of the first kind.

The solution (16) has the singular plane $z = 0$ and the singular axis $\rho = 0$. It is a rotating spiral configuration with exponentially decreasing amplitude of the field for $z \to \pm\infty$. The section $(z = 0) \cap (x^0 = 0)$ of the solution (16) for $m = 1$, $\omega = \lambda = 1$ is shown on Fig. 1.

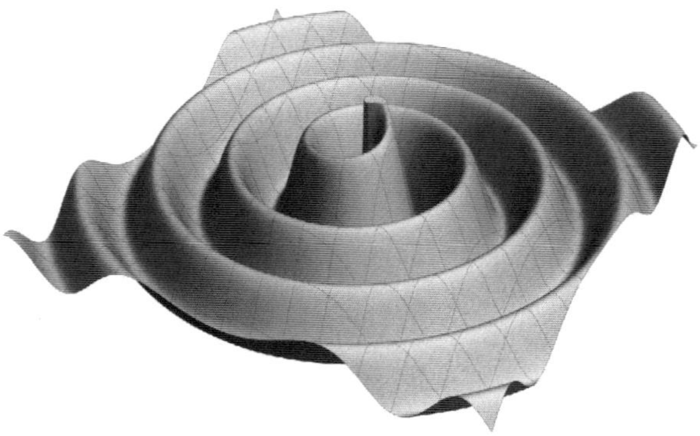

Figure 1. Section $(z = 0) \cap (x^0 = 0)$ of the possible simplest asymptotics (16) ($m = 1$, $\omega = \lambda = 1$) for a galactic soliton.

In view of the equation of motion (12) and expression for $\widetilde{\Gamma}^i_{00}$ (14) we obtain the orbital velocity for an asymptotic function Φ:

$$\frac{\boldsymbol{v}^2}{\rho} = \widetilde{\Gamma}^1_{00} \quad \Longrightarrow \quad \boldsymbol{v} = \sqrt{\rho\,\widetilde{\Gamma}^1_{00}}\,, \quad \widetilde{\Gamma}^1_{00} \geqslant 0\,, \tag{17a}$$

$$\widetilde{\Gamma}^1_{00} \approx \left(\frac{\partial \breve{\tilde{\mathfrak{m}}}_{01}}{\partial x^0} - \frac{1}{2}\frac{\partial \breve{\tilde{\mathfrak{m}}}_{00}}{\partial \rho}\right) \approx 4\pi\,\chi^2\left(\frac{\partial \bar{\mathcal{P}}^1}{\partial x^0} - \frac{1}{2}\frac{\partial \bar{\mathcal{E}}}{\partial \rho}\right). \tag{17b}$$

Here for $\widetilde{\Gamma}^1_{00} < 0$ we have the repulsion from the center and the orbital velocity is absent.

The orbital velocity (17a) for the asymptotic solution (16) on the line $(z = 0) \cap (\varphi = 0) \cap (x^0 = 0)$ for $m = 1$, $\omega = \lambda = 1$ is shown on Fig. 2, and on the plane $(z = 0) \cap (x^0 = 0)$ for $m = 1$, $\omega = \lambda = 1$ is shown on Fig. 3.

As can be seen on Fig. 2, in the region of space, where the orbital velocity exists, its average value has the tendency to be constant. This coincides with experimental data. Also, helical structure shown on Fig. 3 coincides with the images of real galaxies.

It should be noted that for the considered simple asymptotics of galaxy soliton (16), its number of branches or tails equals $2\,m$, where m is the order of twist for the space-time film soliton. As we know, the observable galaxies have even number of branches, that confirms the considered concept of galaxy soliton of space-time film.

Here we have considered only one possible asymptotics of a galactic soliton which coincides with experimental data. But, apparently, for any real galaxy we can discover the appropriate asymptotic solution of the space-time film equation which gives an observable distribution of star density and orbital velocities. This will mean that for this galaxy we have obtained the asymptotics of the appropriate galactic soliton.

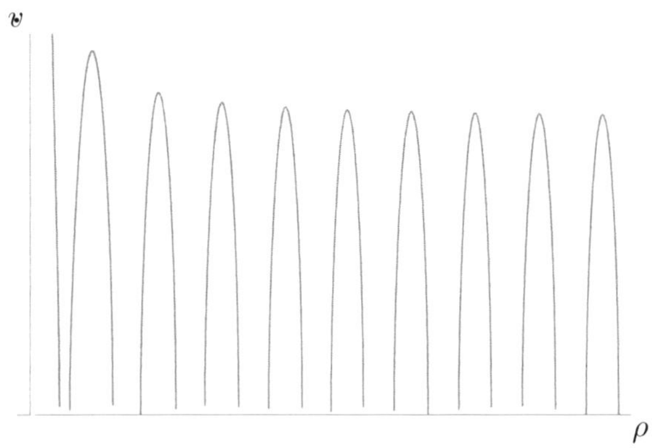

Figure 2. The orbital velocity for the solution (16) on the line $(z = 0) \cap (\varphi = 0) \cap (x^0 = 0)$ for $m = 1$, $\omega = \lambda = 1$.

Figure 3. The orbital velocity for the solution (16) on the plane $(z = 0) \cap (x^0 = 0)$ for $m = 1$, $\omega = \lambda = 1$.

5. Conclusions

Thus we have considered the effect of induced gravitation in the framework of space-time film theory. The approximation of a weak fundamental field for the effective metrics is considered. It is discovered that the choice for the signature of metrics $\{-,+,+,+\}$ in the model formulation provides the gravitational attraction to a region with bigger energy density of the fundamental field.

We introduce the conception of galactic soliton of space-time film. In the framework of this conception, the possible solution of so called dark matter problem and explanation of galactic spiral structure are proposed. We can note the defined agreement of the galaxy soliton conception with the observed data. In particular, this approach gives the observable even number of branches for a spiral galaxy.

References

[1] Chernitskii A A 2018 *J. Phys. Commun.* **2** 105013 (*Preprint* `arXiv:1506.09137`)
[2] Chernitskii A A 1999 *J. High Energy Phys.* JHEP12(1999)010 (*Preprint* `arXiv:hep-th/9911093`)
[3] Chernitskii A A 2005 Born–Infeld equations *Encyclopedia of Nonlinear Science* ed Scott A (New York and London: Routledge) pp 67–69 (*Preprint* `arXiv:hep-th/0509087`)
[4] Chernitskii A A 2020 *J. Phys.: Conf. Ser.* **1557** 012015
[5] Chernitskii A A 2020 *Int. J. Mod. Phys. A* **35** 2040021 (*Preprint* `arXiv:1911.01791`)
[6] Chernitskii A A 2017 *J. Phys.: Conf. Ser.* **938** 012029 (*Preprint* `arXiv:1804.09022`)
[7] Chernitskii A A 2016 *Int. J. Mod. Phys. Conf. Ser.* **41** 1660119 (*Preprint* `arXiv:1808.10266`)
[8] Chernitskii A A 2021 *Universe* **7** 11 (*Preprint* `arXiv:2012.08328`)

On the Schwarzschild solution in TEGR

E D Emtsova[1,2], **M Krššák**[3,4], **A N Petrov**[1] **and A V Toporensky**[1,5]

[1] Sternberg Astronomical Institute, M.V. Lomonosov Moscow State University, Universitetskii pr. 13, Moscow 119992, Russia
[2] Faculty of Physics, Lomonosov Moscow State University, Moscow 119991, Russia
[3] Bangkok Fundamental Physics Group, Department of Physics, Chulalongkorn University, Bangkok 10330, Thailand
[4] Center for Gravitation and Cosmology, College of Physical Science and Technology, Yangzhou University, Yangzhou 225009, China
[5] Kazan Federal University, Kremlevskaya 18, Kazan, 420008, Russia

E-mail: ed.emcova@physics.msu.ru, martin.krssak@gmail.com, alex.petrov55@gmail.com, atopor@rambler.ru

Abstract. Conserved currents, superpotentials and charges for the Schwarzschild black hole in the Teleparallel Equivalent of General Relativity (TEGR) are constructed. We work in the covariant formalism and use the Noether machinery to construct conserved quantities that are covariant/invariant with respect to both coordinate and local Lorentz transformations. The constructed quantities depend on the vector field ξ and we consider two different possibilities, when ξ is chosen as either a timelike Killing vector or a four-velocity of an observer. We analyze and discuss the physical meaning of each choice in different frames: static and freely falling Lemaitre frame. Moreover, a new generalized free-falling frame with an arbitrary initial velocity at infinity is introduced. We derive the inertial spin connection for various tetrads in different frames and find that the "switching-off" gravity method leads to ambiguities.

1. Introduction

Teleparallel approach to gravity became increasingly popular in recent years [1-4]. Naturally, the most popular of these theories is the formulation of the ordinary general relativity itself known as the Teleparallel Equivalent of General Relativity (TEGR) [1,5,6]. The reason why TEGR is interesting, besides being the simplest model, is that one can construct coordinate covariant conserved quantities, in contrast with pseudotensors in the standard general relativity [7].

However, in the original formulation of TEGR, these coordinate covariant conserved quantities are not covariant with respect to local Lorentz rotations in a tangent (tetrad) space [8]. This problem has been solved in [9-11], where, using the language of differential forms, the authors have constructed conserved currents, superpotentials and charges, which are both coordinate covariant and invariant with respect to local Lorentz rotations. Later [12,13], an analogous method was developed in the framework of a more popular tensorial formalism. In this covariant formulation, we have to consider the so-called inertial spin connection (ISC) ${}^{\bullet}A^{a}{}_{b\mu}$, which is not a dynamical quantity, i.e. it does not affect the field equations of TEGR. Nevertheless, it has to be determined since it plays an important role as far as the action and conserved charges derived with it are concerned. One of the methods to determine ISC is a so-called procedure of "switching off" gravity that was originally suggested in [14] and applied later in many studies. In [12, 13] we have formalized this procedure in a more general way and outline it below.

Here, we apply the fully covariant formalism to analyze the Schwarzschild black hole solution in various frames, including the static and free-falling ones. There are some advantages of the fully covariant formalism with respect to non-covariant methods of constructing conserved charges. In particular, the covariant formalism allows us to work with arbitrary tetrads, what is often more convenient than finding the special class of proper tetrads in the non-covariant approach. We then show that the procedure of "switching off" gravity used to determine ISC is not unique and lead to a certain ambiguity since different tetrads can share the same ISC. We generalize our discussion to the case of free-falling observers with a non-zero initial velocity in which we get more physically expected results but the ambiguity still remains and depends on the initial velocity. We introduce a new notion named as a "gauge", which is a pair of a tetrad and a related ISC that correspond to the same physical situation.

These results have been presented at the Moscow PIRT 2021 conference [15] and our recent paper [16]. Compared to [16], we present our results here in a simplified form, such as we use only the static coordinates for our calculations, and focus on the main results.

Here we follow mostly notations used in the book [1]. The Latin indices mean the tetrad components, whereas the Greek indices mean the spacetime components. The tetrad indices are transformed to the spacetime ones and inversely by contracting with the tetrad vectors.

2. Covariant conserved quantities in TEGR

2.1. Lagrangian and field variables of TEGR

TEGR is formulated in tetrad formalism where the fundamental variables are components of the tetrad $h^a{}_\rho$ and the spin connection ${}^\bullet A^i{}_{k\nu}$. The tetrad is related with the metric $g_{\rho\sigma}$ as $g_{\rho\sigma} = \eta_{ab} h^a{}_\rho h^b{}_\sigma$, where η_{ab} is the Minkowski metric in the tangent space, and ${}^\bullet A^i{}_{k\nu}$, in the TEGR case, is the so-called inertial spin connection (ISC) constrained by

$$ {}^\bullet R^a{}_{b\mu\nu} = \partial_\mu {}^\bullet A^a{}_{b\nu} - \partial_\nu {}^\bullet A^a{}_{b\mu} + {}^\bullet A^a{}_{c\mu} {}^\bullet A^c{}_{b\nu} - {}^\bullet A^a{}_{c\nu} {}^\bullet A^c{}_{b\mu} \equiv 0, $$

and can be written as ${}^\bullet A^a{}_{c\mu} = \Lambda^a{}_d(x) \partial_\mu \Lambda_c{}^d(x)$, where $\Lambda^a{}_b(x)$ is a matrix of a local Lorentz transformation.

We can then define the contortion tensor as a difference between ISC ${}^\bullet A^i{}_{k\nu}$ and the Levi-Civita spin connection (L-CSC) ${}^\circ A^i{}_{k\nu}$, i.e. ${}^\bullet K^i{}_{k\nu} = {}^\bullet A^i{}_{k\nu} - {}^\circ A^i{}_{k\nu}$, and write the TEGR Lagrangian as [1]

$$ {}^\bullet L = \frac{h}{2\kappa} \left({}^\bullet K^\rho{}_{\mu\nu} {}^\bullet K_\rho{}^{\nu\mu} - {}^\bullet K^\rho{}_{\mu\rho} {}^\bullet K^{\nu\mu}{}_\nu \right), \tag{1} $$

where $h = \det h^a{}_\rho$ and $\kappa = 8\pi$ (in natural units).

This Lagrangian is equivalent to the Hilbert-Einstein Lagrangian up to a divergence. The variation of the Lagrangian (1) with respect to the tetrad leads to field equations equivalent to Einstein field equations from where the theory gets its name as the Teleparallel Equivalent of General Relativity (TEGR). The variation of the Lagrangian with respect to the spin connection turn out to be trivial [17] and hence ISC is not determined by any field equations. Nevertheless, it plays an important role since the Lagrangian and action do depend on ISC.

The theory defined in this way is covariant under simultaneous local Lorentz transformations of both the tetrad and ISC, i.e. $h'^a{}_\mu = \Lambda^a{}_b h^b{}_\mu$ and ${}^\bullet A'^a{}_{b\mu} = \Lambda^a{}_d {}^\bullet A^d{}_{c\mu} \Lambda_b{}^c + \Lambda^a{}_d \partial_\mu \Lambda_b{}^d$. Due to the particular pure-gauge form of ISC, we can always transform ISC to zero by a transformation with an appropriate $\Lambda^a{}_b(x)$. The tetrad which corresponds to a zero connection is called the *proper* tetrad, following the terminology of [14].

2.2. Covariant conserved quantities in TEGR

Considering invariance of the Lagrangian (1) under diffeomorphisms induced by an arbitrary displacement vector and applying the Noether theorem, one derives the differential conservation law [12,13]:

$$\partial_\rho {}^\bullet J^\rho(\xi) = \overset{\circ}{\nabla}_\rho {}^\bullet J^\rho(\xi) = 0 \tag{2}$$

for the Noether's current ${}^\bullet J^\rho(\xi)$, which is a vector density. Therefore the partial derivative ∂_ρ can be replaced by the Levi-Civita covariant derivative $\overset{\circ}{\nabla}_\rho$. The Klein-Noether identities allow us to express the current through the Noether's superpotential ${}^\bullet J^{\rho\sigma}(\xi)$, which is antisymmetric tensor density,

$$ {}^\bullet J^\rho(\xi) = \partial_\sigma {}^\bullet J^{\rho\sigma}(\xi) = \overset{\circ}{\nabla}_\sigma {}^\bullet J^{\rho\sigma}(\xi). \tag{3}$$

In the case of the TEGR Lagrangain (1), the explicit expression for the superpotential is

$$ {}^\bullet J^{\rho\sigma}(\xi) = \frac{h}{\kappa} {}^\bullet S_\mu{}^{\rho\sigma} \xi^\mu, \tag{4}$$

$$ {}^\bullet S_\mu{}^{\rho\sigma} = {}^\bullet K^{\rho\sigma}{}_\mu + \delta_\mu^\sigma {}^\bullet K^{\pi\rho}{}_\pi - \delta_\mu^\rho {}^\bullet K^{\pi\sigma}{}_\pi . \tag{5}$$

One has to note that both (4) and (5) are spacetime tensors antisymmetric in upper indices and invariant with respect to local Lorentz rotations. The conservation laws (2) and (3) allow us to construct integral conserved quantities defined on hypersurfaces Σ defined by $x^0 = const$. By the standard method in spherical coordinates, $r = x^1$, we obtain the charge

$$P(\xi) = \int_\Sigma dx^3 {}^\bullet J^0(\xi) = \int_{\partial\Sigma} dx^2 {}^\bullet J^{01}(\xi) = \frac{1}{\kappa}\int_{\partial\Sigma} dx^2 h {}^\bullet S_\mu{}^{01}\xi^\mu, \tag{6}$$

where the boundary in a surface integral can be defined by finite $r = r_0$ or $r \to \infty$.

2.3. "Switching off" gravity

In TEGR, ${}^\bullet A^i{}_{k\nu}$ is not a dynamical quantity and hence is not determined by any field equations. However, it does play an important role in definition of conserved charges and hence it needs to be determined. One of the methods to determine is the method of "switching off" gravity that we follow here [14,18,12,13]. This method is based on the idea that in absence of gravity the contortion tensor should vanish ${}^\bullet K^i{}_{k\nu} = {}^\bullet A^i{}_{k\nu} - {}^\circ A^i{}_{k\nu} \to 0$ and hence L-CSC reduces to ISC ${}^\circ A^i{}_{k\nu} \to {}^\bullet A^i{}_{k\nu}$. To define ${}^\bullet A^i{}_{k\nu}$ we follow the steps:

1) For a solution under consideration, we choose a tetrad and calculate L-CSC:

   ${}^\circ A^a{}_{b\sigma} = -h_b{}^\rho \overset{\circ}{\nabla}_\sigma h^a{}_\rho$.

2) From ${}^\circ A^a{}_{b\sigma}$ we construct the related curvature tensor ${}^\circ R^a{}_{b\gamma\delta}$.

3) We find some parameter that allows us to continuously "switch off" gravity and achieve ${}^\circ R^a{}_{b\gamma\delta} = 0$.

4) We then identify ${}^\bullet A^i{}_{k\nu}$ with this "switched-off" L-CSC ${}^\circ A^i{}_{k\nu}$.

It turns out that this method of determining ${}^\bullet A^i{}_{k\nu}$ is not unique since different tetrads can result in the same ISC. In this paper, we analyze and study this problem on the example of the Schwarzschild solution.

3. Mass of the Schwarzschild black hole and free-falling observers

3.1. *Schwarzschild static gauge*

The metric of the Schwarzschild solution is

$$ds^2 = -fdt^2 + f^{-1}dr^2 + r^2\left(d\theta^2 + \sin^2\theta d\phi^2\right), \tag{7}$$

where $f = f(r) = 1 - 2M/r$. A convenient choice of a tetrad is the diagonal form:

$${}^{B}h^{a}{}_{\mu} = diag\left(f^{1/2}, f^{-1/2}, r, r\sin\theta\right). \tag{8}$$

The non-zero components of L-CSC ${}^{\circ}A^{a}{}_{b\sigma}$ are

$${}^{\circ}A^{0}{}_{10} = {}^{\circ}A^{1}{}_{00} = M/r^2,\ {}^{\circ}A^{1}{}_{22} = -{}^{\circ}A^{2}{}_{12} = -(1-2M/r)^{1/2},$$
$${}^{\circ}A^{1}{}_{33} = -{}^{\circ}A^{3}{}_{13} = -\sin\theta(1-2M/r)^{1/2},\ {}^{\circ}A^{2}{}_{33} = -{}^{\circ}A^{3}{}_{23} = -\cos\theta. \tag{9}$$

To obtain ISC ${}^{\bullet}A^{a}{}_{b\sigma}$ we "switch off" gravity by a simple requirement $M = 0$ that reduces (7) to a metric of Minkowski space, and we find

$${}^{\bullet}A^{1}{}_{22} = -{}^{\bullet}A^{2}{}_{12} = -1,\ \ {}^{\bullet}A^{1}{}_{33} = -{}^{\bullet}A^{3}{}_{13} = -\sin\theta,\ \ {}^{\bullet}A^{2}{}_{33} = -{}^{\bullet}A^{3}{}_{23} = -\cos\theta. \tag{10}$$

The difference between (10) and (9) gives us the contortion ${}^{\bullet}K^{i}{}_{k\nu} = {}^{\bullet}A^{i}{}_{k\nu} - {}^{\circ}A^{i}{}_{k\nu}$, from where, using (5), we can find the non-zero components of ${}^{\bullet}S_{\mu}{}^{\rho\sigma}$:

$${}^{\bullet}S_{0}{}^{01} = -{}^{\bullet}S_{0}{}^{10} = \frac{2}{r}\left[f - f^{1/2}\right],\ \ {}^{\bullet}S_{2}{}^{12} = -{}^{\bullet}S_{2}{}^{21} = {}^{\bullet}S_{3}{}^{13} = -{}^{\bullet}S_{3}{}^{31} = -\frac{1}{2r}\left[f - 2f^{1/2}\right]. \tag{11}$$

To obtain the total mass of the Schwarzschild black hole we choose a displacement vector $\xi^{\alpha} = (-1,\ 0,\ 0,\ 0)$, which is both the velocity of the observer at infinity and also a Killing vector. Then using the formula (6) we find

$$P(\xi) = \frac{1}{\kappa}\lim_{r\to\infty}\int_{\partial\Sigma} dx^2 h\, {}^{\bullet}S_{0}{}^{01}\xi^{0} = M, \tag{12}$$

where $\kappa = 8\pi$, $h = r^2\sin\theta$.

It is possible to find a local Lorentz transformation $\left(\Lambda_{Sch}\right)^{a}{}_{b}(x)$ that transforms ISC (10) to zero ${}^{\bullet}A^{a}{}_{c\mu} = 0$. Applying the same local Lorentz transformation to the tetrad (8), we find a new tetrad ${}^{A}h^{a}{}_{\mu} = \left(\Lambda_{Sch}\right)^{a}{}_{b}\, {}^{B}h^{b}{}_{\mu}$ that we call the proper tetrad [14]. By construction, the quantity (12) is both coordinate and locally Lorentz invariant. Thus, the result (12) is the same for the proper tetrad, as well as for any tetrad and ISC obtained from (8) and (10) by arbitrary coordinate transformations or simultaneous local Lorentz rotations. We call this set of pairs of tetrads and related ISCs as the *Schwarzschild static gauge*.

3.2. *Lemaitre gauge*

In order to describe a radially free-falling observer into the Schwarzschild black hole, it is natural to start with the Lemaitre coordinates (τ,ρ,θ,ϕ) that can be defined as [7]:

$$d\rho = dt + \frac{dr}{f\sqrt{1-f}};\ \ d\tau = dt + \frac{dr}{f}\sqrt{1-f}. \tag{13}$$

The metric (7) transforms to

$$ds^2 = -d\tau^2 + \left(1 - f(r)\right)d\rho^2 + r^2\left(d\theta^2 + \sin^2\theta d\phi^2\right), \tag{14}$$

where $r = r(\tau,\rho) = \left[3(\rho-\tau)/2\right]^{2/3}(2M)^{1/3}$. It is then possible to choose a diagonal tetrad for the metric (14) in these new coordinates (13) and follow the same method of determining ISC and calculating conserved charges as we did in the static case. This will result in all quantities being in the Lemaitre coordinates and we then have to transform them to the static coordinates since comparison with other gauges is more convenient when we work in the same coordinates. See for details our paper [16].

Here we follow a simpler method and analyze the free-falling case directly in the static coordinates. In order to obtain a tetrad representing a free-falling observer we consider a boosted tetrad ${}^{C}h^{a}{}_{\mu} = (\Lambda_{boost})^{a}{}_{b}\,{}^{B}h^{b}{}_{\mu}$, where

$$(\Lambda_{boost})^{a}{}_{b} = \begin{bmatrix} \gamma & \gamma\beta & 0 & 0 \\ \gamma\beta & \gamma & 0 & 0 \\ 0 & 0 & 1 & 0 \\ 0 & 0 & 0 & 1 \end{bmatrix}, \tag{15}$$

$\beta = \sqrt{2M/r}$ is a rapidity and $\gamma = 1/\sqrt{1-\beta^2}$. We can then follow the method of switching-off gravity to determine the spin connection. However, note that since the boost depends on M, when we "switch off" gravity $M = 0$, we obtain the same ISC as in the static case (10), i.e.

$$^{\bullet}A^{1}{}_{22} = -{}^{\bullet}A^{2}{}_{12} = -1,\quad {}^{\bullet}A^{1}{}_{33} = -{}^{\bullet}A^{3}{}_{13} = -\sin\theta,\quad {}^{\bullet}A^{2}{}_{33} = -{}^{\bullet}A^{3}{}_{23} = -\cos\theta. \tag{16}$$

This means that we can use the same $\left(\Lambda_{Sch}\right)^{a}{}_{b}(x)$ to obtain the proper tetrad in the free-falling case that we call ${}^{D}h^{a}{}_{\mu} = \left(\Lambda_{Sch}\right)^{a}{}_{b}\,{}^{C}h^{b}{}_{\mu}$.

We can now proceed to calculate the components of the superpotential (5) and we find that the non-zero components are:

$$^{\bullet}S_{0}{}^{01} = -{}^{\bullet}S_{0}{}^{10} = -\frac{4M}{r^2},\quad {}^{\bullet}S_{1}{}^{01} = -{}^{\bullet}S_{1}{}^{10} = -\frac{2}{rf}\sqrt{\frac{2M}{r}},$$

$$^{\bullet}S_{2}{}^{02} = {}^{\bullet}S_{3}{}^{03} = -{}^{\bullet}S_{2}{}^{20} = -{}^{\bullet}S_{3}{}^{30} = -\frac{1}{2rf}\sqrt{\frac{2M}{r}},\quad {}^{\bullet}S_{2}{}^{12} = {}^{\bullet}S_{3}{}^{13} = -{}^{\bullet}S_{2}{}^{21} = -{}^{\bullet}S_{3}{}^{31} = \frac{M}{r^2}. \tag{17}$$

Both the proper tetrad ${}^{D}h^{a}{}_{\mu}$ with zero ISC and the free-falling tetrad ${}^{C}h^{b}{}_{\mu}$ with ISC given by (16) will lead to the same superpotential. We call these combinations of tetrads and ISCs that lead equivalent results for the superpotential (17) as the *Lemaitre gauge*.

In order to calculate the Noether current and conserved charges we consider two distinct choices for the vector field ξ. The first choice is the vector field:

$$\xi^{\alpha} = \left(-f^{-1}, \sqrt{1-f}, 0, 0\right) \tag{18}$$

that represents the velocity of a free-falling observer. Using this vector field we find that the Noether current vanishes

$$^{\bullet}J^{\alpha}\left(\xi\right) = \left(0,0,0,0\right). \tag{19}$$

Naturally, it leads to a vanishing Noether charge (6). This corresponds to the equivalence principle since a free-falling observer is expected to measure a zero gravitational energy. We can also consider the second choice for the vector field ξ and choose it as a Killing vector $\xi^{\alpha} = \left(-1,\ 0,\ 0,\ 0\right)$. In this case we find that the Noether current is non-vanishing and results in the Noether conserved charge

$$P(\xi) = 2M. \tag{20}$$

It is rather difficult to understand this result that seems to be unphysical, what motivates us to support the first choice for the vector field ξ as the velocity of the observer. However, in the next section we demonstrate that it is possible to generalize this situation in a rather natural way by including an initial velocity and obtain a physically meaningful result using the Killing vector.

4. An arbitrary free-falling observer

A free-falling observer with 4-velocity (18) corresponds to the case when its velocity at infinity is zero. It is natural to generalize this by including an initial velocity for the observer. Solving the geodesic equation in a general form for a radially free-falling observer into the Schwarzschild black hole in the static coordinates, one obtains that the observer's velocity is given by [16]:

$$\tilde{\xi}^{a} = \left(-ef^{-1}, \sqrt{e^2 - f}, 0, 0\right) \tag{21}$$

that coincides with (18) when $e = 1$. Here we consider only the case $e > 1$, for a discussion of the case $e < 1$ see [16].

We can then introduce freely falling observer's proper coordinates adopted to (21) and work in these coordinates, construct tetrads and calculate ISC, and then transform the results into static coordinates. See [16] for details. However, it is possible to work in the static coordinates only and find the tetrad representing the arbitrarily free-falling observer with the velocity (21) as

$$^{E}h^{a}{}_{\mu} = (\Lambda_{boost\_e})^{a}{}_{b}\, {}^{B}h^{b}{}_{\mu}, \tag{22}$$

where

$$(\Lambda_{boost\_e})^{a}{}_{b} = \begin{bmatrix} e\gamma & \gamma\sqrt{e^2-1+\beta^2} & 0 & 0 \\ \gamma\sqrt{e^2-1+\beta^2} & e\gamma & 0 & 0 \\ 0 & 0 & 1 & 0 \\ 0 & 0 & 0 & 1 \end{bmatrix}. \tag{23}$$

We can then use the method of "switching off" gravity and obtain ISC corresponding to (22):

$$^{\bullet}A^{0}{}_{22} = {}^{\bullet}A^{2}{}_{02} = -\sqrt{e^2-1},\quad {}^{\bullet}A^{0}{}_{33} = {}^{\bullet}A^{0}{}_{33} = -\sin\theta\sqrt{e^2-1},$$

$$^{\bullet}A^{1}{}_{22} = -{}^{\bullet}A^{2}{}_{12} = -e,\quad {}^{\bullet}A^{1}{}_{33} = -{}^{\bullet}A^{3}{}_{13} = -e\sin\theta,\quad {}^{\bullet}A^{2}{}_{33} = -{}^{\bullet}A^{3}{}_{23} = -\cos\theta. \tag{24}$$

It is possible to find a local Lorentz transformation that transform this ISC to zero. Applying this transformation to the arbitrarily free-falling tetrad $^{E}h^{a}{}_{\mu}$ we obtain the proper tetrad $^{F}h^{a}{}_{\mu}$. Both $^{E}h^{a}{}_{\mu}$ with ISC (24) and the proper tetrad $^{F}h^{a}{}_{\mu}$ represent the same physical situation and hence we can call it the *generalized Lemaitre gauge* or *e*-gauge, for more details see [16].

Now we find the components $^{\bullet}S_{\mu}{}^{\rho\sigma}$ in the *e*-gauge in the static coordinates:

$$^{\bullet}S_{0}{}^{01} = -\frac{2}{r}(A_e - 1 + 2M/r),\quad {}^{\bullet}S_{1}{}^{01} = -\frac{2}{r}\frac{B_e}{1-2M/r},\quad {}^{\bullet}S_{2}{}^{12} = {}^{\bullet}S_{3}{}^{13} = \frac{1}{r}(A_e - 1 + M/r),$$

$$^{\bullet}S_{2}{}^{02} = {}^{\bullet}S_{3}{}^{03} = -\frac{e}{r}\frac{A_e - 1 + M/r}{(1-2M/r)\sqrt{e^2-1+2M/r}}, \tag{25}$$

where one has to add components with opposite signs and antisymmetric in the last indices, and we have introduced a short-handed notation:

$$A_e \equiv e^2 - \sqrt{e^2-1}\sqrt{e^2-1+2M/r}, \quad A_{e=1} \equiv 1,$$

$$B_e \equiv e\left(\sqrt{e^2-1+2M/r} - \sqrt{e^2-1}\right), \quad B_{e=1} \equiv \sqrt{2M/r}. \tag{26}$$

The superpotential (25) reduces to (17) in the case $e = 1$.

We now proceed to calculate the conserved currents and charges using the Noether machinery presented here. We can again consider two choices for the vector field ξ. As the first choice, we consider the velocity of a free-falling observer (21), and using (25), we find the vanishing Noether current

$$^{\bullet}J^{\rho}\left(\tilde{\tilde{\xi}}\right) = (0, \quad 0, \quad 0, \quad 0) \tag{27}$$

and, naturally, a vanishing Noether charge. We can understand this to be a consequence of the equivalence principle. The second choice is the timelike Killing vector $\xi^{\alpha} = (-1, \quad 0, \quad 0, \quad 0)$ that leads to $P(\xi) = M$. This is in contrast with the Lemaitre gauge where we obtained the unphysical result of $2M$ in (20).

5. Discussion

In this paper, we have studied the Noether conserved currents and charges in the covariant formulation of TEGR in the case of the Schwarzschild black hole. We have considered three distinct physical situations by considering the static, Lemaitre, and generalized Lemaitre observers. In each situation, we have considered two choices of the vector ξ in the definition of the Noether conserved quantities: a) as a velocity of the observer, or b) as a time-like Killing vector. In the case of the static observer, both choices lead to the same result and hence we cannot effectively distinguish between them. The case of the Lemaitre free-falling observer revealed the difference and motivated us to suggest that ξ should be chosen as a velocity of the observer to satisfy the equivalence principle, while the other choice, in the charge calculations, leads to an unphysical result **2 *M*.** However, we have found that when we generalize our observer by including some initial velocity, using the observer's proper vector we get the equivalence principle and using the time-like Killing vector we find the conserved charge to be equivalent to the mass of the black hole. Curiously, this happens for an arbitrarily small initial velocity.

The second problem studied here was the problem of calculating ISC, which is left undetermined by the field equations but plays an important role in a definition of conserved charges. We have considered here the method of "switching-off" gravity that is usually used in the TEGR literature. We have demonstrated a certain ambiguity and non-uniqueness in this method since both the static and Lemaitre tetrads correspond to the same ISC, despite representing physically distinct situations. By including the initial velocity this ambiguity seems to be removed for black hole mass calculations, but, in depth of structure, it remains creating the class of gauges depending on the initial velocity, and we distinct this class of ISCs for the generalized Lemaitre observers.

This issue of determining ISC, or, alternatively, the special class of preferred tetrads, is even more important in the case of modified gravity theories, such as the $f(T)$ gravity model, where it affects the field equations of the theory and hence becomes the matter of dynamics. Using our results here, it is possible to show that while the situation in TEGR and $f(T)$ gravity seem to share many similarities in the static case, the free-falling case reveals important differences and demonstrates that the so-called good tetrads in $f(T)$ gravity cannot be always constructed in analogy with the TEGR case [16].

References

[1] Aldrovandi R, Pereira J G 2013 *Teleparallel Gravity: An Introduction* (Dordrecht, Heidelberg, New York, London: Springer)
[2] Cai Y-F, Capozziello S, De Laurentis M and Saridakis E N 2016 *Rept. Prog. Phys.* **79** 106901
[3] Bahamonde S, Bohmer C G and Krššák M 2017 *Phys. Lett.* B **775** 37
[4] Hohmann M, Järv L, Krššák M and Pfeifer C 2018 *Phys. Rev.* D **97** 104042
[5] Maluf J W 2013 *Annalen Phys.* **525** 339
[6] Krššák M, van den Hoogen R, Pereira J, Bohmer C and Coley A 2019 *Class. Quant. Grav.* **36** 183001
[7] Landau L D, Lifshitz E M 1975 *The Classical Theory of Fields* (Oxford: Pergamon Press)
[8] Moller C 1961 *Annals of Physics* **12** 118
[9] Obukhov Y N, Rubilar G F 2006 *Phys. Rev.* D **73** 124017
[10] Obukhov Y N, Rubilar G F 2006 *Phys. Rev.* D **74** 064002
[11] Obukhov Y N, Rubilar G F and Pereira J G 2006 *Phys. Rev.* D **74** 104007
[12] Emtsova E D, Petrov A N and Toporensky A V 2020 *Class. Quantum Grav.* **37** 095006
[13] Emtsova E D, Petrov A N and Toporensky A V 2020 *J. Phys. Conf. Ser.* **1557** 012017
[14] Lucas T G, Obukhov Y N and Pereira J G 2009 *Phys. Rev.* D **80** 064043
[15] Emtsova E D, Krššák M, Petrov A N and Toporensky A V 2021 On the Schwarzschild solution in TEGR *Abstracts of XXII International Scientific Conference Moscow PIRT 5-9 July, 2021* (Moscow: BSTU press) pp 33-34
[16] Emtsova E D, Krššák M, Petrov A N and Toporensky A V 2021 *Eur. Phys. J. C* **81** 743
[17] Krššák M 2017 *Eur. Phys. J. C* **77** 44
[18] Krššák M, Pereira J G 2015 *Eur. Phys. J. C* **75** 519

Acknowledgments

EE, AP and AT have been supported by the Interdisciplinary Scientific and Educational School of Moscow University “Fundamental and Applied Space Research”; MK is supported by the CUniverse research promotion initiative (CUAASC) of the Chulalongkorn University.

Vacuum Polarization and Particle Creation for Two-Horizon Metrics

M L Fil'chenkov and Yu P Laptev

RUDN University, Moscow, Russia

E-mail: fmichael@mail.ru

Abstract. Kerr–Newman and Kottler's metrics with two horizons are considered. Evaporation of Kerr – Newman's horizons in Hawking's effect and Penrose's process as well as de Sitter's horizon decay and Schwarzschild's horizon evaporation for Kottler's metric have been analyzed in terms of an effective temperature, using lifetimes on the horizons. The results are applied to black hole physics and cosmology.

1. Introduction

The metric horizons result in distortion of the spectrum of vacuum zero oscillations followed by vacuum polarization and particle creation [1]. The presence of horizons is related to a loss of information, rise of entropy and appearance of an effective temperature of vacuum and particles being born. Metrics with two horizons will be considered, viz Kerr – Newman's metric [2] describing a rotating charged black hole and Kottler's cosmological metric [3] for inflation models. Horizon lifetimes determining their evolution have been evaluated.

2. Metric Horizons, Their Lifetimes. An Effective Temperature on the Horizons

Kerr – Newman's metric horizons are roots of the equation

$$\Delta^2 = r^2 - \frac{2GMr}{c^2} + \frac{a^2}{c^2} + \frac{GQ^2}{c^2} = 0\,, \tag{1}$$

where r is the radius (in the spherical coordinate system), M the mass, a the specific angular momentum and Q the black hole charge. Since there do not exist massless rotating and / or charged black holes, then the relation

$$Q^2 + \frac{a^2c^2}{G} = \alpha^2 M^2\,, \tag{2}$$

where $\alpha = const$, is valid. Kerr – Newman's metric horizons are written in the form:

$$r_{\pm} = \frac{GM}{c^2}(1 \pm \sqrt{1-A}),\;\; A = \frac{a^2c^2 + GQ^2}{G^2M^2} = \frac{a^2}{G} \leq 1\,, \tag{3}$$

where r_+ is the event horizon, r_- Cauchy's horizon. At $a=0$ Kerr –Newman's axially symmetric metric reduces to Reissner – Nordström's spherically symmetric metric [4, 5] for which $\frac{\Delta^2}{r^2}=g_{00}$, where g_{00} the temporal component of the metric. For Kottler's metric

$$g_{00}=1-\frac{r_g}{r}-\frac{r^2}{r_0^2}, \tag{4}$$

where r_g is Schwarzschild's horizon, r_0 de Sitter's, in the metrics with horizons enters a surface gravity [6]

$$\kappa=\frac{c^2}{2}\left|g'_{00}\left(r_h\right)\right|, \tag{5}$$

equal to the acceleration on the horizon r_h.

The effective temperature on the horizon is given by the formula

$$kT_h=\frac{\hbar\kappa}{2\pi c}. \tag{6}$$

The horizon lifetime

$$\tau_h=\int\frac{dE_h}{P_h}, \tag{7}$$

where E_h is the quantity depending on the form of the metric.

$$P_h=4\pi r_h^2\sigma T_h^4 \tag{8}$$

is the power of horizon evaporation or decay,

$$\sigma=\frac{\pi^2k^4}{60\hbar^3c^2}, \tag{9}$$

Stefan – Bolzmann's constant.

3. Evaporation of Kerr – Newman's Horizons in Hawking's Effect and Penrose's Process

Assume that the charge and angular momentum contribute insignificantly to the horizon, i.e. $A=\frac{\alpha^2}{G}\ll 1$. Then Kerr–Newman's metric has two horizons:

$$r_+=\frac{2GM}{c^2}=r_g, \tag{10}$$

$$r_-=\frac{\alpha^2M}{2c^2}. \tag{11}$$

The ratio of these horizon radii

$$\frac{r_-}{r_+}=\frac{\alpha^2}{4G}\ll 1. \tag{12}$$

The effective temperatures of the particles born on the horizons:

$$kT_+ = \frac{\hbar c^3}{8\pi GM}, \tag{13}$$

$$kT_- = \frac{2\hbar c^3 G}{\pi \alpha^4 M}. \tag{14}$$

The ratio of the temperatures on the horizons

$$\frac{T_-}{T_+} = \frac{16G^2}{\alpha^4} \gg 1. \tag{15}$$

The energies being lost on the horizons

$$E_+ = Mc^2,\ E_- = \frac{\alpha^2 M^2}{r_-} = 2Mc^2. \tag{16}$$

From formulae (12), (15) and (16) we obtain:

$$\frac{\tau_-}{\tau_+} = \frac{1}{2^{11}}\left(\frac{\alpha^2}{G}\right)^6 \ll 1. \tag{17}$$

This means that Cauchy's horizon evaporates considerably faster than the event horizon for a rotating and / or charged black hole, provided that Cauchy's horizon radius is much less than the event horizon one. While evaporating Schwarzschild's horizon in Hawking's effect [7], particles are born. The particles, with a charge opposite to that of the black hole, fall under the horizon diminishing the charge of the latter, which is interpreted as Cauchy's horizon evaporation related as well to a loss of the angular momentum of the black hole due to Penrose's process [8]. Cauchy's horizon evaporation, converts the black hole into Schwarzschild's, which agrees with the results obtained in a manner quite different from ours [9, 10].

4. De Sitter's Horizon Decay and Evaporation of Schwarzschild's for Kottler's Metric

For $r_0 \gg r_g$ the equation $g_{00} = 0$ has two roots $r = r_g$ and $r = r_0$, which are Schwarzschild and de Sitter's horizons respectively.

The effective temperatures on the horizons:
Schwarzschild's with the effective temperature of particles being created

$$kT_H = \frac{\hbar c}{4\pi r_g}, \tag{18}$$

and de Sitter's with the effective temperature of vacuum being polarized

$$kT_{dS} = \frac{\hbar c}{2\pi r_0}. \tag{19}$$

The ratio of the temperature of particles born on Schwarzschild's horizon to de Sitter's vacuum temperature

$$\frac{T_H}{T_{dS}} = \frac{r_0}{2r_g} \gg 1. \tag{20}$$

Using the quantities

$$E_H = \frac{c^4 r_g}{2G}, \quad E_{dS} = \frac{4}{3}\varepsilon_0 r_0^3 = \frac{c^4 r_0}{2G}, \tag{21}$$

where

$$\varepsilon_0 = \frac{3c^4}{8\pi G r_0^2} \tag{22}$$

is de Sitter's vacuum energy density, from formulae (20) – (22) we obtain the ratio of de Sitter's horizon lifetime to that of Schwarzschild's:

$$\frac{\tau_{dS}}{\tau_H} = \frac{1}{2^4}\left(\frac{r_0}{r_g}\right)^3 \gg 1. \tag{23}$$

This means that Schwarzschild's horizons of supermassive black holes at the centres of galaxies evaporate faster than de Sitter's horizon of dark energy, which is a well-known result [11]. However, in case

$$\frac{\tau_{dS}}{\tau_H} < 1, \tag{24}$$

if

$$\frac{r_0}{r_g} < 2^{4/3}, \tag{25}$$

i. e. de Sitter's horizon in the early Universe decays before the evaporation of primordial black holes.

5. Conclusion

The evaporation of Cauchy's horizon before the event one of Kerr – Newman's metric proceeds due to a loss of mass, charge and angular momentum in Hawking's effect and Penrose's process. De Sitter's horizon decay and Schwarzschild's horizon evaporation for Kottler's metric are due to particle creation after the first inflation in the early Universe (Big Bang) and the evaporation of massive black holes in a more stable de Sitter vacuum before the completion of the second inflation (dark energy).

References

[1] Grib A A, Mamaev S G and Mostepanenko V M 1994 *Vacuum Quantum Effects in Strong Fields* (St. Petersburg: Friedmann Lab. Publ.)
[2] Newman E N *et al* 1965 J. *Math. Phys.* **6** 918
[3] Kottler F 1918 *Ann. Phys.* **56** 401
[4] Reissner H 1916 *Ann. Phys.* **50** 15
[5] Nordström G 1918 *Proc. Kon. Ned. Akad. Wt.* **20** 1238
[6] Gibbons G W, Hawking S W 1977 *Phys. Rev.* D **15** 2738
[7] Hawking S W 1975 *Comm. Math. Phys.* **43** 199
[8] Penrose R 1969 *Nuovo Cim.* **1** 252
[9] Gal'tsov D V 1986 *Particles and Fields in the Vicinity of Black Holes* (Moscow: Moscow University Press)
[10] Novikov I D, Frolov V P 1986 *Black Hole Physics* (Moscow: Nauka, Fizmatlit)
[11] Grib A A, Pavlov Yu V 2016 *Grav. Cosmol.* **22** 107

Acknowlegements
This publication was prepared with the support of the "RUDN University Program 5-100".

The novel pushing gravity model and volcanic activity. Is alignment of planets with compact stars a possible cause of natural phenomena?

F Greco[1] and I V Krasnyy[2]

[1] Istituto Nazionale di Geofisica e Vulcanologia, Sezione di Catania, Osservatorio Etneo, Catania, Italy
[2] The State Navigation-Hydrographic Institute (GNINGI), Saint-Petersburg, Russia

E-mail: 9968348@gmail.com; filippo.greco@ingv.it

Abstract. We developed the model, and carried out its discussion at the PIRT-2021 conference, within the framework of the research topic "*External Forcing on Volcanoes and Volcanic Processes: Observations, Analysis and Implications*" announced by the journal "Frontiers in Earth Science" in October 2020. Besides other, external processes considered in this Research Topic included astronomical. In this study, in the category "*Hypothesis and Theory*", we investigate how changes in the position of large bodies of the Solar system can cause natural phenomena, associated with the movement of free masses, such as volcanism, earthquakes and landslides in the lithosphere, as well as various catastrophic events in the atmosphere and hydrosphere. The analysis has shown that the discovered phenomena of celestial bodies' alignments accompanying manifestations of natural phenomena require going beyond the standard cosmological model and clarify the fundamental mechanism of gravity. We propose the novel *Bidirectional Pushing Gravitation* model (BPG), which, in addition to application in Earth Sciences, may occur useful in Astrophysics, Cosmology and Gravitation research.

1. Research goals and methods

The influence at different temporal and spatial scales of external processes on various geological phenomena such as seismicity and volcanism is widely discussed in the scientific literature. Despite it is still debated if the action of external forces (e.g. tidal stress changes) as low as a few kPa may trigger and regulate some natural phenomena [7], the extraordinary cyclicity of some phenomena, suggest an external forcing, such as Earth tides or planetary alignment, may induce changes in the dynamical state of the volcano constituting the ultimate trigger that may lead an active volcano to erupt.

Our research had two major goals - ***a) to verify the existence of the mentioned celestial bodies' alignment phenomena*** and ***b) to suggest a plausible model*** for its explanation, specifically:

- Investigation of Etna volcanic manifestations in order to identify the possible influence of astronomical factors, caused by changes in the position of large celestial bodies of the Solar system. Analysing alignments of the large celestial bodies of Solar system – the seven planets, dwarf planets Pluto and Ceres, the Sun and the Moon relative to Earth observer, coinciding with a wide range of various high-energy natural phenomena in lithosphere, hydrosphere and atmosphere. Discovering the features of alignments and possible force effects on natural phenomena, i.e., the induction of transient gravitational disturbances causing movement of free masses, variation of non-tidal plumb line deviations caused by changes in the position of celestial bodies. Confirmation, refute, or detailing of the popular hypothesis about the phenomenon existence in the extent necessary to build a computational model of the induced gravitational perturbations.

- Proposal of a reasonable mechanism for transmitting gravitational perturbations, caused by changes in the position of large celestial bodies of the Solar system, into the geosphere and a model, which allows, taking into account the identified features, precomputing such perturbations for use in various applications.

To select the time series of registered natural phenomena that are subject to subsequent detailed analysis using ephemeris models, we used various databases volcanic [25] and seismic [26] events, as well as individual mentions of other natural phenomena that have a time reference to the exact or approximate date of their manifestation, and the planetarium program. We employed specialized software for calculating ephemerides and angular separations between celestial bodies Alcyone [28], online services for verification, as well as Excel spreadsheets with the connection of the Swiss Ephemeris library [27] via VBA, as the main tool. The position of celestial bodies at small angular separations and their alignment along several lines in some cases were displayed in the desktop application of the 3D planetarium Solar System Scope [29]. At the time of each event or selected time step, we calculated 220 angular separations for 11 celestial bodies of the Solar System and 15 fixed stellar objects, according to the number of combinations. The spreadsheet allowed us to set the required threshold value (1-5°) of angular separation in order to filter out cases of alignment of celestial bodies along one line.

All the analyzed natural phenomena are associated with the movement of free masses in the geosphere. Data on Etna volcanism were subject to detailed analysis in order to identify the external influence of extraterrestrial factors. The subject for analysis in the lithosphere included large earthquakes, manifestations of volcanism, landslides, mudslides, as well as individual man-made disasters provoked by gravity, "moving stones", and anomalous data from various geophysical observations. In the hydrosphere - floods, manifestations of extreme (rogue) waves in the ocean, tropical storms. In the atmosphere - hurricanes, storms, supercells, tornadoes, wind shear in clear weather.

We did not plan to conduct an in-depth statistical analysis of the alignment effect of celestial bodies that accompanies all the mentioned variety of natural phenomena on Earth, especially taking into account (as will be shown below) that this effect is a) just a necessary condition for such phenomena; b) in general, there is a time shift between alignment and the manifestation of natural phenomena with anticipation or with time lag. Interested researchers can independently use the tools mentioned in this report. This study is a preliminary stage, the purpose of which is to propose a computational model and a method of its parameterization for the pre-calculation of gravitational perturbations from extraterrestrial sources.

2. State of the art

According to generally accepted theories, the influence of planets on terrestrial phenomena is negligible, and is proportional to the gradient of the gravitational field strength, i.e. inversely proportional to the cube of the distance. With regard to the possible mechanism of the influence of extraterrestrial bodies on terrestrial phenomena, only tides are considered, the intensity of which both in the hydrosphere and in the lithosphere significantly decreases with the distance to the disturbing celestial bodies. In the theory of Earth tides, only the lunar and solar harmonic constants (M2, S2, K1, O1) are considered. The gravitational influence of other celestial bodies on tides, as well as on any other terrestrial processes, is considered insignificant.

Some short-lived but significant correlations have been reported between semi-diurnal tides and the frequency of aftershocks in some volcanic regions, such as Mammoth Lakes. The Moon, the Sun and other planets have impact on Earth in the form of perturbations (small changes) of the gravitational field. The relative magnitude of the influence is proportional to the mass of the object and inversely proportional to the third power of its distance from the Earth. The stresses created on Earth by an extraterrestrial mass are proportional to the gradient of the gravitational field $\frac{dg(r)}{dr}$, and NOT the strength of the gravitational field $g(r)$.

$$g(r)=\frac{GMm}{r^2}\text{, thus: }\frac{dg(r)}{dr}=-\frac{2g(r)}{r}=-\frac{2GMm}{r^3} \quad (1)$$

(From the University of California, Berkeley, with the participation of Gary Fuis).

Maurice Allais (1911-2010, French physicist and economist), discovered and investigated the effect of abnormal behavior of pendulums during solar eclipses, named after him [16]. The study of the anomalous behavior of pendulums during solar eclipses was continued by the followers of M. Allais, who also reported the instability of the manifestation of the effect during various eclipses.

Gravitational anomalies are reported during lunar eclipses [18]. Alexander Pugach (1940-2020, GAO of the Academy of Sciences and the National Academy of Sciences of Ukraine), registered the anomalous behavior of torsion scales of his own design ("torsind"), as well as pendulums during syzygies (i.e., during solar and lunar eclipses), during transit of Venus across the solar disk, and when Venus is covered by the Moon.

"Observations were made of the torsion balance and the behavior of the "torsind" at the moments of solar and lunar eclipses, the passage of Venus behind the solar disk, and the eclipse of Venus by the Moon. It is shown that in most cases, the reaction of the devices to these phenomena was either ahead of or behind the actually observed phenomenon" [17].

With the help of a high-precision Lacoste-Romberg gravimeter, continuous and accurate measurements were made during the total solar eclipse of March 9, 1997 in the Mohe region in northeastern China. The presence of two "valleys of gravitational anomalies" with an almost symmetrical decrease in gravity by about 6 ~ 7 microgals at the first and last contact is noted [20].

There are publications about the violation of the synchronization of atomic clocks during solar eclipses, as well as observations that did not confirm this effect [21].

Also, in the context of the proposed mechanism of action of gravitation, it should be mentioned that there are a number of studies and publications claiming that the plasma is pushed out by gravity [19].

A number of researchers report the registration of non-tidal variations of plumb line deviations on dates coinciding with the alignment of celestial bodies [22], [23].

3. The analysis results and conclusions for constructing a model of gravitational perturbations

3.1. Alignment of celestial bodies during natural phenomena occurs near astronomical conjunctions, or oppositions relative to other planets, or compact stars

Almost all the analyzed natural phenomena are associated with the alignment of celestial bodies relative to the observer (a geophysical object, moving masses in the geosphere) at small angular distances (less than 3-5°, see the histogram, Figure 3) from one or several lines connecting two or more celestial bodies with each other, or with some "special points " that repeat for different events on the celestial sphere. Alignment occurs in the forward direction, near astronomical conjunctions (as during solar eclipses), and also (and even more often) in the opposite direction, near astronomical oppositions (as during lunar eclipses).

3.2. Compact stars induce gravitational perturbations when aligned with celestial bodies

The subsequent analysis showed that in the mentioned "special points" on the celestial sphere there are located compact stars with a high density of matter: red dwarfs, white dwarfs, clusters including such stars, neutron stars, as well as other galaxies. Since most of the celestial bodies of the solar system move near the ecliptic plane, the identified "special points" are also grouped near the ecliptic plane. For the primary analysis, we have chosen 15 stellar objects close to the ecliptic plane, near which the solar system bodies are located at the moments of the natural events' manifestation (Table 1).

Table 1. Stellar objects close to the ecliptic plane.

Name	Type	Const.	RA	DEC	Mass, $M_{\odot}$	Radius	Distance
IC 358	lenticular galaxy	Tau	$04^h03^m42.9^s$	+19°53′42″			92,900,222 pc
M1	Crab Nebula, neutron star	Tau	$05^h34^m31.97^s$	+22°00′52.1″	1.4 - 2.0	10 km	2,000 pc
M35	open cluster	Gem	$06^h08^m54.0^s$	+24°20′00″	1,600	11 ly	1,186 pc
M44	open cluster	Cnc	08^h 40.4^m	19°59′	500–600	7.5 ly	160-187 pc
M80	globular cluster	Sco	$16^h17^m02.41^s$	−22°58′33.9″	5.02×10^5	48 ly	10.0 kpc
NGC 4697	elliptical galaxy	Vir	$12^h48^m35.9^s$	−05°48′03″	BH 1.3×10^8		38-50 Mly
QS-Vir	white dwarf and red dwarf binary	Vir	$13^h49^m52.0032^s$	-13°13′37.002″	0.78	0.011 $R_{\odot}$	50.1 pc
AR-Sco	white dwarf and red dwarf binary	Sco	$16^h21^m47.28^s$	−22°53′10.3″	WD 0.81-1.29 RD 0.28-0.45		116 pc
Wolf 28	Van Maanen star, white dwarf	Psc	$00^h49^m09.89841^s$	+05°23′18.9931″	0.67	0.0138 $R^{\odot}$	4.3152 pc
Nu^2Sgr or NGC6717?	binary with WD globular cluster	Sgr	$18^h55^m07.14098^s$ 18^h 55^m 06.04^s	−22°40′16.8185″ −22° 42′ 05.3″			84 pc 7.1 kpc
Ross 154	red dwarf	Sgr	$18^h49^m49.36216^s$	−23°50′10.4291″	0.17	0.24 $R^{\odot}$	2.94 pc
Wolf 359	red dwarf	Leo	$10^h56^m28.99^s$	+07°00′52.0″	0.09	0.16 $R^{\odot}$	2.409 pc
Ross 128	red dwarf	Vir	$11^h47^m44.3969^s$	+00°48′16.4049″	0.168	0.1967 $R^{\odot}$	3.3749 pc
Trappist-1	red dwarf	Aqr	$23^h06^m29.283^s$	−05°02′28.59″	0.0898	0.1192 $R^{\odot}$	12.43 pc
K2-72	red dwarf	Aqr	$22^h18^m29.27^s$	−09°36′44.6″	0.271365	0.330989 $R^{\odot}$	66.56 pc

3.3. The transfer of gravitational perturbations to geosphere occurs through the jerk (derivative of acceleration) with a lag, or ahead of the bodies' alignments

Gravitational perturbations from extra-terrestrial masses on Earth, in addition to the gradient of the gravitational field, are also caused by a change in the strength of the external gravitational field over time, i.e., the derivative of acceleration, kinematic jerk. Figure 1 shows, for example, graphs of accelerations and their derivatives relative to the centre of mass of the Earth, plotted from ephemeris data. The graph clearly shows the obvious fact of the shift of the jerk phase (derivative) relative to the acceleration (primitive). The alignment of the celestial bodies' corresponds to the extremes of perturbing accelerations, however, gravitational perturbations are transmitted to the geosphere through the acceleration derivative - kinematic jerk, and i.e. they have a phase shift relative to the acceleration graph. For this reason, there is a time shift between the moment of alignment of celestial bodies and the manifestation of natural events with the advance or lag of natural phenomena. (Confirmed by the experiments of A. F. Pugach during solar and lunar eclipses, and the covering of Venus by the Moon). This time shift depends on the natural phenomenon size and can comprise intervals from minutes to several weeks.

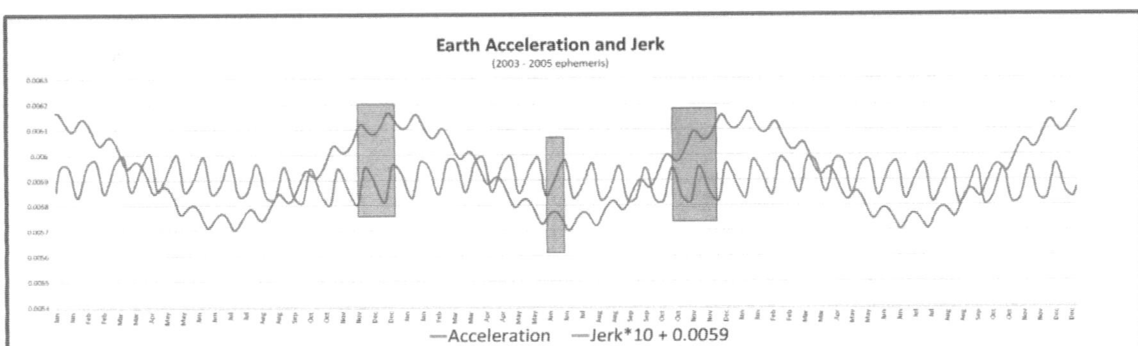

Figure 1. Phase shift between acceleration and kinematic jerk.

3.4. The more powerful natural phenomena correspond to alignment of several celestial bodies, several alignment lines, and a closer angular separations.

We claim that the celestial bodies alignments noted above, in all their diversity, are integral effects accompanying all significant natural phenomena. More powerful natural phenomena manifestations correspond to simultaneous alignment of several celestial bodies along the same line, the presence of several alignment lines, as well as a closer angular separation from astronomical conjunctions, or oppositions.

3.5. Examples of celestial bodies' alignments associated with major natural phenomena.

Figure 4 shows alignments of the celestial bodies for a number of large eruptions of Etna. For ancient eruptions, an unknown date was selected according to the nearest extreme alignment of the celestial bodies.

The Moon and the inner planets move around the celestial sphere faster than other bodies, and under certain circumstances can cause fast-flowing paroxysms of Etna, coinciding with the moments of alignment of the Moon and the planets in terms of the start time and duration, within an accuracy of fractions of hours (Table 2).

Table 2. Etna paroxysms associated with the celestial bodies' alignments.

Dates	Alignment	Reference
10.05.2008	Earth-Moon-Mars	[2]
29.03.2007	Earth-Moon-Saturn	[3]
29.04.2007	Earth-Mars-Uranus	[3]
15.11.2011	Earth-Mercury-Venus	[1]
18.03.2012	Earth-Mercury-Uranus	[1]
04.12.2015	Earth-Moon-Jupiter	[4]
12.01.2011	Mercury-Earth-M35	[5]
19-23.02.2013	Earth-Sun-Neptune	[5]
27-02-2017	Earth-Mars-Uranus	[6]
13-15.04.2017	Earth-Sun-Uranus	[6]
19-21.04.2017	Earth-Mercury-Sun	[6]

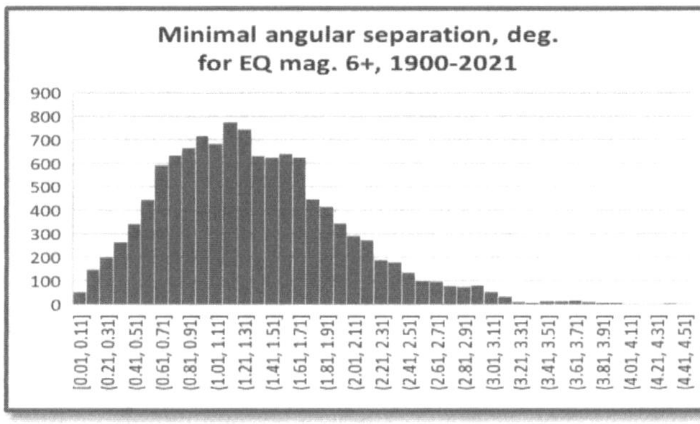

Figure 2. Minimal angular separation for EQ mag. 6+, 1900 – 2021.

We calculated the angular separations between the celestial bodies (Figure 2) relative to each other, and relative to 13 stellar objects at the moments of 11,726 earthquakes of magnitude 6+, filtered and displayed them on the graph (Figure 3) to identify a possible spatial grouping. A similar procedure was performed for three series of randomly generated dates. We were not able to identify significant statistical differences, or a special spatial grouping of the natural series in comparison with random ones. The obvious reason is that the alignment of celestial bodies is just a prercondition, i.e. a prerequisite for the manifestation of an event, just as the presence of clouds is a necessary condition for rain, but not every cloud rains. The sum of the gravitational perturbation vectors takes extreme values only in the case of their corresponding mutual orientation, which cannot be detected by statistical analysis. The second reason is the presence of a time lag of variable length between the alignment of celestial bodies and the event (discussed below). To confirm the statistical dependence, an analysis of the calculated values of gravitational perturbations is required.

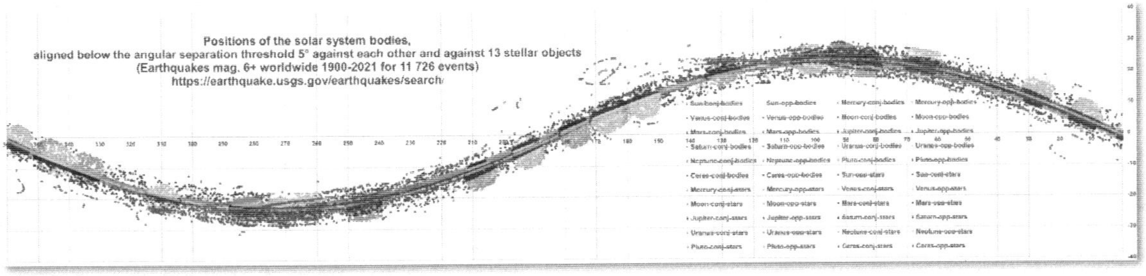

Figure 3. Positions of the solar system bodies aligned below the angular separation threshold 5°, for earthquakes mag. 6 + 1900-2021, 11 726 events (RA°, DEC).

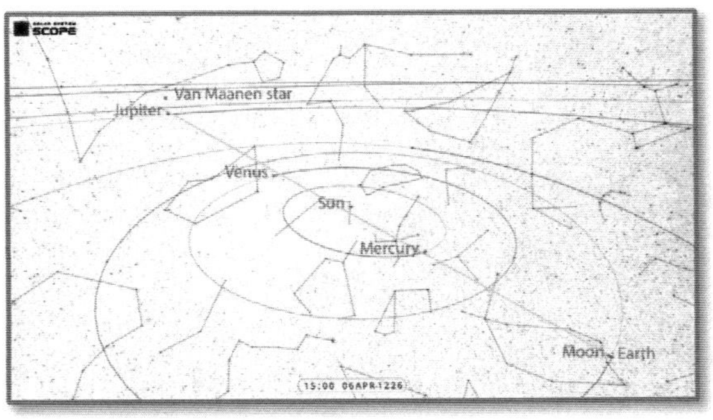

1226 B.C. The first historically confirmed eruption of Etna.

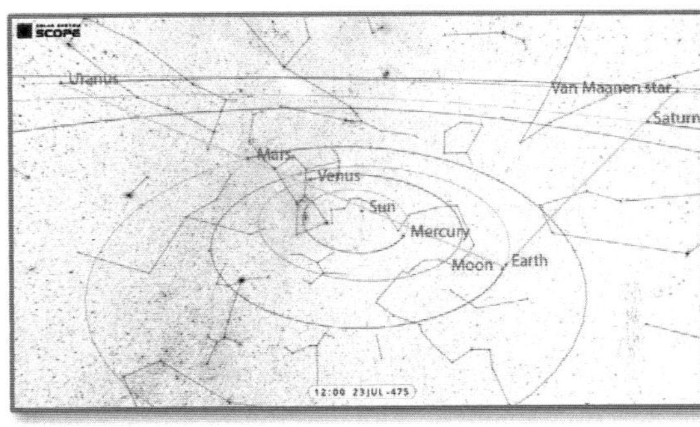

475 B.C. Mount Etna eruption (according to Pindar and Aeschylus).

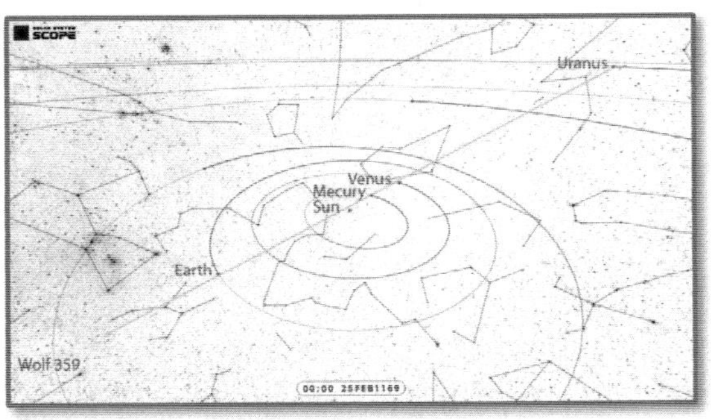

1169 A.D. A powerful eruption of Etna and an earthquake.

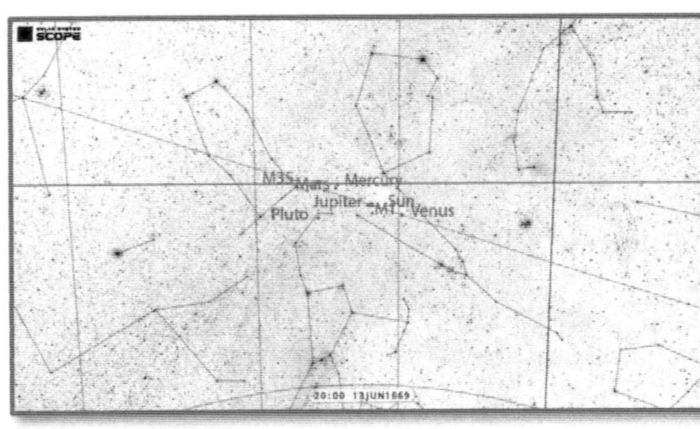

13 June 1669. The eruption of Etna changed the contour of the coast.

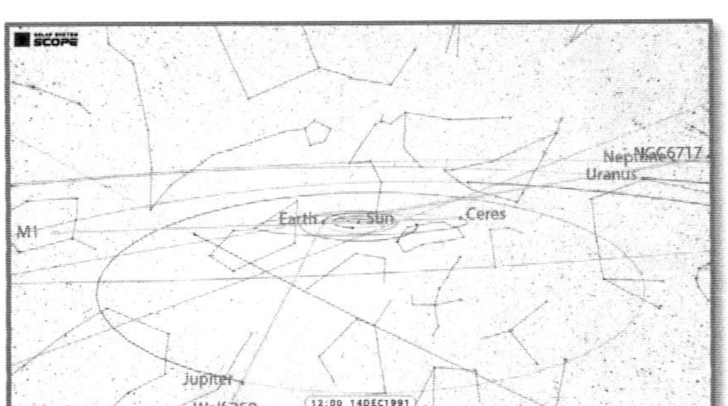

14 December 1991. One of the largest eruptions of Etna.

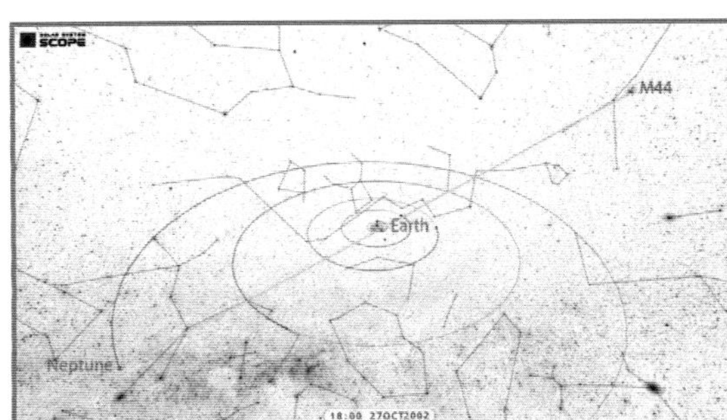

13 June 2001. Etna volcanic activity.

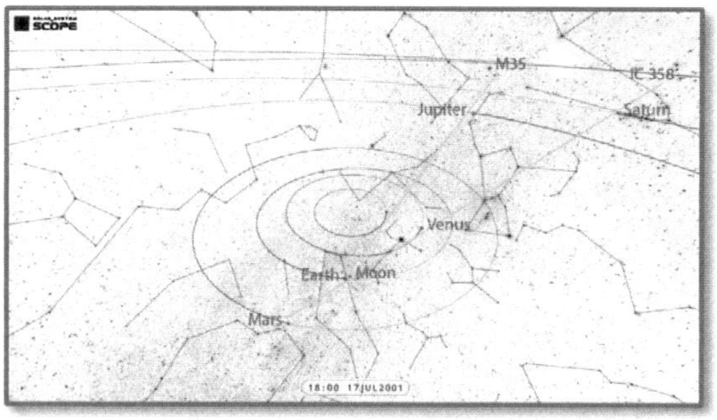

17 July 2001. Etna volcanic activity.

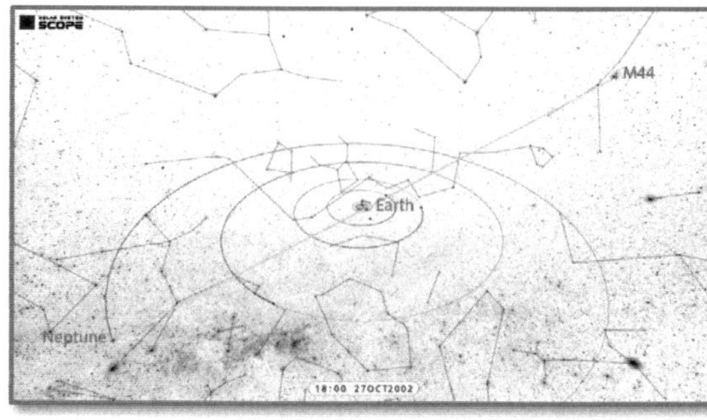

27 October 2002. Etna volcanic activity.

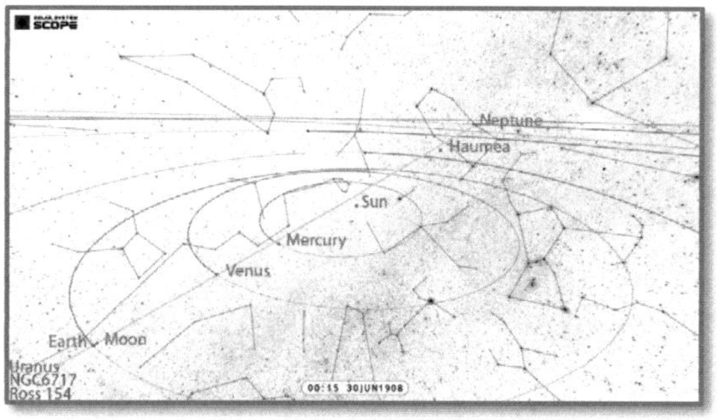

30 June 1908. Tunguska Event, Russia (volcanic version).

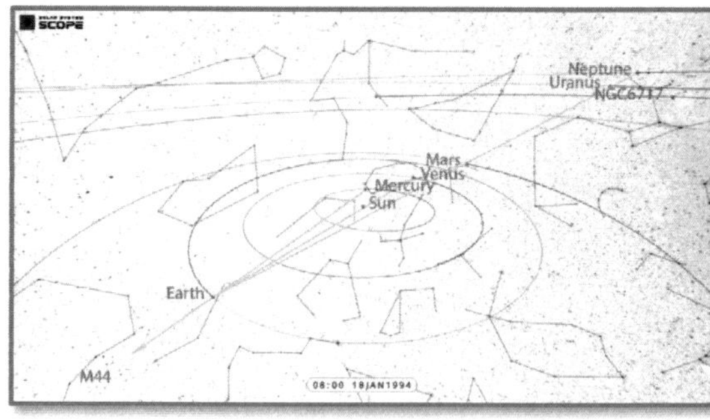

18 January 1994. Event in Cando, Spain (similar to Tunguska).

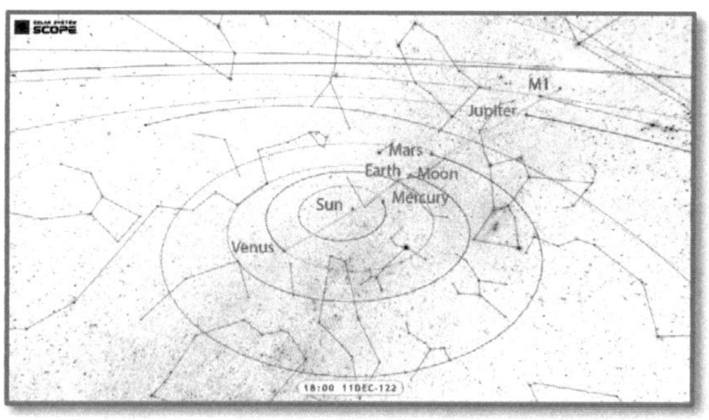

122 B.C. Etna almost wiped out the city of Catania.

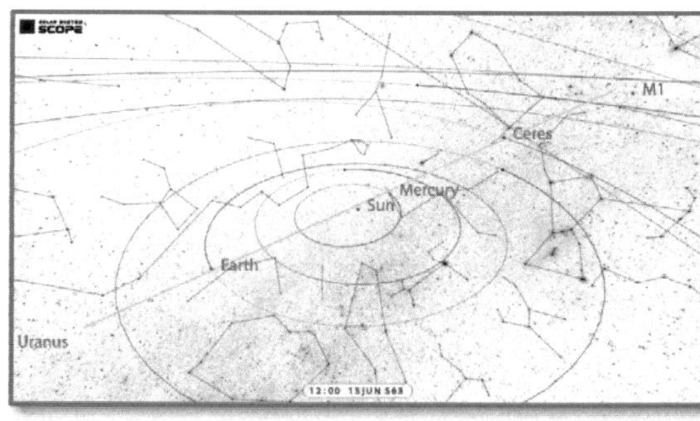

563 A.D. Landslide on Mount Taurendum, caused tsunami on Lake Geneva.

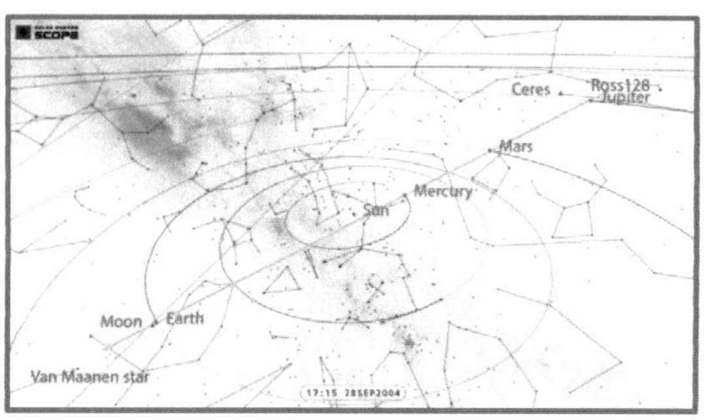

28 September 2004. Parkfield earthquake.

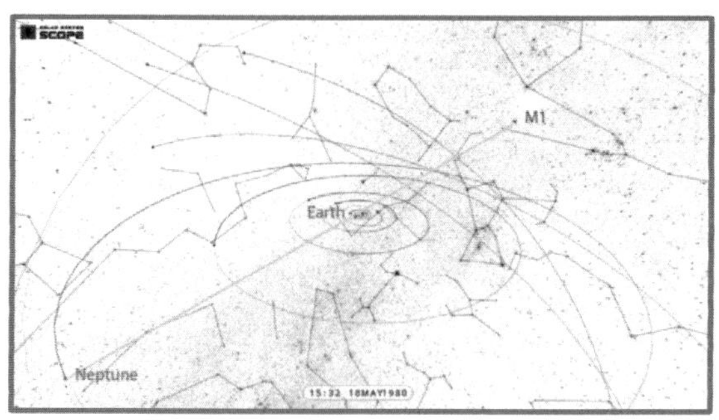

18 May 1980. Eruption of Mount St. Helens.

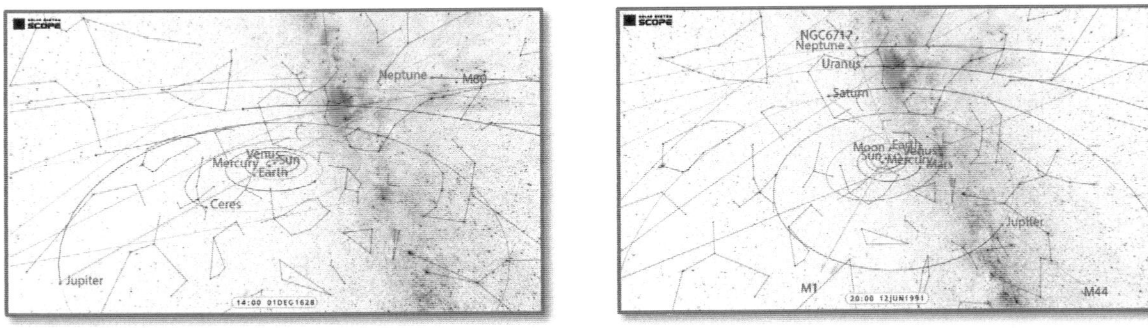

1628 B.C Minoan Eruption (Santorini). 12 June 1991. Pinatubo eruption.

Figure 4. Examples of celestial bodies' alignments, associated with the major natural phenomena.

4. Bidirectional Pushing Gravitation model (BPG)

The conclusions obtained as a result of the analysis of celestial bodies' positions accompanying natural events on Earth do not allow us to propose an appropriate mechanism operating within the framework of the standard cosmological model. Our approach, discussed below, assumes the interpretation of the observed relativistic phenomena of GRT, such as the deflection of geodesic lines near celestial bodies and the Shapiro delay, as gravitationally effective, within the framework of the extension of the law of universal gravitation. The employment of alternative cosmology is dictated by the need to use the bidirectional action of gravitational forces in a closed stationary universe, that is unacceptable in the standard cosmological model, as well as the effect of the "lens of the universe", which ensures the participation of the entire celestial sphere in the gravitational interaction of bodies as an intermediary, without which this interaction cannot occur. The parametrized post-Newtonian formalism (PPN), usually employed for the analysis of alternative theories, is not applicable here, since with the introduction of the apparent gravitational mass, the Einstein equivalence principle for distant bodies is violated.

4.1. Effective gravitational self-lensing and apparent gravitational mass

As is known, the problem of "dark matter", which has appeared since the time of Zwicky's observations, arises due to the discrepancy between the velocities of the circular motion of stars in galaxies and the Kepler curve equation.

$$V(D)=\sqrt{\frac{GM_g}{D}}\,. \tag{2}$$

Here, $\mathbf{M_g}$ – mass of the galaxy, and $\boldsymbol{D}$ – distance to star, and the speeds of the stars decrease proportionally $\sqrt{1/D}$. Instead, the rotation curves of galaxies have plateaus. Thus, to get a plateau on the rotation curves of galaxies, without reducing the velocities of stars, an increase is required in $\boldsymbol{M_g}$ proportionally to D.

Precisely this kind of dependence A. Einstein mentioned in his famous work dated 1936, "Lens-like action of a star by the deviation of light in the gravitational field". He emphasized that the phenomenon of gravitational lensing obeys increase in proportion to $\sqrt{D}$, which in our opinion indicates about the possible participation of this phenomenon in the formation of rotation curves of galaxies

The relativistic deflection of the hyperbolic trajectory of geodesic lines near the solar disk comprises half of the magnitude of this deflection calculated for distant stars (~1.75"). Such a deflection indicates the presence of an observational increase in the geometric properties of the celestial body, in particular, the apparent radius of the Sun, or of any other celestial body. This effect called gravitational self-lensing [14], [15] and considered exclusively as an insignificant optical phenomenon, limited by the laws of geometric optics. However, we believe that in the case of gravitational lensing, in contrast to geometric optics, **the solid angle occupied by the enlarged object on the celestial sphere changes**. As a result, the apparent increase in the geometric properties (including the apparent gravitational radius of the body), under gravitational lensing becomes effective, and violates the law of inverse squares, which lies in the basis of the law of universal gravitation. Thus, the law of universal gravity ceases to be the limiting case for GRT and requires adjustment.

In order to highlight the parameters and constants of the new model, we will mark them with a subscript #.

We introduce the concept of an *apparent gravitational mass,* $\mathrm{M}_{\#} = \mathrm{M}(1+C_{\#} \cdot \mathrm{D})$, which ceases to be invariant and depends on the integral mass self-lensing coefficient of the celestial body $C_{\#}$, determined by its compactness, as well as on the distance D to the observer. The notion of the *apparent gravitational mass* significantly increases the long-range action of gravitation for compact masses or celestial bodies having a compact core and a large integral coefficient of gravitational self-lensing of the mass. This makes it possible to explain the rotation curves of galaxies by the presence of a highly compact central mass without involving the concept of dark matter. The compactness of the central mass in galaxies corresponds to the size of their turning radius, which is determined by the balance of centrifugal forces and gravitational forces. Another task of the introduced concept of the *apparent gravitational mass* is the formation of a distributed repulsive potential of the celestial sphere, as a whole.

In this regard, the question arises on the applicability of Bertrand's theorem [13], which protect the law of universal gravitation from encroachments. J. L. F. Bertrand proved that the law of force, which depends only on distance, and makes its point of application move along an algebraic curve, could be either Hooke's law or the law of universal gravitation. The bodies moving in closed orbits according to the law of universal gravitation must follow elliptical orbits.

4.2. *Bertrand's theorem is not applicable in a curved space*

Generally accepted that violations of the law of universal gravitation can be determined by means of measuring parameters of the orbits using the radar method. However, the length in the international system of units defined basing on the speed of light. According to A. Einstein, the speed of light in a gravitational field decreases due to gravitational time dilation. If we deal with the additional gravity, not caused by the appearance of a third mass between two interacting ones, but due to the very law of gravity, then we must take into account such additional time dilation in radar or laser measurements of the orbits.

Bertrand's theorem applies to Euclidean coordinates. If the law of gravity contains unknown additional terms, depending on the distance, which affect the time dilation in gravitational field, then by affecting the speed of light, this will change the gradation of the distance scale for radar measurements, and we will deal with curved coordinates (without knowing this). Thus, the radar method fundamentally does not allow detecting the presence of additional gravitation in the law. The assumed violations of the ellipticity of orbits for the modified law of gravity canceled out by the corresponding time dilation in gravitational field. This also applies to checking the ellipticity of the orbits for stars orbiting the center of galaxy in the image plane. For a curved space, Bertrand's theorem is inapplicable.

4.3. *The "lens of the universe" effect and bidirectional action of forces in a stationary closed universe*

We employ an alternative cosmological model for a stationary closed universe without inflation. The world space represented by a three-dimensional surface of the hypersphere S^3 with a radius of ~1500 Mpc. Instead of the classical attraction, forces of gravitational repulsion act instantly between the gravitating body and the test mass, both in the forward and in the opposite direction (since the space of the universe, which is a topological manifold, is closed). These repulsive forces are equal in magnitude and mutually compensated. However, in the forward direction, the gravitating mass also shields part of the celestial sphere, blocking the action of repulsive forces applied to the test mass from other distant celestial bodies. Due to the shielding, the compensation for the repulsive forces violated, and attraction emerges. Actually, this is not an attraction of the test body to the gravitating mass, but pushing from the side of the celestial sphere.

Compared to Newtonian gravity, the agents of the interaction change, as well as the direction and number of forces. Due to increase in the range of action for gravitation, and the presence of remote compact masses distributed throughout the entire celestial sphere, the celestial sphere, which has a high gravitational repulsive potential, distributed over its area, becomes the third agent, mediator in the gravitational interaction, without which it cannot occur. This potential pertains to an indefinite set of distant massive bodies in the universe, the long-range gravitation of which is due to the self-lensing of their gravitational mass with distance, as well as with an apparent (but effective) cosmological scaling, due to "lens of the universe" effect (mentioned by J. Wheeler in "Gravitation" [12]).

The closed space of the hypersphere surface (topological manifold S^3) pertains apparent projective scale magnification. For the convenience of insight of the apparent scale magnification in the space of positive curvature, one can lower the dimension of space and consider the surface of the sphere S2 (the Earth's surface). In the azimuthal equidistant projection (Postel's projection), often used to select directions for long-distance radio communication (Figure 6), for an operator located at the pole, an object removed to the equator will have an apparent scale increased in $\pi/2$ times, and at the antipodal point, as well as, when closing the circle, the apparent scale increases asymptotically (Figure 5). It is important to understand that the projective increase is applied to solid angles (because the radius and area of the distant full sphere are reduced). It causes a red shift due to the apparent increase in the size of any cosmologically distant oscillators. This also means that the existence of a time rate declaration for cosmologically distant objects (Just imagine the atomic frequency standards of aliens :).

The increase in redshift with distance due to the lens of the universe effect offers an alternative explanation for the Olbers' paradox instead of the Doppler's effect and eliminates the need for an inflationary model. The entire closed universe acquires the properties of a cosmological black hole, wherein it generates blackbody radiation, interpreted in the standard cosmological model as a cosmic microwave background radiation.

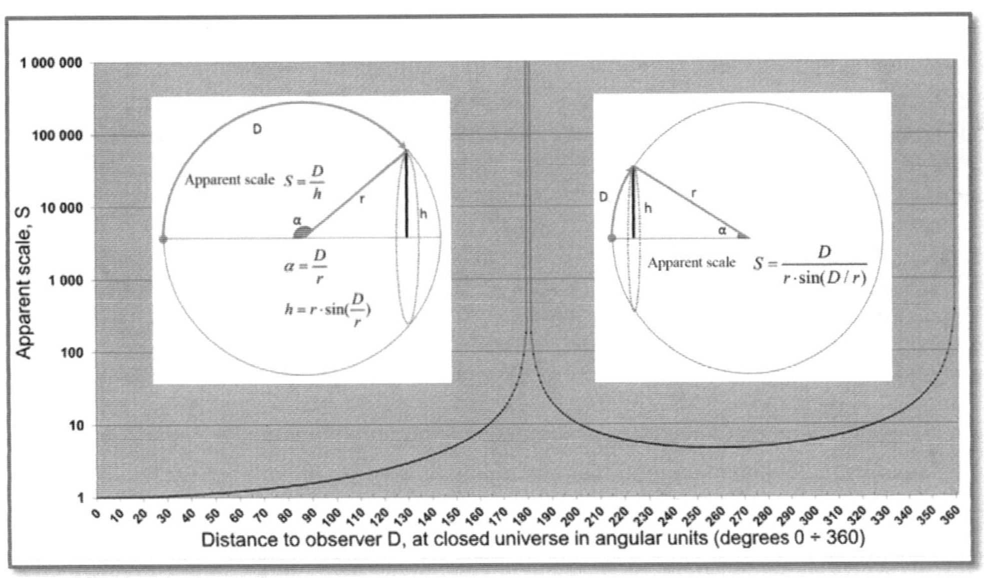

Figure 5. Apparent scale in a stationary closed universe (hypersphere S^3).

In the antipodal zone of the universe, persists an asymptotic increase of the projective scale, causing a strong frequency shift. Thus, with a projective magnification in 10^7 times, the frequencies of the visible range are shifted to the band of radio frequencies of the decimeter range. Accordingly, the apparent increased size of radio galaxies, is taken for their own size.

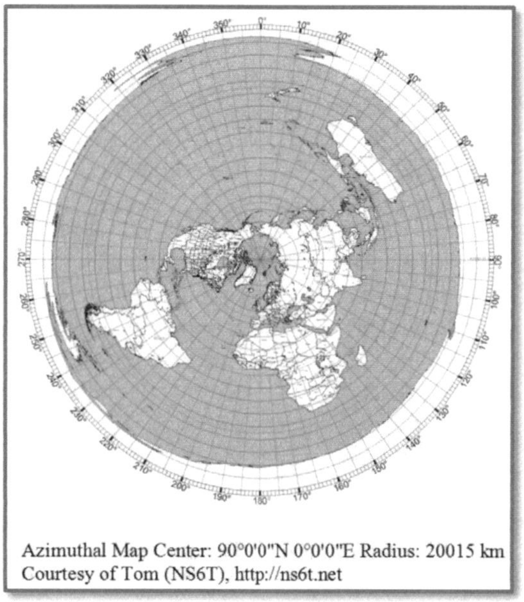

Figure 6. Azimuthal map projection.

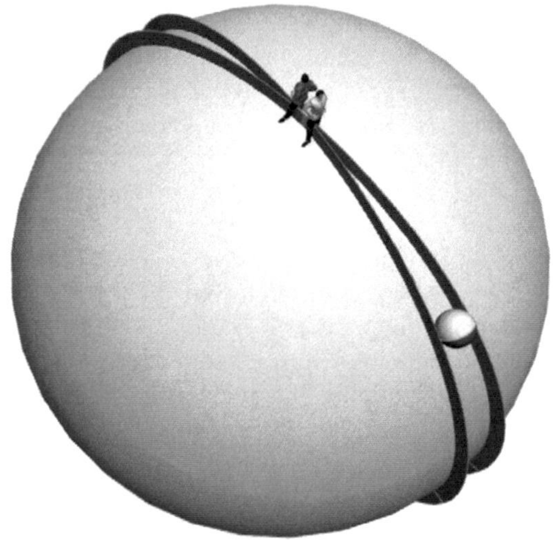

Figure 7. Bidirectional projection in a closed manifold S^2.

If the light propagated instantly, then an observer viewing the Moon in the forward direction could try to "observe" its reverse side in the opposite direction (through the Earth), moreover, of the same angular size, but upside down. However, the asymptotic growth of the projective scale at such a distance appropriately shifts the frequencies of electromagnetic oscillations, which makes this task unsolvable. The geometric relations in a closed space shown in the Figure 7 demonstrate possibility of applying the inverse square law in two opposite directions simultaneously, and give rise to consider the proposed mechanism of action for the compensated pushing gravitational forces plausible.

The lens of the universe effect requires an adjustment of the photometric distance modulus equation (Figure 8, see also [9]). At the same time, a kink appears on the graph, the position of which depends on the presupposed radius of the hypersphere of the universe. **For r = 1500 Mpc**, the beginning of the accelerated darkening of supernovae SN1A (the kink at the graph) falls on the values of the stellar magnitude **m = 21 ÷ 22**, which allows us to choose the radius value and offer an alternative interpretation for the "accelerated expansion of the universe".

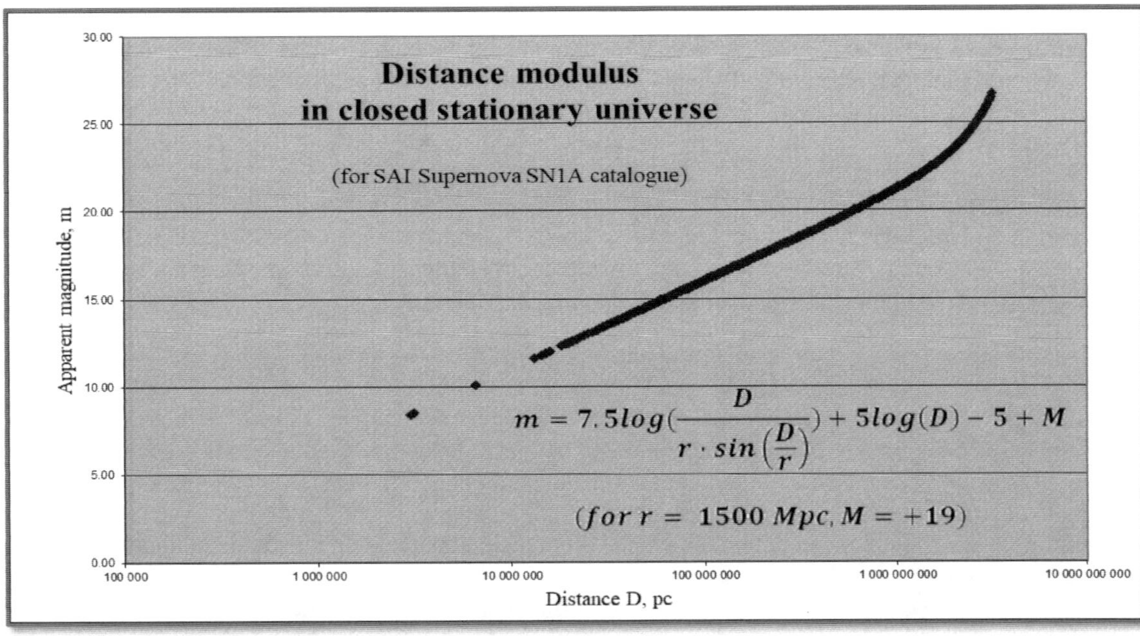

Figure 8. Distance modulus in the stationary closed universe.

4.4. The phenomenon of gravitation is caused by the effect of shielding the celestial sphere repulsion

The question arises: "What is the magnitude of the specific strength of the gravitational repulsive field of the celestial sphere $\mathbf{E}_{\#}$"? For calculations, we will use the following dependencies:

Gravitational field strength: $E = GM / D^2$. Gravitational radius: $r_g = 2\,GM / c^2$. Solid angle occupied by a circle of radius **r**: $\Omega = \pi r^2 / D^2$. Strength of the gravitational field per the solid angle, shielded by the mass on the celestial sphere $E = \Omega E_{\#}$.

After substitutions and transformations, we get: $E_{\#} = E/\Omega = GM/(\pi r^2)$. If we define the radius of the circle of the gravitational shadow as $r_{\#} = \sqrt{r_g / \pi}$, then we get:

$$E_{\#} = \frac{c^2}{2} \text{ (m/s}^2 \text{ per one steradian).} \quad (3)$$

To get rid of the big figures, calculated for a solid angle of 1 sq. second, this value should be divided by $\mathbf{(180 \times 60 \times 60/\pi)^2}$ sq. seconds in one steradian.

Thus, it is possible to propose a universal constant for magnitude of the specific strength of gravitational repulsive field of celestial sphere, applied to the gravitational shadow of the mass, as a circle with radius $r_{\#} = \left(r_g / \pi\right)^{1/2}$, acting in a solid angle:

$$E_{\#} = 1056236.4335137 \text{ (m/s}^2 \text{ per one sq. second of a solid angle).}$$

For an observer on the Earth's surface (D=6 371 000 m), the Earth's gravitational radius $\left(r_g = 0.008857538\ m\right)$ occupies a solid angle $\Omega = 2.18221632\text{E-}16$, which comprises 9.28427648E-06 sq. seconds, thus, the gravitational field strength equals:

$$E = \Omega \cdot E_{\#} = 9.28427648\text{E-}06 \cdot 1.0562364335137\text{E+}06 = 9.81 \text{ m/}s^2.$$

We argue that the very phenomenon of gravity exists due to the effect of shielding the celestial sphere. It is due to this mechanism the inverse square law operates in the gravitational interaction. This case, owing to the presence of a distributed repulsing gravitational potential of the celestial sphere, is radically different from the intended mass shielding effect for two bodies, where strict limits established in the experiments.

4.5. Parameterization of the model. The integral coefficient of gravitational self-lensing of mass.
The angle of the relativistic deflection, and the corresponding magnification coefficient **k (d)** for a thin cylindrical layer is determined by the value of the impact parameter d and the fraction of the body mass (having gravitational radius $\mathbf{r_g}$) located inside the impact parameter.

$$k(d) = 1 + \frac{r_g(d)\cdot D}{d^2}. \quad (4)$$

The total value of the gravitational self-lensing coefficient of the mass $C_{\#}$ of a celestial body can be obtained based on the radial density distribution by means of layer-by-layer integration over a changing impact parameter. For example, the self-lensing coefficient of the Sun obtained by numerical integration for the standard solar model is 1.18% increase of the apparent gravitational mass per 1 AU. For compact celestial bodies with a high average density, the self-lensing coefficient will be high, and the range of their gravity will significantly increase.

For compact stars, such as red dwarfs, white dwarfs, neutron stars, etc., the calculation of self-lensing coefficients can be performed on the basis of the accepted polytropic model for corresponding classes of stars, their known mass and radius. It should be borne in mind that many theoretical models in astrophysics, including data on the masses and radii of stars, as well as on the internal structure of planets, require their verification, and for clusters and other galaxies may be completely absent.

The most reliable way to parameterize the proposed model will be the parameters fitting technique by solving a redundant system of equations. Such a system of equations can be compiled using data on the manifestations of natural events that have a time reference, for selecting the amplitudes of gravitational perturbation vectors induced by corresponding sources of gravity, with known orts of directions to these sources.

4.6. Induction of gravitational perturbations in geosphere due to alignment of celestial bodies
Celestial bodies are affected by compensated powerful forces of repulsive gravity from other celestial bodies distributed over the celestial sphere. However, the repulsive potential of the celestial sphere is not uniform. Outside the solar system, at various distances, there are many compact and massive bodies, as well as their clusters, the repulsive (and at the same time, compensated) gravitation of which has significant values. The inhomogeneity of gravitational potential of the celestial sphere is the main reason that does not allow obtaining exact value of the gravitational constant in different experiments.

The bodies of the solar system, approaching the line of interaction of a distant gravitating mass and a test body, act as gravitational lenses, causing relativistic deflection of the powerful repulsive forces applied to the test body. This deflection induces a vector of gravitational perturbations at the test body, directed orthogonally to the repulsive force.

All the bodies of the solar system, moving along the celestial sphere, are involved in formation of the total vector of disturbances. The impulse to the test body transmitted not by the vectors of total perturbation of the gravitational field strength (accelerations), but by its derivative, known as kinematic jerk.

The relativistic deviation of the direction to the source of gravity also causes an apparent increase in the distance to the source OS' in comparison with OS (Figure 9), as the impact parameter decreases (for sources located further than the deflecting mass) and a corresponding change in the strength of the gravitational field. The Shapiro effect, known as gravitational signal delay, causes similar effective increase in the apparent distance to the source and a change in the field strength. As a result, motion of the deflecting mass breaks compensation of the repulsive forces and creates a kinematic jerk.

The most effective impact on free masses will take place along the plane of horizon of the observer (rotating in space with the Earth). Thus, it is required to calculate the projection of the estimated vector of gravitational disturbances (hodograph of jerks in time window) onto the plane of horizon of the observer. If we integrate the obtained vector field of disturbances in the time window, we can obtain the impulse of force transmitted to the certain geophysical object.

The impact of small gravitational perturbations' (non-tidal variations of plumb line deviation) on the free surface of a large area of liquids (e.g., the sea surface, magma layers of different densities) can cause extreme waves and seiches.

4.7. Relativistic gravitational deflection of geodetic lines

The influence of relativistic gravitational deflection of light by celestial bodies of the solar system on the accuracy of astrometric measurements was studied in detail in the paper by Slava Turyshev and the references given therein [8]. In preparation for the cancelled mission of the space interferometer (SIM), a conclusion is made about the significant contribution to the astrometric observations of the microsecond accuracy of the relativistic gravitational deflection by the monopole gravitational components of all large celestial bodies of the solar system, as well as about the significant value of the quadrupole components of the fields of outer planets of the Solar system. The monopole deflection depends only on the mass of the gravitational lens and the impact parameter (Figure 9).

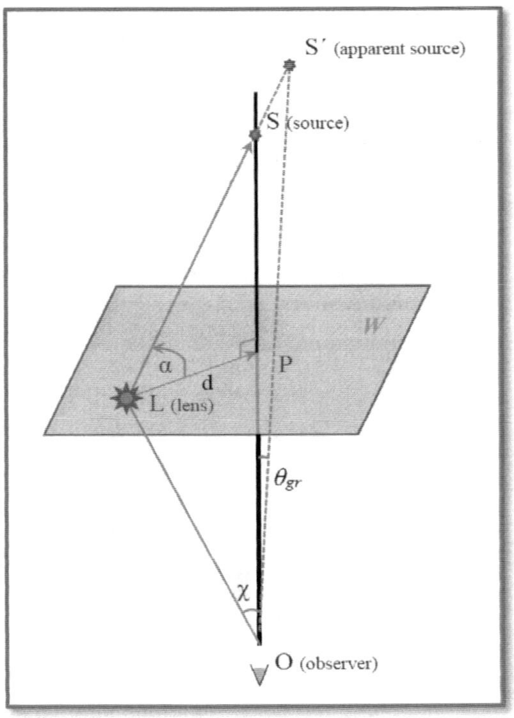

Figure 9. The gravitational deflection geometry.

Gravitational deflection angle θ_{gr} for distant stars depends upon the impact parameter d [8].

$$\theta_{gr} = \frac{4GM}{c^2 d}\cdot\frac{1+\cos\chi}{2}. \tag{5}$$

For the Solar system bodies, **including sources located closer to observer, than a gravitational lens** (deflecting body):

$$\theta_{gr}(\alpha) = \frac{4GM}{c^2 d}\cdot\frac{1+\cos(\alpha)}{2}\cdot(1+\sin(\alpha)); \quad -\frac{\pi}{2}\le\alpha\le+\frac{\pi}{2}. \tag{6}$$

Or, for the convenience of computing:

$$\theta_{gr}(\beta) = \frac{4GM}{c^2 d}\cdot\frac{1+\cos(\alpha)}{2}\cdot(1+\cos(\beta)); \quad 0\le\beta\le\pi. \tag{7}$$

Apparent distance to source OS:

$$D_{s'} = D_S\,\frac{\sin(\beta)}{\sin(\theta_{gr})}. \tag{8}$$

Celestial bodies of the Solar system at small angular separations from other celestial bodies exert a mutual gravitational deflection, and thus can expand the zone of their influence, cascading the transmission of induced gravitational deflection through other celestial bodies acting as gravitational lenses.

When calculating the quadrupole deflection, the second zonal harmonic of the geopotential J2, determined by the polar compression of the planet, will be of significant importance. The corresponding formalism for calculating quadrupole deflection is given in [8].

4.8. The speed of gravity, subatomic structure of matter and ejection of plasma by gravitational field
In the Bidirectional Repulsive Gravitation (BPG) model, the compensated gravitational forces affect instantly, both in the forward and in the opposite direction. This approach requires a justification of the instantaneous action mechanics. We do not use the classical ideas about the necessity of force-carrying particles considered in the early models of Le Sage [11]. The mutual repulsion of masses is a manifestation of the natural property of matter particles to occupy free space. This is simply a choice of the direction of movement due to the collective behavior of matter particles, interpreted as forces. This property is clearly manifested in the behavior of gases, as well as liquids in weightlessness. The behavior of solids obeys more complex laws, which are also implemented in the fundamental structure of matter.

The BPG model is based on the idea that the world space is filled with a medium that transmits interactions and has a limit of fragmentation corresponding to the minimum Planck length ($L_{pl} = 10^{-33}$ cm). Such a property of the medium inevitably means the impossibility of existence of lengths that are not multiples of L_{pl}, and, consequently, the need for quantization of the volume. In fact, such a medium acquires the properties of discrete (cellular) space. The shape of the volume quantum should correspond to the principle of the least action when moving to an adjacent cell. It is proved that a face-centered cubic lattice (FCC-lattice) is the densest regular packing of spheres in three-dimensional space and has the largest number of symmetries. The Voronoi-Delaunay cell for the FCC lattice has the shape of a rhombic dodecahedron. Thus, we can talk about filling the world space with a discrete medium, portions of which have the properties of a perfectly rigid bodies. The only possible metaphor for movement in a dense spatial packing of rhombic dodecahedra can only be the permutation of portions of the medium into neighboring vacant cells of the FCC lattice. This looks similar to the" Game of 15" (Slider-game) rules, but the role of square chips on the plane is performed by rhombic dodecahedra in the three-dimensional space.

Aristotle and the ancient atomists established that a discrete space has three fundamental properties - *isotachy*, *kekinema* and *renovation* [10], [24]. At the subatomic level, the structure of matter can be represented as a three-dimensional cellular automation on a FCC lattice (Figure 10), in which there is a permutation of permanently existing elements of the medium, endowed with the fundamental property of occupying vacant cells of space, and at the same time, competitively implementing this property. In a discrete space, there are only two speeds, 0 (invariantly stationary crystalline ether) and the "speed of light" - the only possible speed of moving elements of the medium to the neighboring cell of the lattice. The entire spectrum of velocities observed in the macrocosm is due to the movement of ensembles of elements, where each element moves with a "speed of light".

The fact that the speed of light is invariant in different reference frames is due to the principle of the *isotachy* of discrete space and the existence of an invariantly stationary ether.

Each cell of the FCC lattice has 12 adjacent faces and 6 adjacent vertices with neighboring cells. If in the" Game of 15 " for one vacant cell, four adjacent chips can competitively strive to occupy it, then on the HCC grid, on the contrary, several vacant cells are required for a single (movable) element in a dense spatial packing. One will need at least three vacant cells adjacent to the three-sided vertex of the rhombic dodecahedron, or five vacant cells adjacent to the four-sided vertex. A complex competitive unidirectional dynamics that emerge generating persistent ensembles of elements, and implementing mechanics of the "arrow of time". As the analysis has shown, in order to build an algorithm for competitive occupation of free cells, it will inevitably be necessary to allow the possibility of accumulation of several elements of the medium in one cell of the FCC-lattice. As a result, we can consider a kind of the weird quantum potential dynamics according to the rules of the cellular automaton. In this model, subatomic particles of matter or energy correspond to vacancies, or to cells with fewer elements of the medium in one cell of the FCC lattice.

Figure 10. Cellular automata using Planck-size FCC-lattice.

4.9. On the closure of a three-dimensional space

The space of the universe is represented to an observer immersed in this space as a closed three-dimensional manifold - the surface of the hypersphere S^3 (in fact, it is a pseudo-hypersphere). Such an observer draws conclusions by evaluating the directions of radiation fluxes and measuring distances using the radar method. Our attempts to construct a model of the closure of the spatial packing of rhombic dodecahedra in the 4th dimension, just as it is done for a hypercube in the architecture of supercomputers by topologically gluing its faces, were unsuccessful, since all the external surfaces of the spatial packing of rhombic dodecahedra are oriented, and do not allow gluing with another external surface. However, a virtual observer located outside the universe will not detect the fourth spatial dimension. For an external observer the trajectories of light rays will be represented as polylines of Planck length segments in Euclidean space, bending towards the cells of space with a lower density of the medium, a kind of a cosmological black-hole in 3D-space.

The instantaneous "impact" of gravitational forces in the described subatomic structure is provided through the contact of neighboring portions of the medium having the properties of absolutely solid bodies. However, along with the *instantaneous impact*, there exists a *contact interaction* in the model under consideration, due to the movement of ensembles of elements, the speed of which corresponds to the property of the *isotachy* of discrete space, and limited by the "speed of light". The described subatomic structure allows the construction of a computer simulation that excludes participation of an immersed observer. A detailed description of the applied space model is beyond the scope of this publication.

4.10. Why hasn't everything collapsed yet?

If the repulsive forces are compensated, and the celestial sphere pushes all the masses to each other, as a result, a collapse should occur. What mechanics resists the pushing of bodies? As noted above, the "lens of the universe" effect causes an asymptotic increase in the observed geometric characteristics of bodies when the circle is closed in the opposite direction. In addition to the apparent sizes of bodies, the sizes of any electromagnetic oscillators, the gravitational radii of bodies, also the observed value of the minimum Planck length increases asymptotically. A quantum threshold arises that restricts transfer of gravitational repulsion forces to ensembles of subatomic elements whose physical size happen to be less than the apparent size of the Planck length. Thus, compensation of the repulsive forces is violated. The plasma is pushed out by the gravitational field [19]. This effect can manifest itself in heating of the solar corona and ionosphere, determine the position of ionopause of the solar system planets, the dynamics of comets' ion tails, astrophysical jets.

5. Conclusions

To give a final answer to the question in the title of the report will be possible only after calculating the gravitational perturbations in accordance with the proposed model. At the moment, we are at the preliminary stage of research and invite interested specialists.

The causal relationship between high-energy natural phenomena and assumed or registered gravitational disturbances accompanying them is interpreted ambiguously. Free masses begin to move when plumb line deviations vary, and vice versa, with the movement of masses, deviations of plumb lines are changing. What is the cause and what is the effect? We have considered the hypothesis about the possibility of transmitting gravitational field perturbations from extraterrestrial masses, and we propose corresponding mechanics and model.

However, computational test of the hypothesis is required. To do this, it is necessary to perform a detailed parameterization of the model (section 4.5). This task will require an interdisciplinary approach of specialists in astrophysics and Earth sciences, performing substantial amount of calculations for selecting model parameters using computational methods, expanding the list of stellar objects, including those, not only near the ecliptic.

It would be possible to check out the model by measuring a spacecraft flight path disturbances of a spacecraft, planning (correction) of space missions for navigation in conditions of celestial bodies' alignments.

In addition to various applied problems in the geosphere (lithosphere, hydrosphere and atmosphere), the application of this model is possible in planetology (volcanism on planets and satellites of the Solar system), in solar physics (flares, CME), ionospheric studies.

The Discussion section will be submitted to the journal Frontiers in Earth Science.

References

[1] Bonaccorso A, Calvari S, Currenti G, Del Negro C, Ganci G, Linde A *et al* 2013 From source to surface: dynamics of Etna's lava fountains investigated by continuous strain, magnetic, ground and satellite thermal data. *Bull. Volcanol.* **75** 690

[2] Bonaccorso A, Cannata A, Corsaro RA, Di Grazia G, Gambino S, Greco F *et al* 2011 Multidisciplinary investigation on a lava fountain preceding a flank eruption: The 10 May 2008 Etna case. *Geochemistry, Geophysics, Geosystems* **12**

[3] Acocella V, Neri M, Behncke B, Bonforte A, Del Negro C and Ganci G 2016 Why Does a Mature Volcano Need New Vents? The Case of the New Southeast Crater at Etna. *Front. Earth Sci.* **4**

[4] Bonforte A, Cannavò F, Gambino S and Guglielmino F 2021 Combining High- and Low-Rate Geodetic Data Analysis for Unveiling Rapid Magma Transfer Feeding a Sequence of Violent Summit Paroxysms at Etna in Late 2015 *Appl. Sci.* **11** 4630

[5] Nunnari G 2021 *Earth Science Informatics* **14** 1121

[6] Viccaro M, Giuffrida M, Zuccarello F, Scandura M, Palano M and Gresta S 2019 Violent paroxysmal activity drives self-feeding magma replenishment at Mt. Etna. *Scientific Reports* **9**

[7] Sottili G, Martino S, Palladino DM, Paciello A and Bozzano F 2007 Effects of tidal stresses on volcanic activity at Mount Etna, Italy. *Geophys. Res. Lett.* **34**

[8] Turyshev V G 2009 Relativistic gravitational deflection of light and its impact on the modeling accuracy for the Space Interferometry Mission. *Astron. Lett.* **35** 215

[9] Heymann Y 2012 Redshift Adjustment to the Distance Modulus *Progress in Physics* **1**

[10] Korukhov V V and Sharypov O V 1995 On the possibility of combining the properties of invariant rest and relative motion on the basis of a new model of space with a minimum length (in Russian) *Philosophy of science* **1** 5

[11] Edwards M 2002 *Pushing Gravity. New perspectives on Le Sage's theory of gravitation.* (Montreal: Apeiron) pp 698-699

[12] Misner C W, Thorne K S, Wheeler J A and Chandrasekhar S 1973 *Gravitation* (San Francisco: Freeman) p 795

[13] Santos F C , Soares V and Tort A C 2011 An English translation of Bertrand's theorem *Latin American Journal of Physics Education* **5** 694
[14] Vargashkin V 2003 The phenomenon of gravitational self-lensing *Proceedings of International Scientific Meeting PIRT-2003* (Moscow: BMSTU press) p 82
[15] Prasanna A R and Ray S 2004 Self-Lensing effects for compact stars and their mass-radius relation *Mod. Phys. Lett.* **19** 2431
[16] Allais M 1999 The 'Allais Effect' and my experiments with the paraconical pendulum (1954-1960) *Memoir C-6083 prepared for NASA (Report)*
[17] Pugach A F 2018 Unexplained Temporal Effects Recorded by The Torsind at Syzygies *J. Adv. Phys.* **15** 5950
[18] Nugraha M G *et al* 2020 Anomalies of earth surface gravity field (g) during total lunar eclipse (TLE) on January 31 and July 28 2018 using video tracker analysis on pendulum harmonic motion. *J. Eng. Sci. Technol.* **15** 1001 http://jestec.taylors.edu.my/Vol%2015%20issue%202%20April%202020/15_2_20.pdf
[19] Dmitriev A L and Nikushchenko E M 2016 Expulsion of Plasma in A Gravitational Field. *Appl. Phys. Res.* **8** 38
[20] Wang Q, Yang X and Tang K 2001 Gravity anomaly during the Mohe total solar eclipse. *Sci. Bull. formerly: Chi. Sci. Bull.* **46** 1833
[21] Zhou S W and Huang B J 1992 Abnormalities of the time comparisons of atomic clocks during the solar eclipses. *Il Nuovo Cimento C.* **15** 133
[22] Barlik M, Bogusz J and Kaczorowski M 2009 Investigations on the field of gravity in Polish tidal laboratories *Reports on Geodesy* **86** 43
[23] Antonov Yu V and Antonova I Yu 2020 *Proceedings of Voronezh State University. Series: Geology* (Voronezh: Voronezh State University)
[24] Vyaltsev A N 1965 *Discrete space-time* (Moscow: Nauka publishers)
[25] https://volcano.si.edu/search_eruption.cfm
[26] https://earthquake.usgs.gov/earthquakes/search/
[27] https://www.astro.com/swisseph/swephinfo_e.htm
[28] https://www.alcyone.de/
[29] https://www.solarsystemscope.com/
[30] www.researchgate.net/profile/Ivan-Krasnyj/projects

Acknowledgements
The authors gratefully acknowledge the support of Professor Giuseppe Nunnari, Università degli Studi di Catania and Professor Innocenzo M. Pinto, University of Sannio at Benevento, Italy.

Null shells and double layers in quadratic gravity

I D Ivanova

Institute for Nuclear Research of the Russian Academy of Sciences, 60th October Anniversary Prospect 7a, 117312 Moscow, Russia

E-mail: pc_mouse@mail.ru

Abstract. For a singular hypersurface of arbitrary type in quadratic gravity motion equations were obtained using only the least action principle. It turned out that the coefficients in the motion equations are zeroed with a combination corresponding to the Gauss-Bonnet term. Therefore it does not create neither double layers nor thin shells. It has been demonstrated that there is no "external pressure" for any type of null singular hypersurface. It turned out that null spherically symmetric singular hupersurfaces in quadratic gravity cannot be a double layer, and only thin shells are possible. The system of motion equations in this case is reduced to one which is expressed through the invariants of spherical geometry along with the Lichnerowicz conditions. Spherically symmetric null thin shells were investigated for spherically symmetric solutions of conformal gravity as applications, in particular, for various vacua and Vaidya-type solutions.

1. Introduction

Quadratic gravity belongs to the class of higher derivative theories analyzed by Ostrogradsky [1], who proved that their Hamiltonian is unbounded from below. Such theories obey Ostrogradsky's instability theorem, which classifies all nondegenerate theories of higher derivatives as Lyapunov unstable. This is a serious problem because such unstable theories usually have negative energy states that are emphatically excluded from quantum field theories. Despite this, recently there has been a surge of interest in theories of higher derivatives for a number of reasons; the main ones are listed below.

Firstly, unlike general relativity higher derivative theories are renormalizable [2]–[5] thus, examination of these theories can provide important clues as to how to quantize gravity. Secondly, several groups of researchers independently discovered [6] – [16] that corrections to the gravitational Lagrangian arising in the one-loop approximation upon renormalization in quantum field theory in curved space include terms that are quadratic in the curvature tensor Riemann and its contractions. Lastly, A.A. Starobinsky [17] used these inevitable corrections to the action of gravity and noticed that a cosmological solution without a singularity, which belongs to the de Sitter type of universe, can be obtained by taking them into account. This led to the creation of an innovative inflation model. Considering all of the above, along with many other reasons, it is clear that the study of quadratic gravity is a topical problem.

Singular hypersurfaces occur in both general relativity and quadratic gravity. These are hypersurfaces on which the curvature tensor has singular components in particular terms proportional to the θ-function or δ-function.

In any theory of gravity singular hypersurfaces are important idealized objects designed to describe the local concentration of matter or energy on a given hypersurface, for example, domain walls, thin layers of matter or gravitational fields, propagation of lightlike matter, gravitational shock waves, matter-vacuum boundaries, caustics, phase transitions in vacuum, etc.

The role of exact solutions in understanding physical phenomena is difficult to overestimate. Since field equations of any theory of gravity are highly nonlinear, search for solutions becomes a very difficult task, therefore, study of singular distributions of matter fields is extremely important.

Einstein's equations on a singular hypersurface were first derived by W. Israel [18]–[20] . In the case of quadratic gravity, the corresponding equations were obtained by J.M.M. Senovilla [21]–[26] . They differ significantly from Israel's equations, primarily in that they contain not only the δ-function, but also its derivative. Thus, they describe not only thin shells, but also so-called double layers.

In the article [27] it was shown that the equations of motion for a singular hypersurface in quadratic gravity can be derived using only the principle of least action. The main advantage of this method is that δ-function derivative does not appear explicitly during computation during the calculations. This paper generalizes the results presented in [27] to lightlike hypersurfaces. In addition, an important special case of spherical symmetry is analyzed for this type of hypersurfaces.

As for applications, null singular hypersurfaces have been investigated for spherically symmetric solutions of conformal gravity, which is a special case of quadratic gravity. The main types of spherically symmetric solutions of conformal gravity were described in article [28]. Here vacuum solutions and Vaidya-type solutions are used.

2. Preliminaries

Adding higher-order curvature terms is a natural generalization of Einstein's theory of gravity. In the first approximation, these corrections give quadratic terms, therefore the action of quadratic gravity in the general case is

$$S_q = -\frac{1}{16\pi}\int_\Omega \sqrt{-g}L_q\, d^4x = \\ = -\frac{1}{16\pi}\int_\Omega \sqrt{-g}\left(\alpha_1 R_{abcd}R^{abcd} + \alpha_2 R_{ab}R^{ab} + \alpha_3 R^2 + \alpha_4 R + \alpha_5\Lambda\right) d^4x. \quad (1)$$

Here g is a determinant of the metric, whose components serve as dynamic variables, α_i are arbitrary constants, $R^a{}_{bcd}$ is a Riemann tensor:

$$R^a{}_{bcd} = \partial_c\Gamma^a_{bd} - \partial_d\Gamma^a_{bc} + \Gamma^a_{ce}\Gamma^e_{bd} - \Gamma^a_{de}\Gamma^e_{bc}.$$

Ricci tensor and scalar curvature are defined as:

$$R_{ab} = R^c{}_{acb},\ R = R^a_a.$$

In what follows the signature $(+,-,-,-)$ is adopted and geometric units where $c = G = 1$ are used.

Consider the four-dimensional spacetime Ω divided into two regions Ω^+ and Ω^- with different geometry by a singular hypersurface Σ_0. A singular hypersurface is a hypersurface on which Riemann curvature tensor has a singular components.

In a coordinate system $\{x^a\}$ continuous in the neighborhood of Σ_0, where the function defining Σ_0 is selected as one of the coordinates, the hypersurface equation is reduced to $n = 0$. In addition, without loss of generality, it can be assumed, that

$$(N_a N^a)|_{\Sigma_0} = (\partial_a n \partial^a n)|_{\Sigma_0} = g^{nn}|_{\Sigma_0} = \varepsilon.$$

Here N^a is a normal to the hypersurface, $\varepsilon = -1$ for the timelike hypersurface, and 1 for the spacelike one. For a lightlike surface:

$$(N_a N^a)|_{\Sigma_0} = (\partial_a n \partial^a n)|_{\Sigma_0} = g^{nn}|_{\Sigma_0} = 0.$$

Taking into account the accepted notation, the metric in the entire spacetime Ω can be formally represented as a sum:

$$g_{ab} = \overset{+}{g}_{ab}(x)\,\theta(n) + \overset{-}{g}_{ab}(x)\,\theta(-n). \tag{2}$$

When differentiating the expression (2), a term with a δ-function appears:

$$\partial_c g_{ab} = \partial_c \overset{+}{g}_{ab}\theta(n) + \partial_c \overset{-}{g}_{ab}\theta(-n) + \delta^n_c \delta(n)[g_{ab}].$$

Hereinafter, δ^n_c is the Kronecker symbol, the brackets $[g_{ab}] = \overset{+}{g}_{ab} - \overset{-}{g}_{ab}$ denote a jump of the corresponding variable on the surface Σ_0.

If $[g_{ab}] \neq 0$ then the δ-function occurs in Christoffel symbols. In this case, δ^2, which is undefined in the theory of generalized functions, appears in the Riemann curvature tensor and in the Lagrangian for both general relativity and quadratic gravity.

Due to coordinate transformations that are different for $\Omega^{\pm}$, it is always possible to make the tensor g_{ab} continuous on Σ_0, this fact was demonstrated, in particular, in the article [29]. Thus for both general relativity and quadratic gravity it is necessary either to work in such a special coordinate system or explicitly require $[g_{ab}] = 0$ in order to avoid the appearance of δ^2 in Lagrangian.

If the metric does not have a jump on Σ_0, from (2) the following expression for the Christoffel symbols is obtained:

$$\Gamma^a_{bc} = \Gamma^{+a}_{bc}\theta(n) + \Gamma^{-a}_{bc}\theta(-n). \tag{3}$$

From which it follows that:

$$R^a_{\ bcd} = R^{+a}_{\ \ bcd}\,\theta(n) + R^{-a}_{\ \ bcd}\,\theta(-n) + (\delta^n_c[\Gamma^a_{bd}] - \delta^n_d[\Gamma^a_{bc}])\,\delta(n). \tag{4}$$

The expression (4) for the Riemann tensor containing the δ-function would be acceptable for the Einstein-Gilbert action and the corresponding equations of motion, because in this case the singular part of the Riemann tensor can be directly related to the surface part of the energy-momentum tensor. This fact was first demonstrated in the papers of W. Israel [18]–[20]. In quadratic gravity, in order to avoid the appearance of undefined functions δ^2 and $\delta(n)\theta(n)$ in the Lagrangian, it is necessary to impose additional restrictions, namely, the Lichnerowicz conditions [30] :

$$[\Gamma^a_{bc}] = 0. \tag{5}$$

Consider the situation when the energy-momentum tensor of matter fields has different behavior in the regions $\Omega^{\ pm}$ and a singular component directly on Σ_0:

$$T^{ab} = S^{ab}\,\delta(n) + T^{+ab}\,\theta(n) + T^{-ab}\,\theta(-n), \tag{6}$$

where the tensor S^{ab} is defined as the surface energy-momentum tensor. In this formulation of the problem, it is assumed that T^{ab} does not contain other singular terms like derivatives of δ-function.

Unlike general relativity the components S^{nn} and S^{ni} of the surface energy-momentum tensor are generally nonzero in quadratic gravity. This fact was first noted in the papers of J.M.M. Senovilla [21]–[26], where S^{nn} and S^{ni} are defined as "external pressure" and "external flow" respectively.

In general relativity the field equations are of the second order in the derivatives of the metric tensor. The appearance of a δ-function in T^{ab} leads to its appearance in the Riemann tensor, then $[\Gamma^a_{bc}] \neq 0$, in this case Σ_0 is called thin shell. Israel's equations describe the relationship between the jumps of the Christoffel symbols and their derivatives with the tensor S^{ab}. If there is only a jump in T^{ab}, then the corresponding jump in curvature describes a shock gravitational wave accompanied by a shock wave in matter.

In quadratic gravity the field equations are of the fourth order in the derivatives of the metric tensor. If certain components of the curvature tensor are continuous on Σ_0, then their second derivatives can contain at most δ-function, which corresponds to the δ-function in T_{ab}, and Σ_0 is a thin shell. If these components of the curvature tensor experience a jump on Σ_0, then their second derivatives contain the δ'-function, and Σ_0 is a double layer. The equations of motion of a double layer in quadratic gravity, as noted above, were first obtained by J.M.M.Senovilla. A jump in curvature describing a gravitational shock wave may or may not be accompanied by a shock wave in the distribution of matter, i.e. in quadratic gravity, a purely gravitational shock wave can exist.

3. Junction conditions in quadratic gravity

The main goal of this study is to find the equations of motion for a singular lightlike hypersurface in quadratic gravity. In order to do this, it is necessary to separate the surface part in the system of equations obtained by varying the action (1) with respect to the inverse metric in the case when the Riemann tensor has the structure (4) and with (6) as the source.

Using the fact that $\theta^2(n) = \theta(n)$ and $\theta(n)\,\theta(-n) = 0$ and substituting the Riemann tensor (4) and its contractions into action (1), the following expression is obtained:

$$S_q = -\frac{1}{16\pi}\int_{\Omega^+}\sqrt{|g|}L_q^+\,d^4x - \frac{1}{16\pi}\int_{\Omega^-}\sqrt{|g|}L_q^-\,d^4x. \tag{7}$$

Here $L_q^+ = \alpha_1 R^+_{abcd}R^{+abcd} + \alpha_2 R^+_{ab}R^{+ab} + \alpha_3 (R^+)^2 + \alpha_4 R^+ + \alpha_5\Lambda$, L_q^- is defined analogically to L_q^+.

From the variation of the action (7) with respect to the inverse metric, one can get [31]:

$$\delta S_q = -\frac{1}{16\pi}\int_{\Omega^+}\sqrt{|g|}\left(H^+_{ab}\delta g^{ab} + \nabla_c V^{+c}\right)d^4x - \frac{1}{16\pi}\int_{\Omega^-}\sqrt{|g|}\left(H^-_{ab}\delta g^{ab} + \nabla_c (V^{-c})\right)d^4x. \tag{8}$$

The following designations are accepted here:

$$\begin{aligned} H^+_{ab} = {} & 2\alpha_1 R^+_{amlp}R_b^{+mlp} - 2(2\alpha_1+\alpha_2)R_d^{+c}R^{+d}_{\ bac} - 4\alpha_1 R_a^{+c}R^+_{bc} + 2\alpha_3 R^+ R^+_{ab} + \alpha_4 R^+_{ab} - \\ & -\frac{1}{2}g_{ab}L_q^+ + (\alpha_2+4\alpha_1)\Box R^+_{ab} + \frac{1}{2}(4\alpha_3+\alpha_2)g_{ab}\Box R^+ - (2\alpha_1+\alpha_2+2\alpha_3)\nabla_a\nabla_b R^+, \end{aligned}$$

$$\begin{aligned} V^{+c} = {} & \left\{(4\alpha_1+\alpha_2)\nabla^c R^{+bd} + \frac{1}{2}(\alpha_2+4\alpha_3)g^{bd}\nabla^c R^+\right\}\delta g_{bd} - \\ & - 2\nabla^b\left\{(2\alpha_1+\alpha_2)R^{+cd} + \alpha_3\, g^{cd}R^+\right\}\delta g_{bd} + (\alpha_4 + 2\alpha_3 R^+)(g^{ab}g^{cd} - g^{ac}g^{bd})\nabla_a\delta g_{bd} + \\ & + \left\{\alpha_2(2g^{cd}R^{+ab} - g^{ac}R^{+bd} - g^{bd}R^{|ac}) - 4\alpha_1 R^{+abcd}\right\}\nabla_a\delta g_{bd}. \end{aligned}$$

The vector V^{-c} and the tensor H^-_{ab} are defined in a similar way.

By definition, to derive the equations of motion using the least action principle a variation with fixed ends should be used, i.e. the values of dynamical variables, metric components in this case, are fixed at the boundary of the entire volume of integration: $\delta g_{ab} = 0$ on $\partial\Omega$. Using the Gauss-Stokes theorem we obtain the following expression for δS_q:

$$\delta S_q = -\frac{1}{16\pi}\int_{\Omega+}\sqrt{|g|}\left(H^{+}_{ab}\delta g^{ab}\right)d^4x - \frac{1}{16\pi}\int_{\Omega-}\sqrt{|g|}\left(H^{-}_{ab}\delta g^{ab}\right)d^4x + \frac{1}{16\pi}\int_{\Sigma_0}[V^c]\,dS_c,$$

where dS_c is a directed surface element.

In this case it is convenient to use the definition of dS_c, which is applicable to both timelike (spacelike) and lightlike hypersurfaces:

$$dS_c = \epsilon N_c\sqrt{|g(x^a(y^i))|}d^3y,$$

where $|g(x^a(y^i))|$ is a modulus of restriction of the metric determinant in the entire spacetime Ω to the Σ_0, $\{y^i\}$ are arbitrary internal coordinates on the hypersurface. As before, $\epsilon = -1$ for a timelike hypersurface and 1 for a spacelike hypersurface, but in addition ϵ is also 1 for a lightlike hypersurface.

Similarly, the variation of the matter action with the energy-momentum tensor (6) is divided into volume and surface parts:

$$\delta S_m = \frac{1}{2}\int_{\Omega+}\sqrt{|g|}\left(T^{+}_{ab}\delta g^{ab}\right)d^4x + \frac{1}{2}\int_{\Omega-}\sqrt{|g|}\left(T^{-}_{ab}\delta g^{ab}\right)d^4x - \\ - \frac{1}{2}\int_{\Sigma_0}\sqrt{|g(x^a(y^i))|}\left(S^{ab}\delta g_{ab}\right)d^3y.$$

According to the least action principle:

$$\delta S = \delta S_q + \delta S_m = 0.$$

From that the system of motion equations is derived:

$$H^{\pm}_{ab} = 8\pi T^{\pm}_{ab}, \tag{9}$$

$$\epsilon[V^c]N_c = 8\pi S^{ab}\delta g_{ab}. \tag{10}$$

The analogue of Israel's equation for quadratic gravity can be obtained directly from (10). The cases of timelike and spacelike hypersurfaces are analyzed, in particular, in the papers [24], [27]. As mentioned above, here the focus is on the lightlike case. However, before moving on to a particular type of hypersurface, the expression (10) can be greatly simplified.

Since the variation of the gravitational part of the action δS_q contains divergence from the vector V^c, it can be argued that the vector V^c itself is determined up to the addition of some divergence-free vector multiplied by an arbitrary constant - $2CU^c$. As demonstrated in the article [31], it is convenient to choose the vector U^c as follows:

$$U^c = -\nabla^b\left(R^{cd} - \frac{1}{2}g^{cd}R\right)\delta g_{bd} + \left(R^{ab}g^{cd} - R^{bc}g^{ad}\right)\nabla_a\delta g_{bd}.$$

In this case, let's put $C = 2\alpha_3$, then:

$$\widetilde{V}^{+c} = V^{+c} + 4\alpha_3 U^{+c} = \left\{(4\alpha_1+\alpha_2)\nabla^c R^{+bd} + \frac{1}{2}(\alpha_2+4\alpha_3)g^{bd}\nabla^c R^{+}\right\}\delta g_{bd} - \\ - 2(2\alpha_1+\alpha_2+2\alpha_3)\nabla^b R^{+cd}\delta g_{bd} + (\alpha_4 + 2\alpha_3 R^{+})(g^{ab}g^{cd} - g^{ac}g^{bd})\nabla_a\delta g_{bd} \\ + \left\{\alpha_2(2g^{cd}R^{+ab} - g^{ac}R^{+bd} - g^{bd}R^{+ac}) - 4\alpha_1 R^{+abcd}\right\}\nabla_a\delta g_{bd} + \\ + 4\alpha_3(R^{+ab}g^{cd} - R^{+bc}g^{ad})\nabla_a\delta g_{bd}\,.$$

As before, the vectors U^{-c}, $\widetilde{V}^{-c}$ are defined similar to U^{+c}, $\widetilde{V}^{+c}$.

In coordinates $\{x^a\}$: $N_c = \delta^n_c$, therefore, replacing $V^{\pm c}$ with $\widetilde{V}^{\pm c}$ in (10) one can get:

$$\epsilon[\widetilde{V}^n] = 8\pi S^{ab}\delta g_{ab}. \tag{11}$$

If the Lichnerowicz conditions are satisfied, then as a consequence: $[\nabla_a \delta g_{bd}] = 0$, so this factor can be taken out of the bracket when calculating the jump $[\widetilde{V}^n]$:

$$\begin{aligned}[\widetilde{V}^n] = &\left\{(4\alpha_1+\alpha_2)[\nabla^n R^{bd}] + \frac{1}{2}(\alpha_2+4\alpha_3)g^{bd}[\nabla^n R] - 2(2\alpha_1+\alpha_2+2\alpha_3)[\nabla^b R^{nd}]\right\}\delta g_{bd} + \\ &+\left\{2\alpha_3[R](g^{ab}g^{nd}-g^{an}g^{bd}) + \alpha_2(2g^{nd}[R^{ab}] - g^{an}[R^{bd}] - g^{bd}[R^{an}])\right\}\nabla_a\delta g_{bd} + \\ &+\left\{4\alpha_3([R^{ab}]g^{nd} - [R^{bn}]g^{ad}) - 4\alpha_1[R^{abnd}]\right\}\nabla_a\delta g_{bd}. \end{aligned}\tag{12}$$

In order to transform the vector $[\widetilde{V}^n]$, let's write out expressions for the jumps of the Riemann tensor components and its contractions, bearing in mind that jumps are present only in those quantities that contain derivatives of the metric with respect to n, starting with the second-order derivatives, since the jumps of the first derivatives of the metric are zero due to the Lichnerowicz conditions.

$$[R_{iklm}] = \frac{1}{2}\left(\delta^n_k\delta^n_l[\partial^2_{nn}g_{im}] + \delta^n_i\delta^n_m[\partial^2_{nn}g_{kl}] - \delta^n_i\delta^n_l[\partial^2_{nn}g_{km}] - \delta^n_k\delta^n_m[\partial^2_{nn}g_{il}]\right), \tag{13}$$

$$R^{ab} = \frac{1}{2}\left(g^{an}g^{bn}g^{cd} + g^{nn}g^{ac}g^{bd} - g^{an}g^{bd}g^{cn} - g^{bn}g^{ad}g^{cn}\right)[\partial^2_{nn}g_{cd}], \tag{14}$$

$$[R] = \left(g^{nn}g^{cd} - g^{cn}g^{dn}\right)[\partial^2_{nn}g_{cd}]. \tag{15}$$

The factor in $[\widetilde{V}^n]$ for $\nabla_a\delta g_{bd}$ (let's denote it A^{abd}) requires more detailed consideration, taking into account the relations (13 - 15), one can get:

$$\begin{aligned}A^{abd} &= -(\alpha_2+4\alpha_3)g^{an}g^{bd}[R^{n\ \ n}_{\ nn}] + (\alpha_2+4\alpha_3)g^{an}(g^{bn}g^{dn} - g^{bd}g^{nn})[R_{nn}] + \\ &\quad + 2(2\alpha_1+2\alpha_3+\alpha_2)g^{nb}g^{nd}[R^{n\ \ a}_{\ nn}] + 4(\alpha_3-\alpha_1)g^{an}g^{dn}[R^{\ bn}_{n\ \ n}] + \\ &\quad + 2(2\alpha_1+2\alpha_3+\alpha_2)g^{nn}g^{nd}[R^{\ b\ a}_{n\ n}] - (\alpha_2+4\alpha_1)g^{nn}g^{na}[R^{\ b\ d}_{n\ n}] = \\ &= -\frac{1}{2}(\alpha_2+4\alpha_3)g^{an}g^{bd}[R] + 2(2\alpha_1+2\alpha_3+\alpha_2)g^{nd}[R^{ab}] - (\alpha_2+4\alpha_1)g^{an}[R^{bd}]. \end{aligned}\tag{16}$$

In order to perform the following operation with A^{abd}, it is necessary to return to the integration of A^{abd} over the hypersurface Σ_0 as a part of $[\widetilde{V}^c]$ present in δS_q:

$$\begin{aligned}&\int_{\Sigma_0}\left(A^{abd}\nabla_a\delta g_{bd}\right)\sqrt{|g(x^a(y^i))|}d^3y = \int_{\Sigma_0}\left(-\nabla_i A^{ibd}\delta g_{bd} + A^{nbd}\nabla_n\delta g_{bd}\right)\sqrt{|g(x^a(y^i))|}d^3y + \\ &+\int_{\Sigma_0}\nabla_i\left(A^{ibd}\delta g_{bd}\right)\sqrt{|g(x^a(y^i))|}d^3y = \int_{\Sigma_0}\left((-\nabla_a A^{abd} + \nabla_n A^{nbd})\delta g_{bd}\right)\sqrt{|g(x^a(y^i))|}d^3y + \\ &\quad + \int_{\Sigma_0}\left(A^{nbd}\nabla_n\delta g_{bd}\right)\sqrt{|g(x^a(y^i))|}d^3y + \int_{\Sigma_0}\nabla_i\left(A^{ibd}\delta g_{bd}\right)\sqrt{|g(x^a(y^i))|}d^3y, \quad i\neq n.\end{aligned}$$

Let's show that the integral $\int_{\Sigma_0}\nabla_i\left(A^{ibd}\delta g_{bd}\right)\sqrt{|g(x^a(y^i))|}d^3y$ is zero. Since the integrand is a complete divergence of some vector field on the hypersurface Σ_0, by virtue of the Gauss-Stokes theorem, this integral reduces to an integral over the boundary of the hypersurface Σ_0, which is a part of the boundary of the entire space-time under consideration $\partial\Omega$, where $\delta g_{bd} = 0$.

With all of the above equation (11) is reduced to:

$$(\alpha_2 + 4\alpha_1)\left\{2[\nabla^n R^{bd}] - g^{nn}[\nabla_n R^{bd}]\right\}\delta g_{bd} + \frac{1}{2}(\alpha_2 + 4\alpha_3)g^{bd}\left\{2[\nabla^n R] - g^{nn}[\nabla_n R]\right\}\delta g_{bd} -$$
$$- (\alpha_2 + 2\alpha_1 + 2\alpha_3)\left\{2g^{ib}[\nabla_i R^{nd}] + g^{nd}[\nabla^b R]\right\}\delta g_{bd} +$$
$$+ \left(-\frac{1}{2}(\alpha_2 + 4\alpha_3)g^{nn}g^{bd}[R] + (2\alpha_1 + 2\alpha_3 + \alpha_2)g^{nd}g^{bn}[R] - (\alpha_2 + 4\alpha_1)g^{nn}[R^{bd}]\right)\nabla_n \delta g_{bd} =$$
$$= 8\pi\epsilon\, S^{ab}\delta g_{ab}, \quad i \neq n. \quad (17)$$

When deriving the equation (17), among other things, the following expression was used:

$$[R^{nb}] = \frac{1}{2}g^{nb}[R], \quad (18)$$

which is a consequence of (13-15).

Let's consider in more detail the multiplier for $\nabla_n \delta g_{bd}$; it can be shown that it is zero if at least one of the indices b or d is equal to n.

$$-\frac{1}{2}(\alpha_2 + 4\alpha_3)g^{nn}g^{bn}[R] + (2\alpha_1 + 2\alpha_3 + \alpha_2)g^{nn}g^{bn}[R] - (\alpha_2 + 4\alpha_1)g^{nn}[R^{bn}] =$$
$$= \frac{1}{2}(\alpha_2 + 4\alpha_3)g^{nn}(g^{bn}[R] - 2[R^{bn}]) = 0.$$

Here the last equality is satisfied due to (18).

As noted in the article [27], on the one hand, the variation of the derivatives of the metric components $\delta(\partial_n g_{ij})$ on Σ_0 is not independent with respect to the variation of the metric components δg_{ij} on the hypersurface, on the other hand, in a sense, the relation between them is arbitrary, since the equations of motion in quadratic gravity are of the fourth order with respect to the derivatives of the metric tensor; therefore, their solutions are not uniquely determined by specifying the initial conditions on the metric components and their first derivatives on some Cauchy hypersurface. Thus, it is necessary to demand:

$$\delta(\partial_n g_{ij}) = B^{kl}_{ij}(y)\delta g_{kl}, \quad i, j, l, k \neq n,$$

where $B^{kl}_{ij}(y)$ are arbitrary functions.

The appearance of arbitrary functions in the equations of motion is due to the implicit presence of δ' and is a marker of a double layer.

Taking into account all of the above, the equations of motion for a double layer of an arbitrary type in quadratic gravity are:

$$2(\alpha_3 - \alpha_1)\left(g^{nn}[\partial^n R] - 2[\nabla^n R^{nn}]\right) + \frac{1}{2}(\alpha_2 + 4\alpha_3)g^{nn}\left(2[\nabla_n R^{nn}] - g^{nn}[\partial_n R]\right) = 8\pi\epsilon S^{nn}, \quad (19)$$

$$(\alpha_2 + 6\alpha_1 - 2\alpha_3)[\nabla^n R^{ni}] + 2(\alpha_3 - \alpha_1)g^{nn}[\nabla_n R^{ni}] + \frac{1}{2}(\alpha_2 + 6\alpha_3 - 2\alpha_1)g^{ni}[\partial^n R] -$$
$$- (\alpha_2 + 2\alpha_1 + 2\alpha_3)\left([\nabla^i R^{nn}] - g^{ni}[\nabla_n R^{nn}] + \frac{1}{2}g^{nn}[\partial^i R] + g^{ni}g^{nj}\Gamma^n_{nj}[R]\right) +$$
$$+ \frac{1}{2}(\alpha_2 + 4\alpha_3)g^{nn}\left(g^{ij}\Gamma^n_{nj}[R] - g^{ni}[\partial_n R]\right) + (\alpha_2 + 4\alpha_1)g^{nn}\Gamma^n_{nj}[R^{ij}] = 8\pi\epsilon S^{ni}, \quad (20)$$

$$(\alpha_2+4\alpha_1)\left\{2[\nabla^n R^{ij}]-g^{nn}[\nabla_n R^{ij}]+g^{nn}\Gamma^j_{nk}[R^{ik}]+g^{nn}\Gamma^i_{nk}[R^{jk}]-g^{nn}B^{ij}_{kl}(y)[R^{kl}]\right\}+$$
$$+\frac{1}{2}(\alpha_2+4\alpha_3)\left\{2g^{ij}[\partial^n R]-g^{nn}g^{ij}[\partial_n R]+g^{nn}g^{ik}\Gamma^j_{nk}[R]+g^{nn}g^{jk}\Gamma^i_{nk}[R]-g^{nn}g^{kl}B^{ij}_{kl}(y)[R]\right\}+$$
$$+(\alpha_2+2\alpha_1+2\alpha_3)\left\{g^{ni}[\nabla_n R^{nj}]+g^{nj}[\nabla_n R^{ni}]-[\nabla^i R^{nj}]-[\nabla^j R^{ni}]-\frac{1}{2}g^{ni}[\partial^j R]-\frac{1}{2}g^{nj}[\partial^i R]\right\}+$$
$$+(\alpha_2+2\alpha_1+2\alpha_3)[R]\left\{-g^{ni}g^{nk}\Gamma^j_{nk}-g^{nj}g^{nk}\Gamma^i_{nk}+g^{nk}g^{nl}B^{ij}_{kl}(y)\right\}=8\pi\epsilon S^{ij},\ i,j,k,l\neq n. \tag{21}$$

For a null double layer, the system of equations (19 - 21) can be greatly simplified:

$$-4(\alpha_3-\alpha_1)g^{ni}[\nabla_i R^{nn}]=-4(\alpha_3-\alpha_1)g^{ni}[\nabla_i(N_aN_bR^{ab})]=$$
$$=-4(\alpha_3-\alpha_1)g^{ni}\partial_i[R^{nn}]=0=8\pi S^{nn}, \tag{22}$$

$$(\alpha_2+6\alpha_1-2\alpha_3)g^{nj}[\nabla_j R^{ni}]+\frac{1}{2}(\alpha_2+6\alpha_3-2\alpha_1)g^{ni}g^{nj}[\partial_j R]-$$
$$-(\alpha_2+2\alpha_1+2\alpha_3)g^{ni}g^{nj}\Gamma^n_{nj}[R]=8\pi S^{ni}, \tag{23}$$

$$2(\alpha_2+4\alpha_1)g^{nk}[\nabla_k R^{ij}]+(\alpha_2+4\alpha_3)g^{nk}g^{ij}[\partial_k R]-$$
$$-(\alpha_2+2\alpha_1+2\alpha_3)\left\{g^{ik}[\nabla_k R^{nj}]+g^{jk}[\nabla_k R^{ni}]+\frac{1}{2}g^{ni}[\partial^j R]+\frac{1}{2}g^{nj}[\partial^i R]\right\}+$$
$$+(\alpha_2+2\alpha_1+2\alpha_3)[R]\left\{-g^{ni}g^{nk}\Gamma^j_{nk}-g^{nj}g^{nk}\Gamma^i_{nk}+g^{nk}g^{nl}B^{ij}_{kl}(y)\right\}=8\pi S^{ij},$$
$$i,j,k,l\neq n. \tag{24}$$

It should be noted that from the equation (22) it follows that S^{nn} is zero in any coordinate system, since it is a scalar:$S^{nn}=N_aN_bS^{ab}$.

During the derivation of motion equations the hypersurface Σ_0 was assumed to be given a priori, but in applications an inverse problem arises when the solutions of the motion equations in the domains $\Omega^{\pm}$ - $g^{\pm}_{ab}$ are known and it is required to find the equation of the hypersurface on which the matching takes place.

Since only equations with S^{ij} on the right-hand side contain arbitrary functions in the case of a double layer they serve to determine $B^{ij}_{kl}(y)$, while equations with S^{nn} and S^{ni} on the right-hand side together with Lichnerowicz conditions are necessary to determine Σ_0 itself.

The picture is different for thin shells. The main criterion that Σ_0 is a thin shell and not a double layer is the absence of arbitrary functions in the motion equations. For the timelike and spacelike cases this criterion is reduced to the absence of jumps in the Ricci tensor: $[R^{bd}]=0$. For the lightlike hypersurface this is the absence of jumps in the scalar curvature $[R]=0$. In this case, the jump $[R_{nn}]$ can be nonzero, since $g^{nn}|_{\Sigma_0}=0$ for a lightlike hypersurface. It can be shown that the absence of the corresponding jumps for any type of thin shell in addition to the absence of arbitrary functions also implies that $S^{nn}=S^{ni}=0$.

Thus the equations of motion for timelike (spacelike) thin shells are a special case of the system (19 - 21) with the additional conditions $[R^{bd}]=0$:

$$S^{nn}=S^{ni}=0,$$
$$(\alpha_2+4\alpha_1)g^{nn}[\nabla_n R^{ij}]+\frac{1}{2}(\alpha_2+4\alpha_3)g^{nn}g^{ij}[\partial_n R]-$$
$$-(\alpha_2+2\alpha_1+2\alpha_3)g^{ni}g^{nj}[\partial_n R]=8\pi\epsilon S^{ij},\quad i,j\neq n. \tag{25}$$

Similarly for null thin shells we get:

$$S^{nn} = S^{ni} = 0,$$

$$2(\alpha_2 + 4\alpha_1) g^{nk} [\nabla_k R^{ij}] -$$
$$- (\alpha_2 + 2\alpha_1 + 2\alpha_3) \left\{ g^{ik} [\nabla_k R^{nj}] + g^{jk} [\nabla_k R^{ni}] + g^{nj} g^{ni} [\partial_n R] \right\} = 8\pi S^{ij},$$
$$i, j, k \neq n. \quad (26)$$

It will also be shown below that for a null hypersurface the condition $[R] = 0$ is satisfied in particular when the normal vector field to Σ_0 is null not only on the hypersurface but also in some neighborhood of the Σ_0.

4. Null case

As noted earlier, in order to avoid the appearance of undefined functions in the Lagrangian, it is necessary to ensure the continuity of the metric components on the hypersurface. This can always be achieved by choosing the appropriate coordinates. For timelike and spacelike hypersurfaces it is usually convenient to use Gaussian normal coordinates. In the paper [29] it was demonstrated that for these types of singular hypersurfaces it is always possible to dock two different Gaussian normal coordinate systems from the $\Omega^{\pm}$ domains on the hypersurface in such a way that the jumps of the metric components are zero.

For lightlike hypersurfaces there are some analogs of Gaussian normal coordinates, such as Gaussian null coordinates (GNC) [32] – [34] or coordinates for foliation of manifold by null hypersurfaces (NSFC) [35]–[37]. Nevertheless, in the present work a more minimalistic approach was used, the one that does not imply additional information regarding the behavior of the normal vector field of the hypersurface outside Σ_0. Using the formalism described in the book of E. Poisson [38], a coordinate system with the necessary properties namely continuity of the metric components on Σ_0 and coordinates themselves in some neighborhood of the hypersurface was constructed. The following is a brief description of this process.

Consider the lightlike hypersurface Σ_0 defined in the domains $\Omega^{\pm}$ with different coordinates $\{x^{\pm}\}$ by the equations: $n^{\pm}(x^{\pm}) = 0$. The normal vector field to a hypersurface is defined as: $N_{\pm a} = \partial_a n^{\pm}(x^{\pm})$. For the sake of convenience, let's temporarily omit the notations $\pm$ and write all the equations in one of the domains.

In the case of a null hypersurface, the norm $N^a N_a$ is zero on Σ_0, so the normal vector is defined up to multiplication by an arbitrary scalar function. Since the vector N^a is null on Σ_0, it is simultaneously tangent to the hypersurface.

Let's show that vector field N^a on the hypersurface is tangent to the family of null geodesics that lie on Σ_0:

$$\nabla_b N_a N^b = \nabla_b \nabla_a n \, \partial^b n = \nabla_a \partial_b n \, \partial^b n = \frac{1}{2} \nabla_a \left(\partial_b n \, \partial^b n \right).$$

The scalar $N^b N_b$ is zero on Σ_0 therefore its gradient is directed along N_a so:

$$\nabla_b N_a N^b = \kappa N_a,$$

for some scalar κ.

Thus, N^a on the hypersurface is a tangent to lightlike geodesics that lie on Σ_0 and are actually its generators.

Let's choose an arbitrary parameter λ on these generators and two additional coordinates - $\theta^A, A = \{2, 3\}$ for marking geodesics, together they form the internal coordinate system $\{y^i\} = \{\lambda, \theta^A\}$ on Σ_0. The parameter λ is affine if the equation $n(x) = const$ defines a whole family of null hypersurfaces, in this case the scalar $N^a N_a$ is zero not only on the hypersurface, but at least in some neighborhood of it.

Let's calculate the induced metric on the hypersurface:

$$ds^2_{\Sigma_0} = g_{ab}\, dx^a(y^i)\, dx^b(y^j) = g_{ab}\, e^a_i\, e^b_j\, dy^i dy^j, \quad e^a_i = \frac{\partial x^a}{\partial y^i}.$$

The vector fields e^a_i are tangent to the curves lying on the hypersurface, so they are orthogonal to the normal on Σ_0:

$$N_a e^a_A = 0, \quad N_a e^a_1 = N_a \frac{\partial x^a}{\partial \lambda} = N_a N^a = 0.$$

These relations also imply that the induced metric on the Σ_0 is effectively two-dimensional:

$$ds^2_{\Sigma_0} = g_{ab}\, e^a_A\, e^b_B\, d\theta^A d\theta^B = \sigma_{AB}\, d\theta^A d\theta^B. \tag{27}$$

The system of vector fields $\{N^a, e^a_A\}$ can be completed to a basis in the restriction of the tangent bundle of the manifold Ω to the hypersurface - $T\Omega|_{\Sigma_0}$ if an auxiliary null vector field l^a with the following properties is found:

$$l^a N_a = 1, \quad l_a e^a_A = 0, \quad l^a l_a = 0.$$

The existence of such a vector, as well as the completeness of the system of vector fields $\{l^a, N^a, e^a_A\}$ in $T\Omega|_{\Sigma_0}$ are demonstrated in the aforementioned book [38], as well as in the article [39], which introduces a generalization for such type of vectors (rigging vector) for singular hypersurfaces of mixed causal character.

Since the vector fields $\{l^a, N^a, e^a_A\}$ form a basis in $T\Omega|_{\Sigma_0}$, it is possible to write completeness relations for the inverse metric:

$$g^{ab} = l^a N^b + l^b N^a + \sigma^{AB} e^a_A e^b_B. \tag{28}$$

Let's choose $\{n, \lambda, \theta^A\}$ as local coordinates in a neighborhood of the hypersurface. In these coordinates on Σ_0 the following relations hold:

$$N_a = \delta^n_a, \quad N^a = \frac{\partial x^a(y^i)}{\partial \lambda} = \delta^a_\lambda, \quad l_\lambda = l^n = 1, \quad l_n = -l^\lambda, \quad l_A = 0. \tag{29}$$

Taking into account (28), in coordinates $\{n, \lambda, \theta^A\}$ the inverse metric has the following structure on the hypersurface:

$$g^{n\lambda} = 1, \quad g^{nn} = g^{nA} = 0, \quad g^{\lambda\lambda} = 2\, l^\lambda, \quad g^{\lambda A} = l^A, \quad g^{AB} = \sigma^{AB}. \tag{30}$$

Then for the tensor g_{ab} inverse to g^{ab} one can get:

$$g_{n\lambda} = 1, \quad g_{\lambda\lambda} = g_{\lambda A} = 0, \quad g_{nn} = -2\, l^\lambda + \sigma_{AB} l^B l^A, \quad g_{nA} = -\sigma_{AB} l^B, \quad g_{AB} = \sigma_{AB}. \tag{31}$$

It should be noted, that the relations (30, 31) are true only on Σ_0, i.e. for $n = 0$. This means, in particular, that the second derivatives with respect to n of $g_{n\lambda}, g_{\lambda\lambda}, g_{\lambda A}$ and their jumps on Σ_0 are generally non-zero. The jumps of the first derivatives with respect to n are taken to be zero by virtue of the Lichnerowicz conditions. A similar kind of metric applies to the neighborhood Σ_0 only if the vector field N^a is null also in some neighborhood of Σ_0, but even in this case $\partial^2_{nn}\, g_{\lambda n}$ is not necessarily zero.

A coordinate system common for $\Omega^\pm$ with the necessary properties is constructed as follows: n^+ and n^- are continuously combined into one coordinate n and y^{+i} or y^{-i} are chosen as the remaining coordinates. The continuity of the metric on Σ_0 is ensured by the relations (30) and the proper selection of functions $y^{+i}(y^{-j})$.

Next, let's proceed to the analysis of the equations (22 - 24, 26) in the coordinates $\{n, \lambda, \theta^A\}$. Let's calculate the jumps present in these equations:

$$[R] = 2\,[R^{n}{}_{nn}{}^{n}] = [\partial^2_{nn} g_{\lambda\lambda}], \quad [\partial_i R] = [\partial_i \partial^2_{nn} g_{\lambda\lambda}], \tag{32}$$

$$[R^{ab}] = \frac{1}{2}\left(\delta^a_\lambda\, g^{bc}\,[\partial^2_{nn} g_{c\lambda}] + \delta^b_\lambda\, g^{ac}\,[\partial^2_{nn} g_{c\lambda}] - \delta^a_\lambda\, \delta^b_\lambda\, g^{cd}\,[\partial^2_{nn} g_{cd}]\right), \tag{33}$$

$$[\nabla_\lambda R^{bd}] = \partial_\lambda [R^{bd}] + \delta^b_\lambda \left(\Gamma^\lambda_{\lambda\lambda}[R^{\lambda d}] + \Gamma^d_{\lambda A}[R^{\lambda A}] + \frac{1}{2}\Gamma^d_{\lambda n}\,[R]\right) + \\ + \delta^d_\lambda \left(\Gamma^\lambda_{\lambda\lambda}[R^{\lambda b}] + \Gamma^b_{\lambda A}[R^{\lambda A}] + \frac{1}{2}\Gamma^b_{\lambda n}\,[R]\right), \tag{34}$$

$$[\nabla_i R^{nd}] = \frac{1}{2}\,\delta^d_\lambda \left(\partial_i [R] + \Gamma^n_{in}\,[R] + 2\,\Gamma^n_{iA}\,[R^{\lambda A}]\right) + \frac{1}{2}\Gamma^d_{i\lambda}\,[R]. \tag{35}$$

Taking into account all of the above, the equations for a null double layer in the coordinates $\{n, \lambda, \theta^A\}$ are:

$$S^{nn} = 0, \tag{36}$$

$$2(\alpha_2 + 4\alpha_1)[\nabla_\lambda R^{\lambda\lambda}] + (\alpha_2 + 4\alpha_3) g^{\lambda\lambda}[\partial_\lambda R] - \\ - (\alpha_2 + 2\alpha_1 + 2\alpha_3)\left(2g^{\lambda i}[\nabla_i R^{\lambda n}] + [\partial^\lambda R] + 2\Gamma^\lambda_{\lambda n}[R] - B^{\lambda\lambda}[R]\right) = 8\pi\, S^{\lambda\lambda}, \tag{37}$$

$$(\alpha_2 - 2\alpha_3 + 6\alpha_1)[\nabla_\lambda R^{\lambda n}] + \frac{1}{2}(\alpha_2 - 2\alpha_1 + 6\alpha_3)[\partial_\lambda R] - (\alpha_2 + 2\alpha_3 + 2\alpha_1)\Gamma^n_{\lambda n}[R] = 8\pi\, S^{n\lambda}, \tag{38}$$

$$2(\alpha_2 + 4\alpha_1)[\nabla_\lambda R^{\lambda A}] + (\alpha_2 + 4\alpha_3) g^{\lambda A}[\partial_\lambda R] - \\ - \frac{1}{2}(\alpha_2 + 2\alpha_1 + 2\alpha_3)\left(2g^{iA}[\nabla_i R^{\lambda n}] + [\partial^A R] + g^{\lambda B}\Gamma^A_{\lambda B}[R] + 2\Gamma^A_{\lambda n}[R] - 2B^{\lambda A}[R]\right) = 8\pi\, S^{\lambda A}, \tag{39}$$

$$(\alpha_2 + 4\alpha_3)\sigma^{AB}[\partial_\lambda R] - (\alpha_2 + 2\alpha_1 + 2\alpha_3)[R](\sigma^{AC}\Gamma^B_{C\lambda} - B^{AB}) = 8\pi\, S^{AB}. \tag{40}$$

Let's also consider an important special case when the vector field N^a is null in some neighborhood of Σ_0. Then it follows from (32 - 35) that:

$$[R] = [R^{\lambda A}] = [\partial_i R] = [\nabla_i R^{nd}] = 0, \tag{41}$$

$$[\nabla_\lambda R^{bd}] = \delta^b_\lambda\, \delta^d_\lambda\, [\partial_\lambda R^{\lambda\lambda}]. \tag{42}$$

Thus, it corresponds to the scenario of a thin shell. In the coordinates $\{n, \lambda, \theta^A\}$ for a null thin shell only one equation of the system (26) remains with a nonzero right-hand side:

$$2(\alpha_2 + 4\alpha_1)[\partial_\lambda R^{\lambda\lambda}] - (\alpha_2 + 2\alpha_1 + 2\alpha_3)[\partial_n R] = 8\pi\, S^{\lambda\lambda}. \tag{43}$$

5. Spherically symmetric case

Let's investigate a special case of null singular hypersurfaces namely spherically symmetric null hypersurfaces separating two spherically symmetric space-times. It will be demonstrated below that such singular hypersurfaces can only be thin shells.

Consider a spherically symmetric metric of the most general form:

$$ds^2 = \gamma_{\alpha\beta}(x)\, dx^\alpha dx^\beta - r^2(x)\, \left(d\theta^2 + \sin^2\theta\, d\phi^2\right) = r^2(x)\left(\widetilde{\gamma}_{\alpha\beta}(x)\, dx^\alpha dx^\beta - d\Omega^2\right) = \\ = r^2(x)\left(d\widetilde{s}_2^2 - d\Omega^2\right), \quad \alpha, \beta = 0, 1. \quad (44)$$

Hereinafter $2+2$ decomposition of the metric in the case of spherical symmetry is used.

For this type of geometry, the following relationships hold:

$$R_{\theta\theta} = 1 + r\sigma + \Delta, \quad R_{\phi\phi} = R_{\theta\theta}\, sin^2\theta, \quad (45)$$

$$R_{\alpha\beta} = \frac{\gamma_{\alpha\beta}}{r^2}\left(\frac{1}{2}\widetilde{R} + \Delta - r\,\sigma\right) - \frac{2}{r}\nabla_\alpha\nabla_\beta\, r, \quad (46)$$

$$R = \frac{1}{r^2}\left(\widetilde{R} - 2 - 6\, r\,\sigma\right), \quad (47)$$

where Δ, σ, $\widetilde{R}$ are invariants of spherical geometry.

$$\Delta = \gamma^{\alpha\beta}\,\partial_\alpha r\,\partial_\beta r, \quad \sigma = \gamma^{\alpha\beta}\,\nabla_\alpha\nabla_\beta\, r\,, \quad (48)$$

$$\widetilde{R} = \frac{1}{\widetilde{\gamma}}\left(-\partial^2_{11}\widetilde{\gamma}_{00} + 2\partial^2_{01}\widetilde{\gamma}_{01} - \partial^2_{00}\widetilde{\gamma}_{11}\right) + \frac{1}{2\widetilde{\gamma}^2}\left(\widetilde{\gamma}_{11}\left(\partial_1\widetilde{\gamma}_{00}\right)^2 + \widetilde{\gamma}_{00}\left(\partial_0\widetilde{\gamma}_{11}\right)^2\right) + \\ + \frac{1}{2\widetilde{\gamma}^2}\left(\partial_1\widetilde{\gamma}_{00}\left(-\partial_0\widetilde{\gamma}_{11}\,\widetilde{\gamma}_{10} + \partial_1\widetilde{\gamma}_{11}\,\widetilde{\gamma}_{00} - 2\partial_1\widetilde{\gamma}_{01}\,\widetilde{\gamma}_{10}\right) + \partial_0\widetilde{\gamma}_{11}\left(\partial_0\widetilde{\gamma}_{00}\,\widetilde{\gamma}_{11} - 2\,\partial_0\widetilde{\gamma}_{10}\,\widetilde{\gamma}_{10}\right)\right) + \\ + \frac{1}{2\widetilde{\gamma}^2}\left(\partial_0\widetilde{\gamma}_{00}\left(\partial_1\widetilde{\gamma}_{11}\,\widetilde{\gamma}_{01} - 2\,\partial_1\widetilde{\gamma}_{10}\,\widetilde{\gamma}_{11}\right) - 2\,\partial_0\widetilde{\gamma}_{10}\left(\partial_1\widetilde{\gamma}_{11}\,\widetilde{\gamma}_{00} - 2\,\partial_1\widetilde{\gamma}_{10}\,\widetilde{\gamma}_{10}\right)\right). \quad (49)$$

Here $\widetilde{R}$ and $\widetilde{\gamma}$ are the determinant and the scalar curvature of the two-dimensional "metric" $\widetilde{\gamma}_{\alpha\beta}$ respectively.

For any spherically symmetric lightlike hypersurface in geometry (44) given by the equation $n(x) = 0$ the following is true:

$$\gamma_{00}\left(\partial^0 n(x)\right)^2 + 2\gamma_{01}\,\partial^0 n(x)\,\partial^0 n(x) + \gamma_{11}\left(\partial^1 n(x)\right)^2 = 0.$$

If this equation can be solved with respect to the variable $\frac{\partial^0 n(x)}{\partial^1 n(x)}$ or the reciprocal of it at any values of x, then the function $n(x)$ potentially defines a whole family of lightlike hypersurfaces $n(x) = const$. Since this is a quadratic equation with respect to the indicated variable, in order for it to always have at least one solution, $-\gamma$ must be nonnegative, but this condition is always satisfied for any Lorentzian spherically symmetric manifold wherever the determinant γ exists. It means that the hypersurface is at least locally a leaf of some null foliation in each of the regions $\Omega^\pm$, and this, as shown in the previous section, is equivalent to the condition $[R] = 0$.

Thus, it has been demonstrated that a lightlike spherically symmetric singular hypersurface in a spherically symmetric geometry can only be a thin shell.

The coordinates $\{n, \lambda, \theta, \phi\}$ defined in the previous section is used to describe a spherically symmetric null thin shell. In these coordinates the metric in the neighborhood of Σ_0 has the form:

$$ds^2 = g_{nn}(n, \lambda)\, dn^2 + 2g_{n\lambda}(n, \lambda)\, dn\, d\lambda - r^2(n, \lambda)\, d\Omega^2. \quad (50)$$

The equation (43) describing the dynamics of a null thin shell in this case can be rewritten using previously defined invariants and their derivatives with respect to n:

$$2(\alpha_2 + 4\alpha_1)[\partial_n \Delta] - (\alpha_2 + 2\alpha_1 + 2\alpha_3)[\partial_n \widetilde{R}] + 4r\,(\alpha_2 + 3\alpha_3 + \alpha_1)[\partial_n \sigma] = 8\pi\, r^2\, S^{\lambda}_{n}. \qquad (51)$$

The Lichnerowicz conditions can also be expressed in terms of invariants:

$$[\Delta] = 0, \quad [\widetilde{R}] = 0. \qquad (52)$$

It needs to be clarified that the conditions (52) ensure the absence of jumps $[\partial_n g_{n\lambda}], [\partial_n r]$, while for a metric of the form (50) it is also necessary to require: $[\partial_n g_{nn}] = 0$. This condition could be satisfied by adding $[\sigma] = 0$ to the Lichnerowicz conditions but there is no need for this, since according to (47) it automatically follows from the continuity of R, r and $\widetilde{R}$ on Σ_0.

6. Conformal gravity

The action of the Weyl conformal gravity is a special case of the action (1) with $\alpha_2 = -2\alpha_1$, $\alpha_3 = \frac{1}{3}\alpha_1$, $\alpha_4 = \alpha_5 = 0$:

$$S_q = -\frac{1}{16\pi}\int_\Omega \sqrt{-g}\alpha_1 \left(R_{abcd}\, R^{abcd} - 2R_{ab}\, R^{ab} + \frac{1}{3}R^2\right) d^4x = \\ = -\frac{1}{16\pi}\int_\Omega \sqrt{-g}\alpha_1 C_{abcd}\, C^{abcd}\, d^4x, \qquad (53)$$

where C_{abcd} is the Weyl tensor, which is defined as:

$$C_{abcd} = R_{abcd} + \frac{1}{2}\left(R_{ad}\, g_{bc} + R_{bc}\, g_{ad} - R_{ac}\, g_{bd} - R_{bd}\, g_{ac}\right) + \frac{1}{6}R\left(g_{ac}\, g_{bd} - g_{ad}\, g_{bc}\right).$$

Therefore, the corresponding equations of motion first obtained in the paper of R. Bach [40] are a special case of (9).

$$\nabla_c \nabla_d\, C^{acbd} + \frac{1}{2}\, C^{acbd}\, R_{cd} = \frac{2\pi}{\alpha_1}\, T^{ab}. \qquad (54)$$

Let's consider spherically symmetric null thin shells separating two spherically symmetric solutions of (54) as applications. As stated earlier, these solutions were presented in the article [28]. Their general form coincides with (44) provided that $r^2(x)$ are arbitrary functions.

Since null thin shells are considered, the general form of the hypersurface equations in the domains $\Omega^\pm$ is known and is determined from the condition of zero interval. The equation (51) together with the Lichnerowicz conditions (52) do not completely fix the equations of the hypersurface, since there are three equations and four unknown functions: $n^\pm(x^\pm), r^\pm(x^\pm)$. Therefore, instead of looking for the exact equations of the hypersurface, let's focus on determining the possibility of matching different spherically symmetric solutions of conformal gravity and the restrictions imposed on them by the Lichnerowicz conditions. It should also be noted that there are cases in wich it is inconvenient to use coordinates $\{n, \lambda, \theta, \phi\}$, thus it is important that conditions (52) are expressed using the invariants of spherical geometry and look the same in any coordinates.

6.1. Vacuum solutions

Let's analyze the case when $\Omega^\pm$ are two spherically symmetric vacuums of conformal gravity. In the aforementioned paper [28] it was demonstrated that in double null coordinates, where:

$$d\widetilde{s}_2^2 = 2\widetilde{H}(u, v)\, du dv, \qquad (55)$$

the Bach equations for vacuum are:

$$\partial^2_{uv}\widetilde{R} = -\frac{\widetilde{H}}{4}\left(\widetilde{R}^2 - 4\right), \tag{56}$$

$$\partial_u\left(ln(\partial_u\widetilde{R})\right) = \partial_u\left(ln(\widetilde{H})\right), \tag{57}$$

$$\partial_v\left(ln(\partial_v\widetilde{R})\right) = \partial_v\left(ln(\widetilde{H})\right). \tag{58}$$

The coordinates $\{n,\lambda\}$ are a special case of double null coordinates, therefore $[\partial^2_{n\lambda}\widetilde{R}] = [-\frac{\widetilde{H}}{4}\left(\widetilde{R}^2-4\right)] = 0$ and $[\partial_n\widetilde{R}] = 0$. Therefore a version of the equation (51) for a thin shell separating two vacuums of conformal gravity is:

$$[\partial_n\Delta] = \frac{2\pi r^2}{\alpha_1}S^{\lambda\lambda}. \tag{59}$$

As noted in the article [28] there are three types of spherically symmetric conformal gravity vacua: two with constant two-dimensional scalar curvature $\widetilde{R} = \mp 2$ and one with variable $\widetilde{R}$.

$$\widetilde{H} = \frac{2}{(u\pm v)^2}, \quad \widetilde{R} = \mp 2. \tag{60}$$

$$\begin{aligned} \widetilde{H} = \frac{1}{2}|A(\widetilde{R}(u,v))|, \quad & A(\widetilde{R}) = \frac{1}{6}\left(\widetilde{R}^3 - 12\widetilde{R} + C_0\right), \\ & \widetilde{R} = \widetilde{R}(v-u) : \int\frac{d\widetilde{R}}{A(\widetilde{R})} = \frac{1}{2}(v-u),\ A(\widetilde{R}) > 0, \\ & \widetilde{R} = \widetilde{R}(v+u) : \int\frac{d\widetilde{R}}{A(\widetilde{R})} = \frac{1}{2}(v+u),\ A(\widetilde{R}) < 0, \end{aligned} \tag{61}$$

where C_0 is some constant.

Since $[\widetilde{R}] = 0$ the following combinations are possible for two vacua: matching a vacuum with a constant $\widetilde{R} = \pm 2$ and a vacuum with a variable $\widetilde{R}$, matching two vacua with variable $\widetilde{R}$, matching two vacua with coinciding constant $\widetilde{R}$.

Let's analyze the situation when Ω^- is a vacuum with a constant $\widetilde{R} = \pm 2$ and Ω^+ is a vacuum with a variable $\widetilde{R}$.

From the condition $[\widetilde{R}] = 0$ it follows that the hypersurface equation in the Ω^+ domain is $n^+(x^+) = \widetilde{R}^+ \mp 2 = 0$. A hypersurface of this type is null only if $C_0 = \pm 16$ when $n^+ = 0$ is an analogue of the double horizon with respect to the variable $\widetilde{R}^+$. In the coordinates $\{n^+, u^+\}$ the two-dimensional "metric" in the Ω^+ domain looks like:

$$d\widetilde{s}_2^{+2} = \frac{1}{6}(n^+)^2(n^+ \pm 6)du^{+2} + 2du^+dn^+. \tag{62}$$

In the domain Ω^- the hypersurface equation is $n^-(x^-) = F^-(v^-) = 0$, where $F^-(v^-)$ is an arbitrary function of v^- (the case when F^- is an arbitrary function of u^- is similar to this one). In the coordinates $\{n^-, u^-\}$ the two-dimensional "metric" in the Ω^- has the following form:

$$d\widetilde{s}_2^{-2} = \frac{4}{(u^- \mp f^-(n^-))^2}\frac{df^-}{dn^-}dn^-du^-, \tag{63}$$

where $f^-(n^-)$ is a function inverse to $F^-(v^-)$.

The continuity of the metric on the hypersurface defines the following connection between coordinates $u^\pm$:

$$r^+(0, u^+) = r^-(0, u^-).$$

The remaining Lichnerowicz condition $[\Delta] = 0$ imposes a restriction on the functions $r^\pm(x^\pm)$:

$$\partial_{n^+} r^+|_{\Sigma_0} = \partial_{\widetilde{R}^+} r^+(2, u^+) = \partial_{n^-} r^-|_{\Sigma_0} = \partial_{n^-} f^-(0)\, \partial_{v^-} r^-(f^-(0), u^-). \tag{64}$$

Let's move on to the analysis of the matching of two vacua with variable $\widetilde{R}$. Without loss of generality the hypersurface Σ_0 can be considered given by the equations: $n^\pm(x^\pm) = F^\pm(u^\pm) = 0$ in $\Omega^\pm$. Here $F^\pm(u^\pm)$ are arbitrary functions of variables $u^\pm$. In this case, it is convenient to choose the coordinates $n^\pm, \widetilde{R}^\pm$:

$$d\widetilde{s}_2^{\pm 2} = A^\pm \left(\frac{df^\pm}{dn^\pm}\right)^2 dn^{\pm 2} + 2sign(A^\pm)\frac{df^\pm}{dn^\pm} dn^+ d\widetilde{R}^\pm. \tag{65}$$

From the continuity of the metric and $\widetilde{R}$ on Σ_0 it follows that $\frac{df^+}{dn^+}(0) = \pm\frac{df^-}{dn^-}(0)$ when $C_0^+ = C_0^-$ .This means that the two given vacua can only differ in functions $r^\pm(n^\pm, \widetilde{R}^\pm)$, but they must also coincide up to the second order in $n^\pm$ on Σ_0:

$$r^+(0, \widetilde{R}^+) = r^-(0, \widetilde{R}^-), \quad \partial_{n^+} r^+(0, \widetilde{R}^+) = \partial_{n^-} r^-(0, \widetilde{R}^-). \tag{66}$$

A similar result, namely a coincidence up to the second derivatives with respect to n of the conformal factor, is obtained by matching two vacua with the same constant $\widetilde{R}$.

6.2. Vaidya-type solution

In this section let's consider the situation when at least one of the domains, namely Ω^+, is a Vaidya-type solution for spherically symmetric conformal gravity also described in the article [28].

In the coordinates $\{\widetilde{R}^+, u^+\}$ the two-dimensional “metric” in the Ω^+ has the following form:

$$d\widetilde{s}_2^{+2} = A^+(\widetilde{R}^+, u^+) du^{+2} + 2du^+ d\widetilde{R}^+, \tag{67}$$

where $A^+(\widetilde{R}^+, u^+) = \frac{1}{6}(\widetilde{R}^{+3} - 12\widetilde{R}^+ + C_0^+(u^+))$, $C_0^+(u^+)$ is an arbitrary function of u^+.

There are two types of null hypersurfaces for this metric. The first type is a hypersurface given by an arbitrary function of the variable u^+, the second type is a hypersurface given by the function $F^+(u^+, \widetilde{R}^+)$ which satisfies the equation:

$$\partial_{u^+} F^+ = \frac{1}{12}(\widetilde{R}^{+3} - 12\widetilde{R}^+ + C_0^+(u^+))\partial_{\widetilde{R}^+} F^+. \tag{68}$$

Let's use the second type of hypersurfaces, since it potentially corresponds to a thin shell with radiation transversal to the hypersurface.

If Ω^- is a vacuum with constant $\widetilde{R}^- = \pm 2$ since $[\widetilde{R}] = 0$ when $n^+(x^+) = \widetilde{R}^+ \mp 2$. It follows from (68) that a hypersurface of this type can be null only if $C_0^+(u^+) = \pm 16$. This corresponds to the degenerate case of the metric (67), which is essentially a vacuum with variable $\widetilde{R}$.

Next let the Ω^- be a vacuum with variable $\widetilde{R}$. Suppose Σ_0 is given by the equation $n^-(x^-) = F^-(u^-) = 0$ in Ω^- and $F^-(u^-)$ is an arbitrary function of u^-. In coordinates $\{n^\pm, \widetilde{R}^\pm\}$ the two-dimensional "metrics” in the $\Omega^\pm$ are:

$$d\widetilde{s}_2^{-2} = A^- \left(\frac{df^-}{dn^-}\right)^2 dn^{-2} + 2sign(A^-)\frac{df^-}{dn^-} dn^- d\widetilde{R}^-, \tag{69}$$

$$d\widetilde{s}_2^{+2} = A^+ \left(\partial_{u+} F^+\right)^{-2} dn^{+2} - 2 \left(\partial_{u+} F^+\right)^{-1} dn^+ d\widetilde{R}^+, \tag{70}$$

where $f^-(n^-)$ is a function inverse to $F^-(u^-)$, $F^+(u^+, \widetilde{R}^+)$ is given by the equation (68).

From the continuity of the metric and $\widetilde{R}$ on Σ_0 it follows that $\frac{df^-}{dn^-}(0) = \pm\partial_{u+}F^+(\widetilde{R}^+, u^+)|_{\Sigma_0}$ and $A^+(\widetilde{R}^+, u^+)|_{\Sigma_0} = A^-(\widetilde{R}^-)|_{\Sigma_0}$. As a result we have: $C_0^+(u^+)|_{\Sigma_0} = C_0^-|_{\Sigma_0} = const$. It means that the hypersurface in Ω^+ is defined by a single variable function $F^+(u^+)$, but this type of null hypersurfaces corresponds to the absence of radiation.

Finally, let's analyze the matching of two Vaidya type solutions. In this case $n^\pm(x^\pm) = F^\pm(u^\pm, \widetilde{R}^\pm)$ are functions defined by the (68). In coordinates $\{n^\pm, \widetilde{R}^\pm\}$ the two-dimensional "metrics" in the $\Omega^\pm$ are:

$$ds_2^{\pm 2} = A^\pm \left(\partial_{u\pm} F^\pm\right)^{-2} dn^{\pm 2} - 2 \left(\partial_{u\pm} F^\pm\right)^{-1} dn^\pm d\widetilde{R}^\pm. \tag{71}$$

From the continuity of the metric and $\widetilde{R}$ on Σ_0 it follows that $\partial_{u-}F^-(\widetilde{R}^-, u^-)|_{\Sigma_0} = \partial_{u+}F^+(\widetilde{R}^+, u^+)|_{\Sigma_0}$ and $A^+(\widetilde{R}^+, u^+)|_{\Sigma_0} = A^-(\widetilde{R}^-, u^-)|_{\Sigma_0}$ hence $C_0^+(u^+)|_{\Sigma_0} = C_0^-(u^-)|_{\Sigma_0}$. It means that the two given solutions can only differ in functions $r^\pm(n^\pm, \widetilde{R}^\pm)$, but they must also coincide up to the second order in $n^\pm$ on Σ_0:

$$r^+(0, \widetilde{R}^+) = r^-(0, \widetilde{R}^-), \quad \partial_{n^+} r^+(0, \widetilde{R}^+) = \partial_{n^-} r^-(0, \widetilde{R}^-). \tag{72}$$

In this case, despite the fact that the hypersurface may be of the second type, it still cannot be interpreted as a shell with radiation.

7. Conclusions

In this concluding section let's discuss the obtained results. Continuing the work started in the article [27], it was shown that for a singular hupersurface of arbitrary type in quadratic gravity equations of motion can be obtained using only the least action principle. It follows from the equations of motion (19-21) that for the ratio of the coefficients corresponding to the Gauss-Bonnet quadratic term: $\alpha_1 = \alpha_3 = -\frac{1}{4}\alpha_2$, $\alpha_4 = \alpha_5 = 0$ all components of the surface energy-momentum tensor are zero. Thus, even if Riemann curvature tensor or its derivatives undergo a jump on the hypersurface the Gauss-Bonnet term does not create neither double layers nor thin shells given that the Lichnerowicz conditions are satisfied.

Analysis of the equations of motion for the null double layer shows that in contrast to the timelike and spacelike cases $S^{nn} = N_a N_b S^{ab} = 0$ in any coordinate system, which means the absence of the so-called "external pressure" for the null double layer in quadratic gravity.

As for thin shells, the main difference between the null case and the others is the condition for the hypersurface to be a thin shell. For a null thin shell is the absence of a jump in scalar curvature, while for the rest the condition is stronger, namely, the absence of jumps in the Ricci tensor.

It was demonstrated that null spherically symmetric singular hupersurfaces in quadratic gravity cannot be a double layer, and only thin shells are possible. In this case the system of equations of motion (26) is reduced to one which is expressed through the invariants of spherical geometry along with the Lichnerowicz conditions.

Spherically symmetric null thin shells were investigated for spherically symmetric solutions of conformal gravity as an applications, in particular, for various vacua and Vaidya-type solutions.

By virtue of Lichnerowicz conditions $[\widetilde{R}] = 0$ therefore, for two vacua, the following combinations are possible: matching a vacuum with a constant $\widetilde{R}$ and a vacuum with a variable $\widetilde{R}$, matching two vacuums with a variable $\widetilde{R}$, matching two vacua with a coinciding constant $\widetilde{R}$. In the first case, the hypersurface is an analogue of the double horizon in the metric with the variable $\widetilde{R}$; in other cases, matching is possible only if the metrics coincide up to a conformal factor.

With the addition of the Vaidya-type solution new possible matchings appear: Vaidya-type metrics with vacuum with variable $\widetilde{R}$, two Vaidya-type metrics. In the first version, the null shell is actually the singular part of the Vaidya-type solution, in the second, they must coincide up to the conformal factor. Moreover the null shell does not emit in both cases presumably due to the fact that the "external flow" is zero.

Acknowledgments
I would like to thank my scientific advisor V. A. Berezin for very valuable discussions.

References

[1] Ostrogradsky M 1850 *Mem. Ac. St. Petersbourg VI* **4** 385
[2] Hooft G, Veltman M J G 1974 *Ann. Inst. H. Poincare Phys. Theor.* A **20** 69
[3] Weinberg S 1974 *Proc. 17th Int. Conf. on High-Energy Physics (London)* vol 3 ed J R Smith (Chilton: Rutherford Laboratory) p 59
[4] Deser S 1975 *Conf. Proc.* C **750926** 229
[5] Stelle K S 1977 *Phys. Rev.* D **16** 953
[6] Utiyama R and DeWitt B S 1962 *J. Math. Phys.* **3** 608
[7] Zel'dovich Ya B 1970 *JETP Lett.* **12** 307
[8] Grib A A, Mamaev S G 1970 *Sov. J. Nucl. Phys.* **10** 722
[9] Zel'dovich Ya B, Pitaevsky L P 1971 *Comm. Math. Phys.* **23** 185
[10] Zel'dovich Ya B, Starobinskii A A 1972 *Sov. Phys. JETP* **34** 1159
[11] L Parker, Fulling S A 1973 *Phys. Rev.* D **7** 2357
[12] Hu B L, Fulling S A, Parker L 1973 *Phys. Rev.* D **8** 2377
[13] Hu B L, Fulling S A, Parker L 1974 *Phys. Rev.* D **10** 3905
[14] Fulling S A, Parker L 1974 *Ann. Phys.* **87** 176
[15] Lukash V N, Starobinskii A A 1974 *Sov. Phys. JETP* **39** 742
[16] Zel'dovich Ya B, Starobinskii A A 1977 *JETP Lett.* **26** 252
[17] Starobinskii A A 1980 *Phys. Lett.* B **91** 99
[18] Israel W 1966 *Il Nuovo Cim.* B **44** 1
[19] Israel W 1967 *Phys. Lett.* A **24** 184
[20] de la Cruz V, Israel W 1967 *Il Nuovo Cim.* A **51** 744
[21] Senovilla J M M 2013 *Phys. Rev.* D **88** 064015
[22] Senovilla J M M 2014 *Class. Quantum Grav.* **31** 072002
[23] Senovilla J M M 2015 *J. Phys.: Conf. Ser.* **600** 012004
[24] Reina B, Senovilla J M M and Vera R 2016 *Class. Quantum Grav.* **33** 105008
[25] Eiroa E F, Figueroa Aguirre G and Senovilla J M M 2017 *Phys. Rev.* D **95** 124021
[26] Senovilla J M M 2018 Equations for general shells *J. High Energ. Phys.* JHEP11(2018)134
[27] Berezin V A, Dokuchaev V I, Eroshenko Yu N and Smirnov A L 2021 *Class. Quantum Grav.* **38** 045014
[28] Berezin V A, Dokuchaev V I, Eroshenko Yu N 2016 Spherically symmetric conformal gravity and "gravitational bubbles" *J. Cosmol. Astropart. Phys.* JCAP01(2016)019
[29] Berezin V A, Kuzmin V A, and Tkachev I I 1987 *Phys. Rev.* D **36** 2919
[30] Lake K 2017 *Gen. Rel. Grav.* **49** 134
[31] Deruelle N, Madore J 2003 On the quasi-linearity of the Einstein- "Gauss-Bonnet" gravity field equations *Preprint* gr-qc/0305004
[32] Moncrief V and Isenberg J 1983 *Commun. Math. Phys.* **89** 387
[33] Friedrich H, Racz I and Wald R M 1999 *Commun. Math. Phys.* **204** 691
[34] Racz I 2007 *Class. Quant. Grav.* **24** 5541
[35] Parattu K, Chakraborty S, Majhi B R *et al* 2016 *Gen. Rel. Grav.* **48** 94
[36] Jezierski J, Kijowski J and Czuchry E 2001 *Rep. Math. Phys.* **46** 399
[37] Padmanabhan T 2011 *Phys. Rev.* D **83** 044048
[38] Poisson E 2004 *A Relativist's Toolkit The Mathematics of Black-Hole Mechanics* (Cambridge: Cambridge University Press)
[39] Mars M and Senovilla J M M 1993 *Class. Quantum Grav.* **10** 1865
[40] Bach R 1921 *Math. Zeit.* **9** 110

Including precision clocks into space-based net as gravitational antennas

G N Izmaïlov and V V Ozolin

Moscow Aviation Institute (National Research University), Russia, 125993 Moscow, Volokolamskoe sh, 4

E-mail: izmailov@mai.ru

Abstract. Here we propose to use a precision clock in a space-based ultra-precise clock network to register sources of low-frequency gravitational waves of cosmic origin in the range of 10-3 ÷ 0.1 Hz. We also show that the method of comparing clocks at inland and intercontinental distances (very long baseline interferometry), originally developed for radio astronomy and geodesy, can be used as a prototype method for recording gravitational waves. Estimates of the measurement accuracy are given. An analyse of precise clocks possibilities for experimental estimates for rotation parameter of Gödel universe and GW recordings is offered, which in particular opens up the prospect of registering circularly polarized gravitational waves. Some new problems of small time intervals registration from general relativity, thermodynamics and quantum mechanics points of view are discussed.

1. Introduction

Registration of gravitational waves of cosmic nature is not only one of the priority fundamental directions of experimental physics and astrophysics, namely the search for a stochastic gravitational background of gravitational waves and dark matter. Such type of registration also has practical implications, for example, in deep space navigation. Instrumentally, the problem of long-term detection of low-frequency signals requires increased reliability of measuring devices, the use of new methods of data collection and algorithms for their processing based on general relativity and quantum mechanics, improvement of operating systems for processing large databases, algorithms for identifying weak signals against the background of various measuring positions of reference sources, operating coordinates of the spacecraft and time intervals.

Unlike SRT spacetime, which can be conventionally represented as a rigid 3D frame structure with preservation of the Euclidean geometry, and with synchronized clocks in the frame nodes [1], the GRT spacetime is considered as a 4D elastic “fabric”. This fabric with an unsynchronized clock can warp and the warp is depending on the condition of the massive body producing the fabric. The degree of deformation of tension (compression), twisting or warping is determined by the movement of the test mass in the vicinity of the central massive body. The stiffness of the "fabric" is large enough, which follows from the Einstein equation [2], which connects the measure of the local curvature of spacetime with the measure of the energy-momentum density $\boldsymbol{T}^{\mu\nu}$. The indices μ and ν run through the values of all coordinates in 4D spacetime. In the linear approximation, the equation for the deviation of space-time from flat $\hat{\boldsymbol{h}}^{\mu\nu}$ looks like

$$\left(-\frac{1}{c^2\partial t^2}+\nabla^2\right)\hat{h}^{\mu\nu}=kT^{\mu\nu}, \tag{1}$$

where the coefficient $k \,\square\, {G}/{c^4} \,\square\, 10^{-42}$ can be interpreted as the rigidity of the vacuum [3]. The huge value of "stiffness" excludes the propagation of longitudinal gravitational waves, leaving transverse or circular waves. For a stationary or slowly moving source with mass density ρ (taking into account the relation $E=mc^2$), the time derivatives disappear, and we can write the equation

$$\nabla^2 h^{\mu\nu}=\frac{G}{c^2}\rho, \tag{2}$$

repeating Poisson's equation in outward form. Far from the source of gravity (in empty space), equation (1) transforms into the equation

$$\left(-\frac{1}{c^2\partial t^2}+\nabla^2\right)h^{\mu\nu}=0. \tag{3}$$

The expression is associated with the equation for the electromagnetic waves in Maxwell's theory. Here the equation describes gravitational shear compression/expansion waves of space-time with deformation amplitude $h^{\mu n}$.

Accelerating masses distort spacetime, generating gravitational waves (GW). When registering GW of a cosmic nature, the experimenters faced a difficult task of registering spatial deformations due to the small deviation of spacetime from the plainness. Suffice to say that the vibration amplitude of the test body position under action of GW from distant sources was less than 10^{-4} sizes of the atomic nucleus. But in 2015, the problem was solved by the LIGO and VIRGO interferometers using [4, 5].

2. Influence of variable gravitational potential on the rate of time passage

It is known that far from the GW source, the field approach can be used to determine the effect of gravitational disturbances on measuring instruments. We will assume that the propagation of the gravitational potential $\phi = {GD}/{r^3}$ (D is the quadrupole moment of a mass system) leads not only to a change in the positions of small masses, but also to a change in their velocities and the rate of time flow [6]. The quadrupole approximation is substantiated by astrophysical observations, such as the change in the orbital period of the binary pulsar PSR 1913 + 16 (accuracy 0.16%) [7], and the analysis of the results of X-ray observations during the cataclysm of binary stars [8].

The experiments of Pound - Rebka and Pound - Snyders [9, 10] confirmed a variation in the frequency of the electromagnetic signal produced by a change in the gravitational potential ϕ. The experimental dependence can be interpreted as a change in recording of the time pace due to ϕ altering.

$${\Delta v}/{v} \cong {gh}/{c^2}, \tag{4}$$

or that $T = f(\phi)$. Recently, researchers using NASA's Chandra X-ray Observatory have confirmed this effect outside the solar system. A gravitational redshift was observed in a two-star system (4U 1916-053) at a distance of 29 kly. ($3.218688\cdot 10^{17}$ m) [11].

The gravitational redshift is significant in the technology for measuring the time delays of the space navigational systems. Because of the effect, clocks on the Earth's surface run slower than clocks on a satellite, as the potential of the gravitational field in whose orbit is lower, requiring the introduction of a correction factor when an object is locked to coordinates on the surface.

It was shown that the GW carries the variable potential of the gravitational field. Therefore, by comparing the time rates recorded by the reference clock (far from the GW source) - T_0 and the clock on the spacecraft exposed to GW (mobile clock) – T_m, one can determine the characteristics of the GW (amplitude, frequency, polarization, direction to the source). Comparison of information from several space-based mobile clocks and laboratory clocks, firstly, will determine the characteristics of the GW:

polarization and direction to the source and, secondly, to develop technologies to increase the effective stability of mobile clocks. The latter will increase the sensitivity of the measuring system to the amplitude of the recorded GW.

Several challenges arise. This is the choice of stable time sources (natural and artificial), the development of a method for measuring the variation of the parameter associated with the time part of the signal, under the action of an alternating gravitational field, and algorithms for data collection and processing. Consequently, with an increase in the precision of mobile clocks, it will become possible to detect gravitational effects by measuring the change in the duty cycle of the pulse train of the mobile clocks. Experimental work on improving atomic clocks for their use in deep space exploration programs is being carried out on a wide scale [12 - 19]. The best error of modern laboratory clocks based on electronic transitions now is about $3.7 \cdot 10^{-18}$ (~ 1 s in 13 billion years) [13]. The current error of the clock on the spacecraft is $\sim 2.3 \cdot 10^{-14} \div 3.2 \cdot 10^{-15}$ [12 - 14, 16 - 18].

A promising way is the practical implementation of nuclear-optical clocks, the stability of which is ensured by the stability of nuclear transitions [19]. It is believed that the accuracy of clocks as a tool increases by an order of magnitude over a decade.

3. Breathing methods for time delays recording in astrophysical experiments

As an effect of the GW action, distortions of spatial intervals arise, as well as compression/expansion of the transit time intervals of light from sources to recorders. This effect served as the basis for the development of methods for registering GWs based on the time delays of electro-magnetic (EM) signals.

One of the methods for determining changes in the parameters of EM signals due to the action of gravitational waves on them is the technique of recording microhertz GWs (sources are massive objects with masses of the order of a million solar masses during a collapse that lasts for many years) using a series of pulses received from an initially regular a sequence of EM pulses from a pulsar is known as the Pulsar Timing method.

The method is based on the assumption that GW contributes to a change in the phase of the signal and makes the sequence irregular. Although the shapes of individual pulses from a given pulsar vary considerably, the shape of the mean profile is fairly stable. Therefore, the time characterizing the moment of arrival of the pulsar pulse is stable, and the change in the time of arrival of the signal can be recorded with an estimate of the error of the pulsars as $\sim 10^{-19}$ per hour [20]. The idea and development of the method belongs to Russian and American astrophysicists [21-24].

Another one is the method of comparing the course of distant atomic clocks, which underlies international timekeeping, global positioning and verification of fundamental physics. This trend is the development of work on the application of very long baseline radio interferometry (VLBI) in astronomy and geodesy [25 - 28].

Fiber optic links allow comparisons of the most accurate optical clocks without degradation over inland distances of up to thousands of kilometers, but intercontinental comparisons remain limited in satellite transmission methods. However, VLBI can overcome this limit and compare distant clocks by observing extragalactic radio sources. Special mobile VLBI stations have been developed that use broadband detection and demonstrate a comparison of two optical clocks located in Italy and Japan, separated by a distance of 9000 km. This system demonstrates performance superior to satellite methods and may pave the way for future long-term stable international clock comparisons [28].

The ensemble of the Event Horizon Telescope network (EHT) of an Earth-sized interferometer was used to measure the size of the emission regions of two supermassive black holes with the largest visible event horizons: SgrA * at the center of the Milky Way and M87 at the center of Virgo A. In both cases, the dimensions correspond to the predicted silhouette caused by the strong lensing of light by the black hole. The addition of key devices in the millimeter and submillimeter wavelength range at high altitudes has opened up the possibility of visualizing such features and determining the dynamic evolution of black hole accretion. The EHT project includes theoretical and simulation research that poses questions rooted in effects at the edge of a black hole that can soon be answered through observation. By integrating existing telescopes with new systems, EHT is leveraging significant global databases to create a ground breaking instrument with the highest possible angular resolution from the Earth's surface [29, 30].

A comparison of the fundamental features of the Pulsar Time Method and VLBI is given in [31]. A few words about the sensitivity of the Doppler method for detecting GW [32, 33]. Since the propagation of the gravitational potential $\varphi = GD/r^3$ leads not only to a change in the positions of small masses, but also to a change in their velocities, the method of Doppler tracking of a spacecraft or a spacecraft complex is used to register GW. With spacetime variations h at a level of $10^{-17} \div 10^{-15}$ and a frequency of $\nu \sim 10^{-4}$ Hz, changes in the spacecraft velocity will be $\Delta\nu \sim 10^{-7} \div 10^{-5}$ m/s at a spacecraft-Earth distance of 1 AU.

4. High-precision space-based clock network

A space-based spacecraft network with precision sensors of displacements or clocks located on them should provide more reliable and accurate characteristics due to the absence of technogenic and geophysical noise compared to ground-based ones [34 - 37]. The implemented projects GRACE [38], GRAIL [39], CHAMP [40], the long-prepared and widely discussed ambitious project LISA [41], the recently considered domestic space project of the orbital gradiometer [42] are analogs of the proposed network.

Consider a network of precision clocks (for example, atomic), which are located at the nodes of a huge space system (see [35, 36] Fig. 1). The diagram shows the location of a clock in a network located in low-earth orbit around the sun. The space operating system of the clocks, covering a large spatial area in Earth's orbit, will form the chronometric space network (CSN). Below we present clock accuracy estimates for event logging and describe possible measurement results and their significance.

For the spacecraft-Earth distance $\Delta L \sim 10^9$ m, the GW amplitude $h \sim 10^{-19} \div 10^{-17}$, the GW frequency $\omega_{\Gamma B} \sim 0.63 \div 63$ rad/s ($\nu \sim 10^{-1} \div 10^{1}$ Hz), we have $\lambda_g = 3 \cdot (10^7 \div 10^9)$ m, therefore $\Delta\nu = h\nu L \sim 3 \cdot (10^{-11} \div 10^{-7})$ m/s; which gives $\Delta\nu/\nu = 10^{-19} \div 10^{-15}$. With a laser clock error of 10^{-19}, the required registration time is about 1 year.

The registered change in the sequence of pulses received from the reference clock will signal the amplitude and frequency of the GW. The delay of signals from different network nodes can be associated with the direction of the gravitational wave front and its polarization. Obtaining other data (for example, the position of the apparatus with the clock and its relative movement) can be used to check some conclusions of general relativity (for example, Shapiro's delays). A functioning CSN will also play the role of a gravitational antenna. As an initial step towards the creation of a CSN, one can limit us to the presence of central precision clocks and ground-based repeaters on satellites. In this scheme (Doppler tracking), the ground station first sends an electromagnetic signal $\nu(t)$, which is compared with a high-quality frequency standard ν_0, then records the mirror response from the spacecraft reproduced by the repeater, $\nu(t-T)$, where T is time travel from a ground station to a spacecraft. Thereafter, the frequency offsets of the transmitted and reference signals are compared. The relative frequency difference $\delta\nu(t) = [\nu(t-T) - \nu(t)]/\nu_0$, ($\nu_0$ is the central frequency of the microwave signal) will be zero if there is no noise, systematic drift, and the GW signal with a wavelength λ is comparable with the dimensions of the detector. Otherwise, the dependence $\delta\nu(t)$ has a three-pulse character (which is a distinctive feature of the method and serves as the basis for signal filtering). According to already known algorithms for calculating the cross-correlation of time residuals from several spacecraft, the result is compared with possible gravitational signals to confirm their registration [33].

We have to point out the difficulties of project implementation:

1. Obtaining long-term and stable readings of the clock on board the spacecraft under observance of cryogenic conditions of measurements
2. The presence of new types of noise (for example, plasma noise in the signal path)
3. The cost of creating and maintaining a CSN can exceed the cost of creating and operating ground-based GW observatories or an array of radio telescopes by tens of times.

5. The principle of registering slow rotations with a clock and some new problems

The spin of a particle can change its behavior after the passage of the GW. Let we suspend a particle in a chamber in a laboratory and measure the speed and direction of its rotation when a gravitational wave packet is sweeping; then measure it again, but after the wave has passed. The difference in the behavior of the spin of the particle would reveal a different kind of memory about the wave, which can be done with the help of post-temporal rotation sensors [34, 42, 43].

The possibility of registering the transient torsion caused by the rotation of a massive body of mass M with angular velocity, which creates a gravimagnetic field with the Lense - Thirring metric, has already been discussed. The metric has the form

$$ds^2 = \left(1 - \frac{2GM}{c^2 r}\right)c^2dt^2 - \left(1 + \frac{2GM}{c^2 r}\right)\left(dx^2 + dy^2 + dz^2\right) - 2\frac{2GS_z}{c^3}\frac{y}{r^3}cdtdx + 2\frac{2GS_z}{c^3}\frac{x}{r^3}cdtdy\ , \quad (5)$$

here the Oz axis is directed along the angular velocity, x, y and z are the current coordinates, r is the distance to the observer, $\mathbf{S} = \mathbf{I\Omega}$ is the angular momentum of the body. To solve an experimental problem: determining the angular velocity of rotation, some schemes were proposed in which centrifugal forces were created. In turn, these forces caused longitudinal stresses in a cylindrical solid sample perpendicular to the $\mathbf{\Omega}$ axis. The stresses can be measured using a composite magnetostrictor-SQUID sensor [44].

However, there is another possibility here. Since the rotation of the clock relative to the center of the Earth under the action of a gravitational wave is relatively slow, it is possible to use the formula of the Lorentzian time dilation when the clock moves on a satellite with a speed of $\Delta v = h\omega_{ГВ}L_{E\text{-}S}$. An estimation is $\Delta\tau \simeq 1/h\omega_{GW}\,(\Delta v/c)^3 \sim 10^{-30}$s, with a measurement time of 10 s.

At the same time, the delay increases as the radius of rotation increases. Therefore, we will consider an experiment involving cosmological distances. Since the time of I. Newton, the cosmological principle has been known, according to which in modern wording an observer located at any point in the Universe and at any moment of time "sees" the Universe as homogeneous and isotropic. In other style, the average values of mass density and pressure obtained after observing large, including galaxies and their clusters, spatial regions (about 250 million light years) and long time intervals, but small in comparison with the size and age of the Universe, are the same in all directions and are the same over time. The statement implies that at any point in space all physical laws are fulfilled in the same way. However, the nature of the principle is more philosophical than physical and has no logical foundation. The reasons for its application were the absence of any significant observed deviations from homogeneity and isotropy in the indicated cosmological limits. After analyzing part of the data of the Planck project (mission), ESA (Planck Mission) determined that the observed deviations of anisotropy are statistically significant and can no longer be ignored [45]. In addition, various other astronomical observations, including the discovery in June 2021 of the structure, the *Giant Arc*, extending approximately 9.2 billion light years and consists of galaxies, galactic clusters, gas, and dust, question the validity of the cosmological principle [46].

Along with the well-known cosmological models that satisfy the requirements of relativistic invariance and have their own privileged time axis, there are other models in which there is no isotropy. These include K. Gödel model. It is believed that at every point in the universe there is some semblance of rotation. This rotation is described by the Gödel metric, which is an exact solution of the Einstein field equations, in which the energy-momentum tensor contains two terms, the first of which represents the density of matter of a uniform distribution of rotating dust particles (which can represent galaxies), and the second one is associated with a nonzero cosmological constant)

$$ds^2 = \frac{1}{2\omega^2}\left[-\left(cdt + e^x dy\right)^2 + dx^2 + \frac{1}{2}e^{2x}dy^2 + dz^2\right], \qquad -\infty\langle t, x, y, z\rangle\infty\ , \quad (6)$$

where ω is a nonzero real constant, which turns out to be the angular velocity of the surrounding dust grains around the y-axis, measured by a "non-rotating" observer located on one of the dust grains. Rotation is reflected in the pace of time, and therefore can be measured with a watch. "Non-rotation" means that the observer does not feel centrifugal forces, but in this coordinate system he will actually rotate around an axis parallel to the *y*-axis. As can be seen, the dust grains remain at constant values of *x*, *y*, and *z*. Their density in this coordinate system increases with increasing x, but their density in their own reference frames is the same everywhere.

If measurements are carried out in the Earth's orbit around the Sun $L \simeq 1.5 \cdot 10^{11}$ m, then with an experiment duration of one year, the time delay will be $\Delta\tau \sim 0.8 \cdot 10^{32}\,\omega^2$, or
the limiting value of the constant ω in Gödel metric, which can be experimentally recorded $\omega \sim 1.2 \cdot 10^{-16}$ 1/s. The measurement error can be reduced by the duration (three to five years) of the mission.

For further reasoning, we will use the following definition of time using a clock "[Time] is considered to be measured by a clock (an ideal periodic process) - [an object] of insignificant spatial extent. The time of the event at the point is defined as the time shown on the clock simultaneously with the event" [47], i.e. we have to deal with the readings of the clock, and not with an abstract definition of time that is not observable. In other words, the clock only measures the time interval between two events, not the absolute time of each of them. This instrumental definition becomes necessary, for example, when we discuss the time it takes for an electron to tunnel through a potential barrier. Another example from classical mechanics - how to establish that two particles began to fall from a certain height (Galileo's experiment) at the same time?

In recent publications [48, 49], a question was discussed that can be formulated in a simplified form as follows. What remains of a black hole after it evaporates?

Theoretical analysis pointed out possible residual phenomena manifested in "homeopathic" deviations of the state of test masses, from the ideal three-dimensional crystal with synchronized clocks assumed in flat space-time conditions, three spatial displacements, three residual angular rotations and three Lorentz transformations (Poincaré group). Those. symmetries of the special theory of relativity (SRT) as asymptotes in the absence of a gravitational field can manifest themselves as one of the sets (groups) of boundary conditions of the equations of the general theory of relativity (GR).

In other words, the vacuum of GR with its symmetries makes possible the existence of a memory matrix that stores this information in the Universe, even after the disappearance of the black hole. These are the shifts of the Poincaré group, which connect the points infinitely distant from the gravitating body. The discovered abundance of symmetries, known as the BMS group, gives empty space-time tremendous latent complexity. The localization of such places in space-time can later serve as a reference point for navigation, similar to observing an overgrown ravine or abandoned mines on topological maps.

Estimates utilizing a toy model with Lense -Thirring metric give us the upper optimistic limit of the residual linear deformation at 5% of the GW already recorded value. Therefore, a multiple series of measurements of time intervals, as well as the position of the end mirrors of the LIGO (VIRGO) interferometer, should indicate the residual displacement.

On the other hand, on the assumption that SRT is not the only asymptote of GR when the gravitational field disappears, it is necessary to discuss the problem of mismatching reference IFRs and introduce the Lorentz transformation into the communication protocol, which connects their reference frames [50].

In conclusion, we would like to note some aspects of measuring time intervals, which will affect the accuracy of timekeeping, and, consequently, the accuracy of recording gravitational waves. These aspects are not directly related to measurement technology and engineering problems, but rather indicate possible connections between the theory of gravity and quantum mechanics.

The first aspect. The point of view of A. Einstein [47], stated above, indicates the clock as a dynamic system that cannot be a stationary object, and the absence of global time. That. it is necessary to introduce the true (proper) time of the observer $d\tau^2 L := -ds^2/c^2$, defined in a curved space-time along its trajectory of motion Γ as $\tau = \int_\Gamma \sqrt{-g_{\mu\nu} dx^\mu dx^\nu}$, $\{\mu,\nu\} = \{0,1,2,3\}$. The insignificance of the spatial extent of clocks (for example, atomic, and even better time crystals) can be interpreted as a quantum dot with many internal degrees of freedom, which moves in a curved spacetime. To measure the time interval, it is necessary to introduce such measures that change covariantly in the group of time transformations acting on the used clock system. Such descriptions of quantum clocks have been developed in the context of quantum metrology. In addition, it is necessary to take into account the influence of the gravitational field on the position of energy levels for a given accuracy of measuring time intervals. Time can then be determined through measurements of quantum systems that serve as clocks. Now we can raise the question of the influence of the GW on the difference in clock readings from a quantum point of view. At the same time, covariant observables in time make it possible to rigorously formulate the time-energy uncertainty relation [51, 52].

The second aspect is how to synchronize the quantum (in terms of size) clocks to a fraction of a tick? What is the impact of the measurement process? It is shown that there are universal limits for the accuracy of measuring time intervals. According to thermodynamics, a watch is a machine that converts one type of energy into another (gravitational into mechanical, mechanical into electrical, nuclear into heat, etc.). That is, any thermodynamically irreversible process can be used as a clock, for example, simply by observing the equilibration process as an indicator of time. Since energy conversion is accompanied by losses, the accuracy of a quantum clock is directly proportional to the created entropy [53]. It was experimentally shown measuring time requires more than just a certain increase in an entropy [54]. It was realized that the marginal relationship between the accuracy of the clock ticks and the entropy produced could be a universal characteristic of timekeeping. The result can make a difference for technologies that depend on accurate timing. A conceptual diagram of a clock for thermodynamic accuracy analysis is shown in fig.1.

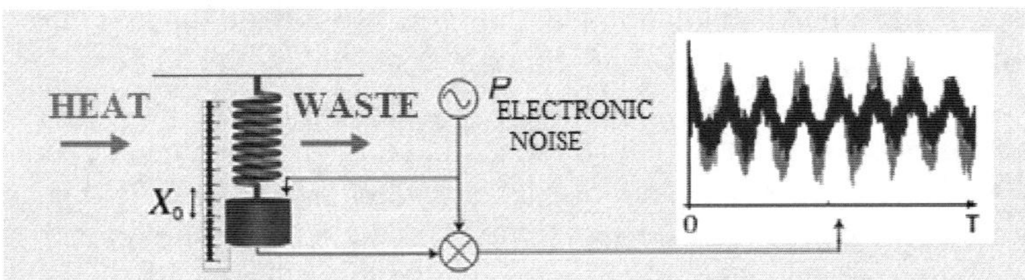

Figure 1. Conceptual clock model from the point of view of thermodynamics in the analysis of the accuracy of measuring time intervals.

For experimental research, physicists have created a simple optomechanical "nano-clock" consisting of a membrane whose deflection is controlled by an electric field. The membrane is driven by heat input to it and electrical work used to measure the displacement. During operation, the watch converts the received energy into a membrane deflection signal and releases excess heat, producing entropy. Thus, by measuring this entropy, we can determine the amount of energy consumed. The implication is that clocks, like other machines such as motors and computers, are subject to a fundamental performance limit determined by the laws of thermodynamics. Although the system was nanoscale, it was too large for quantum mechanical analysis, and physically completely different from the previously studied quantum clock. However, the researchers found a relationship between precision and entropy in the classical clock model, the same as in the quantum clock. The relationship between the clock accuracy and the change in entropy in the experiment is also consistent with the theoretical model of the researchers, confirming that this pattern can be preserved for both classical and quantum models.

Although the researchers have experimentally tested the relationship between entropy and accuracy for only one particular implementation of a classical clock, they argue that the similarity to a quantum result suggests this may be true for any clock. The researchers also propose to redefine the "optimal" watch as one that has the highest possible accuracy with the least entropy dissipation (consistent with the discovered relationship between accuracy and entropy), regardless of the physical details of the watch. The result was the establishment of the dependence of the accuracy of measuring the oscillation period

$$N_{osc} := \frac{(t_{tick})^2}{(\Delta t_{tick})^2}$$

from the entropy produced during the oscillation period – ΔS_{tick}

$$N_{osc} = \frac{2\pi^2}{k_B}\frac{T_C}{T_N}\Delta S_{\text{tick}},$$

where k_B is the Boltzmann constant; T_C is the effective temperature of the amplifier in the refrigerator; T_N is the effective temperature determined by the electronic noise (Johnson noise) during data recording.

Since all clocks use an increase in entropy in one form or another to quantify the passage of time, a better understanding of the measurement of time can provide a new understanding of how heat, work and the arrow of time are related, which will affect the determination of the accuracy limits of the temporal parameters of GW.

The third aspect. One can also raise questions about the minimum distance between two quantum clocks, due to the reaction of space-time to their presence and limiting the joint measurability of time along the nearest world lines [55]; on the influence of the watch masses on the limitation of the measurement time. Among them there is also the question of the quantum entanglement of quantum particles - clocks, since clocks, being, in the final analysis, quantum systems, make it possible to apply the principle of superposition to their states, and, consequently, to the superposition of proper times [56].

The authors of [55] argue that, at least in principle, the proposed scheme can also be used to detect t time dilation effect caused by GW. However, at present, the detection of a gravitational wave according to the proposed scheme is still a difficult task.

For example, the precession of the spin of particles when a magnetic field is applied can be used as a quantum clock. Here we can turn on (turn off) the clock by applying a non-zero (zero) magnetic field. In this case, the accuracy of the two separate clocks will also not be limited by the fit, since we can turn off the two clocks when we measure them. However, in this case, the magnetic field must instead be precisely controlled to the level of the required circuit accuracy, which will be much more difficult than achieving the required accuracy with two naturally defined energy eigenstates, as in a conventional quantum clock circuit.

It is noted that an entangled pair of non-degenerate quantum clocks can be used as a specialized detector to accurately measure the difference in proper times experienced by each constituent quantum clock. In this case, several experiments are considered in which the acceleration effects are mutually compensated. The proposed scheme can be used in precision testing of relativistic time dilation.

In [56], two clocks are considered prepared in localized impulse wave packets moving in Minkowski space. Let clock *B* move with a wave packet with an average localized momentum $\vec{p}_B$, while clock A moves in a superposition of wave packets with a localized average momentum $\vec{p}_A$, and $\vec{p}\,'_A$. It is demonstrated that, on average, quantum clocks *A* measure time dilation relative to clock *B* in accordance with STR special theory of relativity, when the state of their center of mass is localized in momentum space. However, when the state of their center of mass is in a superposition of such states with a localized momentum, then for clocks *A* there additionally takes place the effect of quantum time dilation relative to clock *B*. This quantum time dilation is due to their non-classical state of motion and depends on the parameters that determine the superposition of pulses. An estimate of the order of magnitude of this effect indicates that quantum time dilation can be observed using modern technology.

References

[1] Wald R M 1984 *General Relativity* (Chicago: The University of Chicago Press) p 693
[2] Misner Ch W, Thorn K S and Wheeler J A 1973 *Gravitation* (San Francisco: Freeman W N & Co)
[3] Izmaïlov G N 2011 *Measurement Techniques* **54** 479

[4] Abbott B P (LIGO Scientific Collaboration and Virgo Collaboration) et al 2016 *Phys. Rev. Lett.* **116** 061102
[5] Pustovoit V I, Morozov A N, Gladyshev V O and Izmaïlov G N 2016 *Laser gravitational wave antennae* (Moscow: Bauman Moscow Technical University Press) p 63
[6] Izmaïlov G N 2016 *Measurement Techniques* **59** 560
[7] Weisberg J M and Huang Y 2016 *ApJ* **829** 55 (*Preprint:* astro-ph/1606.02744)
[8] Cordova F A and Mason K O 1984 *Mon.Not.R.Astr.Soc.* **206** 879
[9] Pound R V and Rebka Jr G A 1960 *Phys. Rev. Lett.* **4** 337
[10] Pound R V and Snider J L 1964 *Phys.Rev.Lett.* **13** 539
[11] Trueba N, Miller J M, Fabian A C 2020 (*Preprint*: astro-ph/2008.01083)
[12] https://scitechdaily.com/nasa-deep-space-atomic-clock-mission-extended-in-search-of-future-navigation
[13] Kennedy C J, Oelker E, Robinson J M 2020 *Phys. Rev. Lett.* **125** 201302
[14] McGrew W F, Zhang X, Fasano R J 2018 *Nature* **564** 87
[15] Arnold K J, Kaewuam R, Roy A, Tan T R and Barrett M D 2018 *Nature Communications* **9** 16
[16] Liu L, Lü De-Sh, Wang Yu-Zhu 2018 *Nature Communications* **9** 2760
[17] Ren W, Li T, Liu L 2020 *National Science Review* 7 1828
[18] Cacciapuoti L, Armano M, Much R *et al* 2020 *Eur. Phys. J.* D **74** 164
[19] Sikorsky T, Geist J, Hengstler D 2020 *Phys. Rev. Lett.* **125** 142503
[20] Avramenko A E 2019 *Pulsar: Natural time-space standard* (Moscow: Editorial URSS) p 200
[21] Braginsky V B and Gertzenstein M E 1967 *JETP Letters* **5** 287
[22] Sazhin M V 1978 *Soviet Astronomy* **22** 36-9
[23] Detweiler S 1979 *ApJ.* **234** 1100
[24] Gusev A V, Porayko N K and Rudenko V N 2014 *Gravit. Cosmol.* **20** 290
[25] Pizzocaro M , Sekido M *et al* 2021 *Nature Physics* **17** 223
[26] Sekido M, Takefuji K, Ujihara H *et al* 2019 *ITA-JPN Broadband VLBI Experiment for Optical Clock Comparison*
[27] Marti G E, Ross B, Hutson R B 2018 *Phys. Rev. Lett.* **120** 103201
[28] Kondo T and Takefuji K 2016 *Radio Science* **51** 1686
[29] https://eventhorizontelescope.org/blog/telescopes-unite-unprecedented-observations-famous-black-hole
[30] The Event Horizon Telescope Collaboration 2019 *ApJ* **875** L1 *(Preprint :* astro-ph/1906.11238)
[31] Wang J B, Coles W A, Hobbs G *et al* 2017 *Monthly Notices of the Royal Astronomical Society* **469** 425
[32] Estabrook F B and Wahlquist H D 1975 *Gen. Rel. Grav.* **6** 439
[33] Armstrong J W 2006 *Living Rev. Relativity* **9** 1
[34] Izmailov G N, Ryabov VA, Tskhovrebov A M and Zherikhina L N 2018 *Phys. Astron. Int. J.* **2** 148
[35] Izmailov G N 2019 *J. Phys.: Conf. Ser.* **1348** 012005
[36] Zherikhina L N, Izmailov G N and Ozolin V V 2020 *Al'manac of Modern Metrology* **4** 24
[37] Davies P C W 2004 *Proc. SPIE* 5472, *Noise and Information in Nanoelectronics, Sensors, and Standards* II (Spain)
[38] Klaus A, Heinzel G, Danzmann K *et al* 2019 *Phys. Rev. Lett.* **123** 031101
[39] Lemoine F G *et al* 2014 *Geophys. Res. Lett.* **41** 3382
[40] CHAMP mission status Advances in Space Research 2002 **30** 129
[41] *LISA: Laser Interferometer Space Antena 2017* (pdf). *LISA Consortium* Retrieved 14 Jan. 2018
[42] Milukov V K, Zhirkov A S C, Zharov V E, Ivlev O A *et al* 2020 *Al'manac of Modern Metrology* **4** 296
[43] Touboul P, Willemenot E, Foulon B and Josselin V 1999 *Bollettino di Geofisica Teorica* ed Spplicata **40** 321
[44] Izmaïlov G N *et al* 2018 *J. Phys.: Conf. Ser.* **1051** 012016

[45] https://sci.esa.int/web/planck/-/51551-simple-but-challenging-the-universe-according-to-planck
[46] https://www.newscientist.com/article/2280076-line-of-galaxies-is-so-big-it-breaks-our-understanding-of-the-universe/
[47] Einstein A 1996 *Out of My Later Years* (NewYork: Wings Books)
[48] *Gyros, Clocks, Interferometers...: Testing Relativistic Gravity in Space* 2010 Lecture Notes in Physics (vol 562) ed C Lämmerzahl (Springer Science and Business Media)
[49] Ahmadi M, Smith A R H and Dragan A 2015 *Phys. Rev.* A **92** 062319 (*Preprint*: quant-ph/1508.03341v12015)
[50] Smith A R H and Ahmadi M 2020 *Nature Communications* **11** 5360
[51] Erker P et al 2017 *Phys. Rev.* X **7** 031022
[52] Milburn G J 2020 *Contemp.Phys.* **61** 69
[53] Garrahan J P 2017 *Phys. Rev.* E **95** 032134
[54] Pearson A N *et al* 2021 *Phys. Rev.* X **11** 021029
[55] Hwang W Y, Ahn D, Hwang S W and Han Y D 2002 *Eur. Phys. J.* D **19** 129
[56] Lock M P E and Fuentes I 2019 *Class. Quantum Grav.* **36** 175007 (*Preprint*: quant-ph /1901.08000v3)

Constrains on an uniform model for Dark Matter and Dark Energy

G N Izmaïlov

Moscow Aviation Institute (National Research University), 125993 Moscow Volokolamskoe sh., 4, Russia

E-mail: izmailov@mai.ru

Abstract. In this paper, outlined some of the most important concepts about Dark Matter and methods of their registration, in particular by using SQUIDs, a toy uniform model for Dark Matter and Dark Energy is analyzed. In the frame of the model Dark Matter particles is interpreted as excitations of Dark Energy field. Some constrains are considered. The devices based on SQUID, in particular the SQUID-paramagnetic absorber and the SQUID-magnetostrictor systems, both suitable for investigations of above problems, are considered. Estimates, are carried out within this model, indicate the possibility of experimental detection of the "ether wind" pressure, created by the non-corpuscular incoming flow, corresponding to the galactic orbital motion of the Earth.

1. Introduction

The accelerating expansion of the universe poses challenges for cosmologists. Among the contenders claiming to be the main contributor to uneven and different-temporal expansion at different cosmological distances, we note dark energy (DE), dark matter (DM), dark radiation (DR), and dark gravity (DG). Dark matter (latent mass) is unevenly distributed, thereby explaining the formation of galaxies and their clusters, but leaving the spaces between them free of masses, that opens up the possibility of the action of dark energy. DE is a continuous and homogeneous medium responsible for the accelerated increase in the distances between inhomogeneities. Possible DE-DM interaction leads to the appearance of radiation, part of which have characters that have not yet been detected. A fundamentally different approach to explaining the phenomenon of anomalous expansion is a modification of the laws of gravity, briefly referred to as DG.

The nature of dark matter (DM) is one of the most urgent unsolved problems in modern physics, so physicists persistently try to find signs of elusive dark matter that pulls everything but emits no light. History of observations and theoretical arguments for the DM existence are described in a recent review [1]. Besides large number of independent astrophysical and cosmological observations, such as galaxy rotation curves [2], large structure formation [3], and the observed spectrum of the cosmic microwave background [4-6], indicates that about 26% of the total mass-energy of the Universe is non-baryonic and non-relativistic cold component, which is not subject to electromagnetic interactions. Despite the long history of observations the particle content of DM remains unknown [7-9]. One possible solution would be the existence of new particles beyond the Standard Model (axions, dark photons, sterile neutrinos?).

Among many candidates for DM particles the most favored are the so-called WIMPs (weakly interacting massive particles) with a mass in the GeV to TeV range [10,11]. The WIMPs existences are well motivated, as provide solutions for outstanding issues in both cosmology and particle physics.

These particles appear in extensions to the Standard Model such as supersymmetry with R-parity conservation, and freeze out in the early Universe with an abundance that matches cosmological observations. Now DM particles can be detected in the laboratory experiments via elastic scattering off nuclei, causing recoils of a few keV in energy. However, no experimental indication for a standard WIMP was found in high sensitivity direct search experiments like EDELWEISS [12], XENON100 [13], LUX [14], SuperCDMS [15] for more than a few tens of GeV/c^2 mass.

Thermal relics with masses below 100 GeV/c^2 are also disfavored both from constraints set by Fermi-LAT searches for annihilation signals from Milky Way dwarf galaxies [16], and from the impact that WIMP annihilation would leave on the cosmic microwave background anisotropies [17]. Furthermore search for dark matter particles directly produced in proton-proton collisions in the CERN LHC experiments ATLAS and CMS did not show any indications of such particles so far [18-20].

On the other hand, in [21, 22] some researches insist on in the region of light WIMPs below 30 GeV/c^2, some experiments such as DAMA/LIBRA [21], CoGeNT [22], CRESST [23], CDMS II Si [24] claim or are interpreted as finding dark matter signals.

It is obvious that for observation of low-energy WIMP-induced nuclear recoils, detectors with both low detection threshold and a very low background are needed. The sensitivity of a dark matter direct detection experiment depends on the WIMP mass and on the nature and strength of its coupling to atomic nuclei.

Alternative to the quest of WIMP particles is a search of ultralight DM particles such as neutrino, dark photons or axions. Again, there is no indication that the high-energy neutrinos originating from dark matter annihilations in the Sun, which can escape and be observed by neutrino telescopes, such as Baksan [25], Super-Kamiokande [26], ANTARES [27] and IceCube [28].

When arguing the existence of DM, which exceeds the mass of the visible Universe by about five times, with the help of light particles, a compelling cold dark matter candidate is axion. It has feeble coupling to ordinary matter, moves with non-relativistic energies, and early universe cosmology consistent with dark matter phenomenology. Mass of a pseudoscalar boson – axion is $m_a \sim 10^{-11} \div 10$ eV. This range is probed by the tuple of experimental groups [29, 30].

Axions were introduced as a consequence of an extension of the Standard Model. The expansion, in particular, follows from the desire to preserve CP invariance, which is violated by the absence of an electric dipole moment in the proton, and axions, as particles of a new field that preserves CP invariance, play this role [31–33]. Moreover, new particles have no electric charge. Unexpectedly, it turned out that astrophysical and cosmological problems of the energy balance when calculating the emission budget of stars, the existence of halos around galaxies [34, 35], the cooling processes of white dwarfs [35], and the shortening of the duration of the SN 1987A supernova explosion [36, 37] can also be described more simply, using the existence of axions. (According to estimates [30, 38], the particle density is ~ 10^{21} m^{-3}). Thus, the axions that make up the DM can be detected not only at accelerators, but also in laboratories studying cosmic radiation, since it serves as a source of axions [39, 40].

To describe the axion - photons conversion Primakoff mechanism was proposed [41]. It had been offered to explain the decay of muons. The form of this mechanism was used to count the axion - photons conversion in an external magnetic field and it was chosen as a mechanism which is convenient for an axions registration. In such case the probability of axion-photons conversion is proportional to $(g_{a\gamma\gamma}BL)^2$ where $g_{a\gamma\gamma} \gtrsim 10^{-10}\text{GeV}^{-1}$ is a constant of axion-photons interaction, B is the magnetic field and L is its length [42]. Also the reverse process of transformation of photons into axions is possible. As it can be shown it is possible in the laboratory to use sources of powerful coherent radiation (lasers with intense laser beams ~10^{21}W/cm^2) and with their help simulate the conversion of photons into axions, as well as the reverse process [43]. An additional impetus to research on the conditions for observing axions was the applied problem of detecting the passage of an axion beam through the thickness of matter without a significant weakening of its intensity - the LSTW (light shining through walls) experiment [44, 45]. It was proposed to use a small interaction cross section to establish communication through massive opaque obstacles or very long distances, for example, the Earth-Earth or Earth – Moon.

As noted above, Primakoff mechanism uses low-frequency virtual photons of the magnetic field that resonate with the field of axions. The probability of the axion-photons conversion is $p \sim (g_{a\gamma\gamma}BL)^2$. Due to a very weak interaction the strong enough magnetic fields B ($B \gtrsim 10^2$T) along the entire path L ($\sim 10^2$m), where the conversion takes place for such reactions are required. Because during solar flares on Sun surface there are a lot of photons piercing spacious magnetic field in the protuberance form, then it is possible to observe axion generation by helioscope (IAXO project) [46].

The contemporary paradigm states that DE is a hypothetical all-pervading substance that may be responsible for additional acceleration in Hubble's law of galaxy recession. At the same time we consider DM as a kind of matter with only inertial properties. The DM will keep galaxies as a whole if parts of them break off as they move rapidly. (For example, if Mercury revolves around the Sun in orbit of Pluto at the same speed as in its original orbit, it will leave our system. Thus, in a spiral galaxy, it is DM needed to explain the flattening of the rotation curves). Can we assume that DM has only the inertial property?

A number of recent observations suggest that DM can interact with both baryonic matter, primordial black holes and/or dynamically interactive DE. This appears when 21cm radio signals are analyzed using DM interactions with baryons and/or primary BHs [47]. The possible excess of the 21-centimeter line at the cosmic dawn can be explained also by the active background of the dynamic interactive DE. The active interaction at an early dark energy dominated stage with dynamically interacting dark energy, which manifests itself in the excessive brightness of the 21 cm signal is investigated in [48]. Moreover, in recent astrophysical observations of DM made with the satellite gamma ray telescope Fermi-LAT and DE survey the detection of gamma rays from approximately 45 dwarf spheroidals (dSphs) of the Milky Way in the dense region of galactic centers was reported. Revealing γ-rays indicates processes inside DM and is attributed to the result of self-annihilation of DM, which is another mechanism of activity [49].

Thus, it becomes possible to reveal not only the inertial properties of DM. Both DE and DM play a key role in modern models of galaxy formation and evolution. The discovery of new properties of DM type would provide a clue to the DM-DE interaction.

In recent time some investigators expected the time intervals measurements suppose that they are hint for DM observation [50, 51].

2. A toy uniform model for Dark Matter and Dark Energy

The clues we get from the analysis of astrophysical observations [47-49, 52] can be used to create a unified model in the form DE + DM = DS. In [53] there was considered an approach for DM description, based on the possibility that DM and Dark Energy (DE) are two different aspects of the one cosmological essence, named “Dark Substance”, which is characterized by variable ς. Subsequently, in DS, which has a primary viscosity, but not a form - the original topology - one can describe a property that allows one to construct local maps continuously conjugate to each other and use the smoothness condition (differentiability of smooth forms) to connect them. The reason is that the use of locally flat maps creates the basis for the construction of inertial frames of reference. When constructing, one should use the principle of general relativity - the independence of fundamental systems of physics from the background (DS), which means covariance under a diffeomorphism as a smooth deformation of a spacetime manifold.

The means of offered approach were anticipated from integrated Sachs – Wolfe effect. It points on a correlation between anisotropy of local equilibrium temperature for CMB (i.e. fluctuations of photons) and large scale Universe structure, or in accordance with Einstein GR, of gravitational potential due to the massive particles that do not otherwise interact, i.e. the DM. And both photons and DM are in a primordial soup – DE. So the temperature fluctuation is directly related to the gravitational potential fluctuation by some constant [54]. The influence of scales (hierarchy of scales) can be observed at all levels. Water in the usual macro representation - a continuous, viscous liquid - in the micro representation consists of particles - molecules, which manifests itself in phase transitions. In the Universe, galaxies and galaxy clusters with different astrophysical regularities of the distribution of the gravitational field and the distribution of stars serve as an example, which led to the recognition of the existence of DM and DE. The long-scale structure depends on a matter density, the CMB anisotropy – is defined by fluctuations of a cosmological gravitational potential, which changes the photon frequency, which is consistent with the notion of the absence of scale invariance in an empty Universe. The effect is in a good agreement with the prediction of the concordance ΛCDM cosmology [55-57]. On the other hand, the view of DE as the frame of reference in which DM is located ("river frame + fishes" model) may well be used to represent the DE ↔ DM relationship by using both coordinate transformations and gauge ones [58].

According to an offered model, with origin density of about 300 TeV/m^3, it represents the unperturbed state of "Dark Substance", while its swings or perturbations play the role of elementary DM heavy particles. These particles will be stable if all their decay channels into any combination of other particles are blocked, or also, in our case, if their potential will have local minima, i.e., local traps providing metastable excited states. The density of Hamiltonian with metastable traps can be represented in 2D spacetime, for example, as follows:

$$H = \frac{1}{2}\left(\frac{\partial \varsigma}{\partial x}\right)^2 + \frac{1}{2c^2}\left(\frac{\partial \varsigma}{\partial t}\right)^2 + \alpha\varsigma * \varsigma - \beta \cos\gamma \sqrt{\varsigma * \varsigma}. \quad (1)$$

Here α, β, γ are coefficients. The nonlinear wave equation, corresponding to this Hamiltonian, and describing the dynamics of perturbations to Dark Substance will be like the "quasi-sine-Gordon" equation

$$\frac{\partial^2 \varsigma}{\partial x^2} - \frac{1}{c^2}\frac{\partial^2 \varsigma}{\partial t^2} + \alpha\varsigma * + \frac{1}{2}\beta\gamma\sqrt{\varsigma */\varsigma}\, \sin\gamma \sqrt{\varsigma * \varsigma} = 0. \quad (2)$$

Moreover, the nonlinear potential $\Pi(\varsigma) = \alpha\varsigma * \varsigma - \beta \cos\gamma \sqrt{\varsigma * \varsigma}$ appearing in (1), has the analytic structure very similar to the "parabolic washboard potential" [59] used to describe metastable states in the superconducting ring of a SQUID with one Josephson junction, namely:

$$\Pi(\varsigma) = \alpha|\varsigma|^2 - \beta \cos\gamma\, |\varsigma| \Leftrightarrow E(\Phi) = \frac{1}{2L}\Phi^2 - I_C \Phi_o \cos\frac{2\pi}{\Phi_o}\Phi, \quad (3)$$

where Φ is the magnetic flux piercing the superconducting ring and L is the inductance of the curcuit itself. We represent the profile of the washboard potential in figure 1 of [53]. Some comments about this figure are needed. For small ς (when the disturbance has not yet reached the first trap) folds of the Dark Substance potential have a quasi-harmonic character, and their quanta will have a mass $m = \frac{h}{c}\sqrt{2\alpha + \beta\gamma^2}$. However, such particles may be unstable, and the rate of decay from classical positions will correspond to the viscosity of the Dark Substance [60, 61]. At large amplitudes of ς, the disturbance at some moment will "catch" the lowest energy trap. The trajectory of the oscillation ς will now be a circle in the plane {ς, ς*}, corresponding to a local minimum of potential. The rotation around the circumference of a local minimum is similar to the mechanism of occurrence of massless Goldstone bosons in Weinberg-Salam's model. However, in this example, the mass of excitations ("zero energy") is determined by the height of the bottom of the trap with respect to the main vacuum state ς=0, ς*=0, and will be non-zero. The stability of such excitations, playing (in this example) the role of DM particles, is guaranteed by the height of the wall of the potential well, occurring in the vicinity of the local potential minimum. Due to the model local minimum positions are the metastable states of DM particles. The lower local minimum is associated with a light (or also hot) component of DM, and the top one with a heavy (or also cold) component of DM. The absolute minimum lies in the region of negative energies and is associated with antigravitational properties of DE. For comparison the exemplary shape of potential used in models of spontaneous symmetry breaking is adopted [53].

3. Non-corpuscular "ether wind" and possible registration of its pressure by the SQUID-magnetostrictor system

Therefore, the search of DM particles, as stable moving excitations of Dark Substance, may be intimately connected with the research of the action of the DE non-corpuscular flux on the ordinary matter. By knowing that the free mean path is connected to the cross section of interaction by the relation $\ell^* \approx 1/(\sigma n_A)$, we may say that DE transfers to a slab of material, consisting of ordinary atoms of concentration n_A, with area S and "maximum depth" ℓ, a momentum $q=\ell^* S\rho_{DE}/\upsilon$, where υ is the DM particles speed relative to the substance. In this way, Dark Substance exerts the pressure

$$p^*_{DE}=F/S=(\ell^* S\rho_{DE}/\upsilon)((\ell^*/\upsilon)S)^{-1}=\rho_{DE} \quad (4)$$

on the slab, where ρ_{DE} is an evaluation of DE energy density. Thus, the effective pressure drop across the length ℓ is estimated as $p_{DE}=p^*_{DE}\,\ell/\ell^*=\rho_{DE}\,\ell\sigma n_A$. In accordance with the generally accepted value of the average density of Dark Energy $\rho_{DE}\approx$ 300 TeV/m^3, taking into account the above-obtained "optimistic" estimation of the interaction cross section $\sigma\approx10^{-35}$cm^2, the pressure drop across a one-meter barrier, having a concentration of atoms $n_A\approx3\times10^{22}$cm^{-3}, will be of the order of $p_{DM}(\ell=1\,м)\approx7\times10^{-16}$ Pa.

In order to register such pressure, a dynamometer that performance capable to fix the strength of 10^{-16} N at the end of the cylinder, which dimensions $\ell\times S\approx$ 1m$\times$0.15 m^2, is required. Apparently, the suitable candidate for the role of the super-high-sensitivity dynamometer is the SQUID-magnetostrictor system [62-64] which has been previously supposed to be used in projects for the detection of gravitational waves, etc. (Figure 1).

Ultra-high sensitivity is achieved by means of this system. In fact, high strain-gauge effectiveness of the sensor can be achieved, since it operates on the principle of the reverse magnetostrictive effect, generated, in its turn, by collective quantum solid-state effects [61]. On the other hand, the high ("quantum scale" level) sensitivity of SQUID systems, used as measuring instruments, allows accurate registration of events.

The physical quantity describing the reverse magnetostrictive effect (discovered by Emilio Villari in 1865) in a particular material is the ratio of the internal magnetic induction field to the growth of its outside pressure, i.e., $\Lambda^{-1}=\frac{\partial B}{\partial P}$. For example, an alloy made up of 54% Pt and 46% Fe, with $\mu\approx14000$, will have $\Lambda^{(-1)}\approx10^{-4}$T/Pa (which is basically not a record value). Thus the magnetic response, measured by the SQUID, is related to the force action δF by this expression $\delta\Phi=\Lambda^{(-1)}\delta F$. Accordingly, the capability to register the pressure of non-corpuscular Dark Matter flow, estimated above for $\sigma\approx 10^{-35}cm^2$ at $\delta F\approx10^{-16}$ N, requires an "absolute" (not reduced to the time of the signal accumulation) SQUID sensitivity for the magnetic flux of the order of magnitude of 10^{-20} Wb $\approx5\times10^{-6}\,\Phi_0$. The actual value of a good DC-SQUID is of about $\delta\Phi\approx10^{-6}\,\Phi_0/\sqrt{\text{Hz}}$, which provides the desired sensitivity with a margin of approximately 2 orders of magnitude (at least) due to the possibility of a 3-hour signal accumulation.

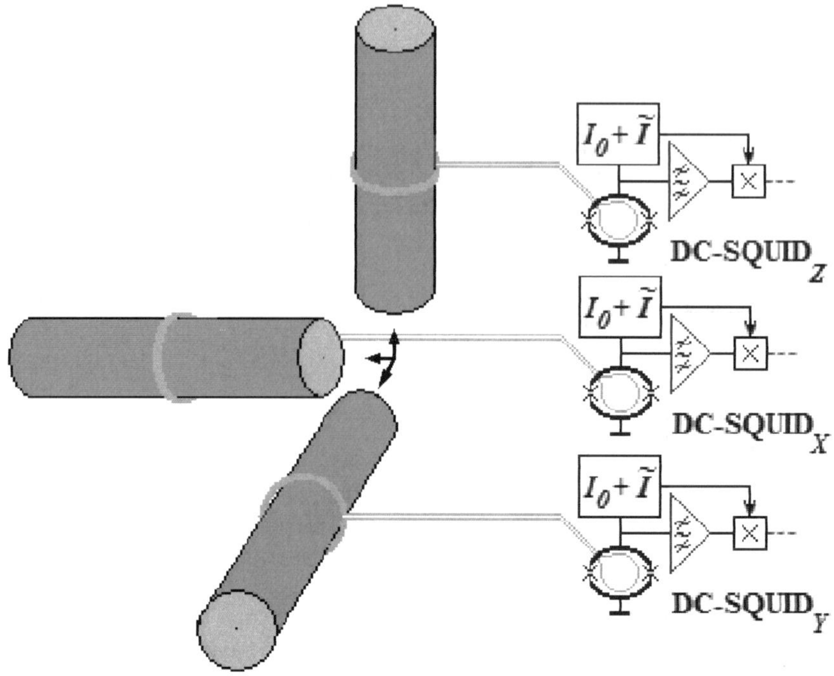

Figure 1. Schematic view of the SQUID-magnetostrictor 3D system for the detection of gravitational waves (the magnetostrictive cylinders are represented in grey) which can be used to determine the action of DE [65].

4. Discussions

In this paper, starting from an introduction about DM and its cosmological properties, in the context of "unifying" trend, clearly dominant in the modern elementary particle physics a simple unimodel is proposed, where it is considered the corpuscular Dark Matter (DM) and non-corpuscular Dark Energy (DE) from single approach. This is the proposed model, in which the DE is an absolutely continuous substance, playing the role of a room for metastable excitations, which can be identified as DM particles.

These excitations, clustering themselves, bind baryonic matter more and more, leaving the intermediate space empty, free for expansion under the action of DE.

From this point of view, we cannot assume DE as a medium suitable for the role of absolute spacetime, relative to which the states of other objects are considered. DE is rather an active medium, one of the manifestations of the activity of which is the effect of the production of DM particles.

The coefficients α, β, γ have some uncertainties need to be reduced in order to satisfy the limiting cosmological models of the classical picture of general relativity. In [66] an unexpected unification between general relativity and QFT was noted due to the superposition of fields, which manifests itself in the existence of entangled states that justify connectivity. From this point of view, the existence of linked DM particles explains the connectivity leading to "emergent geometry". At the same time, it paves the way to theoretical concepts of information.

In modern cosmology, there is an irreconcilable contradiction between the experimentally confirmed position of the accelerating Universe expanding (the Big Bang theory) and the recorded constancy of world constants, leading to an alternative assumption about the constancy of particle density – the Quasi-State-model [67]. Within the framework of the above proposed model, the statement about an increase in the number of particles with time, which leads to the constancy of the density of matter and radiation, is partially confirmed.

Estimates, carried out within that model, indicate the possibility of experimental detection of the "ether wind" pressure, created by the non-corpuscular incoming flow, corresponding to the galactic orbital motion of the Earth. It is argued that these classes of investigations could be performed by using the SQUID-magnetostrictor experimental system. At the same time, the concept of quantizing the continuity of the medium, which was used in the article, should undergo some modification in order to establish the boundaries of possible quantization both in the energy scale and in the spatial one.

It can be noted also that the proposed model admits the existence of negative values of mass (energy) - a long-awaited donate for the theory of gravity.

References

[1] Bertone G and Hooper D 2018 *Rev. Mod. Phys.* **90** 45002 (*Preprint* astro-ph.CO/1605.04909)
[2] Rubin V C, Thonnard N and Ford Jr. W K 1980 *Astrophys. J.* **238** 471
[3] Springel V *et al* 2005 *Nature* **435** 629 (*Preprint* astro-ph/0504097)
[4] (WMAP Collaboration) 2007 Wilkinson Microwave Anisotropy Probe (WMAP) three year results: implications for cosmology *Astrophys. J.* Suppl. **170** 377 (*Preprint* astro-ph/0603449)
[5] (Planck Collaboration) 2014 Planck 2013 results XVI Cosmological parameters *Astron Astrophys* **571** A16 (*Preprint* astro-ph.CO/1303.5076)
[6] Ade P A R *et al* (Planck Collaboration) 2015 Planck 2015 results. XIII Cosmological parameters (*Preprint* astro-ph CO/1502.01589)
[7] Bergstrom L 2000 *Rept.Prog.Phys.* **63** 793
[8] Bertone G, Hooper D and Silk 2005 *J. Phys. Rep.* **405** 279
[9] Olive K A *et al* (Particle Data Group) 2014 *Chinese Phys.* C **38** 090001
[10] Ryabov V A, Tsarev V A and Tskhovrebov A M 2008 *Physics- Uspekhi* **51** 1091
[11] Feng J L 2010 *Ann. Rev. Astron. Astrophys.* **48** 495
[12] Armengaud E *et al* (EDELWEISS Collaboration) 2017 *JINST* **12** P08010 (*Preprint* astro-ph GA/1706.01070)
[13] Aprile E *et al* (XENON100 collaboration) 2012 *Phys. Rev.Lett.* **109** 181301
[14] Akerib D S *et al* (LUX collaboration) 2019 *Phys. Rev. Lett.* **122** 13101 (*Preprint* physics.ins-det/ 1811.11241)
[15] Agnese R *et al* (SuperCDMS collaboration) 2014 Search for Low-Mass Weakly Interacting Massive Particles with Super CDMS *Phys. Rev. Lett.* **112** 241302
[16] Mazziotta M N *et al* (Fermi-LAT Collaboration) 2019 *PoS* ICRC **2020** 531
[17] Ade P A R *et al* (Planck Collaboration) Planck 2015 results. XIII. Cosmological parameters. (*Preprint* astro-ph CO/1502.01589)
[18] Aad G *et al* (ATLAS Collaboration) 2015 *Eur. Phys. J.* C, **75** 299
[19] (ATLAS Collaboration) 2018 *Phys. Rev.* D **98** 032016 (*Preprint* hep-ex/1805.09299)
[20] Khachatryan V *et al* (CMS Collaboration) 2020 *Preprint* hep-ex/2012.04120
[21] Belli P, Bernabei R, Danevich F A *et al.* (DAMA Collaboration) 2019 *Eur. Phys. J.* A **55** 140 (*Preprint* nucl-ex/1908.11458)
[22] Aalseth C E *et al* (CoGeNT collaboration) 2011 *Phys.Rev.Lett.* **106** 131301
[23] Angloher G *et al* (CRESST Collaboration) 2017 *Eur Phys J* C **77** 299 (*Preprint* hep-ex/1612.07662)
[24] Agnese R *et al* (superCDMS Collaboration) 2019 *Phys. Rev.* D **99** 062001
[25] Boliev M M *et al* 2013 *JCAP09* **1309** 019 (*Preprint* astro-ph.HE/1301.1138)
[26] Suzuki Y 2019 *Eur. Phys. J.* C **79** 298
[27] Albert A *et al* 2020 *Phys. Lett.* B **805** 135439
[28] Abbasi *et al* 2021 *Phys.Rev.*D **103** 10 102001 (*Preprint* astro-ph.HE/2101.00610)
[29] Sloan J V *et al* 2016 *Physics of the Dark Universe* **14** 95
[30] Di Luzio L, Giannotti M, Nardi E and Visinelli L 2020 *Phys. Rept.* **870** 1
[31] Peccei R D and Quinn H R 1977 *Phys. Rev. Lett.* **38** 1440
[32] Weinberg S 1978 *Phys. Rev. Lett.* **40** 223
[33] Wilczek F 1978 *Phys. Rev. Lett.* **40** 279
[34] Sikivie P 1983 *Phys. Rev. Lett.* **51** 1415

[35] Giannotti M, Irastorza I, Redondo J and Ringwald A 2016 *JCAP* **05** 057 (*Preprint* astro-ph.HE/1512.08108)
[36] Melendez B Bertolami M M and Althaus L 2012 (*Preprint* hep-ph/1210.0263)
[37] Okun L V, 1982 *Sov. Phys. JETP* **56** 502
[38] Ringwald A 2015 (*Preprint* hep-ph/1506.04259)
[39] Du N *et al* (ADMX Collaboration) 2018 *Phys. Rev. Lett.* **120** 151301
[40] Aprile E *et al* (XENON Collaboration) 2019 *Phys. Rev. Lett.* **123** 251801
[41] Primakoff H. 1951 *Phys. Rev.* **81** 899
[42] Sikivie P, Tanner D and van Bibber K 2007 *Phys. Rev. Lett.* **98** 172002 *(Preprint* hep-ph/0701198*)*
[43] Gorelik V S, Izmailov G N 2011 *Bulletin of LPI (KSF English)* **6** 177
[44] Jaeckel J, Redondo J and Ringwald A 2009 *EPL* **87** 10010
[45] Redondo J and Ringwald A 2011 *Contemporary Physics* **52** 211
[46] Irastorza I G, Pivovaroff M and Kate H T 2014 *CERN Courier*
[47] Halder A, Pandey M. (*Preprint* astro-ph.CO/2101.05228)
[48] Li C, Ren X, Khurshudyan M and Cai Y-F 2020 *Physics Letters* B **801** 135141
[49] Halder A, Banerjee S, Pandey M and Majumdar D 2020 *Mon.Not.Roy.Astron.Soc.* **500** 5589
[50] Kennedy C J *et al* 2020 *Phys. Rev. Lett.* **125** 201302
[51] Pedrozo-Peñafiel E S *et al* 2020 *Nature* **588** 414
[52] Bhattacharya A *et al* 2017 *JCAP* **07** 027 (*Preprint* hep-ph /1706)
[53] Giordano A, Izmailov G N, De Luca R, Ryabov V A, Zherikhina L N and Tskovrebov A M 2016 *(Preprint* astro-ph.HE/1611.04651v1)
[54] Sachs R K, Wolfe A M 1967 *Astrophys. J.* **147** 73
[55] Maeder A 2017 *The Astr. J.* **834** 194
[56] Nishizawa A J 2014 *Progress of Theoretical and Experimental Physics* **6** 06B110
[57] Kovács A 2018 *Monthly Notices of the Royal Astronomical Society* **475** 1777
[58] Roldan O 2018 *JCAP* **03** 014
[59] Barone A, Paternò G 1982 *Physics and applications of the Josephson effect* (New York: J Wiley & Sons) pp 529
[60] Velten H, Schwarz D J, Fabris J C and Zimdahl W 2013 *Phys. Rev.* D **88** 103522
[61] Velten H, Wang J and Meng X 2013 *Phys. Rev.* D **88** 123504
[62] Golovashkin A I *et al* 1994 *JETP letters* **60** 612
[63] Izmailov G N, Zherikhina L N, Ryabov V A and Tskhovrebov A M 2010 *In Dark Energy: Theories, Developments, and Implications* (Nova Sc. Pub.)
[64] Golovashkin A I *et al* 2012 *Quantum Electronics* **42** 1140
[65] Izmaïlov G N *et al* 2018 *J. Phys.: Conf. Ser.* **1051** 012016
[66] Seltmann M 2018 *In Experimental Search for Quantum Gravity* (Cham: Springer Int. Publ. AG) pp 91-104
[67] Narlikar J V, Vishwakarma RG and Burbidge G 2002 (*Preprint* astro-ph/0205064v2)

Geometry and kinematics induced by biquaternionic and twistor structures

V V Kassandrov[1] and N V Markova[2]

[1]Institute of Gravitation and Cosmology, Peoples' Friendship University of Russia, Miklukho-Maklaya Str. 6, Moscow, Russia

[2]Department of Applied Mathematics, Peoples' Friendship University of Russia, Miklukho-Maklaya Str. 6, Moscow, Russia

E-mail: vkassan@sci.pfu.edu.ru

Abstract. The algebra of biquaternions possess a manifestly Lorentz invariant form and induces an extended space-time geometry. We consider the links between this complex pre-geometry and real geometry of the Minkowski space-time. Twistor structures naturally arise in the framework of biquaternionic analysis. Both together, algebraic and twistor structures impose rigid restriction on the transport of singular points of biquaternion-valued fields identified with particle-like formations.

1. Biquaternions as the algebra of extended space-time

The algebra of complex quaternions, *biquaternions*, $\mathbb{B}$ is the most appropriate candidate for the role of a *space-time algebra* (the notion introduced firstly by D. Hestenes in [1]). Indeed, its symmetry (automorphism) group $SO(3,\mathbb{C})$ is isomorphic to the spinor Lorentz group $SL(2,\mathbb{C})$, and its multiplication table can be represented in a manifestly Lorentz invariant form [2, 3]. However, vector space of $\mathbb{B}$ has four *complex* dimensions and does not correspond therefore to the structure of real Minkowski space $\mathbf{M}$. Physical sense of four "superfluous" coordinates (if any) remains indefinite, and we shall concentrate on this problem below.

Nonetheless, biquaternions and $\mathbb{B}$-valued functions, as well as the structure of extended complex space-time $\mathbb{C}\mathbf{M}$ itself, have been exploited in a number of interesting approaches in the field theory and general relativity. Particularly, in GTR complex space-time serves as a source of null shear-free congruences (NSFC) on its "real cut" which are known to determine the structure of Robinson-Trautman metrics. Moreover, consideration of a pointlike source "moving" in a complex extension $\mathbb{C}\mathbf{M}$ results in a number of interesting conclusions about its characteristics and characteristics of particlelike formations – caustics of corresponding NSFC arising on the real cut [4, 5, 6]. A well-known example of this situation is the anomalous Dirac value [7] of the gyromagnetic ratio of the *Kerr singular ring*, the caustic of corresponding NSFC. The latter property has been used in a number of *models of electron*, in particular the "Dirac – Kerr-Newman electron" elaborated by A. Burinskii [8].

As for the biquaternions themselves, there exists a wide variety of works devoted to various applications of them in field theory and physical geometry (for a review see, e.g., [9, 10]). In particular, the quaternionic version of Special Relativity, based on the properties of $\mathbb{B}$ and the concept of 3-dimensional time, has been elaborated in the works of A. Yefremov [11, 12]. There is also a lot of attempts (see [9, 10]) in which physical fields are regarded as $\mathbb{B}$-valued functions subject to a kind of *differentiablity conditions*. The latter, usually constructed in a full analogy with the Cauchy-Riemann (CR) conditions from complex analysis, play the role of the primary field equations.

However, all these generalizations reproduce the linear CR-structure, do not take in account the *non-commutativity* of $\mathbb{B}$ and, moreover, are not mathematically justified (value of a derivative usually depends on the direction of infinitesimal increment, contrary to the complex case).

On the other hand, in our version of non-commutative analysis (see, e.g., [13, 14, 15] and references therein) non-linearity of the differentiability conditions for (bi)quaternionic functions follow directly from the property of $\mathbb{B}$ non-commutativity, so that $\mathbb{B}$-differentiable functions can be actually treated as (self-)interacting physical fields.

Moreover, the system of $\mathbb{B}$-differentiability equations possess natural 2-spinor and twistor structures, and its general solution represents in fact an invariant version of the celebrated Kerr-Penrose theorem [16, 17]. The latter gives a complete algebraic description of the NSFC on the Minkowski $\mathbf{M}$ or Kerr-Shild spaces [16] via twistor generating functions. In the turn, the NSFC define a set of fundamental (both gauge and spinor) fields for which the corresponding vacuum equations are identically satisfied [18]. Singularities of these fields (point-, string- or membrane-like) can be treated as particlelike formations participating in a self-consistent collective dynamics [19]. We obtain thus a non-trivial set of interacting fields and their singular sources whose properties are determined solely by the $\mathbb{B}$-differentiability conditions; this approach has been called "algebrodynamics" [20, 13].

However, the complex space-time geometry defined by the structure of $\mathbb{B}$-algebra embarrass the physical applications of biquaternions and, particularly, the interpretation of results obtained in the framework of the algebrodynamical approach. In our previous works [21, 22] a number of constructions has been presented which, rather naturally, relate complex pre-geometry with real geometry of the Minkowski space or its extensions. These constructions based on $\mathbb{B}$-invariant *bilinear maps* lead to a peculiar kinematics of the particlelike formations; they are described in Section 2.

In Section 3 we impose some additional restrictions on the particles' kinematics motivated by the properties of the primary $\mathbb{B}$-field and its twistor partner. These properties define the transport of twistor field and, in addition, the transport of singulartities, caustic-like structures. In result, one obtains a remarkable classification of the properties of singularities-particles themselves. Some concluding remarks are made in the last Section 4.

2. Real geometry determined by the algebra of biquaternions

Biquaternion algebra $\mathbb{B}$ is isomorphic to the full 2×2 matrix algebra over $\mathbb{C}$, $Mat(2, \mathbb{C})$, so that for any $Z \in \mathbb{B}$ one has, say, the following representation:

$$Z = \mathbf{I}z_0 + \mathbf{\Sigma} \cdot \mathbf{z} = \begin{pmatrix} z_0 + z_3 & z_1 - \imath z_2 \\ z_1 + \imath z_2 & z_0 - z_3 \end{pmatrix}, \tag{1}$$

where $\{z_0, \mathbf{z}\} \in \mathbb{C}$, $\mathbf{I}$ is the unit matrix and $\mathbf{\Sigma} = \{\sigma_a\}, a = 1, 2, 3$ is the triplet of Pauli matrices.

Consider now a Hermitian ($M = M^+$) matrix,

$$M := ZZ^+ = \begin{pmatrix} T + R_3 & R_1 - \imath R_2 \\ R_1 + \imath R_2 & T - R_3 \end{pmatrix} = T + \mathbf{\Sigma} \cdot \mathbf{R}, \tag{2}$$

in which quantities $\{T, \mathbf{R}\}$ depend on the complex coordinates and their conjugates as follows:

$$\begin{aligned} &T := z_0 \bar{z}_0 + \mathbf{z} \cdot \bar{\mathbf{z}}, \quad \mathbf{R} := \mathbf{P} + \mathbf{Q}, \\ &\mathbf{P} = z_0 \bar{\mathbf{z}} + \bar{\mathbf{z}}_0 \mathbf{z}, \quad \mathbf{Q} = \imath\, \mathbf{z} \times \bar{\mathbf{z}} \end{aligned} \tag{3}$$

Bilinear in Z quantities $\{T, \mathbf{R}\}$ could represent the real time and space coordinates, respectively; however, more justified interpretation will be discussed below.

One observes that spatial coordinates are defined as the sum of two different parts, and the two 3D subspaces $\mathbf{P}$ and $\mathbf{Q}$ are easily seen to be orthogonal,

$$\mathbf{P} \cdot \mathbf{Q} = 0. \tag{4}$$

Making now use of the well-known properties of determinants, $\det(ZZ^+) = (\det Z)(\det Z^+) = |\det Z|^2$, one obtains the following remarkable identity [22]:

$$T^2 - \mathbf{R}^2 = S^2, \quad S := |\det Z| = |z_0^2 - \mathbf{z}^2|, \tag{5}$$

which, in account of (3) and the orthogonality relation (4), takes the form

$$T^2 - \mathbf{P}^2 - \mathbf{Q}^2 = S^2. \tag{6}$$

It is worthy to note that, since $S^2 = \Re(S)^2 + \Im(S^2)$, equation (6) represents a remarkable *nine squares identity* closely related to the structure of $\mathbb{B}$-algebra. Of course, identity (6) can be immediately written out in an explicit coordinate form.

Consider now a particular case when the component z_0 is zero. Then equation (5) reduces to the *identity of six squares* for $\mathbf{z}$ of the form

$$\begin{array}{l} t^2 - \mathbf{q}^2 = s^2, \\ t = \mathbf{z} \cdot \bar{\mathbf{z}}, \quad q = \imath\, \mathbf{z} \times \bar{\mathbf{z}}, \quad s = |\mathbf{z}^2|, \end{array} \tag{7}$$

which has been previously obtained and discussed in [21]. Importantly, the identity (7) remains valid, together with (5), if even $z_0 \neq 0$.

Under transformations from the group $SO(3, \mathbb{C})$ which preserve the multiplication law of $\mathbb{B}$-algebra, both $\mathbf{z}^2$ and z_0 are evidently invariant, so that the quantity S (and the reduced one s) can be identified with the *Minkowski interval*. Under *real* 3D-rotations, the quantities T and $\mathbf{R}$ evidently form a scalar and vector, respectively. On the other hand, when the angle of rotation is imaginary, these quantities transform in a way different from the Lorentz transformations. For a reduced case such transformations have been studied in [21].

However, if we consider instead the *left shift* transformation $Z \mapsto AZ$, $A \in SL(2, \mathbb{C})$, then the corresponding Hermitian matrix $M = ZZ^+$ transforms according to the canonical law, $M \mapsto AMA^+$, so that the quantities $\{T, \mathbf{R}\}$ undergo standard Lorentz transformations. Note that the *dual* mapping $Z \mapsto Z^+Z$ and related *right shift* transformation $Z \mapsto ZA$ define in fact the same Lorentz 4-vector $\{T, \mathbf{R}\}$ but for the point in $\mathbb{C}\mathbf{M}^3$ corresponding to complex conjugate coordinates $\mathbf{z} \to \bar{\mathbf{z}}$.

One should, however, bear in mind that the introduced quantities $\{T, \mathbf{R}\}$, being bilinear in Z, are not *translationary invariant*. That is, for two points Z_1 and Z_2 and corresponding real coordinates T_1, T_2 and $\mathbf{R}_1, \mathbf{R}_2$ the *holonomy* conditions $\Delta T = T_2 - T_1$, $\Delta \mathbf{R} = \mathbf{R}_2 - \mathbf{R}_1$ are no longer valid. Here ΔT and $\Delta \mathbf{R}$ are the quantities expressed through the differences of complex coordinates ΔZ and their conjugates ΔZ^+ making use of the same definitions (3).

In view of this fact, one should identify the 4-vector $\{T, \mathbf{R}\}$ with some other quantity attributed to a point of $\mathbb{C}\mathbf{M}^4$, precisely with a set of characteristics of a *material pointlike source*. We shall return to this problem below. As for the coordinates themselves, one can define only their increments $\{\delta T, \delta \mathbf{R}\}$ using corresponding definitions (3), that is:

$$\begin{array}{l} \delta T := dz_0 d\bar{z}_0 + d\mathbf{z} \cdot d\bar{\mathbf{z}}, \quad \delta \mathbf{R} := \delta \mathbf{P} + \delta \mathbf{Q}, \\ \delta \mathbf{P} := dz_0 d\bar{\mathbf{z}} + d\bar{z}_0 d\mathbf{z}, \quad \delta \mathbf{Q} := \imath\, \delta \mathbf{z} \times \delta \bar{\mathbf{z}} \\ \delta S := |dz_0^2 - d\mathbf{z}^2|. \end{array} \tag{8}$$

However, these increments do not constitute full differentials, so that corresponding real coordinates turn our to be *unholonomic*. This property, though seeming rather exotic, leads up to a number of interesting concepts and conclusions ("random time", geometric origin of time irreversibility, effective curvature, etc.) some of which have been discussed in [15].

As the quantities $\{T, \mathbf{R}\}$ themselves, the increments of real coordinates satisfy the identities similar to (5) and (6), that is

$$\delta T^2 - \delta \mathbf{R}^2 = \delta S^2, \tag{9}$$

$$\delta T^2 - \delta \mathbf{P}^2 - \delta \mathbf{Q}^2 = \delta S^2, \tag{10}$$

as well as the orthogonality condition

$$\delta \mathbf{P} \cdot \delta \mathbf{Q} = 0. \tag{11}$$

Note that infinitesimal Minkowski interval δS in (8) is necessarily *non-negative*, so that one deals here with only the *causal* part of the Minkowski-like space-time. On the other hand, the time increment δT, according to the definition (8), is positive definite, $\delta T > 0$, so that one observes here a sort of *time irreversibility*: any displacement in the primary complex space results in an increase of time in the induced real space.

Consider now a pointlike source moving in the primary complex space $\mathbb{C}\mathbf{M}$. Then, for the case of nonzero interval, one can define a real 4-velocity vector ($\mu = 0, 1, 2, 3$)

$$V^\mu := \frac{\delta X^\mu}{\delta S}, \ \delta X^\mu := \{\delta T, \delta \mathbf{R}\}, \tag{12}$$

which, in account of (9), is identically unit,

$$V_\mu V^\mu = \left(\frac{\delta T}{\delta S}\right)^2 - \left(\frac{\delta \mathbf{R}}{\delta S}\right)^2 = 1, \tag{13}$$

while the 3-velocity $\mathbf{V} := \frac{\delta \mathbf{R}}{\delta T}$ is always subluminar,

$$\delta T(1 - V^2) = \delta S > 0, \rightarrow V^2 < 1. \tag{14}$$

Note finally that, in view of the identity (10) and orthogonality condition (11) the full 3-velocity is represented by "partial" velocities

$$\mathbf{u} := \frac{\delta \mathbf{P}}{\delta T}, \quad \mathbf{w} := \frac{\delta \mathbf{Q}}{\delta T} \tag{15}$$

corresponding to the motions in the two real 3-subspaces $\mathbf{P}$ and $\mathbf{Q}$ and mutually orthogonal,

$$V^2 = u^2 + w^2, \quad \mathbf{u} \cdot \mathbf{w} = 0. \tag{16}$$

In the second case of zero interval, $\delta S = 0$, when the complex cone $\det Z = \delta z_0^2 - \delta \mathbf{z}^2 = 0$ is mapped into the future part of the real light cone $\delta T^2 - \delta \mathbf{R}^2 = 0, \ \ \delta T > 0$, the full 3-velocity is equal to the speed of light, $V = 1$, and for partial velocities we have instead of (16):

$$u^2 + w^2 = 1, \quad \mathbf{u} \cdot \mathbf{w} = 0. \tag{17}$$

Remarkably, in this case our $\mathbb{B}$-geometry leads up to the geometry proposed earlier by I. Urusovskii [23, 24, 25] in the framework of his "3+3 treatment of the Special Relativity". From his viewpoint, the speed of light is not only invariant but *universal*, i.e. all material points have one and the same velocity in the full 6D space.

We have seen thus that the $\mathbb{B}$-induced real geometry not only reproduces the principal properties of the Minkowski space-time but impose, in addition, some restrictions on admissible kinematics of pointlike particles. In the next section we shall consider other kinematical relations which arise on the base of $\mathbb{B}$-analysis and corresponding twistor structures.

3. $\mathbb{B}$-analysis, twistors and kinematics in $\mathbb{B}$-induced real space

In the *algebrodynamical program* [20, 13, 14, 15] primary physical fields are $\mathbb{B}$-differentiable functions. Conditions of $\mathbb{B}$-differentiability, a non-commutative generalization of the Cauchy-Riemann conditions from complex analysis, are considered as the primary field equations; in the principal case they reduce to the overdetermined matrix system of equations for differentials,

$$d\xi = \Phi dZ \xi \tag{18}$$

for a 2-spinor $\xi = \{\xi_A(Z)\}$ and complex 4-vector $\Phi = \{\Phi_{AB}(Z)\}$ fields ($A, B = 0, 1$, and Z is the $\mathbb{B}$-coordinate matrix). The structure of general solution of (18) can be easily obtained by the following its transformation

$$d\xi = \Phi d(Z\xi) - (\Phi dZ)\xi, \rightarrow \xi = \xi(\tau), \tag{19}$$

where $\tau := Z\xi$ together with ξ constitute the (projective) *twistor* of complex space $\mathbb{C}\mathbf{M}$. Therefore, in an invariant form general solution of (18) is represented by a system of two algebraic equations [18]

$$\Pi^{(C)}(\xi, Z\xi) = 0, \quad C = 1, 2, \tag{20}$$

with $\Pi^{(C)}$ being two arbitrary analytical functions of four twistor variables. Making use of the projective invariant structure of the incidence relation $\tau = Z\xi$, one can reduce (20) to a simpler form of one equation with one *homogeneous* generating twistor function Π,

$$\Pi(\xi, Z\xi) = 0, \tag{21}$$

from which only the *ratio* G of two spinor components ξ_A can be determined at each fixed point of $\mathbb{C}\mathbf{M}$.

On the Minkowski subspace $\mathbf{M}$ of $\mathbb{C}\mathbf{M}$ represented by Hermitian matrices of coordinates $Z \mapsto X = X^+$, solution of (18) in the form (21) reproduces the celebrated *Kerr-Penrose* theorem [16, 17] completely describing the structure of *null shear-free congruences* (NSFC) on $\mathbf{M}$ mentioned in Section 1. Thus, $\mathbb{B}$-differentiable functions explicitly give rise to a complex extension of the Minkowski twistors and, geometrically, to a complex extension of NSFC.

Caustics of the rays of NSFC correspond to multiple roots of the equation (21) and are defined, therefore, by the condition

$$\frac{d\Pi}{dG} = 0 \tag{22}$$

Structure and dynamics of caustics correspond to those of singularities of fundamental fields defined by NSFC which can be treated as particlelike formations. For more details we refer the reader, say, to [26, 18] and references therein.

Let us consider now the (infinitesimal) transformations of complex coordinates which leave invariant the primary twistor field and the caustic structure. Specifically, from the equation (21) as the *condition of twistor preservation* one obtains $(dZ)\xi = 0$, which in components leads up to the following two real relations for differentials dZ:

$$dy_0 = \mathbf{n} \cdot d\mathbf{y}, \quad d\mathbf{x} = \mathbf{n}\, dx_0 + \mathbf{n} \times d\mathbf{y}, \tag{23}$$

where dx_μ and dy_μ are the real and imaginary parts of the increments of four complex coordinates $dz_\mu = dx_\mu + \imath dy_\mu$. The (projective invariant) *unit director 3-vector* $\mathbf{n}$ is formed from the components of the principal spinor ξ as follows:

$$\mathbf{n} := \frac{\xi^+ \mathbf{\Sigma} \xi}{\xi^+ \xi}, \quad \mathbf{n}^2 \equiv 1. \tag{24}$$

Of course, conditions (23) can be easily inverted,

$$dx_0 = \mathbf{n}\cdot d\mathbf{x}, \quad d\mathbf{y} = \mathbf{n}\, dy_0 - \mathbf{n}\times d\mathbf{x}, \tag{25}$$

and the increments identically satisfy the *complex null cone* condition:

$$dz_0^2 - d\mathbf{z}^2 = 0. \tag{26}$$

It is worthy to note that under transition to the Minkowski space-time, when $Z \mapsto X = X^+$, conditions (23) reduce to $d\mathbf{x} = \mathbf{n}dt$, where $dt := dx_0$ is the real Minkowski time. The latter condition corresponds to the well-known transport of twistor field along a 3-direction (defined by the initial value of the principal spinor) with velocity of light.

The complex cone condition can be equivalently written out in the form of two real conditions:

$$\begin{aligned} dx_0^2 + d\mathbf{y}^2 = dy_0^2 + d\mathbf{x}^2 = \tfrac{1}{2}\delta T, \\ d\mathbf{x}\cdot d\mathbf{y} = dx_0 dy_0. \end{aligned} \tag{27}$$

The last equality in the first equation follows from comparison of the latter with the previous definition of the time increment in (8). Note that the topology of the twistor preserving 6D increment space (actualy, that of the complex null cone (26)), in a consequence of (27), is $S^3\times S^2\times R_+$).

From three equations (27) one can, in particular, express the invariant $|dz_0|^2 = dx_0^2 + dy_0^2$ through the invariants of spacial increments,

$$dx_0^2 + dy_0^2 = \sqrt{(d\mathbf{x}^2 - d\mathbf{y}^2)^2 + (2d\mathbf{x}\cdot d\mathbf{y})^2}, \tag{28}$$

or, after a simple transformation, obtain another important relation:

$$(\delta\mathbf{x}^2 + \delta\mathbf{y}^2)^2 - |2d\mathbf{x}\times d\mathbf{y}|^2 = (dx_0^2 + dy_0^2)^2. \tag{29}$$

Thus, despite the fact that the previously defined complex and real Minkowski intervals in the twistor preserving geometry are both zero, one can, in account of (29), consider as infinitesimal Minkowski interval ds the invariant

$$ds := |dz_0|^2 = dx_0^2 + dy_0^2 \geq 0, \tag{30}$$

while for the redefined time δt and space $\delta\mathbf{r}$ increments of the coordinates take the quantities

$$\delta t := \delta\mathbf{x}^2 + \delta\mathbf{y}^2 \geq 0, \quad \delta\mathbf{r} := 2d\mathbf{x}\times d\mathbf{y}. \tag{31}$$

By this identifications, the principal relation (29) of the twistor preserving geometry takes the Minkowski-like form (again with non-negatively definite increments δt and δs):

$$\delta t^2 - |\delta\mathbf{r}|^2 = \delta s^2 \tag{32}$$

(compare with the reduced identity (7) for corresponding increments in account of the complex null cone condition $dz_0^2 = d\mathbf{z}^2$).

On the other hand, for the full increments of the coordinates (8) one obtains:

$$\begin{aligned} \delta T = \sqrt{(d\mathbf{x}^2 - d\mathbf{y}^2)^2 + (2d\mathbf{x}\cdot d\mathbf{y})^2} + d\mathbf{x}^2 + d\mathbf{y}^2, \\ \delta\mathbf{P} = 2(\mathbf{n}\cdot d\mathbf{x})d\mathbf{x} + (\mathbf{n}\cdot d\mathbf{y})d\mathbf{y}, \quad \delta\mathbf{Q} = 2d\mathbf{x}\times d\mathbf{y}, \end{aligned} \tag{33}$$

and the relations $\delta T^2 - \delta\mathbf{P}^2 - \delta\mathbf{Q}^2 = 0,\ \delta\mathbf{P}\cdot\delta\mathbf{Q} = 0$ are valid for any $d\mathbf{x}, d\mathbf{y}$.

4. Caustic preserving geometry and effective kinematics in real space

Let us now impose, in addition to the twistor preserving condition, the analogous requirement on the caustic structure. Specifically, let us consider the increments of complex coordinates which transfer any caustic point into another one. From the defining equation (22) of the caustic locus, assuming the constancy of twistor field, one obtains the preservation condition for the matrix components:

$$dZ_A^B = \Pi_A \xi^B d\lambda, \quad A, B = 0, 1, \tag{34}$$

where $\Pi_A = d\Pi/d\tau^A$ are the derivatives of generating function Π w.r.t. two of its twistor arguments, and $d\lambda$ is one sole infinitesimal complex parameter.

From (34) it follows that, since the twistor-valued Π_A and ξ_A remain constant, the combined twistor-caustic preserving transport takes place along a *complex null straight line*. If $dz_0 \neq 0$, condition (34) can be equivalently rewritten in a vector form,

$$d\mathbf{z} = \mathbf{m}\, dz_0, \quad \mathbf{m}^2 = 1, \tag{35}$$

where $\mathbf{m}$ is a unit complex 3-vector,

$$\mathbf{m} := \mathbf{a} + \imath\mathbf{b}; \rightarrow \mathbf{a}^2 - \mathbf{b}^2 = 1, \ \mathbf{a}\cdot\mathbf{b} = 0. \tag{36}$$

In the other case, when $dz_0 = 0$, condition (34) gives:

$$d\mathbf{z} = \mathbf{l}\, d\tau, \quad \mathbf{l}^2 = 0, \tag{37}$$

with a null complex 3-vector,

$$\mathbf{l} = \mathbf{c} + \imath\mathbf{d}; \rightarrow \mathbf{c}^2 = \mathbf{d}^2, \quad \mathbf{c}\cdot\mathbf{d} = 0, \tag{38}$$

and an infinitesimal complex parameter $d\tau$.

Calculating then the increments of real temporal and spacial coordinates (8), in the first case (35), one obtains

$$\delta T = \mathbf{a}^2\, ds, \quad \delta\mathbf{P} = \mathbf{a}\, ds, \quad \delta\mathbf{Q} = (\mathbf{a}\times\mathbf{b})ds, \tag{39}$$

while the increment of proper time parameter *turns to be null*, $\delta S = 0$! The increment of real parameter ds is defined by the square of modulus of the zeroth complex coordinate, $ds = 2|dz_0|^2$. Thus, in account of (10), in the induced real space-time singularity moves with the full velocity equal to the speed of light. The latter consists of the two components, $\mathbf{u} = \delta\mathbf{P}/\delta T < 1$ and $\mathbf{w} = \delta\mathbf{Q}/\delta T < 1$ which both are smaller than c and mutually orthogonal.

In the second case (35), when $dz_0 = 0$, one has, respectively:

$$\delta T = \mathbf{c}^2\, ds, \quad \delta\mathbf{P} \equiv 0, \quad \delta\mathbf{Q} = (\mathbf{c}\times\mathbf{d})ds, \tag{40}$$

where $ds = 2|d\tau|^2$, and again the increment of proper time parameter $\delta S = 0$. One observes that singularity again moves with the the speed of light but in this case only one component of velocity is nonzero.

We have already mentioned that the quantities $T, \mathbf{R}$ as well as their increments $\delta T, \delta\mathbf{R}$ are not invariant under translations and unholonomic. Therefore, these (bilinear in complex coordinates $z_0, \mathbf{z}$) quantities should be interpreted not as coordinates but rather as **energy** and **momentum** of a corresponding particlelike formation.

5. Conclusion

Existence of complex pre-geometry seems to be unavoidable in algebrodynamical approach. Indeed, in the framework of the latter, the whole physical geometry and dynamics should be derived solely from the properties of a space-time algebra. However, the only candidate for this role is the algebra of biquaternions with its $SO(3,\mathbb{C}) \sim SL(2,\mathbb{C})$ automorphism group and 4D complex vector space.

Thus, to construct the unified biquaternionic algebrodynamics, one should discover the links between the complex geometry $\mathbb{C}\mathbf{M}$ and real Minkowski space-time $\mathbf{M}$. In the most natural way one can do it via bilinear mapping $Z \mapsto X := ZZ^+$ used in the paper. Then the induced kinematics of particles-singularities turns to be Lorentz invariant. However, quantities X are not translation invariant and correspond rather to the energy-momentum 4-vector of a particle. As for the real coordinates themselves, they can be identified with real parts of complex coordinates while imaginary parts constitute the 4D timelike evolution parameter, see (23). Note that multidimensional time can be the origin of real time irreversibility, of quantum uncertainty etc. [15].

In the paper, we have also studied the effective kinematics of a singular point defined by the requirements of preservation of the primary twistor field and the singularity itself. These conditions arise in the framework of $\mathbb{B}$-analysis and lead to two different types of the transport of a singular point along a straight line in both of which, however, the full real space velocity of transport is equal to the speed of light. Note, however, that for the points on a *focal line* (unique Worldline) [19] the effective kinematics (dynamics) can be much more complicated since the primary spinor field of the congruence G is indefinite on the worldline itself. Complete investigation of this case will be presented elsewhere.

References

[1] Hestenes D 1966 *Space-time algebra* (New York: Gordon and Breach)
[2] Grgin E *Phys. Lett.* B**431** 15
[3] Kassandrov V V and Rizcallah J A 2016 Relativistic algebra of space-time and algebrodynamics *Grav. Cosm.* **22** 230 (*Preprint 1612.02455*)
[4] Kozameh C N and Newman E T 2005 Electromagnetic dipole radiation fields, shear-free congruences and complex center of charge worldlines *Class. Quant. Grav.* **22** 4667 (*Preprint gr-qc/0504093*)
[5] Newman E T 2002 On a classical geometric origin of magnetic moments, spin-angular momentum amd the Dirac gyromagnetic ratio *Phys. Rev.* D **65** 104005 (*Preprint gr-qc/0201055*)
[6] Burinskii A Ya 2003 Complex Kerr geometry and nonstationary Kerr solutions *Phys. Rev.* D **67** 124024 (*Preprint gr-qc/0212048*)
[7] Carter B 1968 *Phys. Rev.* **174** 1559
[8] Burinskii A Ya 2008 The Dirac – Kerr-Newman electron *Grav. Cosm.* **14** 109 (*Preprint hep-th/0507109*)
[9] Gsponer A and Hurni J P 2005 Quaternions in mathematical physics (1): alphabetical bibliography (*Preprint math-ph/0510059*)
[10] Gsponer A 2005 Quaternions in mathematical physics (2): analytical bibliography (*Preprint math-ph/0511092*)
[11] Yefremov A P 1996 *Grav. & Cosm.* **2** 77; **2** 355
[12] Yefremov A P 2005 *Quaternionic Spaces, reference frames and fields* (Moscow: Peoples' Friendship Univ. Press, in Russian)
[13] Kassandrov V V 1995 Biquaternionic electrodynamics and Weyl-Cartan geometry of space-time *Grav. & Cosm.* **1** 216 (*Preprint gr-qc/0007027*)
[14] Kassandrov V V 2007 Quaternionic alalysis and the algebrodynamics *Space-time structure: algebra and geometry* eds. Pavlov D G et al. (Moscow: Lilia-Print) p. 441 (*Preprint 0710.2895*)
[15] Kassandrov V V 2009 Algebrodynamics over complex space and phase extension of the Minkowski geometry *Phys. Atom. Nuclei* **72** 813 (*Preprint 0907.5425*)
[16] Debney G C, Kerr R P and Schild A 1969 *J. Math. Phys.* **10** 1812
[17] Penrose R and Rindler W 1986 *Spinors and space-time, vol.2* (Cambridge: Cambridge Univ. Press)
[18] Kassandrov V V and Rizcallah J A 2017 Maxwell, Yang-Mills, Weyl and eikonal fields defined by any shear-free congruence *Int. J. Geom. Meth. Mod. Phys.* 1750031 (*Preprint 1612.06718*)

[19] Kassandrov V V, Khasanov I Sh and Markova N V 2015 Collective Lorentz invariant dynamics on a single "polynomial" worldline *J. Phys. A: Math. Theor.* **48** 395204 (*Preprint 1501.01606*)

[20] Kassandrov V V 1992 *Algebraic structure of space-time and algebrodynamics* (Moscow: Peoples' Friendship Univ. Press, in Russian)

[21] Kassandrov V V 2005 Algebrodynamics in complex space-time and the complex-quaternionic origin of Minkowski geometry *Grav. Cosm.* **11** 354 (*Preprint gr-qc/0602088*)

[22] Kassandrov V V 2012 Urusovskii's geometry, algebrodynamics and universal quantum-like kinematics in complex space *Proc. Int. Sci. Meet. on Physical Interpretation of Relativity Theory (PIRT-2011)*, eds. M. C. Duffy et al. (Moscow: Bauman Tech. Univ. Press) p. 131 (*Preprint 1209.2377*)

[23] Urusovskii I A 1996 *Zarubezhnaya Radioelectronika* **3** 359

[24] Urusovskii I A 2000 *Zarubezhnaya Radioelectronika* **6** 66

[25] Urusovskii I A 2009 *Proc. Int. Sci. Meet. on Physical Interpretation of Relativity Theory (PIRT-2009)* eds. M. C. Duffy et al. (Moscow: Bauman Tech. Univ. Press) p. 173

[26] Kassandrov V V 2004 Singular sources of Maxwell fields with self-quantized electric charge *Has the last word been said on classical electrodynamics?* eds. A. Chubykalo et al. (Rinton Press) p. 42 (*Preprint physics/0308045*)

Noise factor and reception bandwidth in optoacoustical GW antenna

V A Krysanov

Institute for Nuclear Research RAS, Moscow

E-mail: kv@sai.msu.ru

Abstract. An expression has extracted from the OGRAN project theory, which provides connection between numerical values of noise factor F and achieved displacement resolution and antenna's threshold signal in metric variations. Noise factor and "reception bandwidth" connects across displacement resolution. There is defined analytical expression and numerical value for design displacement resolution (sensitivity) on the base intention $F = 1$. It has appeared that the extracted analytical expression for readout resolution does not correspond to applied Pound-Drever-Hall technique and AURIGA circuitry. This requires an improvement in theoretical design. The achieved resolution value $2 \cdot 10^{-15}$ cm/Hz$^{1/2}$ is matched to the value for metric sensitivity in pulse $h_{min} \simeq 10^{-18}$, which is 15 dB higher than the thermal sensitivity limit.

1. Threshold signal of a bar detector

The effect of a gravitational wave is represented as an equivalent force F_G acting on the test body as the cylindrical acoustic resonator (bar) [1];

$$F_G = h_m M \omega_0^2 L/2. \tag{1}$$

Here h_m is amplitude of metric variation; L, ω_0 and M are length, resonant frequency and equivalent mass of the oscillator.

Force is registered through the mechanical displacement that it causes.

While relaxation time τ_{ref} is long, amplitude of forced oscillations of an oscillator increases linearly; at the end of exposure it reaches the value [1]

$$\Delta A = F_G \tau (2M\omega_0)^{-1},$$

where τ is duration of forecast supernova burst pulse.

Certain antenna "sensitivity bandwidth" appears $\Delta f_S \simeq (1/\tau)$.

Oscillator is in state of permanent thermal Brownian motion; corresponding energy is $k_B T/2$.

Motion has form of a narrow-band stochastic quasi-harmonic process $x(t) = A(t)\cos[\omega_0 t + \varphi(t)]$.

The amplitude and phase change stochastically and slowly. With phase coincidence, signal creates amplitude variation ΔA.

Under strong metric variation affect, thin dissipation effects that determine an oscillator quality factor Q are irrelevant. Consequently, no significant requirements are imposed on accuracy of coincidence of frequency of quasi-harmonic metric signal and resonance frequency of the oscillator ω_0.

During the time τ of signal exposition, stochastic variations of amplitude take place. Its average statistical value is [2]

$$\sigma_B = (2k_B T\tau / M\omega_0^2 \tau_{rel})^{1/2} = (k_B T\tau / QM\omega_0)^{1/2}.$$

Here there is dependence on quality factor.

From comparison of signal and fluctuation variations, threshold signal is determined. This is how a short pulse should detect by a resonant bar antenna.

For general case of arbitrary phase mismatch, the process is represented using quadrature components $x(t) = A_C(t)\cos\omega_0 t + A_S(t)\sin\omega_0 t$. The components A_C and A_S are highlighted by synchronous detection. The algorithm for primary data processing has implemented in the "ULITKA" detector [3,4].

In 2005, development of a full-scale facility OGRAN ($L \simeq 2$ m, $M \simeq 1000$ kg and $f_0 = \omega_0 /2\pi \simeq$ 1300 Hz) has begun [5]. Somewhat earlier, development of the pilot model had begun ($L \simeq 0.5$ m and $M \simeq 50$ kg). In 2004 [6], the first article appeared with the title: «Gravitational wave experiments and Baksan project «OGRAN». The article presents calculation formulas provided by theory (GWA sensitivity analysis), formed in 1996 [7]. In conjunction with predefined parameter values the "theoretical basis" of the OGRAN project had formed. This basis has confirmed in the final article on the pilot model [8].

In 2013, the Moscow adjustment period of full-scale facility OGRAN had completed; the facility has moved to Baksan Neutrino Observatory of the Institute for Nuclear Research of RAS into camera inside the mount. As result, two articles has published [9,10]; in modern view of OGRAN foundation has presented: "Under optimal filtration procedure, the minimum registered perturbation of the OGRAN optical length in the bandwidth $\Delta f_S = (1/\tau)$ is read as" [9]:

$$h_{min} \approx (4/L)\sqrt{(k_B T/M\omega_0^2)(1/Q\omega_0\tau)} \approx 10^{-20}\sqrt{F\Delta f_s}Hz^{-1/2}. \quad (2)$$

Here F is a noise factor; it represents an unique scheme for registration of mechanical displacements and its properties. Further, intended parameters have substituted into the formula: $F = 1$ and $\Delta f_S = 100$ Hz [9]; then forecast of antenna's sensitivity in variations of metrics is $h_{min} \approx 10^{-19}$. The astrophysical forecast $\tau = 10$ ms has become the base parameter. This is intention as of 2014; it is result of evolution of indevour and aspirations, because earlier in 2004, the more optimistic forecast had presented; it was assumed that antenna should register pulses of $\tau = 1$ ms [6]; well-selected set of parameter values in formula for noise factor had allowed to obtain the value $F = 1$. Then expressions (2) for $\Delta f_S = 1000$ Hz provide forecast estimation of $h_{min} = 3\cdot 10^{-19}$. In [6] the forecast some differs numerically because, in particular, parameters of full-scale acoustic oscillator had not known definitely.

Change in theoretical basis in 2014 has led to change in "design sensitivity". This term and numerical value has announced once in [6] for the OGRAN pilot model as 10^{-16} cm/Hz$^{1/2}$. This indicator for the full-scale facility has not announced yet, while this term mentions sometimes and somewhere. In article [10], the similar value of $3\cdot 10^{-16}$ cm/Hz$^{1/2}$ has presented without connection with experimental data. The pilot model had intended to achieve extra high sensitivity of the registration scheme. In it, the calculation of design sensitivity has not opened.

The task of the presented report is to define and calculate design sensitivity for the OGRAN GW antenna to compare it with test result achieved in 2013 [9,10,11].

The expressions (2) has presented as approximate. The analytical expression for spectral density has obtained when changing $\tau \simeq 1/\Delta f_S$. As alternative, it is possible to derive the exact expression of this spectral density and avoid non-exact intermediate analytical transfigurations. So, the expression (1) connects metric variation amplitude and force affecting the oscillator. The Nyquist force is also affected; it is expressed as spectral density: $G_T(f) = 4k_B T H_\mu = 4kTM\omega_0/Q$, where H_μ is oscillator viscosity. When comparing the forces, we find directly the spectral density of antenna's metric limit thermal threshold signal $S_{h0}(f) = (4/L)^2 k_B T(MQ\omega_0^3)^{-1}$. The threshold signal expression takes the view:

$$h_{min} = (S_{h0}\Delta f_S F)^{1/2}. \quad (3)$$

Substituting parameter values, we get the more exact estimation $(S_{h0})^{1/2} = 1.7\cdot 10^{-19}$ Hz$^{-1/2}$ for $Q = 10^5$ [9,10]. It is noticeably different from (2). The exact value is required to specify the resultant value of "reception bandwidth". In article [12] for bar without readout mirrors the first measured value $Q = 1.6\cdot 10^5$ has presented, and the limit threshold value of $(S_{h0})^{1/2} = 1.5\cdot 10^{-19}$ Hz$^{-1/2}$ had calculated.

2. Noise factor in the OGRAN theory

Antenna's registration scheme becomes meaningful when F > 1.

In the OGRAN project, the laser optoelectronic scheme uses Fabry-Perot interferometer; its mirrors has fixed at top ends of the cylinder. The scheme registers interferometer eigenfrequency deviations δν, and small displacements δx are defined by calculation through the ratio ($\nu = c/\lambda$)

$$\frac{\delta x}{L} = \frac{\delta \nu}{\nu}. \quad (4)$$

Fabry-Perot interferometer has used in AURIGA GW detector in the room-temperature version. The full scheme and its description has published and available at the AURIGA website in master's thesis of L.Conti of 1995 [13] and somewhere else in proceedings. This scheme had later presented in the article [14]. In the scheme, Pound-Drever-Hall (PDH) [15,16] technique is applied. Moreover, the article [16] contains formula for maximum achievable resolution of the displacement meter for GW experiment without taking into account laser noise.

In 1996 (soon), the SAI MSU had published the paper [7] with title: "Room-temperature gravitational bar-detector with cryogenic level of sensitivity". Here the theory of article [17] had simplified and elaborated, and term "noise factor" has entered into consideration. Certain optimism in sensitivity presented in the title [7] is because the new PDH technique effectively eliminates noise of laser power technical phenomenological fluctuations, whereas this noise was significant in [17,18]. Afterwards, the presented sensitivity analyze has become the theory of the OGRAN project [6].

MSU has certain groundwork in theoretical investigations on Fabry-Perot application to displacement measurement. There is the article "A combined optical-acoustical gravitational antenna" [17]. The formula for resolution of displacement measurer with a F-P resonator has presented by Braginskii [18]. Moreover, MSU has significant hardware experience [19,20,21,22] in addition to review [5].

Institute of Laser Physics (ILP) RAS has relevant hardware experience too; there is the article with the title [23]: "Use of narrow optical resonances for measuring small displacements and for building gravity-wave detectors". Here the steep slope of the spectral line of methane had used. Moreover, ILP possesses extensive groundwork in Pound-Drever-Hall technology. According to collaborated "Agreement in Intentions", "ILP RAS works out, mounts and adjusts the optoelectronic part of the facility". In fact, ILP has created the meter of small mechanical oscillations. Much earlier, such measurers had created by MSU; now this work became the task of ILP, and SAI MSU accomplishes scientific control.

At MSU, the displacement meters had investigated and developed separately from the probe body [24,25,26]; small vibration were created by a rigid plate with electrostatic force actuator in quasi-static mode. In the OGRAN project, this stage of development has skipped, and the facility should be considered as a whole during customization. As a consequence, SAI MSU and INR RAS has increased essentially their participation in adjustment.

INR represents interest of the Presidium of RAS in the project and from 2006 INR assists hardware adjusters of SAI MSU in study of readout circuitry functioning; there are a number of reports and articles [26,27,28,29]. In report [26], influence of intrinsic laser frequency fluctuations had considered and it had shown how to suppress them; the feedback depth should be increased at least to value $K_f = 1000$ (60 dB) [27]. This led to the need to expand the frequency response of the servo amplifier that controls retuning of laser radiation frequency [30] in order to prevent self-excitation in locked feedback loop. In turn, this has required introduction of electro-optical crystal (EOC) inside the laser cavity as a third, high-speed laser frequency retuning device. This fast control has presented in the functional scheme [31]. Accordingly, the value $K_f = 2000$ has realized [8].

In article [28] there has considered the residual amplitude modulation (RAM) of laser radiation and noise that it induces. Certain efforts have presented at the report [11], thesis [32] and articles [33,34] to suppress this modulation.

Under theoretical consideration of GW antenna in a complex, there have proposed two versions of excess noise origin [35,36].

The OGRAN theory has established the expression for the noise factor as a part of general design procedure [7,6,9]:

$$F = (2M/\tau)\sqrt{G_e/G_T}. \qquad (5)$$

Here G_T is spectral density of bar noise

$$G_T(\omega) = 2kTM\omega_0/\pi Q. \qquad (6)$$

Spectral density G_e of readout "optical noise" [10] is defined as [7, 37, 38]

$$G_e(\omega) = B\omega_0^2(2h\nu/\eta P)(\lambda/2\pi N)^2. \qquad (7)$$

Here N is the "sharpness or the effective number of reflections" in the FP resonator, B is "the phenomenological factor that indicates by how many times laser noise exceeds the Poisson level" [10], η is a photodiode yield. In (7) P is "optical power"; there is not complete certainty in its definition. So, in previous articles [17,18] P is incident power at the interferometer. In [9] it is "effective power". In [39] outside the project, P is the "optical power available at the photodetector". This ambiguity is consequence of incompleteness of optoelectronic circuit functioning theory.

In (5) the noise factor is a function of astrophysical forecast τ. The formula for it has transformed to the "suitable" form [9,10]:

$$F \cong \left(\frac{\sqrt{B}}{N}\right)\left(\frac{1W}{P}\right)^{1/2}\left(\frac{1s}{\tau}\right). \qquad (8)$$

This format simplifies selection of parameter values to calculate the desired value $F = 1$. Substituting parameter values P = 0.01 W, $\tau = 0.01$ s [9,10] and as examples $N = 3\cdot10^3$ and $B = 10$, we find $F = 1$. The formula (8) is convenient for a theorist-designer making general sensitivity forecast; it shows that "the task has a solution". Executor should implement this set of figures in hardware.

In the initial OGRAN article [6] ($\tau = 10^{-3}$ s) one can read: "For the designed OGRAN optical parameters: $P = (1 - 3)$ W, $B \simeq (1 - 10)$, $\lambda = 1.064$ μ, $\eta = 0.8$, $N = (10^3 – 10^4)$ one can find the estimation F $\simeq$ 1." So, in the example presented above, making substitutions $\tau = 0.001$ s and $P = 1$ W, we also get $F = 1$. An acceptable version of initial design set of parameters is formed, which is used below in the formula for calculated (design) resolution value of an optoelectronic meter.

According to reference in intermediate final article [8], "the presented values constitute the theoretical basis of the OGRAN project."

In 2014 intended astrophysical forecast duration τ has changed. Comparison of above selected parameters shows that increase in duration τ in design permits hardware worker reduce significantly power of laser radiation P according to (8). This is result of an attempt to register power of 1 W by creating two complex photodetectors with 16 photodiodes each using light splitting cubes [31,8,10]. This is a response to technical reality. The next parameter to be revised is the laser excess noise factor B.

In addition, perspective is associated with increasing the parameter $N = 1/(1 - R)$ [17]; here R is energy reflectance of interferometer mirrors. For this, there had been acquired mirrors of high quality manufactured by the LAM factory Lyon (France) [40,9,10,33,34]. "A serious problem in the process of OGRAN assembling was a method of mirror installation so as the high quality mirrors require a clear atmosphere conditions" [9]. The clean room as a "laminar box" with dust protection has designed and erected in the BNO INR chamber around the facility in 2015. Results of mirror replacing have presented in articles [33,34].

In the article [8], in sensitivity forecast, calculation there used two base values: $\Delta f_S = 1000$ Hz ($\tau = 1$ ms) and $F = 10$. Also, further there has proposed to decrease in bandwidth towards $\Delta f_S = 10$ Hz, because it should lead to increase in sensitivity. On this way, as the first step we put $\Delta f_S = 100$ Hz; it means $\tau =$ 10 ms. Replacing this time value in the expressions (5), (8) we obtain the value $F = 1$. Thus, we have obtained just the same pair of parameter values on which the sensitivity analysis in articles [9,10] has based. Thus, the transition to desired value $F = 10$ already in 2010 means change of intention preventively.

Further in this way, supposing $\Delta f_S = 10$ Hz [8] we obtain the value of $F = 0.1$. It is a significant departure from reality in forecast. Actually, value of $F \gg 1$ is realized in the facility so far.

The initial sensitivity forecast calculations assume values of $B = (1 \div 10)$; it means practical absence of technical noise in laser radiation. This initial optimism, as indicated above, had based on employment of new advanced and complicated PDH technique.

To take into account significant real excess noise in hardware tests, the conclusion has made that acceptable values of factor B may be expanded in the range: $B = 1 \div 1000$ [9,10]. So, according to the final article on the pilot model [8], excessive technical noise did not allow observing (revealing) thermal noise of the bar. Final estimate of displacement resolution has estimated as $4 \cdot 10^{-16}$ m/Hz$^{-1/2}$ [11]. This level of noise makes it impossible to identify thermal noise at the spectrogram [8]. Further, in the full-scale facility, joint efforts of two institutes succeed in definition and reduction of dominant noise source as of 2010-2012, and thermal noise has revealed quite confidently [41].

In articles [37,38], value range $B = (1 \div 10)$ in the noise factor formula provides a capacious source of perspective again.

3. Latent content of the OGRAN theory

The task is to find connection between two result sensitivity signicatives. The first is the initial threshold signal h_{min} jointed with the noise factor [7]. The second signicative is a novel - the "reception bandwidth". This signicative has introduced in [9,10]. It based on sensitivity result representation in the AURIGA project as a graph at Fig.7 [42] of metric variation spectral density; there the novelty is apportionment the abscissa distance in frequency 10 Hz (865-855) between two points with the ordinate of 10^{-19} Hz$^{-1/2}$.

The reception bandwidth unequivocally relates to the actual achieved displacement resolution.

To define mentioned connection, we should deep into the theory of the OGRAN project to extract the analytical expression for displacement resolution.

There is also a relationship between displacement resolution expression and noise factor expression. Thus, displacement resolution is the third, intermediate significance.

While the noise factor formula is convenient for a design theorist, it is desirable to add to it no less convenient formula for hardware developer measuring resolution directly, because the noise factor formula "with a clear physical sense" is not relevant in test. In particular, one should strive to ensure that traditional correspondence between theoretical introduction and result value inside an article; absence of this connection some signs of theory's inadequacy become not meaningful.

The example of desired correspondence is contained in the fragment of the report [31] on the pilot model. So, we read: "Experiments, had shown sensitivity of $\sim (1 \div 2) \cdot 10^{-14}$ cm/Hz$^{1/2}$ for this setup while theoretical estimation was $\sim 3 \cdot 10^{-15}$ cm/Hz$^{1/2}$. The reason for this mismatch was revealed as technical noise of detection system."

These affirmations allow us to conclude that the required design formula really exists, albeit in a latent form. Moreover, that the presented test result differs moderately from the calculated value.

In the next article [8] there has written: "The equivalent limiting (theoretically expected) spectral density of noise displacements is $0.5 \cdot 10^{-15}$ cm/Hz$^{1/2}$." In both sources, obviously, values of parameters in calculation formula differ. In the message [31], as earlier [6] and later [8], test resultant signicative has presented in the displacement noise format.

In the AURIGA detector, there is significant difference between measured values and calculated ones [42,8].

Meanwhile, the aim of scientific part of instrument development is to achieve coincidence of test result with previously calculated numerical value within limits of measurement errors and accuracy of parameter determination. It is advisable to solve this problem without choosing an appropriate value for parameter B.

Revealing of appropriate resolution formula in OGRAN theory may be considered as the inherent part of general design procedure; the sought formula should connect rigidly with antenna's design sensitivity.

Information about displacement measurement using FP interferometer exists outside the referred above articles. So, the formula for displacement resolution can revealed in the article [17]:

$$\Delta x = \frac{\lambda}{2\pi N}\sqrt{\frac{h\nu}{\eta P}B\Delta f}.$$

Parameter N is not measureable, whereas a generally accepted term "finesse" $\mathcal{F} = \pi\,(1 - R)^{-1}$ is used [16,14]. It determines relative transmission bandwidth of an interferometer at the level of -3 dB. This value has determined experimentally in the OGRAN facility by scanning interferometer transmission function using scanning retune of laser light frequency ν_L with the piezoceramic actuator (driver) [30]. The value $\mathcal{F}$ has announced as 3000 [9,10]. The brief derivation of above finesse expression has implemented in [36]. Accordingly, we get the relation $N = \mathcal{F}/\pi$, and then

$$\Delta x = \frac{\lambda}{2\mathcal{F}}\sqrt{\frac{h\nu}{\eta P}B\Delta f}. \tag{9}$$

Elsewhere, this formula has presented in the article [18]; details of its derivation has presented in monography [2]. Without technical noise ($B = 1$) displacement resolution formula has presented outside the project in the article [39] at page 262. At the beginning of the OGRAN project, the formula (9) without technical noise had presented to ILP RAS as the initial scientific foundation to hardware implementation of the optoelectronic displacement readout; certain optimism with laser noise was due to PDH technique virtue. Later, certain efforts to derive 'readout noise spectral density' in improved form has presented in [12,40].

The project OGRAN theory contains the required analytical expression for displacement noise in latent form. In order to extract it, we should refer to the base article [7]. One can find there the expression for spectral density G_N of the equivalent stochastic force F_N, formally applied to massive body of the equivalent oscillator

$$G_N = G_T + G_f + |Z_\mu|^2 G_e. \tag{10}$$

Here G_f is spectral density of inverse fluctuation force acting by readout circuit; this force represents stochastic component of photon flux pressure.

The third term $|Z_\mu|^2 G_e$ in (10) is output noise of registration circuit adducted to the measured value, in this case, to force. Here Z_μ is the oscillator mechanical impedance: $Z_\mu = M\,[j\omega + 2\delta + \omega_0^2/(j\omega)]$, $2\delta = \omega_0/2Q$. Spectral density G_e contains information about the optical registration scheme, which extremely high resolution is the aim of presenting scientific and engineering development.

It had shown that $G_f \ll G_B$ [7]; this is in contrast to MSU's capacitive displacement meters [19,20,25]. To find displacement noise, we write out the equation of motion

$$M\ddot{x} + \mathrm{H}_\mu \dot{x} + \mathrm{k}_\mu \mathrm{x} = \mathrm{F}_N.$$

Here $x(t)$ is instantaneous coordinate, k_μ is oscillator stiffness.

The solution of this equation "in spectra" allows us to find an analytical expression for spectral density of total displacement noise G_X:

$$G_X(\omega) = G_N/\omega^2|Z_\mu|^2 = G_T/\omega^2|Z_\mu|^2 + G_e/\omega^2 \equiv G_{XT} + G_{XS}. \tag{11}$$

The first term represents thermal displacement noise of the bar. The sense of the second term is spectral density of optoelectronic scheme noise. The frequency range near resonance ($\omega \approx \omega_0$) is of interest here. There is important consequence for determining spectral density of optoelectronic readout noise:

$$G_{XS}(\omega) = G_e(\omega)/\omega_0^2. \tag{12}$$

Thus, the general relationship between displacement resolution and noise factor has determined.

Substituting expressions (6) and (7) into (11), we have obtained the specific noise expression ($\Delta\omega = 2\pi f - \omega_0$):

$$G_X(f) = \frac{4k_BTQ}{M\omega_0^3}\frac{1}{1+(\Delta\omega/\delta)^2} + B\frac{\pi h\nu}{\eta P}\left(\frac{\lambda}{\pi N}\right)^2. \tag{13}$$

It is expression for spectral density of total displacement noise, observed by developer-adjuster in experiment (test) at output spectrogram at the figure 4 in [9] or at the figure 6 in [10]. The first term describes quantitatively thermal noise of the bar as the narrow resonant peak. The second term has no dependence on frequency; it represents spectrum "background". It is the formula for resolution of displacement measurer

$$\Delta x = \frac{\lambda}{\mathcal{F}}\sqrt{\frac{\pi h\nu}{\eta P}B\Delta f}. \quad (14)$$

This expression contains parameters of the design set of parameters. Substituting the mentioned values $N = 3\cdot10^3$ ($\mathcal{F} = 9.5\cdot10^3$), $P = 1$ W, $\eta = 0.7$, $B = 10$, we obtain variant of design sensitivity estimation $\Delta x = 3\cdot10^{-17}$ cm/Hz$^{1/2}$.

The formula (14) has scientific view. However, its engineering significance is reduced by arbitrary defined parameter $B = 1\div1000$ [9,10]. In the AURIGA project "laser power noise spectral density" has measured quite accurately [14]; laser frequency noise has mentioned; but its relative contribution has estimated as negligible [14,42]. It is a consequence of smallness of the "transducer-cavity length".

The formula and the presented theory do not take into account significant energy losses in the interferometer mirrors due to multiple reflections. Phenomenologically this effect has observed and fixed in the facility. It defines sufficient decrease in signal transmission in the readout and accordingly decrease in sensitivity (resolution) of displacement meter. To describe this effect the term "interference contrast" has introduced in [8]. Its measured values are of 0.3 in [9] and of 0.2 in [10]. In addition, we read: "… due to losses in the light guide elements and the interference contrast ~10% the real value of effective power is $P = 0.01$W", whereas "the power of radiation that reached the photodiodes was P = 50 mW."

The attempt to derive analytical dependence of signal transmission and displacement resolution on these energy losses had devoted in the article [12]. The solution that is more adequate has presented in the article [28].

The formula (14) corresponds to the conceptual scheme of Braginskii [18,2]. The resulting formulas for displacement resolution (9), (14) describe Braginskii's conceptual laser scheme as of 1967 too. Slight differences in coefficients [18,39] is due to some uncertainty in interpretation of parameter P and the value $G_T(6)$ in [7,17]. The Pound-Drever-Hall scheme differs significantly. It uses not transmitted beam, but reflected one.

The functional optoelectronic scheme of OGRAN facility has attached to the article [6]. It repeats the AURIGA scheme [13,14]. This scheme has also presented clearly in [8,10,31,37]. In the AURIGA PDH circuit laser radiation is phase modulated by high frequency Ω (~ 10 MHz). The beam reflected from the interferometer falls on photodiode; variable power component has the form $\delta P_{\text{ref}} = D_P(\nu_L - \nu)$ $\sin\Omega t$. Here D_P is a decrement [16,28]. After synchronous detection, a sign alternating discriminator characteristic of the automatic control system of laser frequency is formed. When $\nu_L - \nu = 0$, zero voltage is presented at output of the synchronous detector and further to retune laser frequency. When the feedback loop is locked, the AURIGA circuit implements fast auto-tuning of laser frequency; it exactly follows fast signal variations $\delta\nu$ of interferometer mode frequency (4). In this way frequency manipulated laser radiation is formed; signal and noise encode in laser radiation frequency.

Under condition S/N = 1, the threshold displacement signal in radiation has determined [26]:

$$\Delta x = \frac{\lambda}{8\mathcal{F}}\sqrt{\frac{h\nu}{\eta P_C}\Delta f}. \quad (15)$$

Here P_C is a carrier power spectral component of phase modulated laser radiation at a photodiode; $P_C \approx P/2$ [16]. When $\eta = 1$ expression (15) coincides with the resolution expression in the article [16]. There signal and noise are presented at input of the photodetector as variations of laser radiation power, whereas the formula (15) presents signal and noise containing in laser frequency deviations.

Formulas (14) and (15) present shot noise of photoelectrons and in this sense have single physical meaning. This consideration gives occasion to neglect difference in descriptions of two registration schemes in the OGRAN theory. When focusing on their difference, one can understand the origin of excess noise in the PDH optoelectronic registration scheme [35].

Frequency manipulated radiation is fed to the interferometer of discriminator, second PDH channel. It uses the slope of sign alternating discriminator characteristic. On this slope frequency deviations transforms into voltage variations in output of the facility. The slope is easily measured. Through the slope steepness, the output voltage noise is recalculated (adducted) to frequency noise at test spectrogram and then using (4) - to displacement noise of the facility. The channel discriminator introduces its intrinsic noise; it has not taken into account by the formula (14), and in the noise factor formula. Account of the discriminator's contribution has presented in the report [27].

Regardless analytical expressions, we can use the numerical value of displacement resolution. To connect it with factor F we use the expression (12). The numerous value of noise factor can determined. In addition, we can determine the design value of sensitivity.

4. Consequences and estimates

The readout system registers fast variations in eigenfrequency of the FP resonator, and achieved resultant frequency noise value is available at spectrograms. Noise in this format converts into resultant displacement noise by means of the ratio (4). These two noise dimensions correspond to left and right scales at spectrograms in articles [9,10,38]; where $\delta x/\delta\nu = 0.7\cdot 10^{-12}$ cm/Hz.

The best-achieved resolution can be revealed as background at test low-resolution spectrograms at fig. 6 of [10] and at fig. 3 of [9]. In addition, this important numerical value has pointed out in the text [10]: "The measured level of the spectral density of the total antenna noise (background above which a thermal peak dominates) in the operating antenna range is ~ 0.003 Hz/Hz$^{1/2}$."

Recalculation (4) gives the value of ~ $2\cdot 10^{-15}$ cm/Hz$^{1/2}$. Just the same value has announced in the previous report [11] as a manifestation of ILP RAS competence.

To transform this achieved result into metric format, the absolute and relative height of the thermal peak on the first test low-resolution spectrogram at $f_0 \simeq 1300$ Hz should be determined. We can see on the spectrogram that the peak height is of 0.3 Hz/Hz$^{1/2}$; it corresponds to $2\cdot 10^{-13}$ cm/Hz$^{1/2}$. This absolute height coincides with the value calculated by means of the general formula (13) in the term of the heat peak. The relative height A is of 100 or 40 dB. This important achievement has fixed in the synopsis [42]: "The level of intrinsic fluctuations of the optical displacement sensor has been practically reached, which is two orders of magnitude lower than the level of the thermal acoustic resonance noise."

At two points of resonance curve in (13) thermal noise spectral density is equal to background. Distance between them is Δf_{h0}. Using the expression for shape of the thermal noise curve (13), the bandwidth corresponding the level of - 40 dB is $\Delta f_{h0} = \Delta f_{40} = A\Delta f_0 = Af_0/Q$, where $\Delta f_0 = 0.013$ Hz. Within the specified frequency bandwidth $\Delta f_{h0} = 1.3$ Hz, thermal noise dominates, and limit achievable spectral density is $S_{h0}(f)$.

There is a graph of spectral density of metric variations at the fig. 5 of [9]. This graph is result of processing the experimental data, which has simultaneously represented by the previous spectrum at the fig. 4 as displacement noise. Spectrogram at the fig.5 is connected with spectrogram at the fig. 4 by means of the formula $(G_h)^{1/2} = (G_X)^{1/2}/K_{xh}$, where $K_{xh} \equiv x_m/h_m = Q(L/2)/[1+(\Delta\omega/\delta)^2]^{1/2}$. It is a product of analytical conversion algorithm. In the lowest small part of the spectrum at the fig. 5 (inside frequency bandwidth Δf_{h0}) the spectral density is constant: $S_{h0}(f)$.

Outside the band Δf_{h0}, metric spectral density increases sharply and there are two spectrum sections increasing with distancing from the resonant frequency f_0. This spectrum exactly determined analytically; it is analogous in sense to metric spectrum in the article [42]. The bar thermal peak and this conversion are scientific contribution of SAI MSU into the OGRAN collaboration, whereas ILP provides the unique displacement measurer and its actual noise.

There is decision to extend the bandwidth up to reception bandwidth Δf_{rec}. We are to calculate the distance in frequency between two points with ordinates of 10^{-19} Hz$^{-1/2}$ on the metric graph [9,10]. When using the shape of the resonant curve (13) the distance is determined by relation completing the transformation algorithm:

$$\Delta f_{rec} = \Delta f_{h0}\,[10^{-19}\ \text{Hz}^{-1/2}/(S_{h0})^{1/2}] \simeq 8\ \text{Hz}. \qquad (16)$$

However, articles [9,10] present the value of $\Delta f_{rec} = 4$ Hz. To implement the proper calculation and explain the divergence in figures, we should consider the other, high-resolution spectrogram at fig. 4 [9]. In it, the visual relative peak height is $A \simeq 30$, and then $\Delta f_{h0} = \Delta f_{30} \simeq 0.4$ Hz. In addition, since in articles [9, 10] the approximate value of $(S_{h0})^{1/2} \approx 10^{-20}$ Hz$^{-1/2}$ has established, we obtain using (16) $\Delta f_{rec} \approx 10\Delta f_{h0} = 4$ Hz.

To connect displacement resolution with noise factor we apply expression (12). Spectral density of achieved displacement noise has the numerical value of $G_X(\omega) = 6.3\cdot10^{-35}$ m²/Hz. For the value $\omega_0 = 8.2\cdot10^3$ s^{-1}, we find $G_e(\omega) = 4.2\cdot10^{-27}$ m²/s. For thermal noise we have $G_T(\omega) = 2.2\cdot10^{-19}$ N²/Hz. Then, for $\tau = 10^{-2}$ s according (5) we find $F \simeq 30$. According (3) we find conjugate metric threshold signal $h_{min} \simeq 9\cdot10^{-19}$. It corresponds to $\Delta f_{rec} \simeq 8$ Hz.

Design displacement resolution (sensitivity) we can find when $F = 1$; the estimate is of $0.7\cdot10^{-16}$ cm/Hz$^{1/2}$ as a response to "main technical challenge" [9]. It means A = 3000, $\Delta f_{h0} = 40$ Hz and $\Delta f_{rec} \simeq$ 250 Гц.

Hardware executors should realize this outstanding design resolution value; it considers in the sense that has presented in [6]. This is response to the challenge [9]. For this purpose, it is planned to increase interferometer finesse and reduce the contribution of excess noise.

The intermediate value of signal spectrum bandwidth of $\Delta f_S = 250$ Hz is of interest. It means intermediate value $\tau = 4$ ms [7]; then $F = 2.5$, $h_{min} = 4.3\cdot10^{-19}$. This is moderate distinction of 8 dB from limit thermal threshold.

If the main part of signal pulse energy locates within the bandwidth $\Delta f_{h0} = 40$ Hz ($\tau = 25$ ms), thermal noise dominates, and maximum achievable antenna's sensitivity S_{h0} is realized.

In this case, according to the concept (challenge), a pulse with bandwidth of 100 Hz has detected when the condition $F = 1$ is fulfilled. Some mismatch in band widths has revealed, which is subject to further comprehension.

The article [8] the question is raised for further understanding and elaboration; we read: "... the main question of the pilot experiment is the closeness of the results to the aforementioned theoretical bound." Also, in the article [9] "problems" have pointed out. Similarly, the more significant aspect has arisen in comprehending the OGRAN project theory. Namely, it is shown that executor must reduce the threshold signal of the displacement measurer by 30 times. In this case, value h_{min} changes by the factor of 5.5 (15 dB) from the calculated result of 10^{-18}.to thermal limit $1.7\cdot10^{-19}$. This difference between the current metric threshold signal and the limit one can consider as moderate and acceptable. Similarly, when establishing the reception bandwidth, somewhat arbitrary 6-fold widening of band Δf_{h0} has led to moderate and permissible decrease in sensitivity in comparison with the thermal limit one. The question is to admit the achieved threshold signal in metric of 10^{-18} sufficient to stop making efforts to further improvement the readout scheme. This means recognition of the achieved resolution of readout circuit as satisfactory and the scientific part of the OGRAN project is completed. As an alternative version, the formula (2) is not valid for engineering application for the case $F \neq 1$.

In principle, the noise factor expression as of 1996 requires redefinition to avoid values F < 1. This has done in articles [38,39] and before in report [29].

5. Conclusions

The procedure for preliminary design of sensitivity of bar gravitational antenna is selection of such set of parameter values in the noise factor formula to realize the value $F = 1$ when registering pulse from supernova burst with duration $\tau = 1$ ms [6]. Based on results of tests of the pilot model, significant difficulties have appeared with realization in hardware required values of photodiode power P and excess laser noise factor B. The initial intention has corrected [8,9,10] by increasing pulse duration to value of $\tau = 10$ ms. Then the task of calculating the desired value $F = 1$ has obtained its solution.

Deepening into of the OGRAN project theory [7] the general analytical relationship (12) has revealed between expression for the noise factor and expression for displacement resolution of registration scheme. Regardless of extracted specific expression for resolution, the corresponding numerical relationship has defined; namely, the achieved value of resolution $2\cdot10^{-15}$ cm/Hz$^{1/2}$ [11] has assigned the value of $F = 30$ and value of threshold signal in metric of $h_{min} \simeq 10^{-18}$.

There has determined analytical dependence of design displacement resolution on initial intention of $F=1$ and there has determined its numerical value of $0.7\cdot10^{-16}$ cm/Hz$^{1/2}$. While returning to astrophysical forecast of $\tau = 1$ ms the required resolution of readout is $0.7\cdot10^{-17}$ cm/Hz$^{1/2}$.

The expression for displacement resolution has restored coherence of the OGRAN theory - from initial intention towards engineering realization. The extracted formula does not provide adequate description of features of the registration scheme implemented in hardware. In this way, the analytical expression for value G_e in the formula for noise factor must be much more complex. The absence of connection between theory and experiment allows designer to present inadequate theoretical relations. This may be a reason for difficulties in comprehension of the Pound-Drever-Hall registration scheme operation and, in particular, in determining origin of excessive technical noise in the facility.

References

[1] Rudenko V N and Bichak J 1987 *Gravitational waves in general relativity and methods of their detection - in Russian* (Moscow: Moscow University Press) p 270
[2] Braginsky V B, Manukin A B 1977 *Measurement of weak forces in physics experiments* (Chicago: Univ. of Chicago Press)
[3] Gusev A V, Kulagin V V, Oreshkin S I, Rakhmanov A N, Rudenko V N, Serdobol'skii A V, Tsepkov A N, Tsyganov A V and Motylev A M 1997 *Astronomy Reports* **41** 248
[4] Gavrilyuk Yu M, Gusev A V, Krysanov V A, Kulagin V V, Motylev A M, Oreshkin S I, Rudenko V N, Silin V A and Tsepkov A N 2012 *Astronomy Reports* **56** 638
[5] Rudenko V N 2017 *Phys. Usp.* **60** 830
[6] Bezrukov L B, Popov S M, Rudenko V N, Serdobolskii A V and Skvortsov M N 2004 Astrophysics & Cosmology after Gamow *Proc. Int. Conf.* (*Preprint* gr-qc/0411083v1)
[7] Gusev A V, Kulagin V V and Rudenko V N 1996 *Gravitation & Cosmology* **2** 68
[8] Bezrukov L B, Kvashnin N L, Motylev A M, Oreshkin S I, Popov S M, Rudenko V N, Samoilenko A A, Skvortsov M N, Tsepkov A N, Cheprasov S A and Yudin I S 2010 *Instruments and Experimental Techniques* **53** 423
[9] Bagaev S N, Bezrukov L B, Kvashnin N L, Krysanov V A, Motylev A M, Oreshkin S I, Popov S M, Rudenko V N, Samoilenko A A, Skvortsov M N and Yudin I S 2014 *Rev. Sci. Instrum.* **85** 065114
[10] Bagaev S N, Bezrukov L B, Kvashnin N L, Motylev A M, Oreshkin S I, Popov S M, Rudenko V N, Samoilenko A A, Skvortsov M N and Yudin I S 2015 *Instruments and Experimental Techniques* **58** 257
[11] Vishnyakov V I, Ignatovich S M, Kvashnin N L, Popov S M, Rudenko V N, Samoilenko A A, Skvortsov M N and Yudin I S 2013 *Tech. Digest Int. Symp. MPLP-2013, Novosibirsk, Russia* pp 179
[12] Gusev A V, Rudenko V N, Cheprasov S A and Bassan M 2008 *Class. Quant. Grav.* **25** 055006
[13] Conti L 1995 *Master's thesis*
[14] Conti L, Cerdonio M, Taffarello L, Zendri J P, Ortolan A, Rizzo C, Ruoso G, Prodi G A, Vitale S, Cantatore G and Zavattini E 1998 *Rev. Sci. Instrum.* **69** 554
[15] Drever R W P, Hall J L, Kowalski F V, Hough J, Ford G M, Munley A J and Ward H 1983 *Appl. Phys.* B **31** 97
[16] Black E D 2001 *Am. J. Phys.* **69** 79
[17] Kulagin V V, Polnarev A G and Rudenko V N 1986 *Sov. Phys. JETP* **64** 915
[18] Braginskii V B 1968 *Soviet physics JETP* **26** 831
[19] Braginskii V B, Mitrofanov V P and Rudenko V N 1971 *Prib. Tekh. Eksp.* **4** 241
[20] Krysanov V A, Kuklachov M I and Rudenko V N 1979 *Prib. Tekh. Eksp.* **4** 240
[21] Krysanov V A and Rudenko V N 1984 *Prib. Tekh. Eksp.* **3** 199
[22] Yelfimov S V, Krysanov V A and Rudenko V N 1987 *Preprint* No.32/87 MSU-dept.phys
[23] Bagaev S N, Dychkov A S and Chebotaev V P 1981 *JETP Lett* **33** 79
[24] Braginsky V B, Panov V A, Petnikov VG and Popel'nyuk V D 1977 *Prib. Tekh. Eksp.* **1** 234
[25] Braginsky V B, Panov V A and Popel'nyuk V D 1981 *JETP Letters* **33** 405
[26] Krysanov V A 2007 *Proc. Int. meeting PIRT-2007* (Moscow: BMSTU) 55

[27] Krysanov V A 2008 *Tech. Digest Int. Symp. MPLP, 2008* (Novosibirsk) 168
[28] Krysanov V A 2011 *Gravitation and Cosmology* **17** 97
[29] Krysanov V A 2015 *Proc. Int. meeting PIRT-2015* (Moscow: BMSTU) 230
[30] Okhapkin M V, Skvortsov M N, Belkin A M, Kvashnin N L and Bagayev S N 2002 *Optics Communcations* **203** 359
[31] Rudenko V N, Popov S M, Samoilenko A A, Oreshkin S I and Cheprasov S A 2007 *Proc. Int. meeting PIRT-2007* (Moscow: BMSTU) 49
[32] Yudin I S 2015 *Author's abstract of PhD thesis* (Novosibirsk)
[33] Rudenko V N, Kvashnin N L, Lugovoi A A, Oreshkin S I, Popov S M, Samoylenko A A, Skvortsov M N and Yudin I S 2020 *Physics of Atomic Nuclei* **83** 1682
[34] Rudenko V N, Gavrilyuk Yu M, Gusev A V, Krichevskiy D P, Oreshkin S I, Popov S M and Yudin I S 2020 *Int. J. Modern Phys. A* **35** 2040007
[35] Krysanov V A 2018 *Journal of Physics Conf. Ser.* **1051** 012020
[36] Krysanov V A 2020 *Journal of Physics Conf. Ser.* **1557** 012044
[37] Kulagin V V, Oreshkin S I, Popov S M, Rudenko V N, Skvortsov M N and Yudin I S 2016 *Gravitation and Cosmology* **22** 374
[38] Kulagin V V, Oreshkin S I, Popov S M, Rudenko V N and Yudin I S 2016 *Phys. of Atomic Nuclei* **79** 1552
[39] Grishchuk L, Kulagin V, Rudenko V and Serdobolski A 2005 *Class. Quantum Grav.* **22** 245
[40] Popov S M, Samoilenko A A, Cheprasov S A and Yudin I S 2011 *Gravitation & Cosmology* **17** 94
[41] Bezrukov L B, Kvashnin N L, Motylev A M, Oreshkin S I, Popov S M, Rudenko V N, Samoilenko A A, Skvortsov M N and Yudin I S 2014 *Proc. Int. meeting PIRT-2013* (Moscow: BMSTU) 23
[42] Conti L, De Rosa M, Marin F, Taffarello L and Cerdonio M 2003 *J. Appl. Phys.* **93** 3589

Quantum Nonlocality - Possible Cosmophysical Effects

S N Mayburov
Lebedev Inst. of Physics, Leninsky Prospect 53, Moscow, Russia, RU-117924
E-mail: mayburov@sci.lebedev.ru

Abstract. Mutiple experimental results indicate the existence of cosmophysical effects which influence parameters of nuclear decays and chemical reactions in lab. conditions. In particular, variations of nucleus decay parameters are detected which amplitudes are of the order 10^{-3} and periods of one year, 24 hours or about one month. Similar influence of solar activity on nuclear decays and chemical reactions also was reported. We argue that such deviations from radioactive decay law and other similar effects can be described by novel quantum nonlocality mechanism, different from standard EPR-Bohm nonlocality. Modified Doebner-Goldin model applied for the description of dynamical nonlocal effects concerned with nuclear decays.

1. Introduction

In the last decades, ample bulk of experimental results provided evidence for the existence of long-distance cosmophysical correlations, which presumably can't be explained by standard quantum physics. First observations of this kind were done in the forties by G. Piccardi, who has shown that parameters of some chemical reactions, in particular, their rates correlate in time with solar activity, in particular, with intense solar flares, observed parameter variations are of the order 10^{-1} [1]. Later, similar correlations of chemical reaction parameters with solar activity and other cosmophysical effects were reported by other groups [2, 3]. In the same vein, some recent experiments have found the periodic modulations of nuclear $\alpha-$ and $\beta-$decay parameters of the order of 10^{-3} and with typical periods of one year, one day or about one month [4-6]. Influence of solar activity on some nuclear decay parameters also was observed [5, 7]. These and other related results reviewed below in more detail.

Until now theoretical discussion of these effects had quite restricted character. In particular, oscillations of β-decay rate was hypothized as anomalous interaction of solar neutrino flux with nuclei or seasonal variations of fundamental constants [5]. Yet, neither of these hypothesis can't explain α-decay parameter oscillations of the same order, because nucleus α-decay should be insensitive to neutrino flux or other electro-weak processes. Really, $\alpha-$ and β-decays stipulated by nucleon strong and weak interactions correspondingly. Therefore, observation of parameter oscillations for both decay modes supposes that some universal mechanism independent of particular type of nuclear interactions induces the decay parameter oscillations. Besides, chemical reactions performed via electromagnetic interactions, so it gives, in fact, additional arguments in favor of such hypothetical mechanism universality.

Nowadays, the acknowledged universal theory is quantum mechanics (QM), so it's worth to start from the analysis of its foundations. In this framework, it's notable that QM axiomatics implicates the existence of notorious quantum-mechanical nonlocality. This phenomenon was first formulated as famous EPR-Bohm paradox and later developed as Bell theorem [8, 9]. Now such nonlocal effects confirmed experimentally and applied in quantum communication and computing. In fact, this mechanism permits to realize specific form of nonlocal correlations (NC) or 'action at the distance' between distant parts of quantum system S. Such NC effects obey to strict constraints, in particular, it should not violate causality. NC realization in EPR-Bohm variant demands that initially these S subsystems $S_1, \dots, S_n$ should interact with each other, after such interaction seized and the subsytems departed, such NC conserve correlations of S_i uncertain parameters even at large distance between subsystems. Obviously, such conditions is impossible to fulfill in cosmophysical situations. Meanwhile, it was argued for long that quantum NC can be more general concept than standard QM formalism admits and, in principle, some other NC effects, beyond EPR-Bohm mechanism, can exist [10-12]. In this paper, we develop phenomenological NC model and discuss its possible application to cosmophysical correlation description.

Let's consider first the conditions to which such nonstandard NC mechanism should obey in general. Plainly, beside causality demands, such NC should agree with all standard invariance principles i.e. time, space shift and rotation symmetries. In addition, we'll suppose that NC by itself can't transfer the energy, momentum or orbital momentum between distant objects, such transfer can be performed by conventional fields only. Hence the system average energy, momentum and orbital momentum should not change during such communications. Suppose now that the states of two distant objects S_1, S_2 due to their conjugate NC influence, become to differ from S_1, S_2 states expected after their independent evolution during fixed time interval $\{t_0, t_f\}$. Then, from described assumptions it follows that according to QM rules the initial S_1, S_2 states can't be stationary and non-degenerate, because such states possess the minimal possible energy and only some energy transfer can make them to evolve to another excited state. Hence the only possibility is that S_1, S_2 are degenerate systems, i.e. they have several states with the same energy and during time interval $\{t_0, t_f\}$ evolve from one state to another one. The simple example of such system is the particle with energy level E confined in symmetric double well potential divided by potential wall U_0 with $E < U_0$. Such S_1 has two degenerate orthogonal states g_1, g_2 in these wells with the same energy E, let's suppose that at t_0 it's in state g_1 confined in one well. Thereon, due to under-barrier tunneling it would spread gradually into other well [9], so that it will evolve with the time to some g_1, g_2 superposition. In this case, hypothetic S_2 NC influence on S_1, in principle, can change this state parameters, in particular, resulting g_1, g_2 probabilities at t_f. Such state degeneration is typical for many chemical and nuclear reactions, below we'll consider them in more detail. In this paper, model of such NC processes involving quantum tunneling will be considered. We'll assume that such NC effects can be described analogously to QM evolution equations and construct corresponding Hamiltonians.

2. Experimental motivations

Natural radioactivity law is one of most fundamental laws of modern physics, in accordance with it, nuclear decay parameters are time-invariant and practically independent of environment [13]. First results, indicating the deviations from exponential β-decay rate dependence, were obtained during the precise measurement of ^{32}Si isotope life-time by means of decay exponent fit [4]. Sinusoidal annual oscillations with the amplitude $6 * 10^{-4}$ relative to total decay rate and maxima located at the end of February were found during 5 years of measurements. Since then, the annual oscillations of β-decay rate for different heavy nuclei from Ba to Ra were reported, for most of them the oscillation amplitude is of the order $5 * 10^{-4}$ with its maximum on the average at mid-February [5]. Some other β-decay experiments exclude any decay constant modulations as large as reported ones [14]. Life-time of short-living α-decayed isotope ^{214}Po was measured directly, the annual and daily oscillations with amplitude of the order $9 * 10^{-4}$, with annual maxima at mid-March and daily maxima around 6 a.m. were found during 4 years of measurements [6]. The similar dependence on astronomical cycles was found for some biochemical reaction rates [3]. It was shown also that decay rates of ^{53}Mn, ^{55}Fe e-capture and ^{60}Co β-decay correlate with solar activity, in particular, with intense solar flare moments, observed decay variations are of the order 10^{-3} [5, 7]. As was noticed already, the rates of some chemical reactions demonstrate the similar dependence on solar activity [1, 3].

Individual nuclear decay acts normally are independent of each other [13]; such stochastic processes are called Poissonian and described by Poisson probability distribution [15]. For this distribution, at any time interval dT the dispersion of decay count number $\sigma = N^{\frac{1}{2}}$, where N is average count number per dT. However, some recent results show that under particular external influence this relation can be violated, several experiments of this kind reviewed in [16]. In one of them, α-particles from ^{239}Pu decay detected by Geiger counter located at the distance 2 cm from isotope. Glass volume with diameter 8 cm located at the distance of 10 cm from α-particle beam. In this volume, reaction of $C_{12}H_{22}O_{11}$ disaccharide (sucrose) hydrosolvation performed by dissolving 100 g of sucrose in 400 g of water. Effective reaction time at room temperature is about 15 minutes, during this period the observed dispersion of detector counting rate was reduced by a factor 3, whereas average counting rate practically doesn't change. It assumes that the decay process becomes more regular or sub-Poissonian. In quantum physics, the collective quantum states with sub-Poisson event distribution called squeezed states, they were obtained for optical photon ensembles [17]. Of course, it looks quite surprising that chemical reactions can somehow influence nucleus decay parameters, however, they both involve quantum systems and QM universality, in principle, can provide such possibility via quantum nonlocality.

Experiments of other kind exploited some biochemical and organic-chemical reactions, the example is reaction of ascorbic acid with dichlorphenolindophenol [2, 18]. Authors noticed first that dispersion of these reaction rates can change dramatically from day to day, sometimes by one order of value [19]. Further studies have shown connection of this effect with some cosmophysical factors, like solar activity, solar wind and orbital magnetic field. In particular, reaction rate dispersion becomes maximal during solar activity minima of 11 year solar cycle [2]. Shielding of exploited chemical reactors from external electromagnetic field in steel and permalloy boxes practically don't change these effects, hence such cosmophysical influence can't be transferred by electromagnetic fields. It's notable that such chemical reaction rate distributions aren't Poissonian but rather correspond to flicker-noise regime, for which the dispersion is, in general, essentially larger. If to suppose that all described effects have the same mechanism, then quantum NC seems most reliable candidate for their explanation.

3. Theoretical NC model

Let's discuss possible NC properties for collective systems like the set of noninteracting nuclei or molecules. Experimental results discussed in previous section suppose that NC influence makes their evolution less chaotic and more regular. It's notable that self-ordering is quite general feature of quantum dynamics, examples are crystal lattices or atomic spins in ferromagnetic. Then, it's reasonable to suppose that any large system would gain via NC mechanism its own self-ordering, in particular, making its own evolution more regular and ordered. Besides the self-ordering corresponding to more regular time intervals between events, other forms of system evolution symmetry, in principle, can be induced by NC effects. In particular, enlargement of average time intervals between decay events can be also treated as the self-ordering corresponding to the growth of evolution symmetry, because the event distribution becomes more homogeneous in time. Experimental results reviewed above demonstrate both enlargement and reduction of nucleus life-time induced by cosmophysical factors [1, 6]. Hence under their influence such evolution symmetry can transform in both directions, below only life-time enlargement will be studied.

It's natural to suppose that NC effect for any system of restricted size is proportional to the number of system constituents N_c involved into reactions. For the case of two nearby systems S_1 evolution due to NC effects supposedly results in its own self-ordering and influences S_2 in the similar way and vice versa. It's reasonable to suppose also that the scale of NC effects for decays or chemical reactions will be proportional to the average system process rate. For the case of two nearby systems of which one of them S_1 is large and other one S_2 is small, for them NC effects supposedly realized in master regime, i.e. S_1 can significantly influence S_2 state and make it evolution more ordered, whereas S_2 practically doesn't influence S_1 state evolution. This is characteristic for α-decay experiment described above: for considered set-up the chemical reaction involves about 10^{23} molecules, whereas α-decay isotope contains about 10^{18} unstable nuclei. It can be assumed also that in this case, S_2 self-ordering NC effect is insignificant in comparison with S_1 NC influence. Then, resulting NC effect in S_2 should depend on S_1 evolution properties and S_1, S_2 distance R_{12}.

Here we'll consider simple NC model for the set of N identical, unstable α-decay nucleus $\{A_i\}$. In fact, the similar considerations are applicable to the evolution of arbitrary metastable system, like atom or molecule, yet for nucleus α-decay its description is most simple. Gamow theory of nucleus α-decay supposes that in initial nucleus state, free α-particle already exists inside the nucleus, but its total energy E is smaller than maximal height of potential barrier constituted by nuclear forces and Coulomb potential [20]. Hence α-particle can leave nucleus volume only via quantum tunneling through this barrier. Therefore, alike for double well example, the state energy is the same inside and outside nucleus, and corresponding inside-outside states $\psi_{0,1}$ are degenerate and orthogonal to each other. Hence such degeneration permits, in principle, for some hypothetic NC mechanism to change nucleus decay rate without any energy transfer to α-particle, but just changing the barrier transmission rate. α-particle Hamilton A_i operator

$$H_i = \frac{\vec{P}_i^2}{2m} + U_N(r_i), \tag{1}$$

where m is α-particle mass, $\vec{P_i}$ is its momentum operator, U_N is nucleus barrier potential, r_i is the distance from nucleus center [21]. If at t_0 α-particle was in initial state ψ_0, then the solution for its state $\psi(t)$ in WKB approximation gives for decay probability at given decay moment t for nucleus A_i

$$p_i(t) = \lambda \exp[-\lambda(t - t_0)]. \tag{2}$$

Here p_i is, in fact, the time derivative of total decay probability from t_0 to t; resulting nucleus lifetime is proportional to λ^{-1} . Hence at $t \to \infty$ nucleus state evolves to final state ψ_1 [20]. Typical experimental accuracy of decay time measurement Δt is several nanoseconds [13]. Formally, such measurement described as the sequence of two consequent nucleus state measurements divided at least by Δt interval. If first one shows that nucleus is in ψ_0 state, and next one that it's in the state ψ_1, it means that nucleus decay occurred during this time interval [22]. In QM formalism, a general state of quantum system S described by density matrix ρ, for pure states $\rho = |\psi><\psi|$. If A_1, A_2 nuclei are its components, the partial $A_{1,2}$ density matrixes $\rho_{1,2}$ can be defined. For each A_i it turns out that if some other S components are also measured, then its decay probability would differ from eq. (2) and becomes

$$\gamma_i(t) = \frac{\partial}{\partial t} Tr \rho_i(t) P_1^i, \tag{3}$$

where P_1^i is projector on A_i final state [22].

Formal solution of QM evolution equation for system state prepared at t_0

$$\psi(T) = W(T)\psi(t_0),$$

where integral operator of evolution

$$W(T) = \exp[-\frac{i}{\hbar}\int_{t_0}^{T} H(\tau)d\tau]. \tag{4}$$

Consider now the case $N = 2$, for the system of two independent nuclei A_1, A_2 its initial state $\psi_A = \psi_0^1\psi_0^2$ where formally $\psi_0^{1,2} = \psi_0$. If their evolution is independent, then operator $W(T)$ will be just the product of two right side terms of eq. (4), but it can be also formally written as

$$W(T) = C_t \exp\{-\frac{i}{\hbar(T - t_0)}\int_{t_0}^{T}\int_{t_0}^{T}[H_1(t_1) + H_2(t_2)]dt_1 dt_2\}, \tag{5}$$

where $H_{1,2}$ are $A_{1,2}$ Gamow Hamiltonians with corresponding notations $r_{1,2}$, C_t is time-ordering (chronological) operator [23]. In a sense, here the second integration for each term is dummy giving just multiplier $T - t_0$, yet such ansatz is used here because below NC effects will be treated via time-dependent Hamiltonians for which multiple time parameters is routine approach [23]. Note that here $H_{1,2}$, in fact, don't depend on time, their time parameter just indicates on which of time parameters t_1 or t_2 $H_{1,2}$ integration was truly performed.

4. Nonlinear QM formalism

As was supposed, NC effects should not change the system average energy, however, if the corresponding evolution terms are linear operators, then for α-decay this condition can be violated [24]. It will be shown here that some nonlinear operators satisfy much better to this condition. Interest to nonlinear QM can be dated back to the early days of quantum physics, but at that time it was applied in effective theories describing collective effects. Now it's acknowledged that nonlinear corrections to standard QM Hamiltonian can exist also at fundamental level [25]. In nonlinear QM, particle evolution described by nonlinear Schroedinger equation of the form

$$i\hbar\partial\psi = -\frac{\hbar}{2m}\nabla^2\psi + U(\vec{r},t)\psi + F(\psi,\bar{\psi})\psi, \tag{6}$$

where m is particle mass, U is system potential, F is arbitrary functional of system state. Currently, the most popular and elaborated nonlinear QM model is by Doebner and Goldin (DG) [26, 27]. In its formalism, simple variant of nonlinear term is $F = \frac{\hbar\Gamma}{m}\Phi$ where

$$\Phi = \nabla^2 + \frac{|\nabla\psi|^2}{|\psi|^2} \tag{7}$$

is nonlinear operator, Γ is real or imaginary parameter which, in principle, can depend on time or other external factors, here only real Γ will be exploited. With the notation

$$H^L = -\frac{\hbar}{2m}\nabla^2 + U(\vec{r},t) \tag{8}$$

we abbreviate (6) to $i\hbar\partial\psi = H^L\psi + F\psi$ where in our case, H^L is Gamow Hamiltonian.

Main properties of eq. (6) were studied in [26, 28], for constant Γ they can be summarized as follows: (a) The probability is conserved. (b) The equation is homogeneous. (c) The equation is Euclidian- and time-translation invariant for $U = 0$. (d) Noninteracting particle subsystem remain uncorrelated (separation property). (e) For $U = 0$, plane waves $\psi = \exp[i(\vec{k}_0\vec{r} - \omega t)]$ with $\omega = E\hbar|\vec{k}_0|^2 = 2mE/\hbar$ are solutions both for real and imaginary Γ. Writing $< Q >= \int\bar{\psi}\hat{Q}\psi d^3x$

for operator expectation value, in particular, since $\int \bar{\psi} F \psi d^3x = 0$, the energy functional for solution of (6) is $< i\hbar\partial_t > = < H^L >$. Hence the average system energy would change insignificantly if not at all if F added to initial Hamiltonian, therefore, it advocates DG ansatz application in NC models.

It's notable that nonlinear term F in particle Hamiltonian can modify the particle tunneling through the potential barrier effect. In particular, exact solution of this problem was obtained for rectangular potential barrier, it was shown that the barrier transmission rate depends exponentially on Γ [28]. To calculate corrections to Gamow model, WKB approximation for nonlinear Hamiltonian of (6) can be used [28]. In this ansatz, 3-dimensional α-particle wave function reduced to $\psi = \frac{1}{r}\exp(i\sigma/\hbar)$; function $\sigma(r)$ can be decomposed in $\hbar$ order $\sigma = \sigma_0 + \sigma_1 + ...$ [9, 23]. Given α-particle with energy E, one can find the distances R_0, R_1 from nucleus centre at which $U(R_{0,1}) = E$. Then, for our nonlinear Hamiltonian the equation for σ_0 is

$$(\Lambda - \frac{1}{2m})(\frac{\partial \sigma_0}{\partial r})^2 = E - U(r), \tag{9}$$

where $\Lambda = \frac{2\Gamma}{m}$ for $R_0 \leq r \leq R_1$, $\Lambda = 0$ for $r < R_0$, $r > R_1$. Its solution for $R_0 \leq r \leq R_1$ can be written as

$$\psi = \frac{1}{r}\exp(i\sigma_0/\hbar) = \frac{C_r}{r}\exp[-\frac{1}{\hbar}\int_{R_0}^{r} q(\epsilon)d\epsilon],$$

where C_r is normalization constant,

$$q(r) = \frac{1}{\hbar}\{\frac{2m[U(r) - E]}{1 - 4\Gamma}\}^{\frac{1}{2}}.$$

Account of higher order σ terms doesn't change transmission coefficient which is equal to

$$D = \exp[-\frac{2}{\hbar}\int_{R_0}^{R_1} q(\epsilon)d\epsilon] = \exp[-\frac{\phi}{(1 - 4\Gamma)^{\frac{1}{2}}}] \simeq \exp[-\phi(1 + 2\Gamma)]. \tag{10}$$

Here ϕ is constant, whereas Γ, in principle, can change in time, assuming that its time scale is much larger than the barrier transition time. To calculate nucleus life-time, D multiplied by the number of α-particle kicks into nucleus potential wall per second n_d, so it gives $\lambda = n_d D$ [20], for DG model n_d doesn't depend on F term presence [28].

As was noticed, the growth of average time intervals between events can be considered as the growth of system evolution symmetry. For example, the event distribution of eq. (2) on time half-axe $\{t_0, \infty\}$ becomes maximally homogeneous for $\lambda \to 0$, hence the evolution symmetry rises in this limit. Consider two nucleus systems S_1, S_2 with the average distance between S_1, S_2 elements equal to R_{12}. S_1 is the set of N_0 unstable nuclei $\{A_l\}$ prepared at t_0 with decay probability described by eq. (2). S_2 includes just one unstable nuclide B prepared also at t_0, its evolution normally described by Gamow Hamiltonian H^L ansatz of (8). Its decay constant λ_b, in principle, can differ from λ of eq. (2) if for B $U \neq U_N$ of eq. (4). In such set-up, the system S_1 of N_0 nuclei presumably would induce NC effect in B evolution and also change its own evolution as well. Suppose that all geometric factors of such NC influence on B for given S_1 described by real coefficient $\chi(R_{12})$ which absolute value grows with N_0 and diminishes with R_{12}. Resulting corrections to H^L are supposed to be small and so can be accounted only up to first order of χ . Basing on assumptions discussed above, in particular, that resulting NC effect proportional to S_1 total decay rate, it follows that if no measurements of S_1 states performed, then corrected B Hamiltonian becomes

$$H^d(t) = H^L + \frac{\hbar^2}{m}\chi(R_{12})p_i(t)\Phi, \tag{11}$$

where p_i is of eq. (2), Φ is nonlinear operator of (7) for B α-particle. It means that in this case, $\Gamma = \frac{\bar{h}}{m}\chi p_i(t)$, substituting it in eq. (10), in WKB approximation, it gives B decay probability

$$p'(t) = C_b \exp\{-[(t - t_0)\xi(t)]\}. \quad (12)$$

Here C_b is normalization constant

$$\xi(t) = \exp\{[1 + 2\frac{\bar{h}}{m}\chi(R_{12})p_i(t)] \ln \lambda_b\}. \quad (13)$$

Hence for such ansatz, B Hamiltonian and consequently B life-time depends on S_1 nuclei decay probability at given time moment. Namely, under NC influence resulting B nucleus life-time for $\chi > 0$ becomes larger than initial one. Note that kinematic parameter n_d practically doesn't change in this case. Assuming now that $N > 1$ i.e. S_2 includes not one but many identical B nuclei, on which proposed S_1 NC effect influences independently, then average time intervals between S_2 decays would enlarge. Such S_2 evolution modification can be interpreted as the growth of S_2 evolution symmetry such that resulting decay probability $p'(t)$ becomes more homogeneous in time in comparison with initial B decay probabilty $p_b(t)$.

Now this NC effect can be considered on more fundamental level beyond master regime. Let's suppose that for system S $N_0 = N = 1$ and nuclei A_1, B states described by wave functions ψ_1, ψ_b correspondently. Then for the same initial conditions as above, the system initial wave function $\psi_s = \psi_1\psi_b$, S Hamiltonian in first χ order

$$H_s(t) = H^L + \frac{\hbar^2}{m}\chi(R_{12})p_1(t)\Phi + H_1 + \frac{\bar{h}}{m}\chi(R_{12})p_b(t)\Phi_1, \quad (14)$$

where Φ_1 is is nonlinear operator of (7) for A_1. Solution of evolution equation for such Hamiltonian would give probabilities for A_1, B decays. Resulting A_1, B states are correlated but aren't entangled, so that system state $\psi_s = \psi_1\psi_b$ at any time. Plainly, resulting NC effects will be quite small. Exploited approach to Hamiltonian formulation is, in fact, analogue of well-known Hartry-Fock approximation for electromagnetic interactions in atoms [23].

Now let's consider NC model, which results in sub-Poissonian self-ordering symmetry growth. Let's take two systems S_1, S_2 of N_0, N nuclei, correspondently, with $N \ll N_0$ and the average distance between S_1, S_2 elements equal to R_{12}. As was supposed, due to conjugal NC influence between S_1 elements, its evolution becomes more regular and self-ordered and it induces similar NC influence on S_2 evolution. Due to it, S_2 evolution can differ from the case of independent nuclei and would result in the temporary correlation between S_2 decays. Modified S_2 evolution operator can be chosen from the analogy with squeezed photon production in atomic resonance fluorescence [17]. In that case, the photon production rate is suppressed if the time interval between two consequently produced photons is less than some fixed ΔT. Due to it, the resulting photon registration becomes more regular, and their distribution is sub-Poissonian. Let's start from the simplest case $N = 2$ and A_1, A_2 nuclei prepared at t_0. Let's suppose that S_1 NC influence rate on S_2 elements characterized by some scalar function $k(R_{12})$, which absolute value supposedly diminishes as R_{12} grows and enlarges as N_0 grows. For the simplicity, we assume that NC correlation of A_1, A_2 decay moments is such that their evolution ansatz can be factorized into A_1, A_2 terms. For example, if no measurement of A_1 state is done, then resulting phenomenological A_2 Hamiltonian supposedly becomes

$$H_2^c(T) = H_2 + \frac{\bar{h}}{m}\int_{t_0}^{T} k(R_{12})\varphi(T - t)p_1(t)dt\Phi_2 \quad (15)$$

with $p_1(t)$ is of eq. (2), $\Phi_{1,2}$ are $A_{1,2}$ nonlinear operators of (7) with corresponding notations. φ is causal Green function

$$\varphi(\tau) = \eta(\tau - \nu) - \eta(\tau)$$

its possible dependence on A_1, A_2 distance supposedly accounted in $k(R_{12})$. Thus, corresponding NC time dependence described as the difference of two step functions $\eta(\tau) = \{0, \tau < 0; 1, \tau \geq 0\}$ which is simplest variant of such ansatz [15]. $\nu > 0$ is NC parameter, it corresponds to the time range in which A_1, A_2 decay acts are correlated; ν supposed to be much larger than the barrier transition time [21]. Hence A_2 Hamiltonian H_2^c is time-dependent, at given time moment T it depends on A_1 decay probability during time interval ν previous to T. Analogous modification occurs for A_1 Hamiltonian with corresponding index change. In general, such S_1 NC influence on A_1, A_2 Hamiltonians supposedly results in $W(T)$ nonlinear modification in comparison with eq. (5)

$$W(T) = C_t exp\{-\frac{i}{\hbar(T - t_0)} \int_{t_0}^{T} \int_{t_0}^{T} [H_1(t_1) + H_2(t_2) + (T - t_0)G(t_1, t_2)]dt_1 dt_2\}. \quad (16)$$

Third term in this equality is NC dynamics term, its simple ansatz which supresses nucleus decays at small time intervals between them can be taken as

$$G(t_1, t_2) = \frac{-\hbar}{m} k(R_{12})[\varphi(t_1 - t_2)\gamma_2(t_2)\Phi_1 + \varphi(t_2 - t_1)\gamma_1(t_1)\Phi_2].$$

Note that the second right-side term corresponds to H_2^c Hamiltonian of eq. (15), γ_i is of eq. (3). If no measurement of A_i state was performed for $t_f < T$, then $\gamma_i(t_i) = p_i(t_i)$ of (2). Otherwise, if such measurement was done at some t_f and A_i was found to be in the final state, then for $t_a > t_f$ it follows that $\gamma_i(t_a) = 0$. Thus, for $i = 1, 2$ the nucleus A_i NC term is supposedly proportional to decay rate of neighbour nucleus A_{2-i} of eq. (2). Under these conditions, kinematic factor n_d for α-particle motion inside nucleus changes insignificantly. Then, in WKB approximation for our nonlinear evolution operator the joint $A_{1,2}$ decay probability p_s for $k > 0$ will differ from independent case $p_s(t_1, t_2) = p_1(t_1)p_2(t_2)$ and is equal to

$$p_s(t_1, t_2) = C\lambda^{2+4\theta} \exp[-g(t_1, t_2)(t_1 + t_2 - 2t_0)],$$

where C is normalization constant, analogously to eq. (13)

$$g(t_1, t_2) = \exp[(1 + 2\theta) \ln \lambda], \quad (17)$$

where λ is from eq. (2)

$$\theta = \frac{-\hbar}{m} k(R_{12})[\eta(t_1 - t_2)\varphi(t_1 - t_2)\gamma_2(t_2) + \eta(t_2 - t_1)\varphi(t_2 - t_1)\gamma_1(t_1)].$$

Due to it, if the time interval between two decay acts is less than ν, the nucleus decay rate will be suppressed, and resulting decay event distribution will become more regular. For independent nucleus decays with $N = 2$ described by eq. (5) their joint decay probability corresponds to Poissonian process, whereas NC dynamics term in eq. (16) would transform it to sub-Poissonan one, resulting in less stochastic and more ordered event distribution. Note that in the considered case, A_1, A_2 states are correlated, but aren't entangled. For $N > 2$ the considered NC dynamics term in $W(T)$ would change to $G(t_1, \ldots, t_N)dt_1 \ldots dt_N$ with corresponding integration over N independent time parameters. As the result, for analogous G ansatz the joint decay probability of two arbitrary consequent decays will be suppressed for small time intervals between them, and the decay time distribution of the event sequence will be sub-Poissonian. For all considered effects it can be supposed that analogous metastable system description is applicable also to chemical reactions and other atomic and molecular effects. It's possible also that two considered evolution symmetrization mechanisms, in principle, can coexist and act simultaneously for some systems.

Proposed models exploits NC dynamics based on nonlinear α-decay Hamiltonian, which, in fact, is used as effective Hamiltonian demonstrating how the decay rates can change under NC influence. It was assumed above that NC doesn't change any subsystem average energy, but in our particular model, in fact, it can change slightly. However, it doesn't exclude the possibility that further model development would permit to avoid this difficulty.

5. Conclusion

Considered experimental results and theoretical analysis evidence that novel communication mechanism between distant quantum systems can exist. It's based on new form of QM nonlocality, principally different from well-known EPR-Bohm mechanism. In this paper, such NC effect was studied for the system of metastable states, in particular, unstable α-decay nuclei described by Gamow model. Application of nonlinear Hamiltonian terms for NC description permits to construct consistent dynamical model. Concerning with causality problem for NC communications, at the moment it's still possible to assume that such NC can spread between systems with velocity of light. But even if this spread is instant, it's notable that usually superluminal signalling in QM discussed for one bit yes/no communications [29]. In our case, to define the resulting change of some parameter expectation value or dispersion, one should collect significant event statistics which makes causality violation quite doubtful possibility. In addition, NC dependence on the distance between two systems can be so steep that it would suppress effective superluminal signalling. Note that some nonlinear models by themselves permit superluminal signalling, but that's untrue for our ansatz [27].

Considered QM nonlocality has universal character, so beside nucleus decays, such temporary variations of system parameters can be observed, in principle, for other systems in which metastable states and tunneling play important role. In particular, it's well known that development and functioning of biological systems performed quite consistently even at relatively large distances between their parts. For example, the kidney and liver cells, blood erythrocytes identify and attract the proper partner cells and reject wrong ones at the distance of several microns, which are much larger than the range of chemical forces [30]. Another notorious example is morphogenesis problem, i.e. proportional and optimal growth of plants and organisms. Up to now the mechanism which regulates spontaneous cell division in optimal way at significant distances between them is poorly understood. It's difficult to admit that such long-distance effects can be achieved via chemical messengers only, so it was argued long ago that some other physical mechanism can be responsible for that. Some authors proposed already that such long-distance correlations can be induced by QM nonlocality [31, 32]. However, standard EPR-Bohm mechanism can't transfer signals effectively in dense and warm media, which is characteristic for biological systems. It isn't obvious how NC model considered here can be extended on biological systems which are quite complex, yet there is no direct obstacles for that as well, some possible options discussed in [24].

References

[1] Piccardi G 1962 *The Chemical Basis of Medical Climatology* (Springfield: Charles Thomas)
[2] Shnoll S E 2009 *Cosmophysical Factors in Stochastic Processes* (Stockholm: Svenska fysikarkivet)
[3] Troshichev O A et al 2004 *Adv. Space Res.* **34** 1619
[4] Alburger D et al 1986 *Earth Plan. Science Lett.* **78** 168
[5] Fischbach E et al 2009 *Rev. Space Sci.* **145** 285
[6] Alekseev E et al 2018 *Phys. Part. Nucl.* **49** 557

[7] Bogachev S A et al 2020 *J. Phys.: Conf. Series* **1690** 012028
[8] Jauch J 1968 *Foundations of Quantum Mechanics* (Reading, MA: Addison-Wesly)
[9] Peres A 2002 *Quantum Theory: Concepts and Methods* (New York: Kluwer)
[10] Stapp H P 1997 *Am. J. Phys.* **65** 300
[11] Cramer J 1986 *Rev. Mod. Phys.* **58** 647
[12] Korotaev S M 2011 *Causality and Reversibility in Irreversible Time* (Irvine CA: Scientific Results Publishing)
[13] Martin B 2011 *Nuclear and Particle Physics: An Introduction* (New York: John Wiley & Sons)
[14] Kossert K and Nahl O 2014 *Astrop. Phys.* **55** 33
[15] Korn B and Korn T 1968 *Mathematical handbook* (N-Y: McGraw-Hill)
[16] Parkhomov A G 2009 *Cosmos, Earth, Humankind* (Moscow: Nauka)
[17] Mandel L and Wolf E 1995 *Optical Coherence and Quantum Optics* (Cambridge: University Press)
[18] Shnoll S E 1973 *Conformational oscilations in protein macromolecules* (Biological and Biochemical Oscillators) ed by Chance B (New York: Academic press) pp 667-669
[19] Shnoll S E and Chetverikova E P 1975 *Biochem. Biophys. Acta* **403** 8997
[20] Gamow G 1928 *Zc. Phys.* **51** 204
[21] Newton R 1961 *Ann. of Phys.* **14** 333
[22] Sudbery A 1986 *Quantum Mechanics and Particles of Nature* (Cambridge: Cambridge university press)
[23] Sakurai J J 1994 *Modern Quantum Mechanics* (Reading MA, Addison-Wesly)
[24] Mayburov S N 2022 *Quantum nonlocality and biological coherence* (Ultra-Weak Photon Emission from Biological Systems: Endogenous Biophotonics and Intrinsic Bioluminescence) ed I Volodiaev (Berlin: Springer) to be published
[25] Weinberg S 1989 *Ann. Phys. (N.Y.)* **194** 336
[26] Doebner H and Goldin G 1992 *Phys. Lett.* A **162** 397
[27] Doebner H and Goldin G 1996 *Phys. Rev.* A **54** 3764
[28] Mayburov S N 2021 *Int. J. Theor. Phys.* **60** 630
[29] Blokhintsev D I (1973) *Space and Time in the Microworld* (Berlin: Springer)
[30] Vitiello G 2001 *My Double Unveiled* (Amsterdam: John Benjamins)
[31] Primas H 1982 *Chemistry* **36** 293
[32] Bischof M 2003 *Introduction to integrative biophysics* (Integrative Biophysics) ed Popp F-A and Beloussov L V (Dortrecht: Kluwer Academic Publishers) pp 4-39

Gravitational Radius in view of Existence and Uniqueness Theorem

Boris E Meierovich

P L Kapitza Institute for Physical Problems, 2 Kosygina street, 119334 Moscow, Russia

E-mail: meierovich@mail.ru

Abstract. Talking about a black hole, one has in mind the process of unlimited self-compression of gravitating matter with a mass greater than critical. With a mass greater than the critical one, the elasticity of neutron matter cannot withstand gravitational compression. However, compression cannot be unlimited, because with increasing pressure, neutrons turn into some other "more elementary" particles. These can be bosons of the Standard Model of elementary particles. The wave function of the condensate of neutral bosons at zero temperature is a scalar field. If instead of the constraint $\det g_{ik} < 0$ we use a weaker condition of regularity (all invariants of the metric tensor g_{ik} are finite), then there is a regular static spherically symmetric solution to Klein-Gordon and Einstein equations, claiming to describe the state to which the gravitational collapse leads. With no restriction on total mass. In this solution, the metric component g^{rr} changes its sign twice: $g^{rr}(r) = 0$ at $r=r_g$ and $r=r_h > r_g$. Between these two gravitational radii the signature of the metric tensor g_{ik} is $(+,+,-,-)$. Gravitational radius r_g inside the gravitating body ensures regularity in the center. Within the framework of the phenomenological model "$\lambda\psi^4$", relying on the existence and uniqueness theorem, the main properties of a collapsed black hole are determined. At $r = r_g$ a regular solution to Klein-Gordon and Einstein equations exists, but it is not a unique one. Gravitational radius r_g is the branch point at which, among all possible continuous solutions, we have to choose a proper one, corresponding to the problem under consideration. We are interested in solutions that correspond to a finite mass of a black hole. It turns out that the density value of bosons is constant at $r < r_g$. It depends only on the elasticity of a condensate, and does not depend on the total mass. The energy-momentum tensor at $r \leq r_g$ corresponds to the ultra relativistic equation of state $p = \varepsilon/3$. In addition to the discrete spectrum of static solutions with a mass less than the critical one (where $g^{rr} < 0$ does not change sign), there is a continuous spectrum of equilibrium states with $g^{rr}(r)$ changing sign twice, and with no restriction on mass. Among the states of continuous spectrum, the maximum possible density of bosons depends on the mass of the condensate and on the rest mass of bosons. The rest energy of massive Standard Model bosons is about 100 GeV. In this case, for the black hole in the center of our Milky Way galaxy, the maximum possible density of particles should not exceed $3 \times 10^{81}\ cm^{-3}$.

Spherically symmetric static solutions to the system of Klein-Gordon and Einstein equations, claiming to describe the state of matter which the gravitational collapse leads to, were found and presented in articles [1] and [2]. Getting new solutions became possible when I applied a weaker condition of regularity (all invariants of g_{ik} are finite) instead of the restriction $\det g_{ik} < 0$. In these regular static solutions, the metric component g^{rr} changes its sign twice: $g^{rr}(r) = 0$ at $r = r_g$ and at $r = r_h > r_g$. The signature of the metric tensor g_{ik} is $(+,+,-,-)$ within the spherical layer $r_g < r < r_h$ (between these two gravitational radii). The presence of the gravitational radius r_g inside a gravitating body ensures regularity in the center $r=0$. For a distant observer, the Schwarzschild's gravitational radius r_h is the event horizon.

In the thirties of the last century, in the classical works [3],[4],[5] it was shown that there is a critical mass M_{cr} of matter (of the order of the Sun mass $M_\odot = 2\times10^{33}$g), above which the Fermi elasticity of neutrons cannot withstand gravitational self-compression. Since then, there has been the opinion that objects with more than the critical mass are subject to unlimited self-compression (collapse). It would be so (that is, the collapse would be unlimited), if neutrons remained unchanged in the process of unlimited increase of pressure. In fact, mutual transformations of particles with increasing pressure can slow down and even stop the collapse.

With a further increasing pressure, a strongly compressed matter of dominating elementary particles of the Standard Model, see Figure 1, claims to be "the next stop" after neutrons.

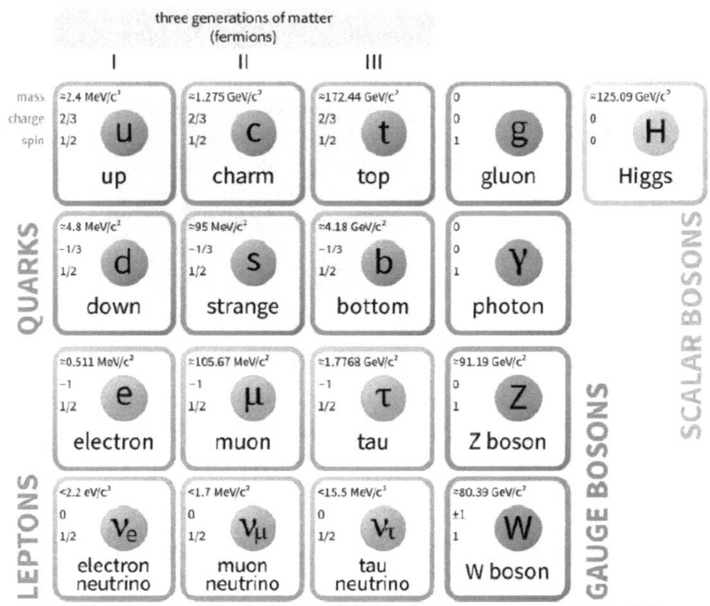

Figure 1. Standard Model of elementary particles [6].

Energetically, the most favorable state of matter is a condensate of massive bosons (gauge bosons Z and W, Higgs' scalar boson H, and also bosonic quasiparticles formed by pairing of fermions due to the Cooper effect [7]).

At zero temperature, the wave function of a Bose-Einstein condensate is a classical scalar field [8], §30. Lagrangian of a complex scalar field ψ in a curved space-time is

$$L = g^{ik}\psi^*_{,i}\psi_{,k} - U(\psi^*\psi), \quad U(0) = 0.$$

With a very large number of particles, the gravitational interaction is predominant. In power series of the potential

$$U(|\psi|^2) = \left(\frac{mc}{\hbar}\right)^2 |\psi|^2 + \frac{1}{2}\lambda|\psi|^4 + \dots, \tag{1}$$

the first term is related to gravity of non-interacting particles (m is the rest mass of a quantum). $\frac{1}{2}\lambda|\psi|^4$ and subsequent terms are corrections including interactions of non-gravitational nature. With no account of these corrections, the wave function $\psi(r)$ diverges logarithmically at $r\to0$ [1]. The phenomenological model "$\lambda\psi^4$", which takes into account the elasticity of a condensate, eliminates the divergence at the center [2]. Earlier, in the article [9], it had been noted that the equilibrium configuration in the model "$\lambda\psi^4$" differs noticeably from the case of non-interacting bosons even when $\lambda \ll 1$.

Being scalars, ψ and $\psi*$ satisfy the Kline-Gordon equation:

$$\frac{1}{\sqrt{det\, g_{ik}}}\left(\sqrt{det\, g_{ik}}\, g^{lm}\psi_{,l}\right)_{,m} = -\frac{\partial U}{\partial|\psi^2|}\psi. \tag{2}$$

Kline-Gordon equation (2) is invariant against changing the sign of $\det g_{ik}$: $\sqrt{-1}$ in numerator and denominator cancel out. A static spherically symmetric scalar field in the state of definite energy E per particle is:

$$\psi_E(x^i) = e^{-iEx^0/\hbar c}\psi(r).$$

$g^{rr}(r)$ is the coefficient at the highest derivative in the Kline-Gordon equation for the radial wave function $\psi(r)$:

$$g^{rr}\psi'' + \left((g^{rr})' + \frac{1}{2}(ln(det\, g_{ik}))' g^{rr}\right)\psi' = \left(\frac{1}{\hbar^2 c^2}(g^{00}E^2 - m^2c^4) - \lambda|\psi|^2\right)\psi. \tag{3}$$

Equation (3) is not defined at the gravitational radii $r = r_g$ and $r = r_h$ where $g^{rr}(r) = 0$.
Kline-Gordon equation (3) together with Einstein equations (see [10], formulae (100.4) and (100.6))

$$(g^{rr})' + \frac{1+g^{rr}}{r} = \kappa r T_0^0, \tag{4}$$

$$g^{rr}\left(\frac{1}{r} - (ln\, g^{00})'\right) + \frac{1}{r} = \kappa r T_r^r \tag{5}$$

constitute a complete system of equations determining a static state of the gravitating Bose-Einstein condensate. Gravitational constant $\kappa = (8\pi k)/c^4 = 2 \times 10^{-48}\, sec^2/\, g \times cm$. Components of the energy-momentum tensor T_0^0 and T_r^r in the Einstein equations (4), (5) are:

$$T_0^0 = \frac{1}{\hbar^2 c^2}(g^{00}E^2 + m^2c^4)|\psi|^2 + \frac{1}{2}\lambda|\psi|^4 - g^{rr}|\psi'|^2, \tag{6}$$

$$T_r^r = \frac{1}{\hbar^2 c^2}(-g^{00}E^2 + m^2c^4)|\psi|^2 + \frac{1}{2}\lambda|\psi|^4 + g^{rr}|\psi'|^2. \tag{7}$$

Energy density (6) and pressure (7) are regular in the center $r = 0$. It follows from Einstein equation (4) that

$$g^{rr}(r) = -1 + O(r^2), \qquad r \to 0. \tag{8}$$

Outside a gravitating body $T_0^0 = 0$ we have the Schwarzshild's solution

$$g^{rr}(r) = -1 + \frac{r_h}{r}, \qquad r \geq r_h. \tag{9}$$

If $T_0^0 = 0$ at $r \geq r_h$, and $T_0^0 \neq 0$ at $r < r_h$, then Schwarzshild's gravitational radius r_h is the event horizon coinciding with the surface of a gravitating body. At $r = r_h$ the derivative

$$\frac{dg^{rr}(r_h)}{dr} = -\frac{1}{r_h}. \tag{10}$$

From equation (5) it follows that on the surface $r = r_h$, where $g^{rr}(r_h) = 0$, the pressure does not vanish:

$$T_r^r(r_h) = \frac{1}{\kappa r_h^2}. \tag{11}$$

I think that this circumstance, along with the constraint $det\, g_{ik} < 0$, was that serious obstacle why, over a hundred years of existence of general relativity, no static solution had been found describing an ultimately compressed black hole. In case of a gravitating degenerate relativistic Fermi-gas, the trouble is that both, density and pressure, vanish on the surface (see [11], problem 3 in the end of §61).
In case of a gravitating degenerate Bose-gas, described by a scalar field, both conditions $T_0^0(r_h) = 0$ and $T_r^r(r_h) \neq 0$ are compatible with the system of equations (3)-(5). The question was how ideal Bose-gas can exist with a finite pressure on the interface with a vacuum. This problem was solved in the article [1]. In reality, the condensate is not in contact with a vacuum. It is in contact with dark matter. The main manifestations of dark matter were described adequately using a longitudinal vector field [12]. Covariant divergence of a longitudinal vector field at the center $\varphi^m_{;m}(0)$ was a free parameter of the theory [12]. It was clear that its non-zero value depends on the interaction of dark matter with a black hole in the center of a galaxy. I draw your attention to the fact that the covariant divergence of a longitudinal vector field

$\varphi^{m}_{;m}(r)$ is a scalar, satisfying the same Klein-Gordon equation (2), though with a different rest mass of a quantum. We can say that the longitudinal vector field is a scalar field "turned inside out". According to the theorem of existence and uniqueness, at the interface $r = r_h$, where $g^{rr}(r_h) = 0$, a solution to equation (2) exists for both fields. However, a solution is not unique. This ambiguity makes it possible to ensure a balance of pressure at the interface of a black hole and dark matter at $r = r_h$. The balance of pressure allowed determining the speed on the plateau of a galaxy rotation curve as a function of the observed mass of a black hole (see [1], formula (68)).

Currently, dark matter is observed due to gravitational interaction only. It follows from Einstein's equation (5) that the pressure balance at the interface $r = r_h$ depends on the observed black hole mass $M = \frac{c^2}{2k} r_h$, $k = 6.67 \times 10^8 \frac{cm^3}{g \times sec^2}$, see (11), regardless of the internal structure of a black hole. It was important to show that there can be a static spherically symmetric solution to Einstein equations with no restriction on mass. As an example, such a solution was found in [1] for a gravitating condensate of non-interacting bosons. However, in case of an ideal Bose-gas (with no elasticity), the condensate wave function diverges logarithmically at the center.

Unfortunately, at the present state of quantum chromodynamics, we don't have an equation of state for ultimately compressed elementary particles of the Standard Model derived from first principles. Current state of the problem, including an interpolation model named "unified equations of state", is considered in the review [13]. The model "$\lambda\psi^4$" (1) takes into account interactions of a non-gravitational nature phenomenologically, including elasticity of the condensate. This model does not take into account that the lifetime of a Standard Model boson falls down with decreasing pressure. Taking into account the elasticity, a necessary condition for the existence of solutions with no restriction on total mass is [2]:

$$\frac{1}{g^{00}(r)} \frac{dg^{00}(r)}{dr} < 0 \quad at \quad r = r_g. \tag{12}$$

Under this condition, in the close vicinity of the upper boundary, numerical solutions with a finite mass practically do not differ from the regular ones at the center. The existence and uniqueness theorem prompts how to find numerical solutions in the whole range of parameters.

To get maximum advantage from the existence and uniqueness theorem, it is necessary to convert equations (3)-(5) to canonical form. In dimensionless variables:

$$x = \frac{mc}{\hbar} r, \quad u(x) = \sqrt{\kappa}\psi(r), \quad w(x) = \frac{\hbar\sqrt{\kappa}}{mc} g^{rr}(r) \frac{d\psi}{dr}, \quad g(x) = g^{rr}(r), \quad h(x) = \frac{E^2 g^{00}(r)}{m^2 c^4}, \tag{13}$$

we have the following system of four first-order equations, resolved with respect to derivatives [2]:

$$du/dx = w/g, \tag{14}$$

$$\frac{dw}{dx} = (h - 1 - \Lambda u^2)\, u - \left(\frac{2}{x} - \frac{x}{g}\left(hu^2 - \frac{w^2}{g}\right)\right) w, \tag{15}$$

$$\frac{dg}{dx} = x\left((h+1)u^2 + \frac{1}{2}\Lambda u^4 - \frac{w^2}{g}\right) - \frac{1+g}{x}, \tag{16}$$

$$\frac{dh}{dx} = \left(\frac{1}{x}\left(1 + \frac{1}{g}\right) - \frac{x}{g}\left((1-h)u^2 + \frac{w^2}{g} + \frac{1}{2}\Lambda u^4\right)\right) h. \tag{17}$$

$g(x) = g^{rr}(r)$ not to be confused with $\det g_{ik}$. In the model "$\lambda\psi^4$", dimensionless parameter

$$\Lambda = \left(\frac{\hbar}{mc}\right)^2 \frac{\lambda}{\kappa} \tag{18}$$

characterizes the elasticity of a condensate. $x_g = \frac{mc}{\hbar} r_g$ is the dimensionless internal gravitational radius; $x_h = \frac{mc}{\hbar} r_h$ is the dimensionless event horizon.

Regularity in the center dictates the following boundary conditions:

$$u(0) = u_0, \quad w(0) = 0, \quad g(0) = -1, \quad h'(0) = 0. \tag{19}$$

Analytical solution to the set of equations (14)-(17) with boundary conditions (19) is

$$u(x) = u_0, \quad w(x) = 0, \quad g(x) = -1 + \frac{1}{3}u_0^2 x^2, \quad h(x) = \frac{1}{3}, \tag{20}$$

provided that there is a balance of density u_0^2 and elasticity Λ :

$$\Lambda u_0^2 = -2/3. \tag{21}$$

In the model "$\lambda\psi^4$", density in the center u_0^2 and internal gravitational radius x_g $\left(x_g^2 = \frac{3}{u_0^2} = 4.5|\Lambda|\right)$ are determined by elasticity Λ of the condensate.

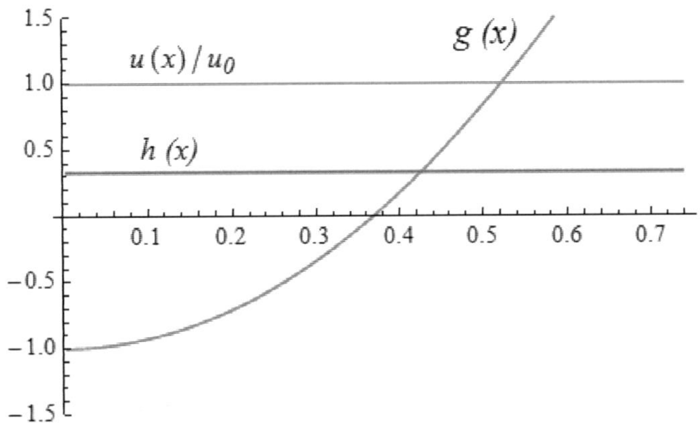

Figure 2. Analytical solution to equations (14)-(17) satisfying initial conditions (19) $u_0^2 = 22$, $x_g = 0.369$. It exists at $0 \le x < \infty$ but it is unique only at $0 \le x < x_g$.

Analytical solution (20) exists in the whole space $0 \le x < \infty$. But it is unique only inside the sphere $0 \le x < x_g$, where $g(x) < 0$. Right-hand sides of equations (14)-(17) are not continuous functions at $x = x_g$ where $g(x) = 0$. In accordance with the theorem of existence and uniqueness (see [14], §3), the density $u^2(x) = u_0^2$ remains constant inside the central sphere $0 \le x < x_g$, regardless of the total mass. Unambiguously

$$u(x_g) = u_0, \quad w(x_g) = 0, \quad g(x_g) = 0, \quad h(x_g) = 1/3. \tag{22}$$

On the central plateau the components (6) and (7) of the energy-momentum tensor

$$T_0^0 = \frac{m^2c^2}{\hbar^2\kappa}u_0^2, \quad T_r^r = \frac{m^2c^2}{3\hbar^2\kappa}u_0^2, \quad 0 \le x \le x_g$$

correspond to ultra relativistic equation of state $p = \varepsilon/3$. The dimensionless metric component $h(x) = 1/3$ also remains constant at $x \le x_g$. As far as $h'(x_g) = 0$, the condition (12) must be specified:

$$\frac{1}{g^{00}(r)}\frac{dg^{00}(r_g + 0)}{dr} < 0.$$

Gravitational radius x_g is the branch point at which, among all possible continuous extensions, we should choose the proper one, corresponding to the problem under consideration. With boundary conditions (22) we would have the same analytical solution (20) at $x > x_g$, Figure 2, describing a condensate which is uniformly spread throughout the whole space. Its total mass would be infinite.

In order to find numerically solutions with finite total mass, coinciding with (22) at $x = x_g$, it is reasonable to search for boundary conditions slightly beyond the gravitational radius $x - x_g = \Delta \to +0$:

$$u(x_g + \Delta) = u_0 - \alpha\Delta^p, \quad g(x_g + \Delta) = \frac{(x_g+\Delta)^2}{x_g^2} - 1 - \gamma\Delta^{1+s}, \quad h(x_g + \Delta) = \frac{1}{3} - \beta\Delta^q. \tag{23}$$

Since the solutions coincide at $x = x_g$, it is sufficient to leave only leading terms at $\Delta \to +0$. In order to find unknowns $\alpha, \beta, \gamma, p > 0, q > 0$, and $s > 0$, one has to submit (23) into equations (14)-(17). With accuracy of leading terms, the equations are satisfied if

$$\alpha(2p^2 - p)\Delta^{p-1} = \sqrt{3}\beta\Delta^q, \tag{24}$$

$$(1+s)\gamma\Delta^s = \frac{3}{x_g}\beta\Delta^q, \tag{25}$$

$$(2q-1)\beta\Delta^q = \frac{1}{3}\gamma\Delta^{1+s}. \tag{26}$$

Due to the arbitrariness of Δ, it follows from (25) that $q = s$. Then relation (26) reduces to $(2q-1)\beta\Delta^q = 0$ at $\Delta \to 0$. Hence, $q = 1/2$ (if $\beta \neq 0$), and $\gamma = 2\beta/x_g$. From (24) we have $p = q + 1 = 3/2$, and $\alpha = \beta/\sqrt{3}$. The constant β remains a free parameter.

In the literature, the system of Einstein and Klein-Gordon equations is widely used to describe hypothetical bosonic stars being in equilibrium with their own gravitational field [9], [15]-[17]. In these and other similar articles, authors usually use the representation $g^{rr} = -e^{-\lambda}$. It imposes the restriction

$$g^{rr} < 0. \tag{27}$$

In this case there is no radius r_0 where $g^{rr}(r_0) = 0$. The Schwarzschild gravitational radius r_h appears only in the asymptotic $g^{rr}(r) = -1 + r_h/r$ which is applicable at $r >> r_h$. In this case, r_h is neither the event horizon, nor the boundary of a gravitating body.

Under the restriction (27) the solution to the eigenvalue problem for the gravitating Bose-Einstein condensate contains a discrete spectrum of eigenvalues $h_n(u_0, x = 0)$ and conjugate eigenfunctions $u_n(x)$ [18]. The continuous parameter $u_0 \equiv u(x = 0)$, on which the star's mass M depends, remains arbitrary. The total mass M of the condensate in the ground state does not exceed the critical value M_{cr}. For bosons with a rest mass $m \sim 100 GeV/c^2$ the critical total mass M_{cr} is about 10^{12}g (a million ton).

With no restriction (27) there are solutions with g^{rr} changing its sign twice. Gravitational radius $r_h > r_g$ where $g^{rr}(r_h) = 0$ exists. $r = r_h$ is the surface of a black hole. r_h is the event horizon for a remote observer. There is no restriction on total mass. The value $h(0) = 1/3$ in (20) is strictly fixed by the condition of regularity in the center. Instead of the discrete eigenvalues $h_n(u_0, 0)$, a continuous parameter $\beta > 0$ appears due to the theorem of existence and uniqueness.

With no restriction (27), there is a one-parametric continuum of regular solutions to the system of equations (14)-(17) with the same boundary conditions (22) at the gravitational radius $x = x_g$. Case $\beta = 0$ corresponds to the analytical solution (20) with infinite total mass M. At $\beta > 0$ we get numerically a continuous spectrum of regular solutions with a finite total mass M. Small deviations of $u(x)$ and $w(x)$ in ambiguous solutions are of the order $\Delta^{3/2}$ at $\Delta \to 0$. Deviations of $g(x)$ and $h(x)$ are proportional to $\Delta^{1/2}$ at $\Delta \to 0$:

$$u(x_g + \Delta) = u_0 - \beta\sqrt{\frac{\Delta^3}{3}}, \quad w(x_g + \Delta) = -\beta u_0 \Delta^{3/2},$$

$$g(x_g + \Delta) = \frac{2}{x_g}\Delta(1 - \beta\sqrt{\Delta}), \quad h(x_g + \Delta) = \frac{1}{3} - \beta\sqrt{\Delta} - 2\beta^2\Delta.$$

With account of only leading deviations ($\sim \beta\sqrt{\Delta}$), boundary conditions at $x = x_g + \Delta$, $\Delta \to +0$ are:

$$u(x_g + \Delta) = u_0, \quad w(x_g + \Delta) = 0, \quad g(x_g + \Delta) = \frac{2\Delta}{x_g}(1 - \beta\sqrt{\Delta}), \quad h(x_g + \Delta) = \frac{1}{3} - \beta\sqrt{\Delta}. \tag{28}$$

The smaller $\beta\sqrt{\Delta} \ll 1$, the more accurate numerical result. Certainly, within the capabilities of a computer.

No matter how small $\beta > 0$, we obtain a solution to equations (14) – (17) with boundary conditions (28) for a black hole with a finite total mass.

Luckily, I managed to find analytical solution (20). It simplified the numerical analysis of the structure and properties of an extremely compressed black hole. Currently, the main characteristic of a black hole is its mass M. Within the frames of the model "$\lambda\psi^4$", varying β within the interval $0 < \beta < \infty$, one can determine the range of possible values of density u_0^2 in the center, and elasticity Λ of the ultimately compressed Bose-Einstein condensate at a fixed mass M.

In order to cover the whole range (from the De Broglie wavelength of 100 GeV bosons to the size of a black hole) on a single graph, it is convenient to use a logarithmic scale along the x axis. And along the y axis also, because of a huge range of variation of the metric component g^{rr}. Using a logarithmic scale along the y axis, I raised $g(x)$ graph two steps up (I show $ln(g(x) + 2)$ instead of $ln\, g\,(x)$ in Figures 3a and 3b), because $g(x)$ changes sign, see Figure 2.

Total mass M is the most reliable parameter of a black hole. In dimensional units, the black hole mass M is connected with the dimensionless event horizon radius x_h by

$$M = \frac{c^2}{2k} r_h = \frac{1}{2}\frac{M_{Pl}^2}{m} x_h. \tag{29}$$

Here $M_{Pl} = \sqrt{\hbar c/k} = 2.177 \times 10^{-5} g$ is the Plank mass, and m is the mass of a boson. Having in mind bosons of the Standard Model with the rest energy about 100 GeV (see Figure 1), I assume a boson mass m to be $1.78 \times 10^{-22} g$ for quantitative estimates. Mass of the black hole in the center of our Milky Way galaxy [19] is

$$M_{MW} = 8.6 \times 10^{39} g \tag{30}$$

The Milky Way black hole's event horizon $r_h = 1.6 \times 10^{12} cm$. In dimensionless units $x_h = \frac{mc}{\hbar} r_h = 6.46 \times 10^{27}$, $ln\, x_h = 64$. Metric function $g(x)$ and dimensionless density $u^2(x)$, corresponding to the Milky Way black hole mass M_{MW} (30), are presented in Figure 3a for a very small $\beta = 0.000045$, and in Figure 3b – for a large $\beta = 15.5$. For $\beta = 0.000045$ the mass (30) corresponds to dimensionless density in the center $u_0^2 = 22$. Gravitational radius $x_g = 0.369$, $ln\, x_g = -0.996$. For $\beta = 15.5$ the density $u_0^2 = 36$, $x_g = 0.289$, $ln\, x_g = -1.242$. Naturally, comparing two cases of different β for the same black hole with mass (30), the event horizon radius is the same: $x_h = 6.46 \times 10^{27}$, $ln\, x_h = 64$.

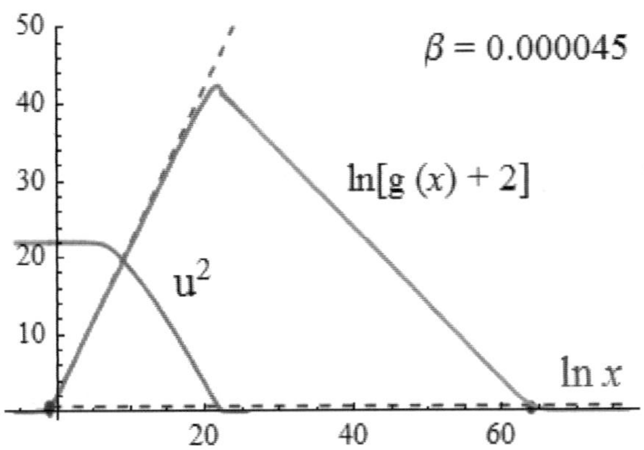

Figure 3a. $\beta = 0.000045$. Dimensionless density in the interval $0 \le x \le x_g$ is $u_0^2 = 22$. Gravitational radius $x_g = 0.369$, $ln\, x_g = -0.996$.

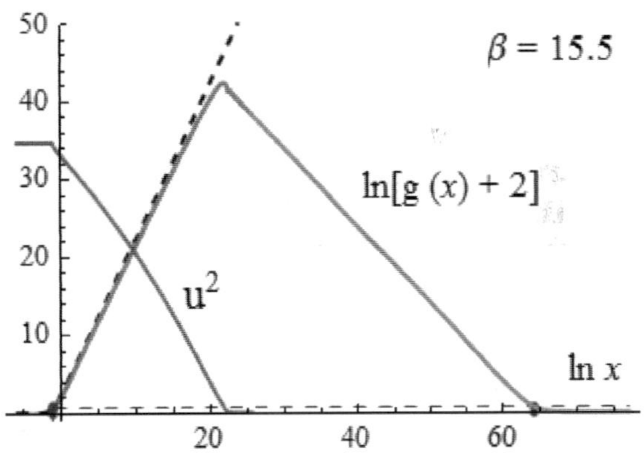

Figure 3b. $\beta = 15.5$. Dimensionless density in the interval $0 \le x \le x_g$ is $u_0^2 = 36$. Gravitational radius $x_g = 0.289$, $ln\, x_g = -1.242$.

Red lines are $g(x) + 2$ in double logarithmic scales. Dashed horizontal lines show the level ln2 where $g(x)$ is zero. Oblique dashed lines correspond to $g(x)$ in Figure 2.

At a fixed mass, the density u_0^2 in the center slightly increases with growing β. Functions $u(x)/u_0$ and $h(x)$ at the fixed mass $M_{MW} = 8.6 \times 10^{39} g$ ($x_h = 6.46 \times 10^{27}$, $ln\, x_h = 64$) are shown in Figure 4a for $\beta = 0.000045$ and in Figure 4b for $\beta = 15.5$.

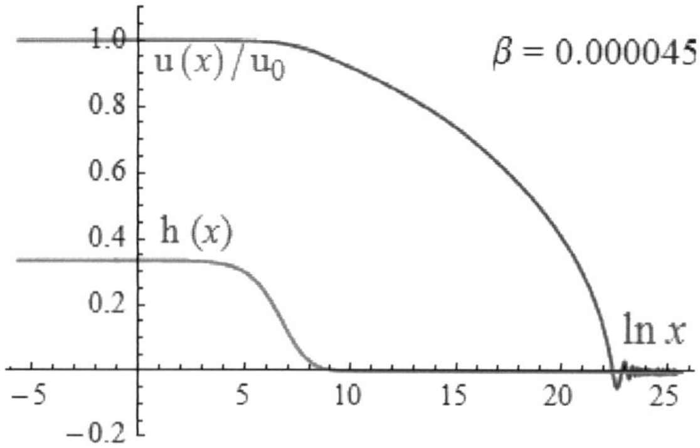

Figure 4a. $\beta = 0.000045$. $u_0^2 = 22$. Gravitational radius $x_g = 0.369$, $\ln x_g = -0.996$.

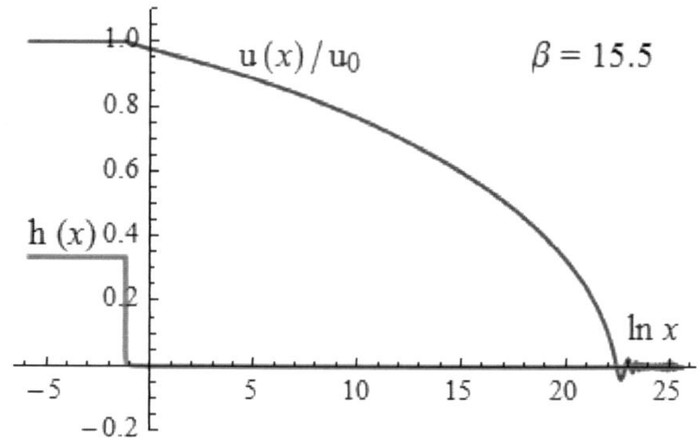

Figure 4b. $\beta = 15.5$, $u_0^2 = 36$. Gravitational radius $x_g = 0.289$, $\ln x_g = -1.242$.

Blue curve $u(x)/u_0$ weakly depends on β. Red curve $h(x)$ turns to a step at large β.
In the center $x \le x_g$ density $u^2(x) = u_0^2$ and metric component $h(x) = 1/3$ remain constant (plateau at $0 \le x < x_g$) regardless of β. This is a consequence of the theorem of existence and uniqueness. At $x > x_g$functions $u(x)$ and $h(x)$ deviate from strait lines in Figure 2 if $\beta > 0$, see Figure 4a. The wave function $u(x)$ decreases with oscillations. The metric component $h(x)$ becomes a smoothly decreasing function. $h(x)$ turns smoothly into a step at $x = x_g$ if $\beta \to \infty$, see Figure 4b. The component $g(x)$ gets positive at $x > x_g$. It turns from a growing function to a decreasing one, and changes sign back to minus at $x > x_h$. $g(x)$ tend to -1 at $x \to \infty$.

It looks reasonable to classify black holes with a mass $M >> M_\odot$ ($M_\odot = 2\times10^{33}\,g$ is the Sun mass) as "heavy" black holes, and those with a mass $M \gtrsim M_\odot$ - as "light" ones. I consider a black hole with the mass $M \gtrsim M_\odot$ as a 'light' black hole for the following reason. The critical mass M_{cr}, above which the collapse of a neutron star is inevitable, is of the order of the Sun mass $M_\odot$. Black holes with masses $M<M_{cr}$ could exist [20]. However, additional elasticity of neutrons as fermions (due to the Pauli principle of exclusivity) is an obstacle for the collapse. At $M>M_{cr}$, elasticity of neutrons is unable to withstand gravitational collapse. The barrier disappears, and matter can be compressed by its own gravitational field to the static state of a Bose condensate. A possibility of super-light black holes with masses $M << M_\odot$ does not contradict the theory of general relativity. Super-light "primordial" black holes could appear at the stage of strong compression of the early Universe [21].

Due to the parametric freedom $0 < \beta < \infty$ the model "$\lambda\psi^4$" allows the existence of black holes with the same mass M, but different elasticity Λ. The dependence of the density $u_0^2(\beta)$ for 'heavy' black holes with masses like the Milky Way's one (30) is shown in Figure 5 (red curve). Blue curve in Figure 5 is the dependence $u_0^2(\beta)$ for 'light' black holes with masses $M\approx M_\odot$.

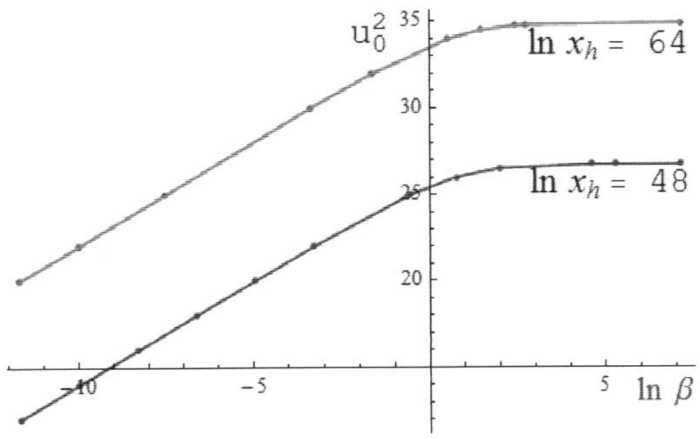

Figure 5. Dependence of the density $u_0^2(\beta)$ in the central plateau of black holes with masses $M = M_{MW}$ (red curve, $ln\, x_h = 64$), and with masses $M \approx M_\odot$ (blue curve, $\ln x_h = 48.76$).

Numerical analysis shows that at $\beta << 1$ density $u_0^2(x_h, \beta)$ in the center grows logarithmically with x_h and β:

$$u_g^2 = u_0^2 = 0.53\, ln\, x_h + 1.2\, ln\, \beta\,, \quad \beta << 1.$$

At $\beta >> 1$ the density $u_0^2(x_h, \beta)$ as a function of β reaches it's maximum $u_{0max}^2(x_h)$. Dependence $u_{0max}^2(x_h)$, found numerically,

$$u_{0max}^2(x_h) = 2.14 + 0.512\, ln\, x_h, \tag{31}$$

is presented in Figure 6.

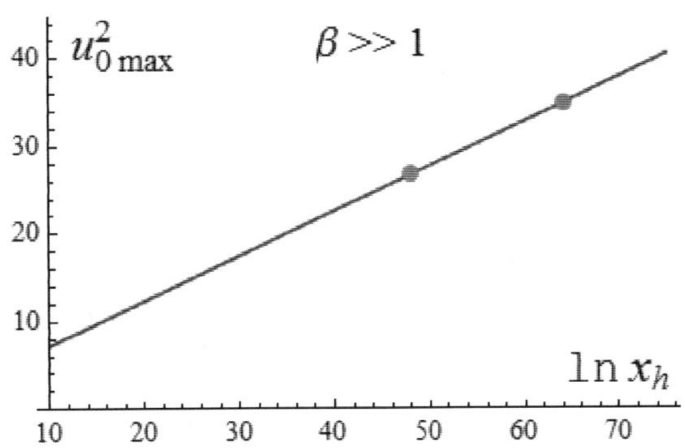

Figure 6. Maximum possible density of a condensate on the central plateau as a function of the total mass in dimensionless units. Two red dots correspond to the 'heavy' black hole in the center of Milky Way, and to a 'light' black hole (with a mass like that of the Sun).

$u_{0max}^2(x_h)$ is the maximum possible density of a condensate at a fixed mass (29). Density $\rho_{0max}(M)$ in dimensional units is

$$\rho_{0max}(M) = \frac{m}{\hbar^2 \kappa} u_{0max}^2(x_h) \frac{1}{cm^3}\ , \ x_h = \frac{2mM}{M_{Pl}^2}.$$

In the model "$\lambda\psi^4$" at a fixed total mass M the maximum density $\rho_{0max}(M)$ depends only on the rest mass m of a boson. For bosons of the Standard Model (rest mass $m \sim 100\ GeV/c^2$) the maximum possible density of the black hole in the center of Milky Way Galaxy (lnx_h=64) is $\rho_{MW_{max}} = 3.1\times10^{81}\ cm^{-3}$.

In the model "$\lambda\psi^4$" regularity in the center dictates a strict balance between density u_0^2 and elasticity Λ of a condensate (21). If, at a fixed mass, the density u_0^2 of a condensate in equilibrium is bounded from above, then, consequently, the elasticity $|\Lambda|$ is restricted from below.

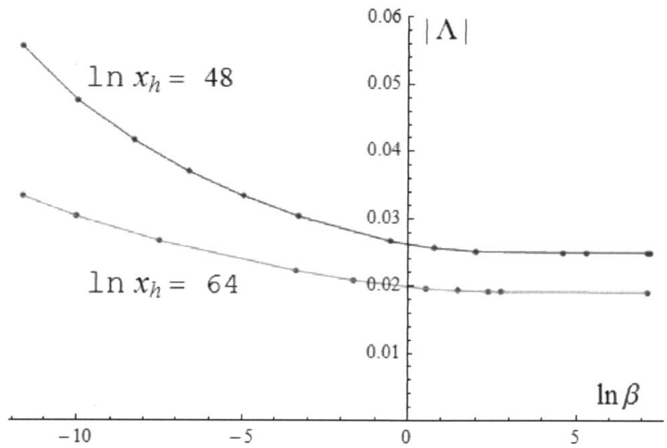

Figure 7. Possible elasticity of a condensate as a function of β at fixed mass. Milky Way's black hole (lnx_h=64) – red line. A 'light' black hole with a mass like the mass of the Sun (lnx_h =48.76) – blue line.

Dimensionless elasticity $|\Lambda|$ of the Milky Way's black hole (lnx_h=64) as a function of β is presented in Figure 7 – red line. Blue curve is $|\Lambda(\beta)|$ of a black hole with the mass like the Sun's one (lnx_h = =48.76). The minimum possible elasticity as a function of the total mass

$$|\Lambda_{min}(x_h)| = \frac{1}{3.21 + 0.768\, ln\, x_h}, x_h = \frac{2mM}{M_{Pl}^2}$$

follows from (31). In the model "$\lambda\psi^4$", a static state is impossible for a gravitating condensate with elasticity weaker than $|\Lambda_{min}(x_h)|$.

Thus, in the model "$\lambda\psi^4$", in addition to the previously known discrete spectrum of static states of the gravitating Bose-Einstein condensate [18], there is also a continuous spectrum of equilibrium states claiming to describe the ultimately compressed condensate as a result of the gravitational collapse. The existence of an internal gravitational radius, at which the uniqueness of solutions to the system of Einstein and Klein-Gordon equations is violated, made it possible to establish some features of a black hole. Within the internal gravitational radius, the condensate density is uniform, and the energy-momentum tensor corresponds to ultrarelativistic equation of state. The density value at the center is determined by elasticity of the condensate. For a given total mass, the range of possible density values in the center is bounded from above. The limiting value of the density depends on the rest mass of the boson m and is proportional to the logarithm of the total mass M. If we proceed from bosons of the Standard Model, then the maximum possible density in the center of the black hole in our Milky Way galaxy is approximately $3 \times 10^{81}\ cm^{-3}$.

References

[1] Meierovich B E 2019 *Universe* **5** 198
[2] Meierovich B E 2020 *Universe* **6** 113
[3] Chandrasekhar S 1931 *Astrophys. J.* **74** 81
[4] Landau L D 1932 *Phys. Zs. Sowjet.* **1** 285
[5] Oppenheimer J R and Volkoff G 1939 *Phys. Rev.* **55** 374
[6] https://en.wikipedia.org/wiki/Elementary_particles.
[7] Cooper L N 1956 *Phys. Rev.* **104** 1189
[8] Lifshits E M, Pitaevski L P 2000 *Course of Theoretical Physics Vol. 9 Statistical Physics Part 2* (Moscow: Fizmatlit)
[9] Colpi M, Shapiro S L and Wasserman I 1986 *Phys. Rev. Lett.* **57** 2485
[10] Landau L D and Lifshitz E M 1995 *Course of Theoretical Physics, Vol. 2: The Classical Theory of Fields* (Moscow: Nauka)
[11] Landau L D and Lifshitz E M 1995 *Course of Theoretical Physics, Vol. 5: Statistical Physics* (Moscow: Nauka)
[12] Meierovich B E 2013 *Physical Review D***10** 103510
[13] Gordon Baum, Tetsuo Hatsuda, Toru Kojo *et al* 2017 *Preprint* Arxiv: 1707.04966v1 [astro-phys. HE]
[14] Pontryagin L S 1961 *Ordinary differential equations* (Moscow: Fizmalit)
[15] Friedberg R, Lee T D and Pang Y 1987 *Phys. Rev. D* **35** 3640
[16] Kaup D J and Klein-Gordon Geon1968 *Phys. Rev.* **172** 1331
[17] Torres D F, Capozziello S and Lambiase G 2000 *Phys. Rev. D* **62** 104012
[18] Meierovich B E 2018 *J. Exp. Theor. Phys.* **127** 889
[19] Gillessen S, Eisenhauer F and Trippe S 2009 *Astrophys. J.* **692** 1075
[20] Zel'dovich Y B 1962 *JETP* **42** 641
[21] Novikov I D, Polnarev A G, Starobinsky A A and Zel'dovich Y B 1979 *Astron. Astrophys.* **20** 104

Spinor vacuum and *C*, *P*, *T* inversions

V V Monakhov and A V Kozhedub

Saint Petersburg State University, St. Petersburg, Russia

E-mail: v.v.monahov@spbu.ru

Abstract. We have developed the theory of Clifford reflections and extended spacetime inversions. This extended spacetime has two additional dimensions associated with the presence of internal degrees of freedom of spinors. Inversions *C*, *P*, and *T* contain not only reflections of the basis Clifford vectors and transformations of basis spinors, but also transformations of the components of vector and spinor quantities. The research is carried out on the basis of algebraic quantum field theory using the superalgebraic representation of spinors as well as the 8-component matrix representation of spinors. We have proved that due to the presence of internal degrees of freedom of spinors, there are two vacua, the vacuum of our Universe and an alternative vacuum. The inversion operators *C* and *T* transform the vacuum into an alternative one, and therefore cannot be operators of the exact symmetry of our Universe.

1. Introduction

Modern theories of spacetime are increasingly based not only on the idea of geometrization proposed by A. Einstein, but also on the idea of algebraization. Modern generalizations of Einstein's general theory of relativity use the theory of spinor bundles taking into account torsion and nonmetricity. However, the quantum theory of the spinor vacuum is still in its infancy. It is hardly possible to construct a realistic quantum theory of spacetime and gravity without understanding the properties of the spinor vacuum.

The modern theory of spinors is the algebraic theory of spinors. It is based on the theory of Clifford algebras [1]-[4]. We develop quantum field theory based on the superalgebraic representation of spinors [5]-[10]. It is an algebraic quantum field theory in which the Dirac spinors are a superposition of Grassmann densities in momentum space and their derivatives [6]-[10]. This theory is a development of the algebraic theory of spinors. Ordinary algebraic theory of spinors [1]-[4] is not based on the canonical anticommutation relations (CAR), although the modern formulation of the algebraic quantum field theory is based on them. In ordinary algebraic spinor theory, the creation and annihilation operators of spinors are elements of a CAR algebra [11] that has nothing to do with the Clifford algebra, of which the spinors are two-valued representations [1]-[4]. And in our theory, the creation and annihilation operators of spinors are elements of the same Clifford algebra as the basis Clifford vectors of spacetime (gamma operators, which are algebraic analogs of Dirac matrices). Moreover, in our theory, we used CAR as a basis for deriving generalized Dirac conjugation formulas [7] and for constructing an explicit expression for the spinor vacuum state vector [6].

Our study of the properties of the spinor vacuum [9]-[10] showed that they differ from the earlier assumptions in the framework of quantum field theory. It turned out that the spinor vacuum is not invariant under the operations of charge conjugation *C* and time inversion *T*. Moreover, it turned out to be invariant with respect to the operation of space inversion *P* (spatial parity) and combinations *CT* and *CPT*. In this article, we refine the results of articles [9]-[10].

2. Superalgebraic representation of spinors

In the theory of superalgebraic spinors, the analogs of Dirac gamma matrices γ_D^μ are gamma operators $\hat{\gamma}^\mu$, formulas for which are given in (1).

$$\begin{aligned}
\hat{\gamma}^0 &= \int d^3p\, [\frac{\partial}{\partial\theta^1(p)}\theta^1(p) + \frac{\partial}{\partial\theta^2(p)}\theta^2(p) + \frac{\partial}{\partial\theta^3(p)}\theta^3(p) + \frac{\partial}{\partial\theta^4(p)}\theta^4(p), \bullet],\\
\hat{\gamma}^1 &= \int d^3p\, [\frac{\partial}{\partial\theta^1(p)}\frac{\partial}{\partial\theta^4(p)} - \theta^4(p)\theta^1(p) + \frac{\partial}{\partial\theta^2(p)}\frac{\partial}{\partial\theta^3(p)} - \theta^3(p)\theta^2(p), \bullet],\\
\hat{\gamma}^2 &= i\int d^3p\, [-\frac{\partial}{\partial\theta^1(p)}\frac{\partial}{\partial\theta^4(p)} - \theta^4(p)\theta^1(p) + \frac{\partial}{\partial\theta^2(p)}\frac{\partial}{\partial\theta^3(p)} + \theta^3(p)\theta^2(p), \bullet],\\
\hat{\gamma}^3 &= \int d^3p\, [\frac{\partial}{\partial\theta^1(p)}\frac{\partial}{\partial\theta^3(p)} - \theta^3(p)\theta^1(p) - \frac{\partial}{\partial\theta^2(p)}\frac{\partial}{\partial\theta^4(p)} + \theta^4(p)\theta^2(p), \bullet],\\
\hat{\gamma}^4 &= i\hat{\gamma}^5 = i\int d^3p\, [\frac{\partial}{\partial\theta^1(p)}\frac{\partial}{\partial\theta^3(p)} + \theta^3(p)\theta^1(p) + \frac{\partial}{\partial\theta^2(p)}\frac{\partial}{\partial\theta^4(p)} + \theta^4(p)\theta^2(p), \bullet],\\
\hat{\gamma}^6 &= i\int d^3p\, [\frac{\partial}{\partial\theta^1(p)}\frac{\partial}{\partial\theta^2(p)} + \theta^2(p)\theta^1(p) - \frac{\partial}{\partial\theta^3(p)}\frac{\partial}{\partial\theta^4(p)} - \theta^4(p)\theta^3(p), \bullet],\\
\hat{\gamma}^7 &= \int d^3p\, [\frac{\partial}{\partial\theta^1(p)}\frac{\partial}{\partial\theta^2(p)} - \theta^2(p)\theta^1(p) + \frac{\partial}{\partial\theta^3(p)}\frac{\partial}{\partial\theta^4(p)} - \theta^4(p)\theta^3(p), \bullet].
\end{aligned} \tag{1}$$

Operator $\hat{A} = [A, \bullet]$ denotes commutator $\hat{A}\Psi = [A,\bullet]\Psi = [A,\Psi] = A\Psi - \Psi A$. Unlike Dirac gamma matrices, gamma operators are commutators. Moreover, they are composite and consist of a combination of Grassmann densities $\theta^a(p)$, $a = 1,2,3,4$, and their derivatives $\frac{\partial}{\partial\theta^a(p)}$ [6]. These densities are generators of CAR-algebra with anticommutation relations (2)

$$\begin{aligned}
&\{\frac{\partial}{\partial\theta^i(p)}, \theta^j(p')\} = \delta_i^j\delta(p-p'),\\
&\{\frac{\partial}{\partial\theta^i(p)}, \frac{\partial}{\partial\theta^j(p')}\} = \{\theta^i(p), \theta^j(p')\} = 0.
\end{aligned} \tag{2}$$

Compared to Dirac's theory, there are two additional gamma operators, $\hat{\gamma}^6$ and $\hat{\gamma}^7$ [7]. They do not refer to spacetime and set the internal degrees of freedom of the spinors [12]-[13].

We obtain the Dirac conjugated spinor $\overline{\Psi}$ from the spinor Ψ using the generalized Dirac conjugation operation [7]

$$\overline{\Psi} = (\hat{\gamma}^0\Psi)^+. \tag{3}$$

The Grassmann densities $\frac{\partial}{\partial\theta^1(0)}$ and $\frac{\partial}{\partial\theta^2(0)}$ are the annihilation operators of spinors with the projections of the spin $s_3 = +\frac{1}{2}$ and $s_3 = -\frac{1}{2}$, respectively. Grassmann densities $\theta^3(0)$ and $\theta^4(0)$ are antispinor creation operators with projections of the spin $s_3 = +\frac{1}{2}$ and $s_3 = -\frac{1}{2}$, respectively.

Lorentz boost $\exp(\hat{\gamma}^{0k}\varphi_k)$ changes operator $b_\alpha(0)=\dfrac{\partial}{\partial\theta^\alpha(0)}$ of annihilation of a spinor (or antispinor) and the operator $\bar{b}_\alpha(0)=\theta^\alpha(0)$ of creation of an antispinor (or spinor) with the momentum p=0 to the operators of annihilation $b_\alpha(p)$ and creation $\bar{b}_\alpha(p)$ with the momentum p

$$b_\alpha(p)=\exp(\hat{\gamma}^{0k}\varphi_k)\frac{\partial}{\partial\theta^\alpha(0)}|_{0\to p},$$
$$\bar{b}_\alpha(p)=\exp(\hat{\gamma}^{0k}\varphi_k)\theta^\alpha(0)|_{0\to p}. \tag{4}$$

Annihilation and creation operators (4) are also generators of the CAR-algebra

$$\{b_i(p),\bar{b}_j(p')\}=\delta_i^j\delta(p-p'),$$
$$\{b_i(p),b_j(p')\}=\{\bar{b}_i(p),\bar{b}_j(p')\}=0. \tag{5}$$

The vacuum state vector

$$\Psi_{\rm V}=\prod_i\Psi_{\rm V}(p_i) \tag{6}$$

is given via the factors $\Psi_{\rm V}(p_i)$ local in the momentum space

$$\Psi_{\rm V}(p_i)=(\Delta^3 p_i)^4\, b_1(p_i)\bar{b}_1(p_i)b_2(p_i)\bar{b}_2(p_i)b_3(p_i)\bar{b}_3(p_i)b_4(p_i)\bar{b}_4(p_i), \tag{7}$$

where $\Delta^3 p_i$ is infinitesimal volume in the momentum space corresponding to the discrete momentum p_i [6], [8], [9].

It is known from the theory of CAR-algebras that there are an infinite number of spinor vacuum state vectors [11]. Moreover, they are all physically equivalent. But only one of them is the vacuum of the Fock representation, for which the operator of the number of particles is defined. At first glance, these statements are contradictory, but they are not. A change in the formula (6) the order of the product of any annihilation operator $b_\alpha(p_i)$ and the corresponding creation operator $\bar{b}_\alpha(p_i)$ gives us another vacuum state vector. But for this vacuum, the operator, which was the annihilation operator for the normal vacuum, becomes the creation operator, and vice versa.

In addition to the spinor vacuum $\Psi_{\rm V}$ of the Fock representation, there is another spinor vacuum that is important for the theory of the *C*, *P*, *T* inversions. It is an alternative spinor vacuum

$$\Psi_{\rm altV}=\prod_i\Psi_{\rm altV}(p_i), \tag{8}$$

where

$$\Psi_{\rm altV}(p_i)=(\Delta^3 p_i)^4\,\bar{b}_1(p_i)b_1(p_i)\bar{b}_2(p_i)b_2(p_i)\bar{b}_3(p_i)b_3(p_i)\bar{b}_4(p_i)b_4(p_i). \tag{9}$$

In it, in contrast to the spinor vacuum $\Psi_{\rm V}$, operators $b_\alpha(p_i)$ are creation operators, and operators $\bar{b}_\alpha(p_i)$ are annihilation operators [9], [10].

3. 8-component matrix representation of spinors

Matrix representation of Grassmann densities [6] is

$$\frac{\partial}{\partial\theta^1(p)}\cong w_1=\begin{pmatrix}1\\0\\0\\0\end{pmatrix};\quad \frac{\partial}{\partial\theta^2(p)}\cong w_2=\begin{pmatrix}0\\1\\0\\0\end{pmatrix};\quad \theta^3(p)\cong w_3=\begin{pmatrix}0\\0\\1\\0\end{pmatrix};\quad \theta^4(p)\cong w_4=\begin{pmatrix}0\\0\\0\\1\end{pmatrix}, \tag{10}$$

where $w1$, $w2$, $w3$, $w4$ are basis matrix spinors. In what follows, for brevity, we will sometimes omit the momentum p as a parameter.

Consider the action of gamma operators (1) on Grassmann densities (10). With the arrow, we indicate the expression into which the gamma operator transforms the corresponding Grassmann density

$$
\begin{aligned}
&\hat{\gamma}^0: \frac{\partial}{\partial\theta^1} \to \frac{\partial}{\partial\theta^1}; \frac{\partial}{\partial\theta^2} \to \frac{\partial}{\partial\theta^2}; \theta^3 \to -\theta^3; \theta^4 \to -\theta^4; \frac{\partial}{\partial\theta^3} \to \frac{\partial}{\partial\theta^3}; \frac{\partial}{\partial\theta^4} \to \frac{\partial}{\partial\theta^4}; \theta^1 \to -\theta^1; \theta^2 \to -\theta^2;\\
&\hat{\gamma}^1: \frac{\partial}{\partial\theta^1} \to -\theta^4; \frac{\partial}{\partial\theta^2} \to -\theta^3; \theta^3 \to \frac{\partial}{\partial\theta^2}; \theta^4 \to \frac{\partial}{\partial\theta^1}; \frac{\partial}{\partial\theta^3} \to \theta^2; \frac{\partial}{\partial\theta^4} \to \theta^1; \theta^1 \to -\frac{\partial}{\partial\theta^4}; \theta^2 \to -\frac{\partial}{\partial\theta^3};\\
&\hat{\gamma}^2: \frac{\partial}{\partial\theta^1} \to -i\theta^4; \frac{\partial}{\partial\theta^2} \to i\theta^3; \theta^3 \to i\frac{\partial}{\partial\theta^2}; \theta^4 \to -i\frac{\partial}{\partial\theta^1}; \frac{\partial}{\partial\theta^3} \to -i\theta^2; \frac{\partial}{\partial\theta^4} \to i\theta^1; \theta^1 \to i\frac{\partial}{\partial\theta^4}; \theta^2 \to -i\frac{\partial}{\partial\theta^3};\\
&\hat{\gamma}^3: \frac{\partial}{\partial\theta^1} \to -\theta^3; \frac{\partial}{\partial\theta^2} \to \theta^4; \theta^3 \to \frac{\partial}{\partial\theta^1}; \theta^4 \to -\frac{\partial}{\partial\theta^2}; \frac{\partial}{\partial\theta^3} \to \theta^1; \frac{\partial}{\partial\theta^4} \to -\theta^2; \theta^1 \to -\frac{\partial}{\partial\theta^3}; \theta^2 \to \frac{\partial}{\partial\theta^4};\\
&\hat{\gamma}^5: \frac{\partial}{\partial\theta^1} \to \theta^3; \frac{\partial}{\partial\theta^2} \to \theta^4; \theta^3 \to \frac{\partial}{\partial\theta^1}; \theta^4 \to \frac{\partial}{\partial\theta^2}; \frac{\partial}{\partial\theta^3} \to -\theta^1; \frac{\partial}{\partial\theta^4} \to -\theta^2; \theta^1 \to -\frac{\partial}{\partial\theta^3}; \theta^2 \to -\frac{\partial}{\partial\theta^4};\\
&\hat{\gamma}^6: \frac{\partial}{\partial\theta^1} \to i\theta^2; \frac{\partial}{\partial\theta^2} \to -i\theta^1; \theta^3 \to i\frac{\partial}{\partial\theta^4}; \theta^4 \to -i\frac{\partial}{\partial\theta^3}; \frac{\partial}{\partial\theta^3} \to -i\theta^4; \frac{\partial}{\partial\theta^4} \to i\theta^3; \theta^1 \to -i\frac{\partial}{\partial\theta^2}; \theta^2 \to i\frac{\partial}{\partial\theta^1};\\
&\hat{\gamma}^7: \frac{\partial}{\partial\theta^1} \to -\theta^2; \frac{\partial}{\partial\theta^2} \to \theta^1; \theta^3 \to -\frac{\partial}{\partial\theta^4}; \theta^4 \to \frac{\partial}{\partial\theta^3}; \frac{\partial}{\partial\theta^3} \to -\theta^4; \frac{\partial}{\partial\theta^4} \to \theta^3; \theta^1 \to -\frac{\partial}{\partial\theta^2}; \theta^2 \to \frac{\partial}{\partial\theta^1}.
\end{aligned}
\tag{11}
$$

Consider the Dirac matrices in the Dirac representation.

$$
\gamma_D^0 = \begin{pmatrix} 1 & 0 & 0 & 0 \\ 0 & 1 & 0 & 0 \\ 0 & 0 & -1 & 0 \\ 0 & 0 & 0 & -1 \end{pmatrix}; \quad
\gamma_D^1 = \begin{pmatrix} 0 & 0 & 0 & 1 \\ 0 & 0 & 1 & 0 \\ 0 & -1 & 0 & 0 \\ -1 & 0 & 0 & 0 \end{pmatrix}; \quad
\gamma_D^2 = \begin{pmatrix} 0 & 0 & 0 & -i \\ 0 & 0 & i & 0 \\ 0 & i & 0 & 0 \\ -i & 0 & 0 & 0 \end{pmatrix};
$$
$$
\gamma_D^3 = \begin{pmatrix} 0 & 0 & 1 & 0 \\ 0 & 0 & 0 & -1 \\ -1 & 0 & 0 & 0 \\ 0 & 1 & 0 & 0 \end{pmatrix}; \quad
\gamma_D^5 = \begin{pmatrix} 0 & 0 & 1 & 0 \\ 0 & 0 & 0 & 1 \\ 1 & 0 & 0 & 0 \\ 0 & 1 & 0 & 0 \end{pmatrix};
\tag{12}
$$

In accordance with (12) and (10), we find that under the action of gamma matrices on the columns, the following correspondence to the transformations of Grassmann densities should be observed

$$
\begin{aligned}
&\gamma_D^0: w_1 \to w_1; \quad w_2 \to w_2; \; w_3 \to -w_3; w_4 \to -w_4;\\
&\gamma_D^1: w_1 \to -w_4; \; w_2 \to -w_3; w_3 \to w_2; \; w_4 \to w_1;\\
&\gamma_D^2: w_1 \to -iw_4; w_2 \to iw_3; \; w_3 \to iw_2; \; w_4 \to -iw_1;\\
&\gamma_D^3: w_1 \to -w_3; \; w_2 \to w_4; \; w_3 \to w_1; \; w_4 \to -w_2;\\
&\gamma_D^5: w_1 \to w_3; \quad w_2 \to w_4; \; w_3 \to w_1; \; w_4 \to w_2.
\end{aligned}
\tag{13}
$$

Comparison of (13) with (11) shows that the matrix representation of gamma operators under their action on Grassmann densities (10) is the Dirac matrices (12)

$$
\hat{\gamma}^\lambda \cong \gamma_D^\lambda, \; \lambda = 0,1,2,3,5. \tag{14}
$$

The situation is more complicated when it comes to finding matrix representations of conjugate spinors. In Dirac's theory, spinors are matrix columns, and Dirac's conjugate spinors are matrix rows. But they cannot be elements of the same algebra. In the theory of Clifford algebras, the so-called algebraic spinors are introduced, whose matrix representations are square matrices [1]-[4]. However, there are a number of problems in the theory of algebraic spinors, and one of them is the problem of conjugate spinors. Since spinors are left ideals of the Clifford algebra, and conjugate spinors are right ideals [3]. Therefore, as in the case of Dirac matrix spinors, they exist in different spaces. Moreover, these spinors have nothing to do with the CAR-algebra of creation and annihilation operators. In the case of a superalgebraic representation of spinors, spinors and conjugated spinors are elements of the CAR-algebra of creation and annihilation operators.

We will consider spinor Ψ_1 and conjugated spinor Φ_2 together as an 8-component column

$$\Psi = \begin{pmatrix} \Psi_1 \\ \Phi_2 \end{pmatrix}. \tag{15}$$

This representation of the Dirac spinors is often used to study the properties of solutions of the Dirac equation [14], [15].

First, let us try to assign $\Phi_2 = \overline{\Psi}_2$ [14]. With such an arrangement of the spinor components, we obtain a correspondence to the basis conjugated spinors-columns

$$\theta^1(p) \cong w'_1 = \begin{pmatrix} 1 \\ 0 \\ 0 \\ 0 \end{pmatrix}; \quad \theta^2(p) \cong w'_2 = \begin{pmatrix} 0 \\ 1 \\ 0 \\ 0 \end{pmatrix}; \quad -\frac{\partial}{\partial\theta^3(p)} \cong w'_3 = \begin{pmatrix} 0 \\ 0 \\ 1 \\ 0 \end{pmatrix}; \quad -\frac{\partial}{\partial\theta^4(p)} \cong w'_4 = \begin{pmatrix} 0 \\ 0 \\ 0 \\ 1 \end{pmatrix}. \tag{16}$$

Then the Dirac matrices will perform the following transformations

$$\begin{aligned}
&\gamma_D^0 : w'_1 \to w'_1; \quad w'_2 \to w'_2; \ w'_3 \to -w'_3; w'_4 \to -w'_4; \\
&\gamma_D^1 : w'_1 \to -w'_4; \ w'_2 \to -w'_3; w'_3 \to w'_2; \ w'_4 \to w'_1; \\
&\gamma_D^2 : w'_1 \to -iw'_4; w'_2 \to iw'_3; \ w'_3 \to iw'_2; \ w'_4 \to -iw'_1; \\
&\gamma_D^3 : w'_1 \to -w'_3; \ w'_2 \to w'_4; \ w'_3 \to w'_1; \ w'_4 \to -w'_2; \\
&\gamma_D^5 : w'_1 \to w'_3; \quad w'_2 \to w'_4; \ w'_3 \to w'_1; \ w'_4 \to w'_2.
\end{aligned} \tag{17}$$

Comparison of (17) with (11) shows that the matrix representation of gamma operators under their action on Grassmann densities (16) is

$$\hat{\gamma}^0 \cong \begin{pmatrix} \gamma_D^0 & 0 \\ 0 & -\gamma_D^0 \end{pmatrix}; \ \hat{\gamma}^1 \cong \begin{pmatrix} \gamma_D^1 & 0 \\ 0 & -\gamma_D^1 \end{pmatrix}; \hat{\gamma}^2 \cong \begin{pmatrix} \gamma_D^2 & 0 \\ 0 & \gamma_D^2 \end{pmatrix}; \ \hat{\gamma}^3 \cong \begin{pmatrix} \gamma_D^3 & 0 \\ 0 & -\gamma_D^3 \end{pmatrix}; \ \hat{\gamma}^5 \cong \begin{pmatrix} \gamma_D^5 & 0 \\ 0 & \gamma_D^5 \end{pmatrix}. \tag{18}$$

Second, we can try to use *CPT*-conjugate spinors as columns of conjugated spinors. Let us try to assign $\Phi_2 = \Psi'_{2,CPT}$

$$\Psi = \begin{pmatrix} \Psi_1 \\ \Psi'_{2,CPT} \end{pmatrix}. \tag{19}$$

According to [9] and [10],

$$\Psi'_{CPT} = \pm R_{-q} R_{-x^\mu} \hat{\gamma}^{26} \Psi^*, \tag{20}$$

where R_{-q} changes to the opposite value q of the electric charge, operator R_{-x^μ} changes to opposite coordinates x^μ, and $\hat{\gamma}^{26} = \frac{1}{2}(\hat{\gamma}^2\hat{\gamma}^6 - \hat{\gamma}^6\hat{\gamma}^2)$. Wherein

$$\hat{\gamma}^{26} = \int d^3p \, [\frac{\partial}{\partial\theta^3(p)}\theta^1(p) + \frac{\partial}{\partial\theta^4(p)}\theta^2(p) - \frac{\partial}{\partial\theta^1(p)}\theta^3(p) - \frac{\partial}{\partial\theta^2(p)}\theta^4(p) \ ,\bullet], \tag{21}$$

$$\begin{aligned}
&\hat{\gamma}^{26} : \frac{\partial}{\partial\theta^1} \to \frac{\partial}{\partial\theta^3}; \frac{\partial}{\partial\theta^2} \to \frac{\partial}{\partial\theta^4}; \ \theta^3 \to -\theta^1; \theta^4 \to -\theta^2; \\
&\theta^1 \to \theta^3; \theta^2 \to \theta^4; \frac{\partial}{\partial\theta^3} \to -\frac{\partial}{\partial\theta^1}; \frac{\partial}{\partial\theta^4} \to -\frac{\partial}{\partial\theta^2}.
\end{aligned} \tag{22}$$

Let us choose the minus sign in (20). We obtain a correspondence to the basis conjugated spinors-columns

$$-\frac{\partial}{\partial\theta^3(p)} \cong w'_1 = \begin{pmatrix}1\\0\\0\\0\end{pmatrix}; \quad -\frac{\partial}{\partial\theta^4(p)} \cong w'_2 = \begin{pmatrix}0\\1\\0\\0\end{pmatrix}; \quad \theta^1(p) \cong w'_3 = \begin{pmatrix}0\\0\\1\\0\end{pmatrix}; \quad \theta^2(p) \cong w'_4 = \begin{pmatrix}0\\0\\0\\1\end{pmatrix}. \tag{23}$$

Similar to the previous consideration, we get the matrix representation of gamma operators under their action on Grassmann densities (23)

$$\hat{\gamma}^0 \cong \begin{pmatrix}\gamma_D^0 & 0\\ 0 & \gamma_D^0\end{pmatrix}; \quad \hat{\gamma}^1 \cong \begin{pmatrix}\gamma_D^1 & 0\\ 0 & \gamma_D^1\end{pmatrix}; \hat{\gamma}^2 \cong \begin{pmatrix}\gamma_D^2 & 0\\ 0 & -\gamma_D^2\end{pmatrix}; \hat{\gamma}^3 \cong \begin{pmatrix}\gamma_D^3 & 0\\ 0 & \gamma_D^3\end{pmatrix}; \hat{\gamma}^5 \cong \begin{pmatrix}\gamma_D^5 & 0\\ 0 & \gamma_D^5\end{pmatrix}. \tag{24}$$

At first glance, it seems that from relations (17) and (23) it follows that conjugated spinors in the matrix representation are transformed according to a law that differs from the law of transformation of spinors. However, it is not. It can be noted that the difference between the basis conjugated spinors in the two considered conjugation variants is only in the order of the basis spinors in the basis. As the basis conjugated spinors, we can choose any linear combination of them that does not violate the positive frequency of the annihilation operators and the negative frequency of the creation operators. Let us choose as the conjugated spinor a linear combination corresponding to the 4-component column $i\gamma_D^5\gamma_D^2\Psi'_{2,CPT} = \Psi_{2,CPT}$. Then the basis spinors of the conjugated spinors-columns are

$$-\frac{\partial}{\partial\theta^4(p)} \cong w'_1 = \begin{pmatrix}1\\0\\0\\0\end{pmatrix}; \quad \frac{\partial}{\partial\theta^3(p)} \cong w'_2 = \begin{pmatrix}0\\1\\0\\0\end{pmatrix}; \quad -\theta^2(p) \cong w'_3 = \begin{pmatrix}0\\0\\1\\0\end{pmatrix}; \quad \theta^1(p) \cong w'_4 = \begin{pmatrix}0\\0\\0\\1\end{pmatrix}, \tag{25}$$

and the 8-component spinor

$$\Psi = \begin{pmatrix}\Psi_1\\ i\gamma_D^5\gamma_D^2\Psi'_{2,CPT}\end{pmatrix}. \tag{26}$$

is transformed by gamma operators in accordance with the relations

$$\hat{\gamma}^0 \cong \begin{pmatrix}\gamma_D^0 & 0\\ 0 & \gamma_D^0\end{pmatrix}; \quad \hat{\gamma}^1 \cong \begin{pmatrix}\gamma_D^1 & 0\\ 0 & \gamma_D^1\end{pmatrix}; \hat{\gamma}^2 \cong \begin{pmatrix}\gamma_D^2 & 0\\ 0 & \gamma_D^2\end{pmatrix}; \hat{\gamma}^3 \cong \begin{pmatrix}\gamma_D^3 & 0\\ 0 & \gamma_D^3\end{pmatrix}; \hat{\gamma}^5 \cong \begin{pmatrix}\gamma_D^5 & 0\\ 0 & -\gamma_D^5\end{pmatrix}. \tag{27}$$

Thus, when choosing the Grassmann densities (25) as the basis conjugated spinors, we obtain for the conjugated spinors the same transformation formulas as for the spinors themselves, but in this case the matrix γ_D^5 changes sign. That is, the "right" spinor $\Psi_{1,\mathrm{R}} = \frac{1+\gamma_D^5}{2}\Psi_1$ corresponds to the "left" antispinor $\Psi_{2,\mathrm{L}} = \frac{1-\gamma_D^5}{2} i\gamma_D^5\gamma_D^2\Psi'_{2,CPT}$.

Note that in [15] the 8-component spinor satisfying the Dirac equation was written in the form

$$\Psi = \begin{pmatrix}\Psi_1\\ \Psi_2^c\end{pmatrix} = \begin{pmatrix}\Psi_1\\ i\gamma_D^2\Psi_2^*\end{pmatrix}. \tag{28}$$

The formula for the charge conjugated spinor in the matrix representation

$$\Psi_2^c(x) = i\gamma_D^2\Psi_2^*(x) \tag{29}$$

differs by a factor γ_D^5 from the *CPT*-conjugated spinor formula in the matrix representation

$$\Psi_{2,CPT}(x') = i\gamma_D^5\gamma_D^2\Psi'_{2,CPT}(x'). \tag{30}$$

This factor arises due to the fact that during *CPT* conjugation, the sign of spatial coordinates and coordinates along the time axis changes, and therefore $x'^{\mu} = -x^{\mu}$ and $\partial'_{\mu} = -\partial_{\mu}$.

Note the fundamental difference between the operators $\hat{\gamma}^{\mu}$ and γ_D^{μ} [10]. Operators γ_D^{μ} have two different causes. First, there are matrix representations of the operators $\hat{\gamma}^{\mu}$. Second, these are matrix operators arising from the replacement of the basis of conjugate spinors, which does not change the physical quantities themselves. In our case, the choice of basis (25) consists of a change in the order of the terms and a change in the sign of two of the four basis spinors. Spinors are specified up to a sign, but the relative signs of their components matter.

Therefore, when considering matrix representations of superalgebraic spinors, it is necessary to distinguish between matrices that represent gamma operators and matrices associated with the choice of basis spinors. In the superalgebraic representation of spinors, in addition to gamma operators $\hat{\gamma}^{\mu}$, there are Lorentz-invariant operators $\hat{\Gamma}^{\mu}$ composed of the creation and annihilation operators of spinors and antispinors [8]. As shown in [10], it is these operators that transform the basis of spinors and conjugated spinors. It is their matrix representation that corresponds to the matrices associated with the choice of basis spinors.

4. Action of spatial parity P, time inversion T, and charge conjugation C operators on the spinor vacuum

In [9] and [10], we derived formulas for operators of reversal operator R, spatial parity (inversion) P, time inversion (reversal) T and charge conjugation C

$$R = R_{\hat{\gamma}^{05}} R_{\hat{\gamma}^{26}} (\bullet)^T , \tag{31}$$

$$P = R_{-x^k} R_{i\hat{\gamma}^0} R_{\hat{\gamma}^{67}} = R_{-x^k} R_{\hat{\gamma}^0 \hat{Q}} , \tag{32}$$

$$T = R R_{-x^0} R_{\hat{\gamma}^1 \hat{\gamma}^3} (\bullet)^* = R_{-x^0} R_{\hat{\gamma}^7} (\bullet)^+ , \tag{33}$$

$$C = R R_{-q} R_{i\hat{\gamma}^{56}} = R_{-q} R_{-i\hat{\gamma}^{02}} (\bullet)^T . \tag{34}$$

In [9], we made a statement about the action of operators (31) - (34) on the spinor vacuum (6) and the alternative vacuum (8)

$$\begin{aligned} R\Psi_{\mathrm{V}} &= \Psi_{\mathrm{altV}} , \\ R\Psi_{\mathrm{altV}} &= \Psi_{\mathrm{V}} , \end{aligned} \tag{35}$$

$$P\Psi_{\mathrm{V}} = \Psi_{\mathrm{V}} , \tag{36}$$

$$T\Psi_{V} = \Psi_{\mathrm{altV}} , \tag{37}$$

$$C\Psi_{V} = \Psi_{\mathrm{altV}} . \tag{38}$$

However, the proofs of these statements, with the exception of formula (35), were either absent or were not completely correct. Below we will correctly prove formulas (35)-(38). First, let us indicate how the operators R, P, T, C and the factors R_{-x^k}, R_{-x^0}, R_{-q} included in them change the spatial momentum p, on which the Grassmann densities depend [9]

$$\begin{aligned} R \quad &: p \to p, \\ P, R_{-x^k} &: p \to -p, \\ T, R_{-x^0} &: p \to -p, \\ C, R_{-q} &: p \to p. \end{aligned} \tag{39}$$

Consider the action of the operator P on each factor $b_{<l>}(p_i)\bar{b}_{<l>}(p_i)$ in $\Psi_{\mathrm{V}}(p_i)$ given in (7). The brackets <l> mean that there is no automatic summation at index l. The proof is very close to the insufficiently correct proof we made in [9]. But, in contrast to [9], the action on $b_{<l>}(p_i)\bar{b}_{<l>}(p_i)$ the operator $R_{-x^k} R_{\hat{\gamma}^0 \hat{Q}}$, not the operator $R_{\hat{\gamma}^0 \hat{Q}}$, is considered.

$$
\begin{aligned}
&Pb_{<l>}(p_i)\bar{b}_{<l>}(p_i) = R_{-x^k} R_{\hat{\gamma}^0\hat{Q}}(b_{<l>}(p_i)\bar{b}_{<l>}(p_i)) = (R_{-x^k}\hat{\gamma}^0\hat{Q}b_{<l>}(p_i))(R_{-x^k}\hat{\gamma}^0\hat{Q}\bar{b}_{<l>}(p_i)) = \\
&= -(R_{-x^k}\hat{\gamma}^0 b_{<l>}(p_i))(R_{-x^k}\hat{\gamma}^0 \bar{b}_{<l>}(p_i)) = -(\hat{\gamma}^0 R_{-x^k} b_{<l>}(p_i))(\hat{\gamma}^0 R_{-x^k} \bar{b}_{<l>}(p_i)) = \\
&= -\left(\hat{\gamma}^0 \exp(\hat{\gamma}^{0k}\varphi_k) R_{-x^k} \frac{\partial}{\partial\theta^{<l>}(p_i)}\right)\left(\hat{\gamma}^0 R_{-x^k}\left(\hat{\gamma}^0 \exp(\hat{\gamma}^{0j}\varphi_j)\frac{\partial}{\partial\theta^{<l>}(p_i)}\right)^+\right) = \\
&= -\left(\exp(-\hat{\gamma}^{0k}\varphi_k) R_{-x^k}\hat{\gamma}^0 \frac{\partial}{\partial\theta^{<l>}(p_i)}\right)\left(R_{-x^k}\hat{\gamma}^0\left(\exp(-\hat{\gamma}^{0j}\varphi_j)\hat{\gamma}^0 \frac{\partial}{\partial\theta^{<l>}(p_i)}\right)^+\right) = \\
&= \left(\exp(-\hat{\gamma}^{0k}\varphi_k)\frac{\partial}{\partial\theta^{<l>}(-p_i)}\right)\left(\left(\hat{\gamma}^0 \exp(-\hat{\gamma}^{0j}\varphi_j)\frac{\partial}{\partial\theta^{<l>}(-p_i)}\right)^+\right) = \\
&= b_{<l>}(-p_i)\bar{b}_{<l>}(-p_i).
\end{aligned} \tag{40}
$$

It follows from (40) that $P\Psi_{\mathrm{V}}(p_i) = \Psi_{\mathrm{V}}(-p_i)$. Then

$$
P\Psi_{\mathrm{V}} = \prod_i (P\Psi_{\mathrm{V}}(p_i)) = \prod_i \Psi_{\mathrm{V}}(-p_i) = \prod_i \Psi_{\mathrm{V}}(p_i) = \Psi_{\mathrm{V}}. \tag{41}
$$

Statement (36) is proved.

Now consider the action of the time inversion operator (33) on Ψ_{V}. To do this, first consider the action of the operator T on $b_{<l>}(p_i)\bar{b}_{<l>}(p_i)$.

$$
\begin{aligned}
&Tb_{<l>}(p_i)\bar{b}_{<l>}(p_i) = R_{-x^0}R_{\hat{\gamma}^7}(b_{<l>}(p_i)\bar{b}_{<l>}(p_i))^+ = \\
&= (R_{-x^0}\hat{\gamma}^7\hat{\gamma}^0 b_{<l>}(p_i))(R_{-x^0}\hat{\gamma}^7 b_{<l>}(p_i)^+) = \\
&= \left(R_{-x^0}\hat{\gamma}^7\hat{\gamma}^0\exp(\hat{\gamma}^{0k}\varphi_k)\frac{\partial}{\partial\theta^{<l>}(p_i)}\right)\left(R_{-x^0}\hat{\gamma}^7\left(\exp(\hat{\gamma}^{0j}\varphi_j)\frac{\partial}{\partial\theta^{<l>}(p_i)}\right)^+\right) = \\
&= \left(R_{-x^0}\hat{\gamma}^7\hat{\gamma}^0\exp(\hat{\gamma}^{0k}\varphi_k)\frac{\partial}{\partial\theta^{<l>}(p_i)}\right)\left(R_{-x^0}\hat{\gamma}^7\exp(-\hat{\gamma}^{0j}\varphi_j)\theta^{<l>}(p_i)\right) = \\
&= \left(R_{-x^0}\exp(-\hat{\gamma}^{0k}\varphi_k)\hat{\gamma}^7\hat{\gamma}^0\frac{\partial}{\partial\theta^{<l>}(p_i)}\right)\left(R_{-x^0}\exp(-\hat{\gamma}^{0j}\varphi_j)\hat{\gamma}^7\theta^{<l>}(p_i)\right) = \\
&= \left(\exp(-\hat{\gamma}^{0k}\varphi_k)\hat{\gamma}^7\frac{\partial}{\partial\theta^{<l>}(-p_i)}\right)\left(\exp(-\hat{\gamma}^{0j}\varphi_j)\hat{\gamma}^7\theta^{<l>}(-p_i)\right)
\end{aligned} \tag{42}
$$

According to (11), operator $\hat{\gamma}^7$ transforms $\dfrac{\partial}{\partial\theta^l(-p_i)}$ into $\pm\theta^a(-p_i)$ and $\theta^l(-p_i)$ into $\pm\dfrac{\partial}{\partial\theta^a(-p_i)}$, where the plus sign for a equal to 1 and 3, and the minus sign for a equal to 2 and 4.

Therefore

$$
\begin{aligned}
&Tb_{<l>}(p_i)\bar{b}_{<l>}(p_i) = \pm\left(\exp(-\hat{\gamma}^{0j}\varphi_j)\theta^{<a>}(-p_i)\right)\left(\exp(-\hat{\gamma}^{0k}\varphi_k)\frac{\partial}{\partial\theta^{<a>}(-p_i)}\right) = \\
&= \pm\bar{b}_{<a>}(-p_i)b_{<a>}(-p_i).
\end{aligned} \tag{43}
$$

In $\Psi_{\mathrm{V}}(p_i)$, the number of factors with a minus sign is even, and we get that $\Psi_{\mathrm{V}}(p_i)$ is transformed by the operator T into $\Psi_{\mathrm{altV}}(-p_i)$. As a result

$$T\Psi_{\rm V}=\prod_i (T\Psi_{\rm V}(p_i))=\prod_i \Psi_{\rm altV}(-p_i)=\prod_i \Psi_{\rm altV}(p_i)=\Psi_{\rm altV}\,. \quad (44)$$

Statement (37) is proved.

The action of the charge conjugation operator (34) on $\Psi_{\rm V}$ is considered in a similar way. First, consider the action of the operator C on $b_{<l>}(p_i)\bar{b}_{<l>}(p_i)$.

$$C b_{<l>}(p_i)\bar{b}_{<l>}(p_i)=RR_{-q}R_{i\hat{\gamma}^{56}}b_{<l>}(p_i)\bar{b}_{<l>}(p_i)=(i\hat{\gamma}^{56}\bar{b}_{<l>}(p_i))(i\hat{\gamma}^{56}b_{<l>}(p_i))=$$
$$=\left(i\hat{\gamma}^{56}\exp(\hat{\gamma}^{0j}\varphi_j)\theta^{<l>}(p_i)\right)\left(i\hat{\gamma}^{56}\exp(\hat{\gamma}^{0k}\varphi_k)\frac{\partial}{\partial\theta^{<l>}(p_i)}\right)= \quad (45)$$
$$=\left(\exp(\hat{\gamma}^{0j}\varphi_j)i\hat{\gamma}^{56}\theta^{<l>}(p_i)\right)\left(\exp(\hat{\gamma}^{0k}\varphi_k)i\hat{\gamma}^{56}\frac{\partial}{\partial\theta^{<l>}(p_i)}\right).$$

From (11) we obtain that

$$i\hat{\gamma}^{56}:\frac{\partial}{\partial\theta^1}\to\frac{\partial}{\partial\theta^4};\frac{\partial}{\partial\theta^2}\to-\frac{\partial}{\partial\theta^3};\theta^3\to\theta^2;\theta^4\to-\theta^1;\frac{\partial}{\partial\theta^3}\to\frac{\partial}{\partial\theta^2};\frac{\partial}{\partial\theta^4}\to-\frac{\partial}{\partial\theta^1};\theta^1\to\theta^4;\theta^2\to-\theta^3. \quad (46)$$

It follows from (46) that $i\hat{\gamma}^{56}\theta^l(p_i)=\pm\theta^l(p_i)$ and $i\hat{\gamma}^{56}\frac{\partial}{\partial\theta^l(p_i)}=\pm\frac{\partial}{\partial\theta^k(p_i)}$. Therefore

$$C\Psi_{\rm V}(p_i)=\Psi_{\rm altV}(p_i)\,. \quad (47)$$

Taking into account (6)-(9), we obtain

$$C\Psi_{\rm V}=\Psi_{\rm altV}\,. \quad (48)$$

Statement (38) is proved.

5. Conclusions

We have derived formulas for 8-component matrix representation of spinors and carried out a correct proof that the spatial parity operator P leaves the spinor vacuum invariant, and the operators of time inversion T and charge conjugation C transform the spinor vacuum into an alternative one.

References

[1] Lounesto P 2001 *Clifford Algebras and Spinors* (Cambridge: Cambridge University Press)
[2] Shirokov D S 2018 *Proc. of the Nineteenth Int. Conf. on Geometry, Integrability and Quantization* (Varna: Bulgarian Academy of Sciences) pp 11–53
[3] Marchuk N G, Shirokov D S 2021 *Theory of Clifford algebras and spinors* (Moscow: URSS)
[4] Floerchinger S 2021 *Universe* **7** 168
[5] Pavšič M 2012 *Advances in Applied Clifford Algebras* **22** 449
[6] Monakhov V 2018 *J. of Physics: Conf. Series* **1051** 012023
[7] Monakhov V 2019 *Theor. and Math. Physics* **200** 1026
[8] Monakhov V 2019 *Universe* **5** 162
[9] Monakhov V 2021 *Universe* **7** 124
[10] Monakhov V, Kozhedub A 2021 *Geom. Integrability & Quantization* **22** 165
[11] Gårding L, Wightman A 1954 *Proc. Nat. Acad. Sci. USA* **40** p617
[12] Monakhov V 2020 *J. of Physics: Conf. Series* **1557** 12031
[13] Monakhov V 2020 *Bull. of Russian Acad. of Sciences: Physics* **84** 1216
[14] Krivskii I Y, Simulik V M 1992 *Theor. Math. Physics* **90** 265
[15] Fushchich V, Nikitin A 1994 *Symmetries of equations of quantum mechanics* (New York: Allerton Press)

R+RG gravity with maximal Noether symmetry

Yu E Pokrovsky

National Research Center "Kurchatov Institute", Moscow, 123182, Russia

E-mail: yury_pokrovsky@mail.ru

Abstract. A Noether symmetric, 3rd order polynomial in the Riemann curvature tensor $R_{\alpha\beta\mu\nu}$ extension of the General Relativity (GR) without cosmological constant (R+RG gravity) is suggested and discussed as a possible fundamental theory of gravity in 4-dimensional space-time with the geometric part of the Lagrangian to be $L_{R+RG} = \frac{\sqrt{-g}}{2\kappa} R\left(1 + \frac{G}{G_P}\right)$. Here $\kappa = \frac{8\pi G_N}{c^4}$ is the Einstein constant, $g = \det\left(g^{\mu\nu}\right)$, $g^{\mu\nu}$ - the metric tensor, G_N - the Newton constant, c - the speed of light, $R = R_{\mu\nu}{}^{\mu\nu}$ - the Ricci scalar, $G = R^2 - 4R_{\mu\nu}R^{\mu\nu} + R_{\alpha\beta\mu\nu}R^{\alpha\beta\mu\nu}$ - the Gauss-Bonnet topological invariant, and G_P – a new constant of the gravitational self-interaction to model the cosmological bounce, inflation, accelerated expansion of the Universe, etc. The best fit to the Baryon Acoustic Oscillations data for the Hubble parameter $H(z)$ at the redshifts $z < 2.36$ leads to $G_P^{1/4} = (0.557 \pm 0.014)Tpc^{-1}$ with the mean square weighted deviation from the data about 3 times smaller than for the standard cosmological (ΛCDM) model. Due to the self-gravitating term $\sim RG$ the respective Einstein equation in the R+RG gravity contains the additional (tachyonic in the past and now) scalar (spin = 0) graviton and the perfect geometric fluid tensor with pressure- and matter- like terms equal to the respective terms in the ΛCDM model at $|z| \ll 1$. Some predictions of this R+RG gravity for the Universe are also done.

1. Introduction

There are three well known fundamental interactions in the Nature: the strong, electroweak, and gravitational interactions defined in the classical limit by the principle of least action with fundamental Lagrangians L being polynomials of degree $n \geq 3$ in the respective fundamental fields [1], except the General Relativity (GR) where contribution of the Riemann curvature tensor $R_{\mu\nu\rho\sigma}$ is a polynomial of the first degree (n=1) only [2]: $L_{GR} = \frac{\sqrt{-g}}{2\kappa} R$, where $\kappa = 8\pi G_N c^{-4}$ is the Einstein constant, $g = \det\left(g^{\mu\nu}\right)$ - the determinant of the metric tensor $g^{\mu\nu}$, G_N - the Newton constant, c - the speed of light, and $R = R_{\mu\nu}{}^{\mu\nu}$ is the Ricci curvature scalar. Other curvature scalars explored in the action may be quadratic in $R_{\mu\nu\rho\sigma}$ [3]-[5] ($R^2, R_{\mu\nu}R^{\mu\nu}, R_{\mu\nu\rho\sigma}R^{\mu\nu\rho\sigma}$), cubic [6]-[7] ($R^3$, $R R_{\mu\nu}R^{\mu\nu}$, $R R_{\mu\nu\rho\sigma}R^{\mu\nu\rho\sigma}$, $R^{\nu}_{\mu}R^{\rho}_{\nu}R^{\mu}_{\rho}$, $R_{\mu\nu\rho\sigma}R^{\mu\rho}R^{\nu\sigma}$, $R_{\mu\nu\rho\sigma}R^{\mu c\nu}{}_{\gamma}R^{\sigma\gamma}$, $R_{\mu\nu}{}^{\rho\sigma}R_{\rho\sigma}{}^{\gamma\beta}R_{\gamma\beta}{}^{\mu\nu}$, $R_{\mu\nu}^{\ \ \rho\sigma}R_{\rho\delta}^{\ \ \gamma\beta}R_{\gamma\delta}^{\ \ \mu\nu}$), etc. In particular, by use of the Gauss-Bonnet topological invariant $G = R^2 - 4R_{\mu\nu}R^{\mu\nu} + R_{\alpha\beta\mu\nu}R^{\alpha\beta\mu\nu}$, a possible extension of GR in order to take into account all curvature components of $R_{\mu\nu\rho\sigma}$ may be realized by the replacement $R \to F(R,G)$ in the Einstein-Hilbert action (see [8] and references therein).

Then there is a problem to constrain the function $F(R,G)$ by necessary physical conditions: In particular, by unnatural singularities, mathematical and physical symmetries, and experimental data. In this way the Noether symmetry is the most important and necessary condition to introduce canonical variables and develop a theory based on a suitable Lagrangian with the respective Hamiltonian and conservation lows (for the physically maximal application of Noether symmetry see [8] and references therein). Polynomiality of this Lagrangian at fundamentally minimal ($n=3$) degree in $R_{\mu\nu\rho\sigma}$, good agreement with experimental data and adequacy to the string theory [9], considered as the leading candidate for a "theory of everything" [10], may lead to the most fundamental theory of gravity, called below as R+RG gravity, suggested and discussed in the next sections of this paper.

2. F(R,G) gravity with the Noether symmetry and polynomiality in Riemann curvature tensor

The most comprehensive study for $F(R,G)$ extension of GR has been done in [8]. In particular, different forms of $F(R,G)$ has been detail classified, and a lot of possible solutions with different Noether symmetries are obtained and presented in [8]. The particular solution $F(R,G) = f_1(G) + R f_2(G)$ has been selected from [8] in this paper as the only possibility for the Noether symmetric extension of GR. Then, assuming that $F(R,G)$ is a polynomial of the minimal fundamental degree ($n=3$) in $R_{\mu\nu\rho\sigma}$, there may be $f_1(G) = -2\Lambda + G/R_P$, and $f_2(G) = G/G_P$ with 3 constants: Λ, R_P, G_P. Because the linear in G term, being the topological invariant, don't contribute to the respective equations of motions in the 4-dimensional space-time, the minimal extension of GR with the full budget of the Riemann curvature tensor components may be of the following form:

$$F(R,G) = -2\Lambda + R\left(1 + \frac{G}{G_P}\right), \tag{1}$$

where Λ is the cosmological constant (note, that, according to the string theory [9], Λ is expected to be zero), and G_P is a new constant of gravitational self-interaction in addition to the Newton constant G_N of gravitational attraction.

The respective Einstein equation for this theory is:

$$R_{ij} - \frac{1}{2} g_{ij} R = \kappa T_{ij}^{Matter} + \Sigma_{ij}, \tag{2}$$

where $\Sigma_{ij} = \kappa\left((\mu + \pi)u_i u_j + \pi g_{ij}\right)$ [11][10] - arises from equation (1). As well, the geometric inflation and bounce [7] are possible in this theory instead of the Big Bang singularity in the standard cosmological (ΛCDM) model.

There are no known rigorous theoretical conditions to exclude Λ (the only theoretical motivations for $\Lambda = 0$ are concerned with the string theory [9]). Note, that at $G_P \to \infty$ equation (1) is the geometric part of the ΛCDM model. On the other hand at $\Lambda = 0$ the $\frac{RG}{G_P}$ term may be approximated by the effective cosmological constant Λ^{eff} and effective Dark Matter (DM) density near the recent time (redshift $z = 0$): $\Lambda^{eff} + \kappa c^2 \rho_{DM}^{eff} (1+z)^3$. Therefore only cosmological data at $z \cdot 1$ would resolve the problem of cosmological constant: $\Lambda = 0$ or not.

3. The R+RG gravity motivated by the cosmological data

In order to compare this theory with cosmological data in assumptions of homogeneous, isotropic, and spatially flat Universe the equation (2) can be written for the spatially flat FRW metrics:

$$ds^2 = -dt^2 + a(t)^2 \left(dx^2 + dy^2 + dz^2\right). \tag{3}$$

Then the 1st Friedmann equation has the following form:
$\dot{a} = aH$,

$$H^2 = \frac{\Lambda}{3} + \frac{\rho_{m0}}{3}\left(\frac{a_0}{a}\right)^3 + \frac{\rho_{r0}}{3}\left(\frac{a_0}{a}\right)^4 + 24\frac{H^6}{G_P}\left(1 - 6\frac{\dot{H}}{H^2} - \frac{\dot{H}^2}{H^4} - 2\frac{\ddot{H}}{H^3}\right), \tag{4}$$

with initial conditions: $a(0) = a_0$, $H(0) = H_0$, $\dot{H}(0) = -(1+q_0)H_0^2$,

where $q_0 = q(0)$, $q(z) = -1 - \frac{\dot{H}}{H^2}$ is the deceleration parameter, ρ_{m0} and ρ_{r0}. are the ordinary matter and radiation densities at $z = 0$.

Numerical calculations of $H(z)$, where $\left(z = -1 + \frac{a_0}{a}\right)$, by use of equati4), and the best fit of G_P, Λ, and H_0, q_0 (at ρ_{m0}, ρ_{r0} taken from [12]) to the cosmological data for $H(z)$ (see Figure 1) lead to the mean square weighted deviation from the data about 3 times smaller than for the ΛCDM model. For the best fit the following values for the fundamental constants $G_P^{1/4} = (0.557 \pm 0.014)Tpc^{-1}$, $\Lambda = (0.0000 \pm 0.0003)G_P^{1/2}$, and the initial conditions $H_0 = (68.15 \pm 1.8) km\, s^{-1} Mpc^{-1} \approx 0.41 c\, G_P^{1/4}$, $q_0 = -0.50 \pm 0.05$ are obtained with the deceleration-acceleration transition predicted to be at $z_{da} = 0.71$. These values are in good agreement with the string theory [9], and the Planck 2018 results [12]: $H_0 = (67.4 \pm 0.5) km\, s^{-1} Mpc^{-1}$, $q_0 = -0.50 \pm 0.05$.

Note, that the obtained value of $\Lambda = 0$ is expected in the string theory [9]. This indicates that the Lagrangian function (1) may be simplified by $\Lambda \equiv 0$, and the new fundamental Lagrangian of the gravity (let us call this new theory of gravity as R+RG gravity) may be only of the form

$$L_{R+RG} = \frac{\sqrt{-g}}{2\kappa} R \left(1 + \frac{G}{G_P}\right). \tag{5}$$

In this case the geometric part of the energy density is

$\mu_{GF} = 72\frac{H^6}{G_P}\left(1 - 6\frac{\dot{H}}{H^2} - \frac{\dot{H}^2}{H^4} - 2\frac{\ddot{H}}{H^3}\right)$, and the geometric part of the pressure is

$$\pi_{GF} = -72\frac{H^6}{G_P}\left(1 - 4\frac{\dot{H}}{H^2} - 9\frac{\dot{H}^2}{H^4} - \frac{2}{3}\frac{\dot{H}^3}{H^6} - 4\left(1 + \frac{2}{3}\frac{\dot{H}}{H^2}\right)\frac{\ddot{H}}{H^3} - \frac{2}{3}\frac{\dddot{H}}{H^4}\right).$$

In this theory the cosmological perfect Geometric Fluid (GF) is well defined by the term $\frac{RG}{G_P}$ with the new constant G_P of gravitational self-interaction. This perfect GF may be an analog of the Quintessence [17] and etc. widely discussed in the literature.

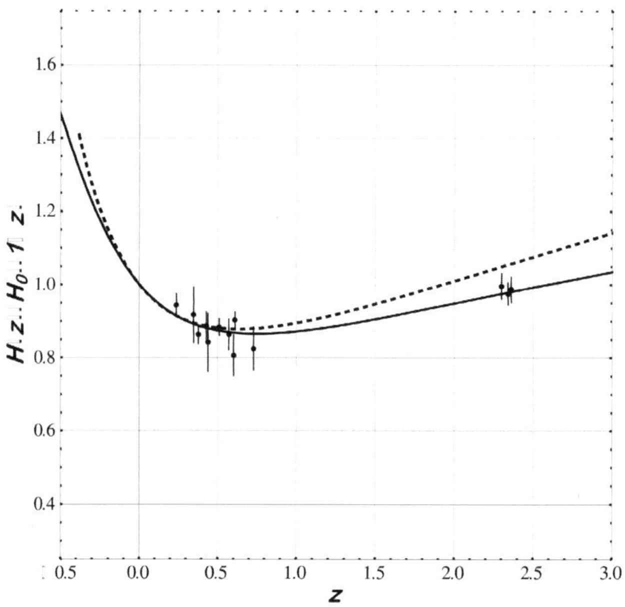

Figure 1. Comparison of R+RG-gravity (solid line) and ΛCDM model (dashed line) with the data selected from [13]-[16] for the Hubble parameter $H(z)$ measured by use the Baryon Acoustic Oscillation (BAO) method, in combination with a Cosmic Microwave Background (CMB) prior on the sound horizon scale (in respect with the minimal experimental errors).

Note that $H(z)$ in the R+RG gravity is very close to the ΛCDM model at $z \approx 0$ (Figure 1). Moreover, ΛCDM model can be derived as an exact limit of R+RG gravity at $|z| \ll 1$.

The R+RG gravity contain one massless tensor (spin=2) graviton, and one massive scalar (spin = 0) graviton. Square of the mass of the scalar graviton in the R+RG gravity is derived of the following form $M_{SG}^2 = -3H^2\left(1+\frac{1}{2}\frac{\dot{H}}{H^2}\right)^{-1}\left(1+\frac{\dot{H}}{H^2}+\frac{1}{3}\frac{\dot{H}^2}{H^4}-\frac{1}{72}\frac{G_P}{H^4}\right)$. It depends on z, is negative in the past and now (tachyon), and it is predicted to became massless in the nearest cosmological future with $M_{SG}^2 > 0$ later. This may lead to significant strengthening of gravitational attraction in future with attendant effects and phenomena.

4. Conclusion

The new fundamental Lagrangian of the pure geometric gravity (the R+RG gravity) $L_{R+RG} = \frac{\sqrt{-g}}{2\kappa} R\left(1+\frac{G}{G_P}\right)$ is suggested and discussed as possibly the most fundamental extension of the General Relativity. It is shown that the standard cosmological (ΛCDM) model is the very good approximation of the R+RG gravity at the small $|z| \ll 1$ only. As well the R+RG gravity predicts that there is tachyonic scalar graviton $M_{SG}^2 < 0$ that became massless in the nearest cosmological future with $M_{SG}^2 > 0$ later. Instead of the Big Bang singularity, the geometric inflation and bounce are possible in the R+RG gravity, and their detailed study is the subject of the future papers.

References

[1] Bansal S, Evnin O and Mkrtchyan K 2021 Polynomial duality-symmetric lagrangians for free p-forms *Eur. Phys. J. C* **81** 257
[2] Palese M and Vitolo R 2001 On a class of polynomial Lagrangians *arXiv:math-ph* 0111019
[3] Stelle K S 1977 Renormalization of higher-derivative quantum gravity *Phys. Rev. D* **16** 953
[4] Stelle K S 1978 Classical Gravity with Higher Derivatives *Gen. Rel. Grav.* **9** 353
[5] Starobinski A A 1980 A new type of isotropic cosmological models without singularity *Phys. Lett. B* **91** 99
[6] Bueno P and Cano P A 2016 Einsteinian cubic gravity *Phys. Rev. D* **94** 104005
[7] Arciniegaa G,.Edelsteina J D and Jaimeb L G 2020 Towards geometric inflation: The cubic case *Phys. Lett. B* **802** 135272
[8] Dialektopoulos K F and Capozziello S 2018 Noether Symmetries as a geometric criterion to select theories of gravity *Int. J. Geom. Methods Mod. Phys.* **15** 1840007
[9] Mukhi S 2011 String theory: a perspective over the last 25 years *Class. Quant. Grav.* **28** 153001
[10] Hawking S W 2002 *The Theory of Everything: The and Fate of the Universe Origin* (New Millennium Press)
[11] Capozziello S, Mantica S A and Molinari L G 2019 Cosmological perfect fluids in Gauss–Bonnet gravity *Int. J. Geom. Methods Mod. Phys.* **16** 1950133
[12] Planck Collaboration Aghanim N, Akrami Y, Ashdown M *et al* 2020 Planck 2018 results VI Cosmological parameters *A&A* **641**
[13] Farooq O, Madiyar F R, Crandall S and Ratra B 2017 Hubble parameter measurement constraints on the redshift of the deceleration-acceleration transition, dynamical dark energy, and space curvature *ApJ* **835** 26
[14] BOSS Collaboration Alam S, Ata M, Bailey S *et al* 2017 The clustering of galaxies in the completed SDSS-III Baryon Oscillation Spectroscopic Survey: cosmological analysis of the DR12 galaxy sample *MNRAS* **470** 2617
[15] Gaztañaga E, Cabré A and Hui L 2009 Clustering of luminous red galaxies – IV Baryon acoustic peak in the line-of-sight direction and a direct measurement of H(z) *MNRAS* **399** 1663
[16] SDSS-III Collaboration Delubac T, Bautista J E, Busca N G *et al* 2015 Baryon acoustic oscillations in the Lyα forest of BOSS DR11 quasars *A&A* **574** A59
[17] Tsujikawa S 2013 Quintessence: a review *Class. Quantum Grav.* **30** 214003

Journal of Physics: Conference Series 2081 (2021) 012029 doi:10.1088/1742-6596/2081/1/012029

Multiworld motives by closed time-like curves

A V Shepelin[1], A M Rostom[2], V A Tomilin[2,†] and L V Il'ichov[2]

[1] Novosibirsk State University, 630090 Novosibirsk, Russia

[2] Institute of Automation and Electrometry SB RAS, 630090 Novosibirsk, Russia

E-mail: † 8342tomilin@gmail.com

Abstract. We propose a new model, entitled S-CTC, for description of quantum systems in the presence of CTC – closed time-like curves. The model is based on the viewpoint on any quantum state as an observer's state of knowledge of the system preparation procedure. We compare and contrast our S-CTC model with D-CTC and P-CTC models and show that S-CTC shares special quantum features with both D-CTC and P-CTC. As far as the interaction of the quantum system with itself coming from the future concerns, S-CTC is formally equivalent to P-CTC. On the other hand, when calculating outcome probabilities for a measurement within the time interval between the entrance and exit of CTC, S-CTC becomes equivalent to D-CTC. Both these models require the concept of alternative realities (worlds) where different measurement outcomes are recorded and alternative connections of these realities by CTC.

1. Introduction

Presently, the key scientific approach to reality consists of two drastically different fundamental theories: quantum physics and geometrical physics (General Relativity). One of the main problems nowadays is incorporating these theories into a single framework. In search of such a framework, the 'borderline' problems, e.g. field quantization near the black hole's event horizon [1], that require instruments of the both theories can be regarded as a kind of 'metatheoretical colliders' for Quantum Mechanics and General Relativity. Quantum models with closed time-like curves (CTC) focus attention on this point in perhaps the best way. The assumption of existence of particles with closed world lines is supported by the well-known solution of Einstein's equations, e.g. the Gödel's solution for rotating Universe [2].

However, the existence of CTC in the real world has long been a controversial issue. In general, quantum models with causal loops could potentially strengthen the scepticism if their predictions were in strong disagreement with the well-established views. The value of quantum physics as a referee regarding this question depends on unambiguity and correctness of its application to quite a non-standard problem that causes paradoxes even in its classical version [3]. Quantum theory in presence of CTC is ambiguous, which is evident from existence of different approaches to the problem. The D-CTC approach [4] developed by Deutsch is perhaps the most well-known. He suggested the space-time setup that became a de facto standard for dealing with quantum aspects of CTC. Fig.1 shows a part of a 2D Minkowski space-time with an attached 'corridor' (not shown on the scheme) connecting the lower edge of the 'entrance' cut at $t = \tau$ with the upper edge of the 'exit' cut at $t = \tau$. The latter is in the absolute past (in terms of causal structure of Minkowski space) with regards to the 'entrance', which potentially allows closing particles' world lines, thus forming CTC. We will further use abbreviation TM (time machine) as a synonym of CTC. Hence, a particle entering TM at $t = \tau$ almost instantly (according to its own clock) appears at its past at $t = 0$.

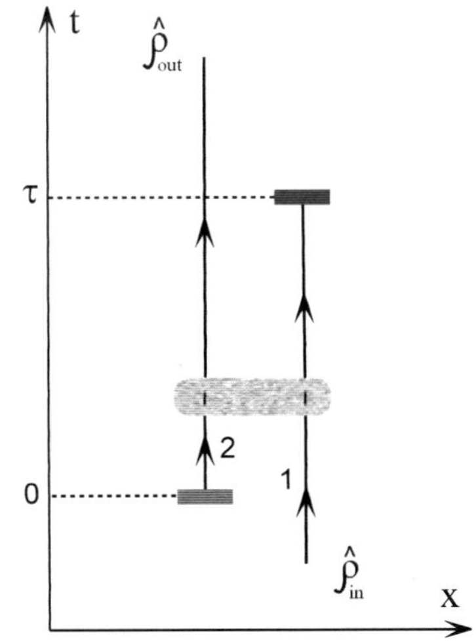

Figure 1. The space-time scheme of quantum models with CTC due to Deutsch.

The distinctive feature of the Deutsch's approach is ontic view on the state of the quantum system. Its density matrix appears in the model as a real physical entity for which local self-consistency conditions hold. Deutsch's model is based on a boundary condition that identifies the exit density matrix with the one at the entrance of TM. This matrix appears to be a solution of a consistency equation and depends nonlinearly on the initial density matrix of the system, $\hat{\varrho}_{in}$, prepared in the absolute past of the TM. Consequently, $\hat{\varrho}_{out}$, the state in the absolute future of TM, also gets a nonlinearly dependence on $\hat{\varrho}_{in}$. This dependence, non-trivial and unusual for quantum mechanics, is due to interaction between the system and its version that moved through CTC during the time interval $(0, \tau)$. The space-time region of interaction is shown at Fig.1 (gray). Note that it is the interaction that gives rise to the classical TM paradoxes [3, 5].

Recently, a model known as P-CTC was proposed [6, 7], in which the particle's state is transferred to the past via quantum teleportation. In Fig.2, the preparation of entangled pair of particles (2,3) plays the role of the TM's exit, while the entrance is represented by post-selection of the pair (1,3) in the same entangled state. The dashed fragments closing the world lines of the particles allow interpreting the transfer to the past upon successful post-selection (P-CTC) as a one-particle evolution – the world line of particle 3 is considered to be the world line of particle 1 moving backwards in time.

Both D-CTC and P-CTC predict non-trivial possibilities for processing quantum information. Despite the highly vague feasibility perspectives, TMs are important for information theory (both classical and quantum) by clarifying the nature of its tools [8]. For example, it was proved that the classical computer with access to the D-CTC resource is able to solve PSPACE-complete problems [9]. Other notable investigations of CTC in quantum information theory include development of novel communication protocols [10], information encoding [11], studies of state cloning [12].

The P-CTC model compared to D-CTC does not offer exceptional possibilities for quantum information processing. P-CTC demonstrates a more 'mild' nonlinear dependence of $\hat{\varrho}_{out}$ on $\hat{\varrho}_{in}$. On the other hand, as noted above, the quantum teleportation and, consequently, P-CTC, lacks real change of the space-time topology.

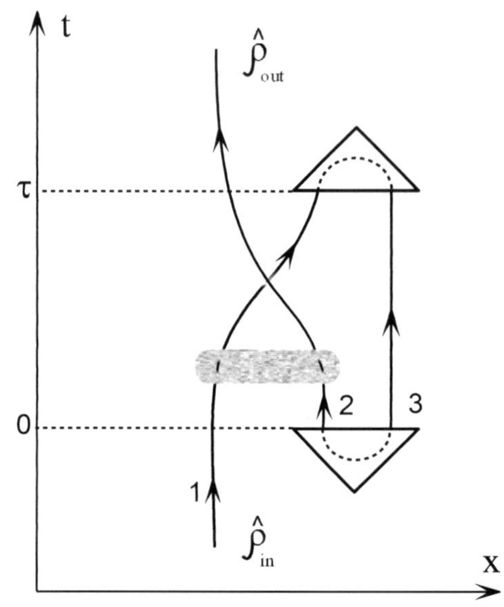

Figure 2. The scheme of P-CTC model.

The fact that P-CTC is less radical than D-CTC does not give advantage to the latter. It was shown that in D-CTC quantum states cannot be purified, while in P-CTC this is possible [13]. Moreover, unlike D-CTC, P-CTC is compatible with the formalism of path integrals [14], which allows it to effectively describe quantum fields in the curved space-time.

D-CTC and P-CTC were examined in terms of the so-termed quantum causal analysis [15]. This shed light on intrinsic character of the types of self-consistency in these models. These types appeared quite different in their nature.

In the present work, we propose a novel quantum model with CTC, its features being in a certain sense in between those of D-CTC and P-CTC. The new model resembles D-CTC by assuming a non-trivial topology of space-time. However, its nonlinearity and the way it treats the effects of interaction between the system and its older version coming through CTC are similar to P-CTC. On the contrary, when describing a measurement in between the entrance and exit of the CTC loop, the new model shares attributes with D-CTC. In both these cases, one needs accounting alternative realities with different measurement outcomes and averaging over different connections of these realities via CTC.

2. Model

Two guiding principles are invoked in order to formulate the new model of quantum evolution in the presence of CTC. In the first place is the identity condition for inner state of the system entering TM at $t = \tau$ and exiting it at $t = 0$. In D-CTC it is reflected in a special self-consistency equation. Its solution yields an explicit form of $\hat{\varrho}_{CTC}$ at the entrance and at the exit as well. Our second guiding consideration is the radically different interpretation of the quantum state notion, in which the quantum state is treated as a state of knowledge possessed by a dweller of Minkowski space-time. Such an observer, who prepared the system 1, has no knowledge of the past of the system 2. These two points may serve as a foundation for a formal theoretical model.

Let us note that the Dirac notations for ket-vector $|\Psi\rangle$ (and bra-vector $\langle\Phi|$) imply a certain time direction. The ket-vector symbolizes preparation of the system state, and dynamical transformation of this knowledge is given by unitary evolution operators written to the left. This allows associating a ket-vector with the direction of time flow from right to left, and vice versa for bra-vector[1]. A somewhat similar idea involving two mutually opposite time directions $\langle\Phi|\,|\Psi\rangle$ forms the basis of time-symmetric quantum formalism [17].

The same considerations allow to 'juxtapose' the two opposite time directions starting from the same time moment by an arbitrary operator, for instance by the unit operator

$$\hat{1} = \sum_k |k\rangle\langle k|. \tag{1}$$

If the preparation procedure is completely unknown, the system state is given by the unit operator. It is then reasonable to use it for the system exiting TM at $t = 0$. The red color for ket and the blue one for bra in (1) show their correspondence to different time moments. It is convenient to represent the resulting picture (in the absence of interaction) horizontally (Fig. 3).

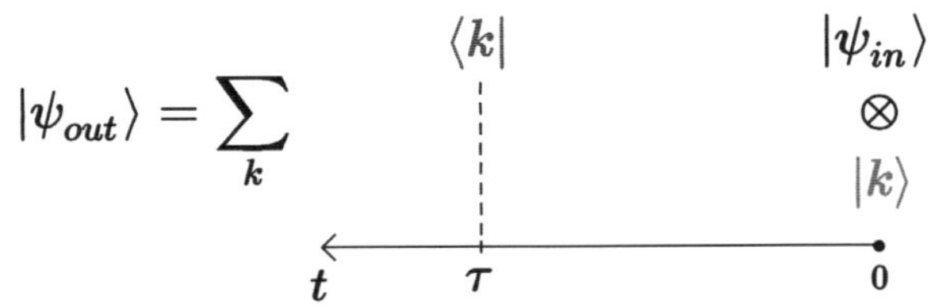

Figure 3. The splitting of the ket-bra structure of (1) illustrating relation between $|\psi_{in}\rangle$ and $|\psi_{out}\rangle$ in the absence of interaction.

By placing ket- and bra-parts of (1) in different sides of (3), both of the guiding principles of our theory are satisfied: on the one hand, the observer is of total ignorance with regard to the preparation of the second system's state; on the other hand, an explicit coincidence of the entrance and exit states of TM is granted. The initially indivisible essence, (1) is now split into two parts. For this reason the new model will from now on be called S-CTC ('S' stands for 'splitting').

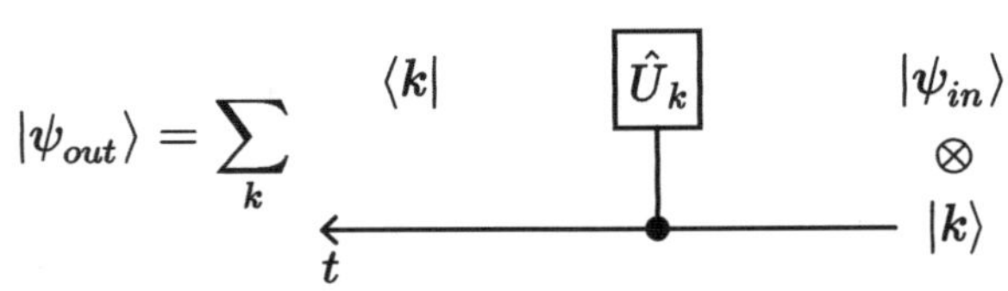

Figure 4. The scheme of Fig. 3 added with interaction between the system and its later version (system 2) via controlled operation represented by a set of unitaries $\hat{U}_k$.

Quantum evolution in the presence of CTC with interaction between 1 and 2 has several peculiarities. Consider a special case of a unitary operation applied to 1 and controlled by the state of 2. Such an interaction, CONTROL-U, is typical for quantum information. It is defined in a a fixed 'interaction' basis of 2. For simplicity, we will use the same basis for (1) when

[1] We are using the visual representation of quantum processes developed in [16].

studying the scheme presented in Fig. 4, but the result is eventually insensitive to the choice of basis. It follows that

$$|\psi_{out}\rangle = \sum_k |k\rangle\langle k|\hat{U}_k|\psi_{in}\rangle. \tag{2}$$

Introducing operator

$$\hat{W} \doteq \sum_k |k\rangle\langle k|\hat{U}_k, \tag{3}$$

we get

$$\langle\psi_{in}|\hat{W}^\dagger\hat{W}|\psi_{in}\rangle = \langle\psi_{out}|\psi_{out}\rangle. \tag{4}$$

This is the probability that the act of interaction will happen, or, equivalently, CTC get realised (the probability of system 2 to appear).

The result of interaction between 1 and 2 in P-CTC appears identical to (2). Indeed, quantum teleportation scenario implies preparation of entangled state of 2 and 3, in addition to $|\psi_{in}\rangle_1$:

$$|\Psi\rangle_{2,3} = \sum_k |k\rangle_2 \otimes |k\rangle_3, \tag{5}$$

followed by post-selection of 1 and 3 in the same state $|\Psi\rangle_{1,3}$. In the case of successful post-selection,

$$|\psi_{out}\rangle_2 = \sum_{k,k'} \Big(\langle k'|_1 \otimes \langle k'|_3\Big)\Big(\hat{U}_k|\psi_{in}\rangle_1 \otimes |k\rangle_2 \otimes |k\rangle_3\Big) = \sum_k \langle k|\hat{U}_k|\psi_{in}\rangle \otimes |k\rangle_2, \tag{6}$$

which is exactly the same as (2).

In what follows, we use the simplest qubit model for systems ($k = 0, 1$) and controlled phase shift as interaction. In this case $\hat{U}_0 = \hat{1}$ and

$$\hat{U}_1 = |0\rangle\langle 0| + e^{\imath\varphi}|1\rangle\langle 1|. \tag{7}$$

Then $\hat{W} = \hat{U}_1$ and CTC realization probability is 1.

3. Measurement in CTC

Now there is a measurement of qubit 2 after it exits TM but before it interacts with qubit 1. Eigenstates $|\psi_\sigma\rangle$ ($\sigma = \pm$) of the measured observable are related to the 'interaction basis' via amplitudes α and β:

$$|\psi_+\rangle = \alpha|0\rangle + \beta|1\rangle, \;\; |\psi_-\rangle = -\beta^*|0\rangle + \alpha^*|1\rangle, \tag{8}$$

where $|\alpha|^2 + |\beta|^2 = 1$.

We are interested in the probabilities $p(\sigma)$ of measurement outcomes. Let us first calculate them according to properly modified D-CTC self-consistency conditions for states $\hat{\varrho}_{CTC}$ at the entrance and exit of TM [4]. We denote these states by $\hat{\varrho}_\pm$, relating them to possible measurement outcomes:

$$p_I(\sigma)\hat{\varrho}_\sigma(I) = \sum_{k=0,1} \langle k|\hat{P}_\sigma\hat{\varrho}_\sigma(I)\hat{P}_\sigma|k\rangle\hat{U}_k\hat{\varrho}_{in}\hat{U}_k^\dagger. \tag{9}$$

Here $\hat{P}_\sigma = |\psi_\sigma\rangle\langle\psi_\sigma|$. We introduced the additional index I which implicitly reflects the way of making the states consistent. Specifically, the consistency condition is satisfied independently for each of two alternative realities (worlds $W_\pm$), in which the measurement outcomes are recorded (Fig.5).

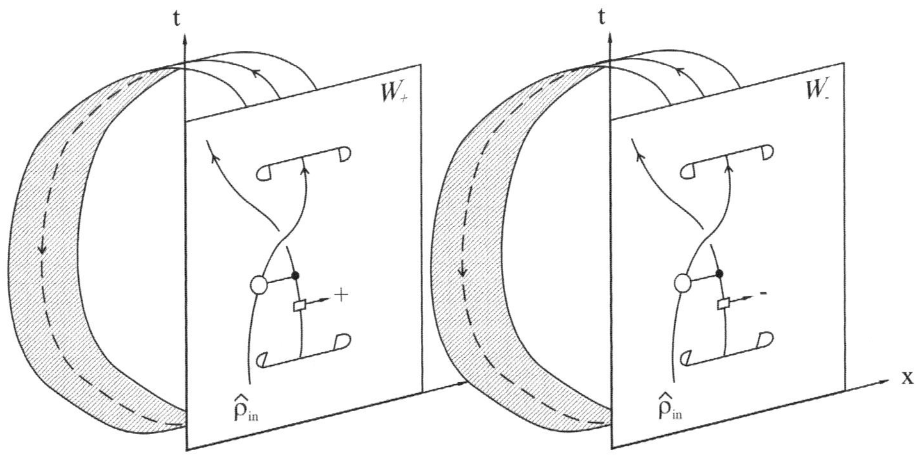

Figure 5. Two alternative worlds $W_\pm$ with CTC where measurement outcomes $\pm$ are recorded (connection by the variant I).

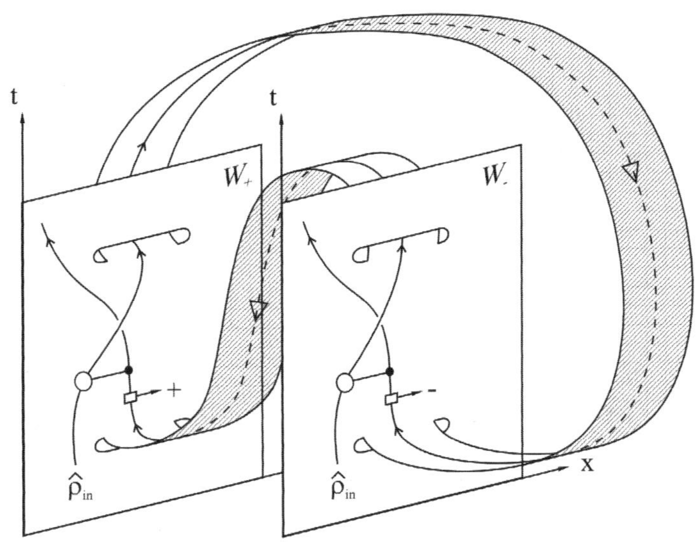

Figure 6. The same worlds connected by the variant *II*.

It appears that for the outcome probabilities $p_I(\pm)$ there is a problem: generally $p_I(+) + p_I(-) \neq 1$, i.e. $p_I(+)$ and $p_I(-)$ can not be viewed as probabilities. One can try a different way to satisfy consistency condition (variant *II*). In this case the corresponding equations are coupled(see Fig.6):

$$p_{II}(+)\hat{\varrho}_-(II) = \sum_{k=0,1} \langle k|\hat{P}_+\hat{\varrho}_+(II)\hat{P}_+|k\rangle \hat{U}_k\hat{\varrho}_{in}\hat{U}_k^\dagger, \tag{10}$$

$$p_{II}(-)\hat{\varrho}_+(II) = \sum_{k=0,1} \langle k|\hat{P}_-\hat{\varrho}_-(II)\hat{P}_-|k\rangle \hat{U}_k\hat{\varrho}_{in}\hat{U}_k^\dagger.$$

Again, $p_{II}(+) + p_{II}(-) \neq 1$. However,

$$\frac{1}{2}\Big(p_I(+) + p_{II}(+)\Big) + \frac{1}{2}\Big(p_I(-) + p_{II}(-)\Big) = 1. \tag{11}$$

This suggests interpreting the two terms in (11) as measurement outcome probabilities from the point of view of the observer – dweller of Minkowski space:

$$Prob^{(\sigma)}_{D-CTC} = \frac{1}{2}\Big(p_I(\sigma) + p_{II}(\sigma)\Big). \tag{12}$$

The observer's complete ignorance of the connection variant used is reflected by 1/2 probabilities.

Let us now consider outcome probabilities in S-CTC. Like in D-CTC, a similar dilemma of connecting entrances and exits of TM in the worlds W_+ and W_- arises. The scheme for the

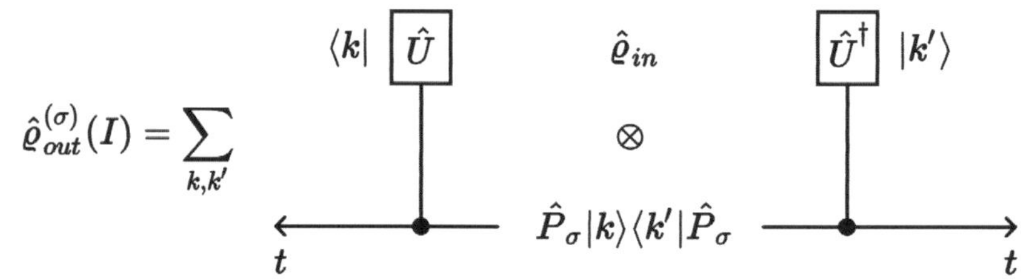

Figure 7. The scheme of measurement in S-CTC with connection by the variant I.

variant I is illustrated in Fig.7 and yields

$$\hat{\varrho}^{(\sigma)}_{out}(I) = \sum_{k,k'} |k\rangle\langle k|\hat{P}_\sigma \hat{U}_k \hat{\varrho}_{in} \hat{U}^\dagger_{k'} \hat{P}_\sigma |k'\rangle\langle k'| \tag{13}$$

with the quasi-probabilities similar to D-CTC:

$$Tr\, \hat{\varrho}^{(\sigma)}_{out}(I) = p_I(\sigma). \tag{14}$$

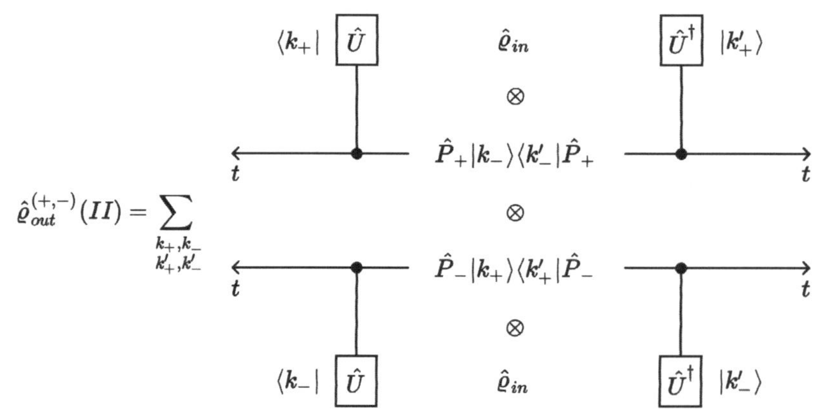

Figure 8. The scheme of measurement in S-CTC with connection by the variant II.

For connection by the variant II, there is a correlated state for the worlds W_+ and W_- (Fig. 8)

$$\hat{\varrho}_{out}^{(+,-)}(II) = \sum_{k_+,k'_+}\sum_{k_-,k'_-} |k_+\rangle\langle k_+|\hat{P}_+\hat{U}_{k_-}\hat{\varrho}_{in}\hat{U}^\dagger_{k'_-}\hat{P}_+|k'_+\rangle\langle k'_+| \otimes |k_-\rangle\langle k_-|\hat{P}_-\hat{U}_{k_+}\hat{\varrho}_{in}\hat{U}^\dagger_{k'_+}\hat{P}_-|k'_-\rangle\langle k'_-|. \tag{15}$$

By tracing, one gets the product of the same quasi-probabilities which appeared in (10):

$$Tr\,\hat{\varrho}_{out}^{(+,-)}(II) = p_{II}(+)p_{II}(-). \tag{16}$$

One can introduce density matrices $\hat{\varrho}_{out}^{(\sigma)}(II)$ according to

$$\hat{\varrho}_{out}^{(+)}(II) \doteq Tr_2\frac{\hat{\varrho}_{out}^{(+,-)}(II)}{p_{II}(-)};\ \ \hat{\varrho}_{out}^{(-)}(II) \doteq Tr_1\frac{\hat{\varrho}_{out}^{(+,-)}(II)}{p_{II}(+)} \tag{17}$$

which ensures the coincidence between S-CTC and D-CTC with respect to outcome probabilities for both of the connection types.

The expression (12) can be generalised up to

$$\hat{\varrho}_{out}^{(\sigma)} = \frac{1}{2}\left(\hat{\varrho}_{out}^{(\sigma)}(I) + \hat{\varrho}_{out}^{(\sigma)}(II)\right), \tag{18}$$

thus introducing the final state of the qubit. In contrast to the outcome probabilities the final states in D-CTC and S-CTC prove to be quite different.

4. Joint probability distribution

Consideration of purely quantum-mechanical effects, such as quantum entanglement, can provide an understanding of a deeper connection between the theories under study.

So, we now move to a scheme which involves two-party measurements performed by experimenters Alice (A) and Bob (B). As before, one can not dispense with the two topologies (Fig. 9, 10). The experimenters prepare a pair of particles in the entangled state $|\psi_i\rangle$, and hold one of the particles for themselves. Also, both researchers have measuring devices ($\hat{P}_a$ and $\hat{P}_b$), and Alice additionally has an access to CTC with the $\hat{U}$ interaction. It is proposed to find joint outcome probabilities in this scheme and determine their nature (whether they are full-fledged probabilities or pseudo-probabilities).

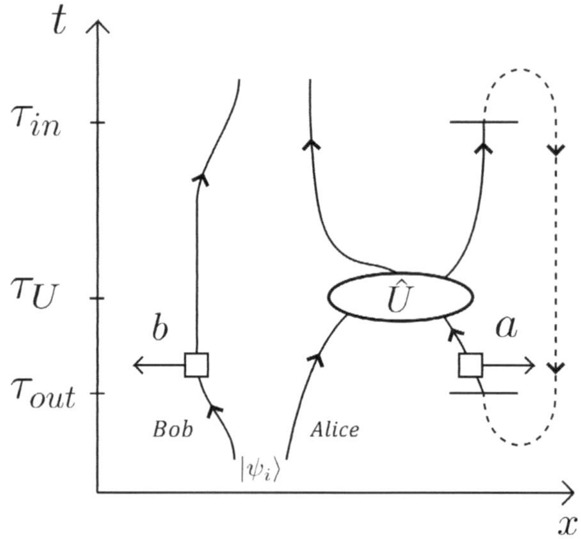

Figure 9. Physical scheme of the CTC experiment with a joint probability distribution (topology I).

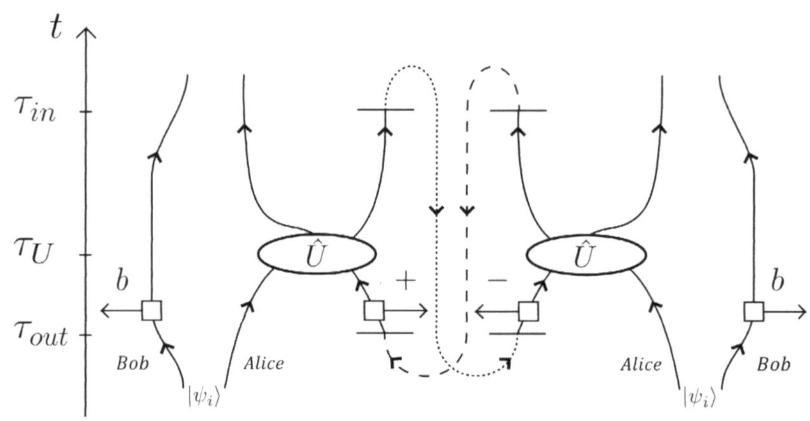

Figure 10. Physical scheme of the CTC experiment with a joint probability distribution (topology II).

For certainty, we will assume that the entangled particles are in a singlet state

$$|\psi_i\rangle = \frac{1}{\sqrt{2}} \left(|0\rangle_A \otimes |1\rangle_B - |1\rangle_A \otimes |0\rangle_B \right). \tag{19}$$

The Alice's and Bob's measurement settings are denoted by α and β. The corresponding measurement bases are

$$|+;\alpha\rangle = \alpha_0|0\rangle_A + \alpha_1|1\rangle_A, \tag{20}$$

$$|-;\alpha\rangle = -\alpha_1^*|0\rangle_A + \alpha_0^*|1\rangle_A, \tag{21}$$

$$|+;\beta\rangle = \beta_0|0\rangle_B + \beta_1|1\rangle_B, \tag{22}$$

$$|-;\beta\rangle = -\beta_1^*|0\rangle_A + \beta_0^*|1\rangle_B, \tag{23}$$

where $\alpha_{0,1}$, $\beta_{0,1}$ – complex amplitudes. The outcomes of the measurements will be denoted by $a = \pm$ and $b = \pm$. Then, the corresponding projectors are expressed in the standard way:

$$\hat{P}_\sigma(\varepsilon) = |\sigma;\varepsilon\rangle\langle\sigma;\varepsilon|, \tag{24}$$

where $\sigma = \pm$, $\varepsilon \in \{\alpha, \beta\}$. As before, we will choose the interaction as a controlled phase shift (7) and will be interested in the joint outcome probabilities $p(a, b|I)(\alpha, \beta)$ and $p(a, b|II)(\alpha, \beta)$ (here the explicit dependence on the measurement settings is shown).

Equivalent computational schemes for the S-CTC model are shown in the Figures 11, 12.

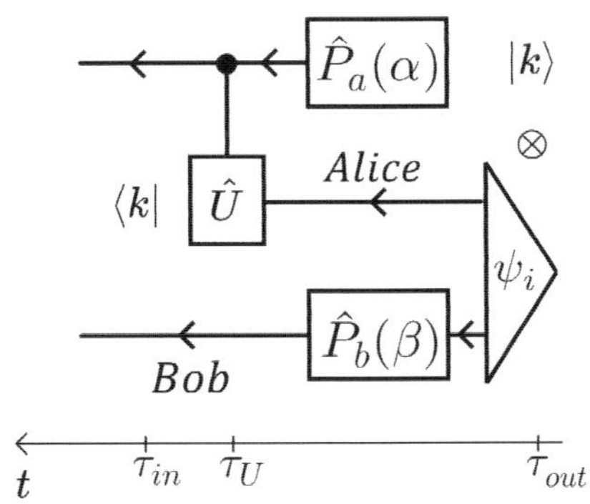

Figure 11. Computational scheme of the S-CTC experiment with two measurements (topology I).

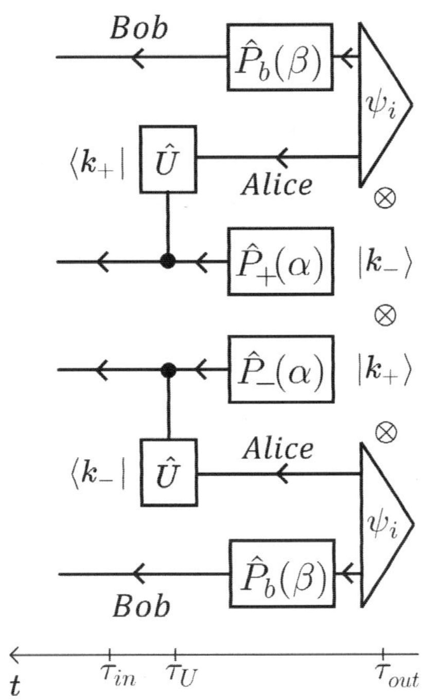

Figure 12. Computational scheme of the S-CTC experiment with with two measurements (topology II).

Based on them, we can now write the final states:

$$|\psi_{f,I}^{(a,b)}(\alpha,\beta)\rangle = \Big(|k\rangle_A\langle k|\hat{P}_a(\alpha)\hat{U}_k \otimes \hat{P}_b(\beta)\Big)\,|\psi_i\rangle, \tag{25}$$

$$\begin{aligned}|\psi_{f,II}^{(+,-,b)}(\alpha,\beta)\rangle = &\Big(|k_+\rangle_A\langle k_-|\hat{P}_-(\alpha)\hat{U}_{k_+} \otimes \hat{P}_b(\beta)\Big)\,|\psi_i\rangle\otimes\\ &\otimes\Big(|k_-\rangle_A\langle k_+|\hat{P}_+(\alpha)\hat{U}_{k_-} \otimes \hat{P}_b(\beta)\Big)\,|\psi_i\rangle,\end{aligned} \tag{26}$$

where $|\psi_{f,I}^{(a,b)}(\alpha,\beta)\rangle$ is the final two-qubit state for the first topology I, $|\psi_{f,II}^{(+,-,b)}(\alpha,\beta)\rangle$ is the final four-qubit state for the second topology II, the symbols $+$ and $-$ refer to Alice's outcomes.

Desired values $p_I(a,b|\alpha,\beta)$ and $p_{II}(a,b|\alpha,\beta)$ expressed in terms of (25) and (26), consequently:

$$p_I(a,b|\alpha,\beta) = \langle\psi_{f,I}^{(a,b)}(\alpha,\beta)|\psi_{f,I}^{(a,b)}(\alpha,\beta)\rangle \tag{27}$$

$$p_{II}(+,b|\alpha,\beta)\cdot p_{II}(-,b|\alpha,\beta) = \langle\psi_{f,II}^{(+,-,b)}(\alpha,\beta)|\psi_{f,II}^{(+,-,b)}(\alpha,\beta)\rangle. \tag{28}$$

The expression (27) is simplified to the form

$$\begin{aligned}&p_I(a,b|\alpha,\beta) =\\ &= \frac{1}{2}\Big[\langle 0|\hat{P}_a(\alpha)|0\rangle\langle 1|\hat{P}_b(\beta)|1\rangle + \langle 1|\hat{P}_a(\alpha)|1\rangle\langle 0|\hat{P}_b(\beta)|0\rangle -\\ &- \langle 0|\hat{P}_a(\alpha)|1\rangle\Big(\langle 0|\hat{P}_a(\alpha)|0\rangle + e^{i\varphi}\langle 1|\hat{P}_a(\alpha)|1\rangle\Big)\langle 1|\hat{P}_b(\beta)|0\rangle -\\ &- \langle 1|\hat{P}_a(\alpha)|0\rangle\Big(\langle 0|\hat{P}_a(\alpha)|0\rangle + e^{-i\varphi}\langle 1|\hat{P}_a(\alpha)|1\rangle\Big)\langle 0|\hat{P}_b(\beta)|1\rangle\Big].\end{aligned} \tag{29}$$

The expression $p_{II}(a,b|\alpha,\beta)$ is a decomposition of (28), finding which in the general case is not a trivial task. However, it is possible to notice some regularity in the pseudo-probabilities obtained earlier and 'guess' the proper structure, using (29):

$$\begin{aligned}&p_{II}(a,b|\alpha,\beta) =\\ &= \frac{1}{2}\Big[\langle 0|\hat{P}_a(\alpha)|0\rangle\langle 1|\hat{P}_b(\beta)|1\rangle + \langle 1|\hat{P}_a(\alpha)|1\rangle\langle 0|\hat{P}_b(\beta)|0\rangle -\\ &- \langle 0|\hat{P}_a(\alpha)|1\rangle\Big(\langle 1|\hat{P}_a(\alpha)|1\rangle + e^{i\varphi}\langle 0|\hat{P}_a(\alpha)|0\rangle\Big)\langle 1|\hat{P}_b(\beta)|0\rangle -\\ &- \langle 1|\hat{P}_a(\alpha)|0\rangle\Big(\langle 1|\hat{P}_a(\alpha)|1\rangle + e^{-i\varphi}\langle 0|\hat{P}_a(\alpha)|0\rangle\Big)\langle 0|\hat{P}_b(\beta)|1\rangle\Big].\end{aligned} \tag{30}$$

A bit longer evaluation in D-CTC model confirms this result. One can also check that the said joint distributions are probabilities indeed:

$$\sum_{a,b} p(a,b|\tau)(\alpha,\beta) = 1 \tag{31}$$

for $\tau = I, II$ and arbitrary choice of α and β, thus being free from inconsistencies discussed above.

From the joint probabilities distributions the one-party (Bob) probability distributions are readily obtained:

$$\pi_\mathcal{T}(b|\alpha,\beta) = \sum_a p_\mathcal{T}(a,b|\alpha,\beta), \tag{32}$$

$$\begin{aligned}\pi_I(b|\alpha,\beta) = &-\frac{1}{2}\langle 1|\hat{P}_b(\beta)|0\rangle \left(|\alpha_0|^2 - |\alpha_1|^2\right)\alpha_0\alpha_1^* \left(1-e^{i\varphi}\right) - \\ &-\frac{1}{2}\langle 0|\hat{P}_b(\beta)|1\rangle \left(|\alpha_0|^2 - |\alpha_1|^2\right)\alpha_1\alpha_0^* \left(1-e^{-i\varphi}\right) + \frac{1}{2},\end{aligned} \tag{33}$$

$$\begin{aligned}\pi_{II}(b|\alpha,\beta) = &-\frac{1}{2}\langle 1|\hat{P}_b(\beta)|0\rangle \left(|\alpha_1|^2 - |\alpha_0|^2\right)\alpha_0\alpha_1^* \left(1-e^{i\varphi}\right) - \\ &-\frac{1}{2}\langle 0|\hat{P}_b(\beta)|1\rangle \left(|\alpha_1|^2 - |\alpha_0|^2\right)\alpha_1\alpha_0^* \left(1-e^{-i\varphi}\right) + \frac{1}{2}.\end{aligned} \tag{34}$$

Expressions (33) and (34) indicate the presence of the effect of 'superluminal telegraph' – Alice, choosing the type of her measurement, has the opportunity to influence the outcome of Bob's measurement. Due to the non-physical nature of such an effect, it is necessary to find the conditions for 'restoring physicality'. And this turns out to be possible when taking into account two topologies with equal weights:

$$\frac{1}{2}\pi_I(b|\alpha,\beta) + \frac{1}{2}\pi_{II}(b|\alpha,\beta) = \frac{1}{2}. \tag{35}$$

That is, the randomness of the implementation of a particular topology deprives Alice of the opportunity to influence the statistics of measurement outcomes in Bob's laboratory.

5. Discussion and Conclusion

The comparative study of outcome probabilities in CTC models can not be complete without considering the case of P-CTC. Assume that in Fig.2, immediately after preparation of $|\Psi\rangle_{2,3}$ at $t = 0$, the qubit 2 is subject to the measurement in the basis (8). In P-CTC there is neither need nor possibility to take into account the type II of connecting $W_\pm$. For this reason $\hat{\varrho}_{out}^{(\sigma)}$ relates to $\hat{\varrho}_{in}$ by (13). Outcome probabilities are given by the ABL rule [18]. Since the trace of (13) is equal to $p_I(\sigma)$, one gets

$$Prob_{P-CTC}^{(\sigma)} = \frac{p_I(\sigma)}{p_I(+) + p_I(-)}. \tag{36}$$

Dissimilarity from (12) is evident.

As previously noted, the quantities $p(\sigma|I)$ and $p(\sigma|II)$ are not probabilities in general. We suppose, however, that they may enter Bayesian expressions. *A priori*, the variants I and II are equiprobable for the external observer. Upon learning the measurement outcome σ, the probabilities get updated:

$$p(I|\sigma) = \frac{p_I(\sigma)}{p_I(\sigma) + p_{II}(\sigma)}, \quad p(II|\sigma) = \frac{p_{II}(\sigma)}{p_I(\sigma) + p_{II}(\sigma)}. \tag{37}$$

The new probabilities find application if the same TM is used in experiment with another qubit B along with the measured qubit A. The B system is not measured, but it interacts with its later version as pictured in Fig.1. Let it be a controlled phase shift $\hat{V}$ as for the qubit A. The final state of B in S-CTC model is

$$\hat{\varrho}_{B-out}(I) = \sum_{k,k'} |k\rangle\langle k|\hat{V}_k \hat{\varrho}_{B-in} \hat{V}_{k'}^\dagger |k'\rangle\langle k'| \tag{38}$$

for the first variant and

$$\hat{\varrho}_{B-out}(II) = \sum_{k,k',l} |k\rangle\langle k|\hat{V}_l\hat{\varrho}_{B-in}\hat{V}_l^\dagger|k'\rangle\langle k'| \cdot \langle l|\hat{V}_k\hat{\varrho}_{B-in}\hat{V}_{k'}^\dagger|l\rangle \tag{39}$$

for the second one. If measuring the first qubit results in outcome σ, the final state of B is

$$\hat{\varrho}^{(\sigma)}_{B-out} = p(I|\sigma)\hat{\varrho}_{B-out}(I) + p(II|\sigma)\hat{\varrho}_{B-out}(II). \tag{40}$$

Note that A and B did not interact directly, but $\hat{\varrho}^{(\sigma)}_{B-out}$ is sensitive to the fact of measurement on A and depends on its outcome. Therefore, in S-CTC (as well as in D-CTC) a system can influence the evolution of another one by means of additional measurement-induced modification of space-time topology.

Finally, let us outline the similarities and differences in D-CTC, P-CTC and S-CTC. In absence of measurement, but in the presence of interaction between different versions of the system, P-CTC and S-CTC predict the same final states $\hat{\varrho}_{out}$ given the same $\hat{\varrho}_{in}$. In case of measurement, apart from final state $\hat{\varrho}^{(\sigma)}_{out}$ for the outcome σ one needs to know its probability. Despite the radical difference, D-CTC and S-CTC give identical probabilities. In both cases, for obtaining physically meaningful result, it has proved the necessity to account different realities W_σ attached to different outcomes. This necessity appears not in the context of interpretation issues, but as an indispensable technical element of a formal construction. The probabilities predicted by P-CTC are different from D-CTC and S-CTC. The final states are different in all three models.

The research was financially supported by the state order (project AAAA-A17-117052210003-4, the internal FASO number 0319-2016-0002) at the Institute of Automation and Electrometry SB RAS.

References

[1] Horowitz G and Maldacena J 2004 *J. High Energy Phys.* **2004** 8
[2] Gödel K 1949 *Rev. Mod. Phys.* **21** 447
[3] Allen J M A 2019 Reality, Causality, and Quantum Theory *Preprint* arXiv:1901.01618
[4] Deutsch D 1991 *Phys. Rev.* D **44** 3197
[5] Pfarr J 2010 *Found.-Phys.* **40** 1326
[6] Lloyd S *et al.* 2011 *Phys. Rev. Lett.* **106** 040403
[7] Svetlichny G 2011 *Int. J. Theor. Phys.* **50** 3903
[8] Barrett J 2007 *Phys. Rev.* A **75** 032304
[9] Brun T A 2003 *Found. Phys. Lett.* **16** 245
[10] Durham I T 2008 Quantum communication on closed time-like curves *Preprint* arXiv:0803.3287
[11] Brun T, Harrington J and Wilde M 2009 *Phys. Rev. Lett.* **102** 210402
[12] Ahn D, Myers C R, Ralph T C and Mann R B 2013 *Phys. Rev.* A **88** 022332
[13] Jung E and Park D 2016 *Quantum Inf. Process.* **15** 3055
[14] Lloyd S, Maccone L, Garcia-Patron R, Giovannetti V and Shikano Y 2011 *Phys. Rev.* D **84** 025007
[15] Korotaev S M and Kiktenko E O 2015 *Physica Scripta* **90** 085101
[16] Coecke B 2010 *Contemp. Phys.* **51** 59
[17] Aharonov Y andVaidman L 2008 *Lect. Notes Phys.* **734** 399
[18] Aharonov Y, Bergmann P G and Lebowitz J L 1964 *Phys. Rev.* **134** B1410

Examples of Calabi-Yau threefolds with small Hodge numbers

A O Shishanin[1,2]

[1] Bauman Moscow State Technical University, Moscow, Russia;
[2] Lomonosov Moscow State University, Moscow, Russia

E-mail: shishandr@rambler.ru

Abstract. We observe some suitable examples of Calabi-Yau threefolds for heterotic superstring compactifications. It is reasonable to seek CY threefolds with Euler characteristic equals ±6 because of generation's number. Hosotani mechanism for violations of the gauge group by the Wilson loops requires such CY space has a non-trivial fundamental group. These spaces can be obtained by factoring the complete intersection Calabi-Yau spaces by the free action of some discrete group. Also we shortly discuss cases when discrete groups act with fixed point sets.

1. Introduction

Calabi - Yau (CY) spaces and structures associated with them are one of the most important objects of study in modern mathematical physics. CY spaces appear naturally in areas such as two-dimensional supersymmetric field theories, conformal field theory, topological strings, and mirror symmetry [1]. In string theory [2], [3], [4], [5] CY spaces arise as objects for obtaining reasonable compactifications.

Candelas et all [6] had studied the compactification of $E_8 \times E_8$ heterotic string on $M \times X$. Here M is 4d Minkowski space and X is some six-dimensional manifold. They wondered when such a compactification leads to 4d $N = 1$ supersymmetric field theory. They demonstrated [6] that supersymmetry requires the existence of a covariant constant spinor on X. The presence of such spinors gives the *SU(3)* holonomy group on X. For any n-dimensional Riemann manifold holonomy group is subgroup of $SO(2n)$. If this manifold is Kähler then holonomy group is subgroup of $U(n)$. For Calabi-Yau manifolds maximal holonomy group reduces to $SU(n)$.

The embedding of the spin connection in the heterotic string leads to the violation of the gauge group from $E_8 \times E_8$ to $E_6 \times E_8$. Adjoint representation of E_8 with dimension 248 is reducible to the group $SU(3) \times E_6$ as

$$248 = (8,1) + (1,78) + (3,27) + (\bar{3},\overline{27}).$$

The generation number of elementary particles equals difference between the number for generation of particles (3,27) and the number for anti-generation of particles $(\bar{3},\overline{27})$. These numbers (3,27) and $(\bar{3},\overline{27})$ equal $h^{2,1}$ and $h^{1,1}$ respectively. Thus it turns out that the number for generations of elementary particles is half of Euler characteristic of X $|h^{2,1} - h^{1,1}| = |\chi|/2$. Then quest of CY threefolds with $\chi = \pm 6$ represents significant interest because of the generation number is expected to be no more than 3. The simplest and most important approach to break E_6 gauge group is Hosotani mechanism [7], [8]. The Hosotani mechanism uses Wilsonian loops. It requires the fundamental group $\pi(X)$ of CY threefold X to be non-trivial.

2. General information about Calabi-Yau spaces

There are many definitions of Calabi-Yau spaces. It is mathematically correct to treat of Calabi-Yau spaces as Kähler manifolds with a trivial canonical class $K = 0$. This means that the holomorphic form of the highest degree Ω has no poles and zeros anywhere. Physically Calabi-Yau spaces are Ricci- flat (zero first Chern class) Kähler-Einstein manifolds.

The low-dimensional examples of Calabi-Yau spaces are torus and famous *K3*-surface. There are many examples of K3-surface a quartic in $\mathbb{C}P^3$, the intersection of a quadric and a qubic in $\mathbb{C}P^4$, the intersection of three quadrics in $\mathbb{C}P^5$, a Kummer surface and so on. One important example of CY manifolds is elliptic fibrations. *K3*-surface can be obtain, for instance, as elliptic fibration on weighted projective space

$$y^2 = x^3 + f(z)x + g(z).$$

If *f(z)* is polynomial degree 8, *g(z)* is polynomial degree 12, *x* has weight 4 and *y* has weight 6 then this equation defines smooth K3-surface. Also we can obtain some K3-fibration Calabi-Yau threefold if polynomials *f(z,w)* and *g(z,w)* have coordinates *z* and *w* on *K3*-surface.

Let recall that Hodge numbers are dimensions of Dolbeault cohomology groups

$$h^{p,q} = dim\, H_{\bar{\partial}}^{p,q}.$$

For complex manifolds, the numbers $h^{p,q}$ consist some table called the Hodge diamond. The Hodge diamond for CY threefold is defined as

$$\begin{array}{ccccccc} & & & 1 & & & \\ & & 0 & & 0 & & \\ & 0 & & h^{1,1} & & 0 & \\ 1 & & h^{2,1} & & h^{2,1} & & 1 \\ & 0 & & h^{1,1} & & 0 & \\ & & 0 & & 0 & & \\ & & & 1 & & & \end{array}.$$

Hence the Euler characteristic χ equals $2(h^{1,1} - h^{2,1})$. Numbers $h^{1,1}$ are called Kähler moduli and $h^{2,1}$ are complex moduli. Mirror manifold for CY threefold with Hodge numbers $(h^{1,1}, h^{2,1})$ has Hodge numbers $(h^{2,1}, h^{1,1})$.

The most known example of CY threefold is quintic in $\mathbb{C}P^4$. Let $(x_0, x_1, x_2, x_3, x_4)$ homogeneous coordinates in $\mathbb{C}P^4$ then the quintic equation has form

$$x_0^5 + x_1^5 + x_2^5 + x_3^5 + x_4^5 = 0$$

with $h^{1,1} = 1$, $h^{2,1} = 101$ and $\chi = -200$.

There are few examples of complete intersections CY (CICY) threefolds in $\mathbb{C}P^N$ for some *N*. $\mathbb{C}P_N[d_1, \dots, d_k]_\chi$ denotes complete intersection of k homogeneous polynomials with d_i degrees in $\mathbb{C}P^N$. Then the total Chern class is given by adjunction formula

$$c = \frac{(1+J)^{N+1}}{\prod_{i=1}^{k}(1+d_i J)}, \tag{1}$$

where J is some 2-form obtained by normalizing the Kähler form. For CY threefolds, the condition zero first Chern class $c_1 = 0$ gives expression

$$\sum_{i=1}^{k} d_i = k + 1.$$

There are five nonlinear examples of such manifolds the quintic $\mathbb{C}P_4[5]_{-200}$ and $\mathbb{C}P_5[2{,}4]_{-176}$, $\mathbb{C}P_5[3{,}3]_{-144}$, $\mathbb{C}P_6[2{,}2{,}3]_{-144}$, $\mathbb{C}P_7[2{,}2{,}2{,}2]_{-128}$.

Hübsch notation [9] for CICY threefold is defined *N* polynomials in $\mathbb{C}P^{i_1} \times \mathbb{C}P^{i_2} \times \dots \times \mathbb{C}P^{i_n}$

$$\begin{matrix} P^{i_1} \\ \vdots \\ P^{i_n} \end{matrix} \begin{bmatrix} a_{11} & a_{12} & \dots & a_{1N} \\ \vdots & \dots & \dots & \vdots \\ a_{n1} & a_{n2} & \dots & a_{nN} \end{bmatrix}^{h^{1,1},h^{2,1}}_{\chi} .$$

Each column of this table corresponds polynomial in $\mathbb{C}P^{i_1} \times \mathbb{C}P^{i_2} \times \dots \times \mathbb{C}P^{i_n}$.

The easiest way to lower the Hodge numbers is factorizing manifold X_0 by some discrete group *G*. If G has free action on *X* (without fixed points) then the Euler characteristic of quotient manifold $X = X_0/G$ equals

$$\chi(X_0) = \frac{\chi(X_0)}{|G|},$$

where $|G|$ is order of *G*. This equality is true for spaces of odd dimension, as in our case. It can be obtained using the Atiyah-Bott fixed point formula. It is necessary to check the transformation of the form of highest degree Ω [10]. Number $h^{2,1}$ is usually easy to find. Then the number $h^{1,1}$ of *X* is given by

$$h^{1,1} = h^{2,1} + \frac{\chi(X_0)}{2|G|}.$$

For example, the quintic is factorized by $\mathbb{Z}_5 \times \mathbb{Z}_5$ has Euler characteristic -8.
Such way the first example of CY threefold with Euler characteristics $|\chi| = 6$ was constructed by S.T. Yau in 1985 [11]. This CY threefold with $\chi = -6$ is called Tian-Yau space. Let us denote quotient complete intersection Calabi-Yau manifold as QCICY.

3. Tian-Yau space and its twins

There exist complete intersection K_0 in $\mathbb{C}P^3 \times \mathbb{C}P^3$ with table

$$\begin{matrix} P^3 \\ P^3 \end{matrix} \begin{bmatrix} 1 & 3 & 0 \\ 1 & 0 & 3 \end{bmatrix}^{14,23}_{-18} .$$

Let us denote (x_0, x_1, x_2, x_3) four homogeneous coordinates of the first $\mathbb{C}P^3$ and for the second $\mathbb{C}P^3$ homogeneous coordinates (y_0, y_1, y_2, y_3). K_0 is given by polynomials

$$x_0^3 + x_1^3 + x_2^3 + x_3^3 = 0,$$
$$y_0^3 + y_1^3 + y_2^3 + y_3^3 = 0,$$
$$x_0 y_0 + x_1 y_1 + x_3 y_3 + x_4 y_4 = 0.$$

There is a group $\mathbb{Z}_3$ with free action. This action is given by $g \in \mathbb{Z}_3$

$$g: (x_0, x_1, x_2, x_3, y_0, y_1, y_2, y_3) \to (x_0, \alpha^2 x_1, \alpha x_2, \alpha x_3, y_0, \alpha y_1, \alpha^2 y_2, \alpha^2 y_3),$$

where $\alpha^3 = 1, \alpha \neq 1$. The Tian-Yau space $K_1 = K_0/\mathbb{Z}_3$ has Hodge numbers (6,9) and Hodge diamond

$$\begin{matrix} & & & 1 & & & \\ & & 0 & & 0 & & \\ & 0 & & 6 & & 0 & \\ 1 & & 9 & & 9 & & 1 \\ & 0 & & 6 & & 0 & \\ & & 0 & & 0 & & \\ & & & 1 & & & \end{matrix} . \quad (2)$$

An explanation of obtaining Hodge numbers for Tian-Yau space can be found in [12]. The fundamental group of Tian-Yau space is $\mathbb{Z}_3$.
Schimmrigk had found another example of factorized CICY with same Hodge numbers and fundamental group [13]. Consider complete intersection in $\mathbb{C}P^2 \times \mathbb{C}P^3$ with table

$$\begin{matrix} P^2 \\ P^3 \end{matrix} \begin{bmatrix} 3 & 0 \\ 1 & 3 \end{bmatrix}^{8,35}_{-54} .$$

This CICY can be factorized by $A \times B$ group. Both group are isomorphic $\mathbb{Z}_3$. Action of group A is specified by element g_1 of order 3

$$g_1: \mathbb{C}P^2 \times \mathbb{C}P^3 \to \mathbb{C}P^2 \times \mathbb{C}P^3,$$

$$g_1 = \begin{pmatrix} 0 & 1 & 0 \\ 0 & 0 & 1 \\ 1 & 0 & 0 \end{pmatrix} \oplus \begin{pmatrix} 0 & 1 & 0 & 0 \\ 0 & 0 & 1 & 0 \\ 1 & 0 & 0 & 0 \\ 0 & 0 & 0 & 1 \end{pmatrix}.$$

This action is free. The group B acts only in $\mathbb{C}P^2$

$$g_2 : (x_0, x_1, x_2, y_0, y_1, y_2, y_3) \to (x_0, \alpha x_1, \alpha^2 x_2, y_0, y_1, y_2, y_3).$$

Here there are three fixed point sets which define three invariant tori

$$(1,0,0) \times \{y_1^3 + y_2^3 + y_3^3 = 0\},$$
$$(0,1,0) \times \{y_0^3 + y_2^3 + y_3^3 = 0\},$$
$$(0,0,1) \times \{y_0^3 + y_1^3 + y_3^3 = 0\}.$$

It is possible to resolve singularities without leaving class of CY threefolds [11].
There is other sample of CY threefold with same Hodge diamond (2). Let us consider bicubic L

$$L = \begin{matrix} P^2 \\ P^2 \end{matrix} \begin{bmatrix} 3 \\ 3 \end{bmatrix}^{2,83}_{-162}.$$

L is factorized by three groups with fixed point acting groups G_1,G_2 and free acting group G_3 Resolved $\widehat{L/G_1 \times G_2 \times G_3}$ has Hodge diamond and fundamental group as Tian-Yau and Schimmrigk spaces.

4. More modern examples of QCICY

Many examples of "three generation" manifolds with $\chi = \pm 6$ was constructed in [15]. Two more samples were built by Braun, Candelas and Davies [16]. CICY is denoted in [16] as $Y^{8,44}$. This manifold is defined by following Hübsch table

$$\begin{matrix} P^2 \\ P^2 \\ P^2 \\ P^2 \end{matrix} \begin{bmatrix} 1 & 1 & 1 & 0 & 0 \\ 0 & 0 & 1 & 1 & 1 \\ 1 & 1 & 1 & 0 & 0 \\ 0 & 0 & 1 & 1 & 1 \end{bmatrix}^{8,44}_{-72}.$$

Two groups of order 12 act freely on $Y^{8,44}$. One group is abelian $\mathbb{Z}_{12}$. And other group is non-Abelian dicyclic group Dic_3 (semidirect product $\mathbb{Z}_3$ and $\mathbb{Z}_4$). The quotient manifolds for both these groups have Hodge numbers $(h^{1,1}, h^{2,1}) = (1,4)$. Fundamental groups these manifolds are $\mathbb{Z}_{12}$ and Dic_3 respectively.
A huge number of examples of CY threefolds with small Hodge numbers are given in the table article [17]. Many examples with a non-trivial fundamental group provide quotients for complete intersections of four quadrics $\mathbb{C}P_7[2,2,2,2]_{-128}$.

Relevant way to change Hodge number for Calabi-Yau spaces is conifold transition [18], [19], [20]. Usual conifold transition shifts the Hodge numbers as follows

$$\delta(h^{1,1}, h^{2,1}) = (1, -1).$$

The conifold transition does not change the fundamental group. However, there is a hyperconifold transition [21], [22] that can change the fundamental group.
In this text we have given some examples CY threefolds that are interesting for compactifications of heterotic string theory. Mirror symmetry [1] is an actual way to study Calabi-Yau spaces. Euler characteristic for mirror manifold is same. Therefore, from the modern point of view, it is important to explore mirror manifolds with small Hodge numbers.

Acknowledges
The author is grateful to Maxim Leenson for valuable discussions.

References

[1] Hori K, Katz S, Klemm A, Pandharipande R, Thomas R, Vafa C, Vakil R and Zaslow E 2003 *Mirror symmetry* (American Mathematical Society)
[2] Polchinski J 1998 *String Theory Vol. II: Superstring Theory and Beyond* (Cambridge University Press)
[3] Becker K, Becker M and Schwarz J H 2007 *String Theory and M-theory: A Modern Introduction* (Cambridge University Press)
[4] Green M, Schwarz J H and Witten E 1987 *Superstring Theory. Vol 2.* (Cambridge University Press)
[5] Greene B 1996 String theory on Calabi-Yau manifolds *e-Print hep-th/*9702155
[6] Candelas F, Horowitz G T, Strominger A and Witten E 1985 *Nucl. Phys.* B **258** 46
[7] Hosotani Y 1983 *Phys. Lett.* B **126** 309
[8] Witten E 1985 *Nucl. Phys.* B **258** 75
[9] Hübsch T 1992 *Calabi-Yau manifolds. A besriary for physicists.* (World Scientific Pub)
[10] Davies R 2011 *Adv. High Energy Phys.* **2011** 901898
[11] Yau S T 1985 *Compact three dimensional Kahler manifolds with zero Ricci curvature in Proceedings of a symposium on anomalies geometry topology* (World Scientific Pub)
[12] Greene B, Kirklin K, Miron P and Ross G 1986 *Nucl. Phys.* B **278** 667
[13] Schimmrigk R 1987 *Phys. Lett.* B **193** 175
[14] Candelas P, de la Ossa X, He YH and Szendroi B 2008 *Adv. Theor. Math. Phys.* **12** 429
[15] Candelas P and Davies R 2010 *Fortsch. Phys.* **58** 383
[16] Braun V, Candelas P and Davies R 2010 *Fortsch. Phys.* **58** 467
[17] Candelas P, Constantin A and Mishra C 2018 *Fortsch. Phys.* **66** 6
[18] Reid M 1987 *Mathematische Annalen* **278** 329
[19] Candelas P, Dale A M, Lutken A, Schimmrigk and Rolf 1988 *Nuclear Physics B* **298** 493
[20] Green P and Hübsch T 1988 *Comm. in Math. Phys.* **119** 431
[21] Davies R 2010 *Adv. Theor. Math. Phys.* **14** 965
[22] Davies R 2011 *Nuclear Physics B* **850** 214

Completely geometric theory I: gravitation

S V Siparov

State University of Civil Aviation, St. Petersburg, 196210, RF

E-mail: sergey@siparov.ru

Abstract. In order to describe motions in arbitrary physical systems, a geometric approach is proposed, in which the goal is not to find the Lagrangian, but to find the geometry of space that models the experimental space. The requirement that the observed trajectory coincides with the geodesic makes it possible to use geometric identity to find the equation for the metric. As a result, it is possible to give an interpretation of a number of observations that do not have such in the existing theories.

When constructing the dynamics, Newton relied on the observations of Kepler, who discovered the elliptical form of the trajectories of the planets. Along with Galileo's principle of relativity, this led to the formulation of the equation of motion in the form of the second law of dynamics, $m\frac{d\vec{v}}{dt}=\vec{F}$. Having determined the force with the help of trajectories, Newton considered his task completed. If the force of gravity is considered to be the gradient of some potential, then the field equation corresponding to the observed trajectories is the Poisson equation. Therefore, Newton's theory of gravity is reduced to the following equations

$$m\frac{d\vec{v}}{dt}=-\nabla\varphi, \tag{1}$$

$$\Delta\varphi=\rho(r), \tag{2}$$

where $\vec{v}=\frac{d\vec{r}}{dt}$, and $\rho(r)$ is the source distribution density, setting which one can solve the direct problem and find the trajectory.

Along with the well-known successes, Newton's theory of gravity incorrectly describes i) the precession of the Mercury orbit (missing 43'' in a century) and ii) the deflection of a light beam by a massive body (the angle differs by half).

2. If the field potential is known, then the trajectory can be found using Hamilton's principle. The variation method leads to the equation of motion in the form of the Euler-Lagrange equation containing this potential. Turning to geometrodynamics, we see that a trajectory which is a solution to the Euler-Lagrange equation coincides with the geodesic equation on a surface with a given curvature. This equation has the form

$$\frac{d^2x^i}{ds^2}+\Gamma^i{}_{kl}\frac{dx^k}{ds}\frac{dx^l}{ds}=0, \tag{3}$$

where $\Gamma^i{}_{kl}$ are the connectivity coefficients defined by the choice of metric (i.e. of geometry), s is a natural parameter along the trajectory. In order to find the metric, and hence the curvature at each point, the field equation is needed. Einstein obtained it by equating the zero divergence of the energy-momentum tensor (conservation law) to the zero divergence of a combination of geometric parameters that had the desired structure. The resulting equation of the gravitational field has the form

$$R^{ik} - \frac{1}{2} g^{ik} R = -\kappa T^{ik} \tag{4}$$

By specifying the energy-momentum tensor, within the framework of the direct problem, one can solve equation (4), substitute the found metric into equation (3), and find the trajectory. And vice versa, by specifying the trajectory, one can solve the inverse problem and find the components of the energy-momentum tensor. Thus, Einstein's theory of gravity (GRT) is reduced to equations (3-4).

The use of geometrodynamics made it possible to cope with the problems i)-ii) of Newton's theory, as well as predict the gravitational redshift, subsequently found in experiment. However, on the galactic scale, GRT ran into problems: iii) in spiral galaxies, the rotation curves turned out to be flat, and iv) in some gravitational lenses, the detected light refraction turned out to be significantly larger than the calculated one.

3. In electrodynamics, the equation of motion contains the Lorentz force,

$$m\frac{d\vec{v}}{dt} = q\{\vec{E} + [\vec{v}, \vec{H}]\}, \tag{5}$$

which includes two types of measured forces: $\vec{E} = -\frac{\partial \varphi_e}{\partial \vec{r}}$ and $\vec{H} = rot\vec{A}$, where φ_e and $\vec{A}$ are scalar and vector potentials. The electromagnetic field equations (Maxwell's equations) in convenient units have the form

$$\begin{aligned} &\frac{\partial \vec{H}}{\partial t} + rot\vec{E} = 0 \\ &div\vec{H} = 0 \end{aligned} \tag{6}$$

$$\begin{aligned} &rot\vec{H} - \frac{\partial \vec{E}}{\partial t} = \vec{j} \\ &div\vec{E} = \rho \end{aligned} . \tag{7}$$

They were obtained as a result of generalization of observations related to measurements of the strengths of electric and magnetic fields and the introduction of the concept of "displacement current", which follows from the continuity equation. This introduction became possible due to the identical equality of two zeros: zero, which is on the right in the continuity equation, and zero, which is equal to the divergence of the rotor. Equations (6-7) can be transformed into d'Alembert equations for the corresponding potentials:

$$\begin{aligned} &\Delta\varphi_e - \frac{\partial^2 \varphi_e}{\partial t^2} = -\rho \\ &\Delta\vec{A} - \frac{\partial^2 \vec{A}}{\partial t^2} = -\vec{j} \end{aligned} \tag{8}$$

or

$$\Delta A^i - \frac{\partial^2 A^i}{\partial t^2} = -I^i, \tag{9}$$

where A_i are the components of the 4-potential (φ_e, $\vec{A}$), $I^0 = \rho; \vec{I} \equiv \vec{j}$. Thus, equation (9) is the electromagnetic field equation.

It is important to note that in Maxwell's theory based on Faraday's ideas about lines of force, the sources on the right-hand sides of equations (7-9) are auxiliary quantities and are actually functions of strengths [1]. However, due to the complexity of solving the complete direct problem, the goal of which is to find the trajectory, the sources' distribution is usually assumed to be known. If the sources are given, equations (9) allow you to find the potentials, then calculate the strengths and find the trajectory with equation (5).

1. Construction of a completely geometric theory

In the middle of the 19th century, Clifford pointed out that there is no experimental possibility of distinguishing the effect of physical fields from the manifestation of the geometry with which the experimental space is described. Therefore, a variety of physical ideas and results that made it possible to obtain the above equations of motion and field equations can be applied to the construction of a completely geometric theory that is not tied to a specific physical problem. For this, it is necessary to obtain analogs of the equation of motion and of the field equation. The first, of course, is the geodesic equation for a space with a given geometry. Taking into account the possibility of the existence of two types of forces, we will consider space not only curved, but also anisotropic. Then the geodesic equation (3) should be generalized, and we should take into account the possible dependence of metric on the derivative with respect to the natural parameter, i.e. put $g_{ik} = g_{ik}(x, y)$, where $y^i = \frac{dx^i}{ds}$. This greatly complicates the geometric aspect of the matter. For the sake of simplicity, let us set $g_{ik}(x, y) = \eta_{ik} + \varepsilon_{ik}(x, y); \eta_{ik} = diag\{-1,1,1,1\}$, where $\varepsilon_{ik} = \sigma\varsigma_{ik}; \sigma << 1$. Then the generalized geodesic takes the form

$$\frac{dy^i}{ds} + \left(\Gamma^i{}_{lk} + \frac{1}{2}\eta^{it}\frac{\partial^2 \varepsilon_{kl}}{\partial x^j \partial y^t} y^j\right) y^k y^l = 0. \tag{10}$$

Then we use the well-known [2,3] simplifying assumptions for the geodesic and obtain

$$\frac{dy^i}{ds} + \Gamma^i{}_{00} + \frac{1}{2}\eta^{ik}\frac{\partial^2 \varepsilon_{00}}{\partial x^j \partial y^k} y^j = 0. \tag{11}$$

Thus, the only significant component of the metric tensor that remains in the equations for the considered case of small curvature and small anisotropy is ε_{00}. Let us transform the third term so as to select the "potential" and "solenoid" parts. Then the spatial 3D-section of equation (11) takes the form

$$\frac{d\vec{v}}{dt} = -\nabla\varepsilon_{00} + \nabla(\vec{v}, \frac{\partial\varepsilon_{00}}{\partial\vec{v}}) + [\vec{v}, rot\frac{\partial\varepsilon_{00}}{\partial\vec{v}}]. \tag{12}$$

Now let us find the field equation. The well-known expression $F_{ik} = \frac{\partial A_k}{\partial x^i} - \frac{\partial A_i}{\partial x^k}$ for the tensor of electromagnetic field, where A_i is the 4-potential, satisfies the identity $\frac{\partial F_{ij}}{\partial x^k} + \frac{\partial F_{jk}}{\partial x^i} + \frac{\partial F_{ki}}{\partial x^j} = 0$. Now we note that this identity has not a specific, but a universal geometric character and holds for any covariant anti-symmetric tensor of the second rank. The consequences of this fact are as follows. Let F_{ik} be an arbitrary anti-symmetric tensor, g_{ik} be the metric (fundamental tensor) of the modeling space, so that $F^{ij} = g^{ik}g^{jm}F_{mk}$. If we introduce the notation $\vec{F}^{(*)} = (F_{12}, F_{31}, F_{10})$ and $\vec{F}^{(**)} = (-F_{30}, F_{20}, -F_{23})$, as well as $I^i = \frac{\partial F^{ij}}{\partial x^j}$ and $I^0 \equiv \rho; I^1 \equiv j_x; I^2 \equiv j_y; I^3 \equiv j_z$, then the two geometric identities

$$\frac{\partial F_{ij}}{\partial x^k} + \frac{\partial F_{jk}}{\partial x^i} + \frac{\partial F_{ki}}{\partial x^j} = 0; \frac{\partial F^{ij}}{\partial x^k} + \frac{\partial F^{jk}}{\partial x^i} + \frac{\partial F^{ki}}{\partial x^j} = 0, \tag{13}$$

lead to the equations

$$\begin{aligned} &\frac{\partial\vec{F}^{(**)}}{\partial t} + rot\vec{F}^{(*)} = 0 \\ &div\vec{F}^{(**)} = 0 \end{aligned} \tag{14}$$

and

$$rot\vec{F}^{(**)} - \frac{\partial \vec{F}^{(*)}}{\partial t} = \vec{j},$$
$$div\vec{F}^{(*)} = \rho \tag{15}$$

that coincide in form with equations (6-7), but do not have any pre-assigned physical meaning. This allows to remain within the framework of exclusively geometric constructions in all further reasoning.

Instead of electromagnetic field tensor, we choose a kinematical tensor in the form $F_{ik} = B_{k,i} - B_{i,k}$, where $B_{k,i} = \frac{\partial y_k}{\partial x^i}$. It is an anti-symmetric 4x4 matrix, the lower right corner of which is occupied by the components of the curl of the velocity, and the first row and column are the acceleration components. We introduce $\vec{F}^{(*)} = -\nabla \varphi_g$, ($\Delta\varphi_g = -\rho$) and $\vec{F}^{(**)} = rot\vec{U}_g$, ($\Delta\vec{U}_g = -\vec{j}$), where φ_g and $\vec{U}_g$ are "metric potentials", determined exclusively by the metric. Then, similarly to equation (9), we obtain the equation of the "metric field"

$$\Delta D^i - \frac{\partial^2 D^i}{\partial {x_0}^2} = -I^i, \tag{16}$$

where D_i is the metrical 4-potential.

The equation of motion (10) (geodesic) and the equation of the metric field (16) represent a closed system of geometric equations based on identity and suitable for solving both direct and inverse problems. The theory based on them is completely geometric, because no additional physical quantities are initially included into it. The equations of the form (14-16) can be solved with the help of efficient mathematical apparatus developed in electrodynamics.

2. Geometric theory and classical physics

Consider the applications of the constructed geometric theory to physics.

a) We use the notation from electrodynamics and put $\varepsilon_{00} \equiv \varphi_e$, $\frac{\partial \varepsilon_{00}}{\partial \vec{v}} \equiv \vec{A}$, ($rot\vec{A} \equiv \vec{H}$). Then the geodesic equation (12) takes the form of the equation of motion of the test (charged) particle

$$\frac{d}{dt}\vec{v} = -\nabla \varphi_e + (\vec{v}\cdot\nabla)\vec{A} + [\vec{v}, rot\vec{A}]. \tag{17}$$

Choose $\vec{A} = \mathrm{Re}\{\vec{A}_0 e^{i(\vec{k}\vec{r}-\omega t)}\}$ and make replacements $\vec{E}_{(a)} = ik\vec{A}$ and $\vec{H}_{(a)} = i[\vec{k}, \vec{A}]$. Then, for the quantity $\Psi_{(a)} = E_{(a)} + iH_{(a)}$, the pair of homogeneous Maxwell equations (14) obtained from the geometric identity can be transformed into the equation

$$i\frac{\partial \Psi_{(a)}}{\partial t} = rot\Psi_{(a)}. \tag{18}$$

Solving equation (18) we can find the fields in the region free from sources, and then use equation (17) to obtain the trajectory of a probe body. The stationary equation corresponding to equation (18)

$$rot\psi_{(a)} = \lambda\psi_{(a)} \tag{19}$$

has eigenvalues of the following form [4]

$$\lambda_{n,m} = \rho_{m,n} {R_A}^{-1}, \tag{20}$$

where R_A is the radius of the (spherical) domain of definition, and $\rho_{m,n}$ are the zeros of the Bessel function of half-integer order. Their arrangement is similar to the arrangement of the zeros of a sinusoid. The eigenfunctions of equation (19) are expressed in terms of spherical functions. In plasma physics, they are called force-free fields, and this fully corresponds to the geometric approach under consideration. The appearance of Bessel functions is typical for various wave problems. In this case, the

Bessel beam (wave field) does neither refract nor diffract, and, being partially obscured by an obstacle, self-regenerates. Thus, the geometric approach applied to electrodynamics leads to physically significant consequences.

b) Now use the notation from hydrodynamics and put $\frac{\partial \varepsilon_{00}}{\partial y^i} \equiv u_i$, $rotu_i \equiv \Omega_i$, $\varphi = \varepsilon_{00} - (\vec{v}, \vec{u})$.

Then equation (12), after rearranging and multiplying and dividing the right-hand side of the equation by a constant value ρ_0, takes the form

$$\frac{\partial \vec{v}}{\partial t} + (\vec{v}, \nabla)\vec{u} + 2[\vec{\Omega}, \vec{v}] = -\frac{1}{\rho_0}\nabla\{\rho_0\varepsilon_{00} - (\vec{v}, \vec{u})\rho_0\} \quad .(21)$$

If we consider vector $\vec{u}$ characterizing the anisotropy of the modeling space as the flow velocity of an incompressible fluid, then, assuming the gradient of the two terms on the right-hand side to be the effective pressure, we obtain the Euler equation for a (sufficiently large) fluid region rotating as a whole with angular velocity $\vec{\Omega}$. Accordingly, equations (21) have both potential and wave solutions. Indeed, assuming the velocity of a liquid particle to be equal to the velocity of its environment, $\vec{v} \cong \vec{u}$, we will seek solution of the equation (21) in the form

$$\vec{v} = \vec{V}e^{i((\vec{k}, \vec{r}) - \omega t)} \,. \quad (22)$$

The amplitude $\vec{V}$ will be considered small so that the term $(\vec{v}, \nabla)\vec{u} \to 0$ can be neglected. Then

$$\frac{\partial \vec{v}}{\partial t} + 2[\vec{\Omega}, \vec{v}] = -\nabla\varphi \,. \quad (23)$$

Let us choose the local Oz axis parallel to the vector $\vec{\Omega}$, and apply the *rot* operator to both sides of equation (23), taking into account that $rot[\vec{\Omega}, \vec{v}] = \vec{\Omega} div\vec{v} - (\vec{\Omega}, \nabla)\vec{v} = -(\vec{\Omega}, \nabla)\vec{v}$. We get [5]

$$\frac{\partial}{\partial t} rot\vec{v} = 2\Omega \frac{\partial \vec{v}}{\partial z} \,. \quad (24)$$

In such a wave, the velocity vector $\vec{v}$ of a liquid particle retains its value and changes only in direction. In classical hydrodynamics, such waves are generated by the "Coriolis force" and are called inertial waves. However, in geometric theory there are no Coriolis forces, but only a metric tensor that describes the anisotropic space in which the motion occurs. In this space, the zero component of the metric is a wave in accordance with the definition $\vec{v} \cong \vec{u}$, i.e.

$$\varepsilon_{00}(v) = \frac{1}{2}v^2 = \frac{V^2}{2c^2}e^{2i((\vec{k}, \vec{r}) - \omega t)} \,. \quad (25)$$

Thus, the desired small correction to the metric has a wave character.

3. Geometric theory and gravitation: interpretation of observations

Applying a completely geometric approach to the theory of gravity, it is natural, as in Newton's theory, to start with the inverse problem, i.e. "Do not invent hypotheses" and, knowing the trajectory, find a suitable metric that satisfies equation (12). For illustration, consider two limiting cases. In the first, only the potential contribution of the anisotropy of the modeling space is taken into account

$$\frac{d\vec{v}}{dt} = -\nabla\left\{\varepsilon_{00} - (\vec{v}, \frac{\partial \varepsilon_{00}}{\partial \vec{v}})\right\} \quad (26)$$

and in the second, only the solenoid contribution of the anisotropy is taken into account

$$\frac{d\vec{v}}{dt} = -\nabla\varepsilon_{00} + [\vec{v}, rot\frac{\partial \varepsilon_{00}}{\partial \vec{v}}] \,. \quad (27)$$

Let us compare equations (26, 27) with the motion of a test body in a spiral galaxy. The simplest model of a spiral galaxy is the "center + current" (C+C) system [6-10], in which all the motion is concentrated along a circle, in the vicinity of which there is also a test body. So, we will assume that on the periphery of the galaxy, the motion occurs in a circle. Then in both left-hand sides of equation (26) and equation (27) there is a centripetal acceleration. The first terms on the right-hand sides correspond to the approximate equations of general relativity, leading to Newton's equation. In order to save this result, choose $\varepsilon_{00}(r) \sim \frac{C_1}{r}$. (Under physical interpretation, C_1 will become the Schwarzschild radius, $C_1 = r_S$). The second terms in equations (26, 27) arise from the geometric approach and are associated with the dependence of the acceleration on the speed of both the test body itself and other surrounding bodies, which can also be selected as test ones. We will talk about the "mass current density" in the vicinity of the test body and denote $\frac{\partial \varepsilon_{00}}{\partial \vec{v}} \equiv \vec{u}$. We emphasize that this terminology is formal in nature, and no substance is initially considered here.

If it turns out that the first and second terms on the right-hand side of equation (26) make the same contribution, then the classical GRT will not be enough for an adequate description of the motion. Assuming $\vec{u} \sim \vec{v}$, we find the condition under which this will happen

$$\frac{GM}{r} \sim v^2. \quad (28)$$

Then if the galaxy contains 10^{11} stars with a mass of the Sun of 10^{30} kg, and the radius of the galaxy is 10^{20} m, then according to (28), both terms will give the same contribution if the velocity of the test body is 10^5 m/s. It is just this value that corresponds to observations, therefore, GRT is not enough to describe galactic phenomena.

Let us now estimate the second term on the right-hand side of equation (27). To do this, we introduce formal notation $rot\frac{\partial \varepsilon_{00}}{\partial \vec{v}} = rot\vec{u} (\equiv rot\vec{A} \equiv \vec{H})$ and use the geometric analogue of the Biot-Savart-Laplace law. Then we get $|[\vec{v}, rot\frac{\partial \varepsilon_{00}}{\partial \vec{v}}]| \sim v\frac{C_2}{r}$, and from equation (27) it follows

$$v^2 = \frac{C_1}{r} + vC_2. \quad (29)$$

With $r \to \infty$, this equation has two roots: $v_1 = 0$, which corresponds to both GRT and Newton's theory, and $v_2 = C_2 = const$, which corresponds to observations. Thus, in a completely geometric theory which in this case becomes anisotropic geometrodynamics (AGD), a spiral galaxy has flat rotation curves, and problem iii) does not arise.

When studying gravitational lensing within the framework of the geometric theory, it turns out that if a spiral galaxy-lens has a certain orientation, namely, with its plane parallel to the direction of the beam from the source to the observer, Figure 1a, then the refraction will also depend on the rotation speed of this galaxy, which leads to a significant increase in the angle of refraction, Figure 1b, [11]. This removes problem iv).

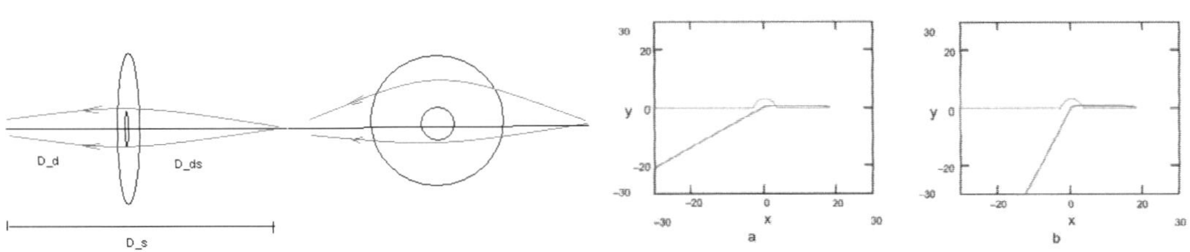

a) b)

Figure 1. a) Orientations of a spiral galaxy-lens; b) Refraction on a Coulomb centre and on a C+C system.

Thus, the current situation with (gravitational) dark matter resembles the situation with (electromagnetic) ether at the turn of the 19th and 20th centuries. Now observations that contradict GRT are explained using a completely geometric theory, and the introduction of dark matter turns out to be superfluous. All circumstances considered as proof of its existence (the Bullet cluster, possibility of the primary nucleosynthesis, etc.) require not matter as such, but additional interaction, which manifests itself in the geometric theory in the form of additional acceleration.

Let us find the value C_2 arising in geometric theory. Simple estimates give: $C_2 \sim M/T$, with $M \sim R^2$, where R is the apparent radius of a spiral galaxy, the mass M of which is concentrated in the plane of the disk, and T is the orbital period corresponding to the periphery. Kepler's law implies $T \sim R^{3/2}$, therefore $C_2 \sim R^{1/2}$. On the other hand, the luminosity L is also proportional to the area of the galaxy, $L \sim R^2$. Therefore, $R \sim L^{1/2}$, and finally

$$v \sim L^{1/4}. \tag{30}$$

This corresponds to the Tully-Fisher's law known from observations, and it also has no theoretical explanation in GRT. Exponents other than ¼ found in Faber-Jackson's law for elliptical galaxies can be obtained from a more accurate calculation using numerical methods.

Thus, the correction to the Minkowski metric obtained for the C+C system within the framework of the geometric approach can be chosen in the form

$$\varepsilon_{00}(r,v) = \frac{GM}{c^2 r} + \frac{V^2}{2c^2} e^{2i((\vec{k},\vec{r})-\omega t)}. \tag{31}$$

The corresponding trajectory of the test body is not flat and has the form of a helix wound on a circle along which the motion occurs.

In addition to the results mentioned above, the constructed geometric theory or AGD makes it possible to obtain trajectories of motion corresponding to the strange shape of the distribution of cold gas in the center of the Milky Way (in the form of the ∞ sign), obtained by the observatory of the Herschel mission [12], Figure 2. The equations

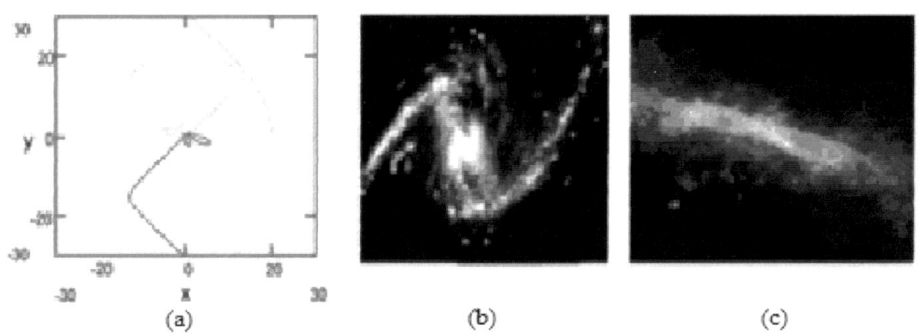

Figure 2. (a) Numerically calculated trajectories of the equal parts explosion in the C+C system; (b) Galaxy NHG-1365 (Hubble telescope, NASA) (free access); (c) Distribution of the cold gas in the Milky Way center (Herschel observ.) (free access).

of geometric theory also allow us to propose a new interpretation of the linear Hubble's law. Since now inertia and gravity are completely indistinguishable, the motion of distant stars across the line of sight of an arbitrary observer imparts to them a centripetal acceleration, which is linearly dependent on the distance to them. This acceleration corresponds to the "centrifugal" force of inertia, and thus the detected redshift is not cosmological, but gravitational one.

References

[1] Tamm I E 1976 *Foundations of the theory of electricity* (Moscow: Nauka)
[2] Einstein A 1979 *Foundations of the general theory of relativity* In: *Albert Einstein and the Theory of Gravity* (Moscow: Mir)
[3] Siparov S 2011 *Introduction to the Anisotropic Geometrodynamics* (London-New Jersey-Singapore: World Scientific)
[4] Sax R S 2013 *Ufa mathematical journal* **5** 63
[5] Landau L D and Lifshits E M 1986 *Hydrodynamics* (Moscow Nauka)
[6] Siparov S V 2008 *HNGPh* **10** 64 (*Preprint* gr-qc/0809.1817v3).
[7] Siparov S V 2009 HNGPh **12** 140 (Preprint gen-ph/1001.1501)
[8] Siparov S V 2009 *AIP Conf. Proc. Proc. 4-th Int. Gamow Conf.* Odessa 2010 **1206** Ed Chakrabarti S K, Bisnovatyi-Kogan G S and Zhuk A I (New York: Melville) p 152
[9] Siparov S V 2010 BSG *Proc. Proc. Int. Conf. DGDS-2009* (Geom.Balkan Press) p 190
[10] Siparov S V 2010 *AIP Conf. Proc. Proc.Int. Conf. "Mathematics and Astronomy"* (New York: Melville) p 222
[11] Siparov S V 2010 *Proc. 10th Int. Gamov Conf.* Odessa-2010 p 71
[12] Siparov S V 2012 *Space, time and fundamental interactions* **1** 99

Completely geometric theory II: basics of quantum mechanics

S V Siparov

State University of Civil Aviation, St. Petersburg, 196210, RF

E-mail: sergey@siparov.ru

Abstract. The geometrical approach suggested earlier, makes it possible to investigate the regions both in mega- and micro worlds that cannot be accessed by the direct observation. This makes it possible not only to interpret the basic experiments of quantum mechanics in a new way but also to escape the paradoxes stemming from the wave function introduction. It also gives perspectives to adjust the quantum mechanics and the relativity theory.

In the previous paper [1], a completely geometric theory was constructed, the main result of which was the geodesic equation

$$\frac{dy^i}{ds}+\left(\Gamma^i{}_{lk}+\frac{1}{2}\eta^{it}\frac{\partial^2\varepsilon_{kl}}{\partial x^j\partial y^t}y^j\right)y^k y^l=0 \qquad (1)$$

for a weakly curved and weakly anisotropic space with metric $g_{ik}(x,y)=\eta_{ik}+\varepsilon_{ik}(x,y);\eta_{ik}=diag\{-1,1,1,1\}$, where $\varepsilon_{ik}=\sigma\varsigma_{ik};\sigma<<1$, and the metric field equation

$$\Delta D^i-\frac{\partial^2 D^i}{\partial x_0{}^2}=-I^i \qquad (2)$$

for the kinematical tensor $F_{ik}=\frac{\partial y_k}{\partial x^i}-\frac{\partial y_i}{\partial x^k};y^i=\frac{\partial x^i}{\partial s}$.

The possibility and meaning of such a construction follows from Clifford's idea of the experimental indistinguishability of physical and geometric fields. This idea was partially reflected in Einstein's general relativity, in which the principle of equivalence was adopted, concerning the experimental indistinguishability of inertial forces and gravitational forces. It turned out [1] that the geometric approach allows one to present the main results of electrodynamics and hydrodynamics in a new language. Moreover, with its help it is possible [1] to cope with the difficulties encountered by general relativity on the galactic scale[1], as well as to give an interpretation to a number of observations that are absent in modern theory.

In this regard, it seems promising to further develop the approach, which is based on a statement analogous to the principle of least action: *there exists a geometry suitable for describing phenomena in experimental space such that all the motions of test bodies occur along the corresponding geodesics*. At the same time, within the framework of the geometric approach, it is natural to assume that the stationary

[1] Flat rotation curves of spiral galaxies and the anomalous refraction of light in gravitational lenses led to the introduction of the notion of dark matter that is self-contradictory. The decades' search for the particles of dark matter also failed.

state is determined not by the principle of minimum potential energy, but by the *principle of minimum occupied volume*.

1. Principles of quantum mechanics and its basic experiments

Quantum mechanics arose as a result of the impossibility of interpreting all experiments either only from the wave or only from the corpuscular point of view. Let's list the main of these experiments:

1. Wave interpretation of a heated body radiation spectrum led to the "ultraviolet catastrophe". Planck put forward the hypothesis of discrete quanta of radiation, which made it possible to successfully describe the experiment.

2. The regularities of spectral terms and the result of Rutherford's experiment on the scattering of α-particles allowed Bohr to formulate postulates about the existence of stationary states and about transitions between them.

3. The results of experiments on the scattering of electrons by crystals led de Broglie to the hypothesis that the electron has wave properties.

A self-consistent interpretation of these circumstances was proposed by Schrödinger, who refused to describe the microsystem and its behavior using classical concepts. Instead, he introduced the concept of a "wave function" satisfying a certain differential equation. The properties of this equation made its solutions satisfactorily describe the experimental results.

We see that at each stage of the quantum mechanics formation, formal assumptions of a paradoxical nature were made. Wave-corpuscle dualism, the absence of a visual meaning of the wave function and the impossibility of reconciling its instantaneous reduction with the relativity theory, as well as communicating physical meaning to the concepts of probability theory (probability waves, probability amplitude) constitute a conceptual problem of quantum mechanics.

2. Geometric approach

Let's apply a geometric approach to interpret the above experiments. After simplifications and transformations, the three-dimensional section of equation (1) will take the form

$$\frac{d\vec{v}}{dt} = -\frac{c^2}{2}\nabla\varepsilon_{00} + \nabla(\vec{v}, \frac{c^2}{2}\frac{\partial\varepsilon_{00}}{\partial\vec{v}}) + [\vec{v}, \vec{\Omega}], \qquad (3)$$

where $\vec{\Omega} \equiv \frac{c^2}{2} rot \frac{\partial\varepsilon_{00}}{\partial\vec{v}}$. Note that the gradients of scalars on the right-hand side of the equation are polar vectors, while the cross product is an axial vector. However, one of the factors of the vector product is also the axial vector (curl), so equation (3) retains its meaning.

If the observed acceleration is absent, then $\frac{d\vec{v}}{dt} = 0$, and the trivial solution is $\varepsilon_{00}(x, y) = const$, which reduces to the Minkowski metric for the flat space and corresponds to uniform rectilinear motion. However, equation (3) can have another solution. We will search for $\vec{v}$ in the form

$$\vec{v} = \vec{V}\exp i\{(\vec{k}, \vec{r}) - \omega t\}, \qquad (4)$$

and the correction ε_{00} to the metric tensor component in the form

$$\varepsilon_{00} = \frac{1}{2}\frac{v^2}{c^2} = \frac{1}{2}\frac{V^2}{c^2}\exp 2i\{(\vec{k}, \vec{r}) - \omega t\}, \qquad (5)$$

in both cases, referring to the real parts. Then

$$\nabla\varepsilon_{00} = 2i\vec{k}\varepsilon_{00}$$

$$\nabla(\vec{v}, \frac{c^2}{2}\frac{\partial\varepsilon_{00}}{\partial\vec{v}}) = c^2 2i\vec{k}\varepsilon_{00}, \qquad (6)$$

$$\vec{\Omega} = \frac{i}{2}[\vec{V}, \vec{k}]\exp\{i\{(\vec{k}, \vec{r}) - \omega t\}\},$$

and equation (3) gives

$$\frac{d\vec{v}}{dt} = ic^2\left\{\vec{w} + [\vec{w}, [\vec{w}, \vec{k}]]\right\}\varepsilon_{00}, \tag{7}$$

where $\vec{w} = \frac{\vec{V}}{V}$, and the dispersion relation has the form

$$\omega\vec{w} = -\frac{1}{2}V\left\{\vec{k} + [\vec{w}, [\vec{w}, \vec{k}]]\right\}\exp i\{(\vec{k}, \vec{r}) - \omega t\}. \tag{8}$$

Thus, when a weakly curved, weakly anisotropic space is used to simulate the experimental space in regions inaccessible for direct observation, the trajectory of free motion of a particle is a generalized helix. This circumstance makes it possible to naturally use the concept of phase to describe the motion of a particle. The radius R and the pitch λ of such a helical line are

$$R = \frac{v\sin(\vec{v}, \vec{\Omega})}{|\vec{\Omega}|}, \lambda = 2\pi\frac{v\cos(\vec{v}, \vec{\Omega})}{|\vec{\Omega}|}. \tag{9}$$

In general case, the axis of the helix may not be straight. If the axis of such a helical line is closed, then a selected finite area of motion appears. In the simplest case, such a closed line is a circle and the trajectory will lie on the surface of the torus. Note that the surface of a torus embedded in a 4-dimensional space has zero curvature, which has an additional important meaning.

3. Diffraction of electrons

Figure 1a. depicts Young's experiment for light waves. The position z_{max} of the interference maximum numbered n is determined by the expression $z_{max} = \frac{n\lambda L}{d}$, i.e.

$$d \cdot tg\varphi = n\lambda. \tag{10}$$

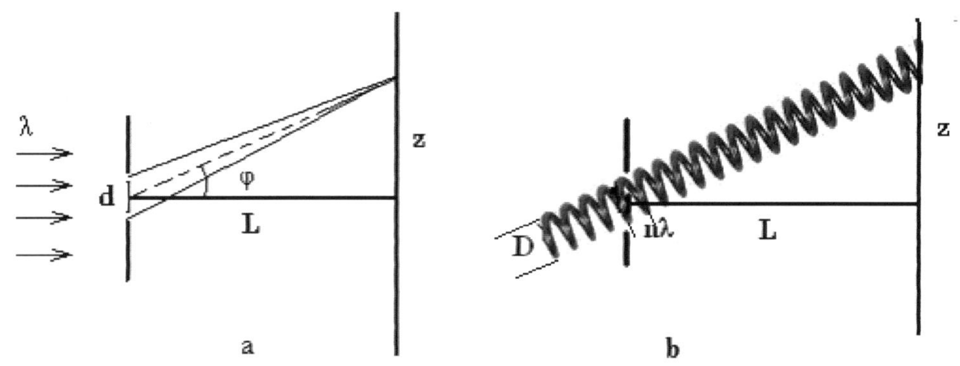

Figure 1. (a) Young's experiment; (b) Electron's trajectory in an anisotropic space.

Sequential transmission of single electrons through the crystal diffraction grating also leads to the illumination of the screen only in certain directions. The experimenter controls the device of the electron gun, the parameters of the crystal and the relative position of the gun, obstacle and screen, and also measures the location of the traces of electrons entering the screen. Everything that happens in the experimental space is one or another interpretation that explains the results. If we assume that the electron has wave properties, then the wave corresponding to each individual electron passes through the lattice, interacts with itself and, finally, is localized on the screen, forming a diffraction pattern. What is this wave?

Suppose that the electron has no wave properties, but the experimental space is slightly curved and slightly anisotropic. Then the electron moves along a helical line. Figure 1b shows an example of such a trajectory when an electron moves through a pair of slits, as in Young's experiment. Obviously, the electron can move in only such a direction towards the screen, the sine of which is equal to $\sin\varphi = \dfrac{n\lambda}{d}$, i.e.

$$d \cdot \sin\varphi = n\lambda, \tag{11}$$

where λ is the pitch of the helix and d is the distance between the slits. The particle will not be able to move in all other directions, because the phase of its passage through the second slit will differ from the phase of the passage of the first slit; the particle will not enter the second slit of the obstacle and, thus, will not reach the screen. At small angles typical for experimental verification, the sine differs little from the tangent, therefore, the interpretation of the experiment using formula (11) instead of formula (10) will remain satisfactory, and the geometric approach will adequately describe the results of the experiment. Replacing $d = \dfrac{D}{\cos\varphi}$, where $D = 2R$ is the helix diameter, we obtain the expression for the position of the maxima

$$D \cdot tg\varphi = n\lambda, \tag{12}$$

allowing to evaluate the parameters of the electron's trajectory.

4. Model of the atom

Bohr's semiclassical theory, which arose as a result of the interpretation of the results of Rutherford's experiments and spectroscopic observations of hydrogen emission, provided for the existence of stationary states and transitions between them. Let us combine Bohr's second postulate with the generalized Balmer formula and write down the expression for the energy of the emitted quantum

$$| E_n - E_m | = C^{(*)} \left| \frac{n^2 - m^2}{n^2 m^2} \right|; n, m = 1,2,3...; n \neq m, \tag{13}$$

where $C^{(*)}$ is proportional to the Rydberg constant.

Let the metric be such that the axis of the helix is a closed circle with radius R_A which is greater than the radius R of the helix. Then the particle will move along the surface of a torus. We will consider trajectories stationary if they are closed. Then

$$n_1 2\pi R_A = n_2 \lambda, \tag{14}$$

where n_1 and n_2 are integers, and in this case

$$tg(\vec{v}, \vec{\Omega}) = \frac{n_1}{n_2} \tag{15}$$

is a rational number. This means that the characteristic linear size of the region, on the surface of which the trajectory of the particle is located during its finite motion, cannot be less than $2R$. Numerical estimates [2], consistent with independent experiments, give a reasonable estimate for the size of such an atom. Then the discrete states will correspond to the volumes of the tori with closed trajectories on the surface. The volume of the torus is

$$V = 2\pi^2 R_A R^2. \tag{16}$$

Substitute here the expression for R from equation (9) and use formulas (4-6). Then, taking into account equation (15), we obtain

$$V = 2\pi^2 R_A \frac{v^2 \sin^2(\vec{v}, \vec{\Omega})}{|\vec{\Omega}|^2} = 2\pi^2 R_A \frac{v^2}{|\vec{\Omega}|^2} \frac{n_1^2}{n_1^2 + n_2^2}. \tag{17}$$

For simplicity, let us assume that the number of large revolutions for different states is the same and equal to 1, and the state is determined only by the number of small revolutions, i.e. $n_1 = 1; n_2 \equiv n$, and we get

$$V_n = 2\pi^2 R_A \frac{v^2}{|\vec{\Omega}|^2} \frac{1}{1+n^2}. \tag{18}$$

Then

$$|V_n - V_m| = C^{(**)} \left| \frac{n^2 - m^2}{(1+n^2)(1+m^2)} \right|; n, m = 0,1,2,...; n \neq m. \tag{19}$$

Comparing the obtained expression with formula (13), we see that the regularity of the arrangement of the observed spectral lines remained unchanged. Thus, as in the previous section, this group of experiments also admits an adequate geometric interpretation.

The constant present in the last formula has the form

$$C^{(**)} = 2\pi^2 R_A \frac{v^2}{|\Omega|^2} = 2\pi^2 R_A \frac{4}{|[\vec{w}, \vec{k}]|^2} = 2R_A \lambda^2 \tag{20}$$

and is related both to the expression for the correction to the metric and to the size of the atom.

Further analysis can take into account the different number of revolutions along the great circumference of the torus; deformation of the torus and the formation of various nodes; two different directions of movement for each stationary state. All this provides opportunities for introducing additional quantum numbers if necessary.

5. The Planck formula

The Planck formula follows from the Boltzmann law on the distribution of states by energy, according to which the probability of a state with number n is equal to $P_n = \frac{\exp(-E_n / kT)}{\sum_n \exp(-E_n / kT)}$. Then the average energy will be equal to $\langle E \rangle = \sum_n P_n E_n$.

Within the framework of the geometric approach, the state corresponds to the occupied volume, and the change in state is accompanied by the appearance or disappearance of one or several elementary toroidal vortices of the metric having an elementary volume

$$V_0 = 2\pi^2 R_A \frac{v^2}{|\vec{\Omega}|^2}. \tag{21}$$

The thermodynamic quantity kT, in accordance with its kinematic meaning, is determined by the square of the velocity

$$kT = bv^2. \tag{22}$$

Then, denoting $x \equiv \frac{2\pi^2 R_A}{b|\Omega|^2}$, we get

$$\langle V \rangle = V_0 \frac{\sum_n n \exp(-nx)}{\sum_n \exp(-nx)} = \frac{V_0}{\exp(x) - 1}, \tag{23}$$

as in the usual Planck formula. Since, according to equation (18), in the geometric approach, the volume V_n is inversely proportional to the squares of integers, the formula will not change its usual form.

6. Equation of state

Thus, the geometric theory satisfactorily describes the basic experiments of quantum mechanics without invoking the ideas of wave-particle duality. As is known, Schrödinger did not derive his equation, but postulated it on the basis of an analogy between action in Jacobi theory and the phase of a wave in wave theory. In this case, the wave function used to describe the physical situation had no visual meaning, and the measured quantities served as parameters of the equation.

A typical example of the use of the Schrödinger equation is the well-known problem of a harmonic oscillator and its energy levels [3]. The one-dimensional stationary Schrödinger equation for a linear oscillator illustrates the well-known approach to describing the state of an atom. It has the form

$$\frac{d^2\psi}{dx^2}+\frac{2m}{\hbar^2}(E-\frac{m\omega^2x^2}{2})\psi=0 \tag{24}$$

and the solution

$$\psi_n(x)=\left(\frac{m\omega}{\pi\hbar}\right)^{1/4}\frac{1}{\sqrt{2^n n!}}\exp(-\frac{m\omega}{2\hbar}x^2)H_n(x\sqrt{\frac{m\omega}{\hbar}}), \tag{25}$$

where

$$H_n(\xi)=(-1)^n\exp(\xi^2)\frac{d^n\exp(-\xi^2)}{d\xi^n} \tag{26}$$

are Hermite polynomials. The eigenvalues of equation (24), (in physical applications identified with energy levels), have the form

$$E_n=(n+\frac{1}{2})\hbar\omega;n=0,1,2.... \tag{27}$$

In the geometric approach, the state of an atom is characterized by the volume of the region available for the finite motion of the particle, which, in turn, is described by the geodesic equation. Using the kinematical tensor according to [1], instead of Maxwell's equations, one can obtain the geometric equation

$$i\frac{\partial\Psi_{(a)}}{\partial t}=c\cdot rot\Psi_{(a)}, \tag{28}$$

where $\Psi_{(a)}=E+iH$, $\vec{E}_{(a)}=-\nabla\varphi$ and $\vec{H}_{(a)}=rot\vec{A}$, and the potentials have a metric meaning

$$\frac{c^2}{2}\varepsilon_{00}\equiv\varphi,\ \frac{c^2}{2}\frac{\partial\varepsilon_{00}}{\partial\vec{v}}\equiv\frac{1}{c}\vec{A}. \tag{29}$$

Besides, $\vec{E}_{(a)}=ik\vec{A}$ and $\vec{H}_{(a)}=i[\vec{k},\vec{A}]$, where $\vec{A}=\mathrm{Re}\{\vec{A}_0e^{i(\vec{k}\vec{r}-\omega t)}\}$, and the wave vector is $\vec{k}=\frac{\omega}{c}\vec{n}$.

Then, to describe the stationary state, one can use the stationary equation

$$rot\psi_{(a)}=\lambda\psi_{(a)}, \tag{30}$$

whose eigenvalues are proportional to the zeros of the Bessel function of half-integer order [4]. They are also located at an equal distance from each other (see Figure 2).

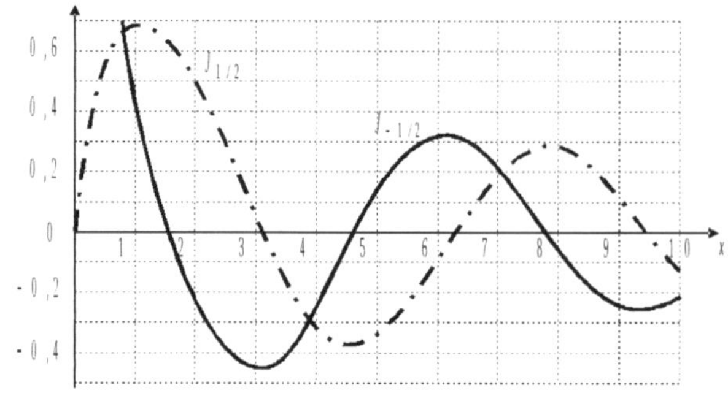

Figure 2. Location of the zeros of the Bessel function of half-integer order.

For the components of the eigenfunctions of equation (30), written in a spherical coordinate system, the following expressions take place [4]

$$\begin{aligned}(\psi_r)_\kappa &= c_\kappa(\lambda_{n,m}r)^{-1}\varsigma_n(\lambda_{n,m}r)Y^k{}_n(\theta,\varphi)\\ (\psi_\varphi + i\psi_\theta)_\kappa &= c_\kappa(\lambda_{n,m}r)^{-1}\varsigma_n(\lambda_{n,m}r)\Phi_n(\lambda_{n,m}r)HY^k{}_n(\theta,\varphi)\end{aligned}, \quad (31)$$

where

$$\varsigma_n(z) = \sqrt{\frac{\pi}{2z}}J_{n+\frac{1}{2}}(z)$$

$$\Phi_n(\lambda_{n,m}r) = \int_0^r \exp\{i\lambda_{n,m}(r-t)\}\varsigma_n(\lambda_{n,m}t)t^{-1}dt \quad (32)$$

$$HY^k{}_n(\theta,\varphi) = (\sin^{-1}\theta\partial_\varphi + i\partial_\theta)Y^k{}_n(\theta,\varphi)$$

$Y^k{}_n(\theta,\varphi)$ - real spherical functions, c_κ - rational numbers. These formulas are suitable for describing the motion of a curled fluid flow in a ball, there the flow velocity is $\psi_\kappa(x)$ at n = 1,2,3, ..., and its vorticity $rot\,\psi_\kappa$, equal to $\lambda_{n,m}\psi_\kappa$, is nonzero at each point of the ball. According to Arnold's theorem [5], almost all flow streamlines of an ideal fluid that satisfy equation (30) are wound either on cylinders or on tori. The rest of the flows can have streamlines with a very complex topology. Both these statements are consistent with the assumptions mentioned in Section 4.

Thus, the use of the geometric approach to describe both experimental results and fundamental properties concerning the microworld does not contradict the known results and has a number of advantages.

References

[1] Siparov S V 2021 *Proc. Conf. PIRT-2021* (Moscow: BMSTU)
[2] Siparov S V 2015 *Proc. Conf. PIRT-2015* (Moscow: BMSTU) pp 483-501
[3] Landau L D and Lifshits E M 1963 *Quantum mechanics* (Moscow: Gosudarstvennoe Izdatel'stvo Fiziko-matematichskoi literatury)
[4] Sax R S 2013 *Ufa mathematical journal* **5** 63
[5] Arnold V I 1965 *C. R. Acad. Sci. Paris*. **261** 17

On the force caused by a null Einstein-Maxwell field with the plane symmetry

V N Timofeev

Department of Physics and Chemistry, Saint Petersburg State University of Civil Aviation, 38 Pilotov Street, Saint Petersburg 196210, Russia

E-mail: WTimoff@yandex.ru

Abstract. The article shows that a large flat platform with a constant current, which flows over its surface, accelerates time. It is also shown that if an alternating current flows along the surface of a flat platform while creating a null electromagnetic field then a force repelling from the platform acts on the test particle located near it. This force has no gravitational nature and arises as a result of the curvature of space-time by the electromagnetic field of a flat platform with an alternating current.

1. Problem statement

In this paper, the following task is solved: to find out whether a flat platform, along the surface of which an electric current flows, creates a force whose nature is different from that of gravity. If this force exists, then find the direction of the force relative to the flat platform and the dependence on the current flowing along the surface of the platform.

2. The algorithm for solving the problem

Let the electric current flow along the conductive surface of the platform along the direction of the z-axis. To find out whether the flat platform creates the current-controlled force, let us apply the following algorithm of actions:

1) Find the solution to the equations of Einstein – Maxwell field describing a flat platform with your current.
2) Since we are talking about the man-made platform, which is unlikely to create a noticeable gravity by its weight, let us move on to the limit of small mass, so that when you turn off the current, the metric becomes flat.
3) Drawing on the behaviour of test particle we find out whether the force acts on it from the side of the platform.

3. The requirements for the metric for the plane with a current

Let us consider that the platform is quite large, and it can be envisioned as a boundless plane within the framework of the problem solved. Therefore, to solve the problem, it is necessary to know the type of the metric for the plane with a current. The metric should depend on two parameters I and σ, where I is the current flowing along the plane surface band of unit width (linear current density), σ is the surface mass density.

Choose the "Cartesian coordinate system" (t,x,y,z) so that the coordinate plane (y,z) coincides with the plane with current, and the direction of the z-axis with the direction of the current I.
The desired metric must meet the following conditions:

1) In the vacuum case $I=0$ the metric must turn into Taub's plane-symmetric vacuum solution.
2) For $I=0$ and $\sigma=0$, the metric should become flat:
$$ds^2 = dt^2 - dx^2 - dy^2 - dz^2.$$
3) The current density in this problem has only one non-zero component:
$$j^i = \left(0;0;0;\ j^3\right),\ where\ j^3 = I\delta(x). \tag{1}$$
4) In the classical case, the electromagnetic field outside the plane with alternating current will be a plane electromagnetic wave. Therefore, in this problem, we require the field of the plane with alternating current to be null. This means that the invariant
$$J = F^*_{\ kl} F^{*kl}, \tag{2}$$

where $F^*_{lk} = \frac{1}{2}\left(F_{lk} + i\tilde{F}_{lk}\right)$, $\tilde{F}_{lk} = \frac{1}{2}\sqrt{-g}\,\varepsilon_{lkmn}F^{mn}$ equals zero.

4. Taub's Metric

There are various types of writing this metric in the literature [1-4]. In this paper, we will choose the Taub's metric in the form in which it is presented in [5]:
$$ds^2 = \frac{1}{\sqrt{1-Mx}}\left(dt^2 - dx^2\right) - \left(1-Mx\right)\left(dy^2 + dz^2\right), \tag{3}$$
where M is a constant.

Let us determine the dependence of M on the surface mass density σ. Considering Mx small and taking into account only the first approximation of this value, we rewrite the metric component g_{00} in the form:
$$g_{00} \approx 1 + \frac{M}{2}x.$$
Whence it follows that the Newtonian gravitational potential of a massive boundless flat platform has the form:
$$\varphi_{grav} = \frac{M}{4}x,$$
and the intensity of the gravitational field is equal to
$$g_{grav} = \frac{d\varphi_{grav}}{dx} = -\frac{1}{4}M.$$
As is known from classical mechanics, the gravitational strength of the field of a boundless flat layer is equal to:
$$g_{grav} = -2\pi G\sigma,$$
where G is the gravitational constant, σ is the surface mass density.

Then for the value M we find:
$$M = 8\pi G\sigma.$$
The constraint $\left(x < \frac{1}{M}\right)$ in the metric is a consequence of the fact that the plane is infinite. The problem under consideration is local: $Mx \ll 1$. Therefore, this restriction does not affect the final result, and the chosen type of the Taub's metric is adequate for solving the task.

5. The general form of the metric. The null tetrad and the NP equations

As is known, the plane symmetry is given by three space-like Killing vectors [1]:

$$\xi_1 = \partial_y,\ \xi_2 = \partial_z,\ \xi_3 = y\partial_z - z\partial_y,$$

but for a plane with current, the y and z directions are not equal. Therefore, we can assume with greater confidence that there are only two Killing vectors:

$$\xi_1 = \partial_y,\ \xi_2 = \partial_z.$$

In GR, two space-like vectors define solutions with cylindrical symmetry. Therefore, we will choose the desired metric in the Weyl form.:

$$ds^2 = e^{-2U+2K}\left(dt^2 - dx^2\right) - \left(1 - Mx\right)^2 e^{-2U} dy^2 - e^{2U} dz^2, \qquad (4)$$

where $U = U\left(x,t\right),\ K = K\left(x,t\right)$.

In order to find out the metric needed to solve the problem we apply the Newman-Penrose formalism. Let's choose a null tetrad $\left(\boldsymbol{l}, \boldsymbol{n}, \boldsymbol{m}, \bar{\boldsymbol{m}}\right)$ in the form:

$$\begin{aligned} l^i &= \left(1,1,0,0\right), \\ n^i &= \left(n^0, n^1, 0, 0\right), \\ m^i &= \left(0,0,m^2,m^3\right), \\ \bar{m}^i &= \left(0,0,m^2,-m^3\right). \end{aligned} \qquad (5)$$

Taking into account the fact that the vector $\bar{\boldsymbol{m}}$ is complex-conjugate to the vector $\boldsymbol{m}$, it follows from the Weyl metric that the equalities are fulfilled for the components:

$$n^0 = -n^1,\ m^2 m^3 = \frac{i}{2\left(1 - Mx\right)}. \qquad (6)$$

For the selected tetrad, the spin coefficients have the form:

$$\alpha = \beta = \gamma = \tau = \pi = v = k = 0$$

$$\rho = \frac{M}{2\left(1 - Mx\right)},\ \varepsilon = \varepsilon^*,\ \tilde{\sigma} = \tilde{\sigma}^*,\ \mu = \mu^*,\ \lambda = \lambda^*.$$

From the equations of the Newman-Penrose formalism for the tetrad components and the spin coefficients we have (NP equations):

$$n^1 = -\frac{2}{M}\left(1 - Mx\right)\mu, \qquad (7)$$

$$Dm^2 = \left(\frac{M}{2\left(1 - Mx\right)} + \tilde{\sigma}\right) m^2, \qquad (8)$$

$$\Delta m^2 = -\left(\mu + \lambda\right) m^2. \qquad (9)$$

$$\tilde{\sigma}^2 - \frac{M\varepsilon}{\left(1 - Mx\right)} + \Phi_{00} - \frac{M^2}{4\left(1 - Mx\right)^2} = 0, \qquad (10)$$

$$-\Delta\varepsilon = \Psi_2, \qquad (11)$$

$$D\lambda = \frac{\lambda M}{2\left(1 - Mx\right)} + \mu\tilde{\sigma} - 2\varepsilon\lambda + \Phi_{20}, \qquad (12)$$

$$D\mu = \frac{\mu M}{(1 - Mx)} - 2\mu\varepsilon, \tag{13}$$

$$-\frac{\mu M}{2(1 - Mx)} + \lambda\tilde{\sigma} + \Psi_2 = 0, \tag{14}$$

$$-\Delta\mu = \mu^2 + \lambda^2 + \Phi_{22}, \tag{15}$$

$$-\Delta\tilde{\sigma} = \mu\tilde{\sigma} + \frac{\lambda M}{2(1 - Mx)} + \Phi_{02}, \tag{16}$$

where $D = \partial_t + \partial_x$, $\Delta = n^0\partial_t + n^1\partial_x$.

Here $\Phi_{ab} = \kappa_0 \phi_a \phi_b^*$, where ϕ_a is complex scalar, the electromagnetic field, $\kappa_0 = 8\pi G$ is the Einstein's constant.

Due to the plane symmetry, the electromagnetic field is polarized linearly and will be given by the electromagnetic field tensor, whose nonzero covariant components are:

$$F_{03}, \quad F_{13}.$$

The scalars of the electromagnetic field ϕ_a are equal to:

$$\begin{aligned} \phi_0 &= F_{ij} l^i m^j = (F_{03} + F_{13}) m^3, \\ \phi_1 &= \frac{1}{2} F_{ij} \left(l^i n^j + \bar{m}^i m^j \right) = 0, \\ \phi_2 &= F_{ij} \bar{m}^i n^j = (-F_{03} + F_{13}) m^3 n^1. \end{aligned} \tag{17}$$

As a result, Maxwell's equations written in terms of these scalars have the form:

$$D\phi_2 + \lambda\phi_0 - (\rho - 2\varepsilon)\phi_2 = -2\pi j^{\hat{3}}, \tag{18}$$

$$-\Delta\phi_0 - \mu\phi_0 + \tilde{\sigma}\phi_2 = -2\pi j^{\hat{4}}, \tag{19}$$

where $j^{\hat{3}} = m_i j^i$, $j^{\hat{4}} = \bar{m}_i j^i$ are the tetrad components of the 4-vector of the current density (1).

6. The Stokes' Theorem for the plane with a current

Let us find the magnetic field of the boundless plane with a current. Let us take a parallelepiped $S \times T$, where $S = \{x \times y \mid x \in [-a; a], y \in [-b; b]\}$ is a rectangle which is perpendicular to the z-axis, $T = [t_1, t_2]$ is a time interval. Let an electric current $\mathcal{J}$ flow through the rectangle S.

First, consider the case when a constant electric current flows through the conductor $\mathcal{J}$. From the left and right sides of Maxwell's equation

$$F^{ik}_{;k} = -4\pi j^i$$

take the integral over the parallelepiped $S \times T$:

$$\int F^{ik}_{;k} dS_i = -4\pi \int j^i dS_i, \tag{20}$$

where $dS_i = -\frac{1}{6}\varepsilon_{ijkl}\sqrt{-g}\,dS^{jkl}$. The integral on the right side of the equality (19)

$$\int j^i dS_i \equiv \int j^3 dS_3$$

is equal to the change in the charge that flows through the rectangle S over a period of time $\Delta t = t_2 - t_1$:

$$\Delta q=\int j^3 dS_3=-\frac{1}{6}\int j^3\varepsilon_{3jkl}\sqrt{-g}\,dS^{jkl}=\int j^3\sqrt{-g}\,dxdydt=\int_{t_1}^{t_2}\mathcal{J}dt.$$

To calculate the integral $\int F^{ik}_{;k}dS_i$, we apply the Stokes theorem [6]:

$$\int F^{ik}_{;k}dS_i=\frac{1}{2}\oint F^{ik}df^*_{ik}, \tag{21}$$

where $df^*_{ik}=\frac{1}{2}\varepsilon_{iklm}\sqrt{-g}\,df^{lm}$ is the tensor dual to the bivector of the infinitesimal element of the two-dimensional surface $df^{lm}=dx^l dx'^k-dx^k dx'^l$.

Given that the two faces of the parallelepiped passing through the points $x=-a$ and $x=a$ are parallel to the coordinate plane ty the integral on the right side (21) will be equal to

$$\frac{1}{2}\oint F^{ik}df^*_{ik}=\int F^{13}\varepsilon_{13lm}\sqrt{-g}\,df^{lm}=-2\int F^{13}\sqrt{-g}\,dtdy=-4b\int_{t_1}^{t_2}F^{13}\sqrt{-g}\,dt$$

As a result, the equality (20) takes the form:

$$-4b\int_{t_1}^{t_2}F^{13}\sqrt{-g}\,dt=-4\pi\int_{t_1}^{t_2}\mathcal{J}dt.$$

Whence it follows that the nonzero contravariant component of the electromagnetic field tensor is equal to:

$$F^{13}=\frac{2\pi I}{\sqrt{-g}}, \tag{22}$$

where $I=\frac{\mathcal{J}}{2b}$ is the current flowing through a band of unit width.

Now consider the case, an alternating current with a linear density flows along the plane $I=I(t)$. It follows from the energy-momentum tensor for the electromagnetic field

$$T^{ik}=-g_{lm}F^{il}F^{km}+\frac{1}{4}g^{ik}F_{lm}F^{lm}$$

that the Poynting vector for the desired metric has the form:

$$\vec{S}=\left(T^{01},T^{02},T^{03}\right)=\left(\frac{1}{2m^3\left(m^3\right)^*}F^{03}F^{13};0;0\right).$$

We assume that $\vec{S}$ is directed from the plane. This is possible when the values F^{03} and F^{13} have the same signs.

In accordance with requirement 4, the invariant (2) is zero:

$$J=F^*_{kl}F^{*kl}=16\phi_0\phi_2=0$$

from where, for the specified direction of the Poynting vector, we have:

$$\phi_0=0\Leftrightarrow F^{03}=F^{13}. \tag{23}$$

Outside the plane, we get from Maxwell's equations:

$$\partial_t\left(\sqrt{-g}\,F^{13}\right)+\partial_x\left(\sqrt{-g}\,F^{13}\right)=0, \tag{24}$$

$$\partial_t F_{13}+\partial_x F_{13}=0. \tag{25}$$

From equation (24) follows

$$\partial_t\left(\sqrt{g_{22}/g_{33}}F_{13}\right)+\partial_x\left(\sqrt{g_{22}/g_{33}}\,F_{13}\right)=0.$$

Comparing it with (25), we find: $g_{22}=g_{33}$. Then it follows from (4) that for a plane with an alternating current creating a null electromagnetic field around itself, the equation holds:

$$g_{22}=g_{33}=-\frac{1}{1-Mx}. \tag{26}$$

While integrating the equation (25), the nonzero contravariant components of the electromagnetic field tensor are obtained in the form:

$$F^{13}=F^{03}=\frac{2\pi I(u)}{\sqrt{-g}}, \tag{27}$$

where $u=t-x$.

7. A plane with a direct current. The acceleration of time

The solution for a plane with a direct current is known and presented in [7], but in this solution it is not clear how the linear current density I and the surface mass density σ are included in the metric. Their location in the metric is the key point in solving the problem. In order to find out this, we will carry out the process of integrating the NP equations.

For the field of a plane with a direct current, the nonzero component of the tensor F^{ij} is F^{13}. The field will be non-null, since for the invariant (2) we have

$$J=16\phi_0\phi_2\neq 0,$$

where $\phi_0=F_{13}m^3$, $\phi_2=-F_{13}n^1\bar{m}^3$. These scalars are related by the equality

$$\phi_2=n^1\phi_0=-\frac{2}{M}(1-Mx)\,\mu\phi_0. \tag{28}$$

From (22) we have

$$I=\frac{1}{2\pi}F^{13}\sqrt{-g}=-\frac{1}{\pi}(1-Mx)m^3\phi_0=Const. \tag{29}$$

As a consequence of (28), from the equations (18) and (19) the equations follow:

$$\lambda=-2(1-Mx)\,\mu\sigma,$$

$$\partial_x\phi_0=\tilde{\sigma}\phi_0+\frac{M}{2(1-Mx)}\phi_0,$$

From the equations (7)-(16) we get:

$$\partial_x\varepsilon=\frac{M^2}{4(1-Mx)^2}-\tilde{\sigma}^2,$$

$$\partial_x\sigma=\frac{M}{(1-Mx)}\tilde{\sigma}-\Phi_{00},$$

$$\Phi_{00}=\frac{M^2}{4(1-Mx)^2}-\tilde{\sigma}^2-\frac{M}{(1-Mx)}\varepsilon,$$

$$\partial_x\mu=\frac{M}{(1-Mx)}\mu-2\mu\varepsilon,$$

$$\partial_x m^3=-\tilde{\sigma}m^3+\frac{M}{2(1-Mx)}m^3.$$

Given the equality (29), integrating these equations yields:

$$\phi_0 = \frac{i}{\sqrt{2}}\frac{2\pi I}{\sqrt{1-Mx}\left(C(1-Mx)+1\right)},$$

$$\varepsilon = -\frac{3M}{4(1-Mx)} + \frac{M}{(1-Mx)\left(C(1-Mx)+1\right)},$$

$$\tilde{\sigma} = \frac{M}{(1-Mx)} - \frac{M}{(1-Mx)\left(C(1-Mx)+1\right)},$$

$$\mu = \frac{M}{4\sqrt{1-Mx}\left(C(1-Mx)+1\right)^2},$$

$$m^3 = i\frac{\left(C(1-Mx)+1\right)}{\sqrt{2}\sqrt{1-Mx}}.$$

As a result, while using the equalities (6) and (7), we get the metric in the form:

$$ds^2 = \frac{\left[C(1-Mx)+1\right]^2}{\sqrt{1-Mx}}\left(dt^2 - dx^2\right) - (1-Mx)\left[C(1-Mx)+1\right]^2 dy^2 - $$
$$ - \frac{(1-Mx)}{\left[C(1-Mx)+1\right]^2} dz^2, \qquad (30)$$

where $C = \frac{\pi I^2}{4G\sigma^2}$, $M = 8\pi G\sigma$, I is the linear current density.

The metric (30) has the following properties:

1) When the current is switched off $(I = 0)$, the metric turns into the Taub's vacuum metric (3).
2) When $I = 0$ and $M = 0$, the metric becomes flat.
3) It follows from (22) that the magnetic field is given by the non-zero covariant component of the tensor of the electromagnetic field:

$$F_{13} = \frac{2\pi I}{\left(C(1-Mx)+1\right)^2}.$$

In the case when I and M are small, the magnetic field strength is equal to:

$$\vec{H} = 2\pi I\vec{j},$$

where $\vec{j}$ – unit vector along the z-axis. From where, in SI units, we get the magnetic field induction in the form:

$$\vec{B} = \frac{\mu_0 I}{2}\vec{j},$$

where μ_0 is the magnetic constant. This expression coincides with the classical value of the magnetic field induction, which is created by a boundless plane with a direct current.

From the metric it can be seen that in the case of a light platform $(Mx \ll 1)$ a flat platform with a direct current does not create any tangible force directed along or against the x-axis and having a non-gravitational nature. In this case, the electric current causes the deformation of space-time. As a result, when a direct current flowing along the surface of the plane is turned on the acceleration of time occurs:

$$\Delta t_0 = \Delta t\left(1+\frac{\pi I^2}{4G\sigma^2}\right),$$

Δt_0 is proper time.

8. The metric of a plane with an alternating current

Let us consider the case when an alternating current flows through the plane. It follows from (23):

$$\phi_0 = 0.$$

For convenience, let us use the coordinate system (u, r, y, z), where $u = t - r$. In these coordinates

$$n^0 = \frac{4}{M}(1-Mx)\mu,\ n^1 = -\frac{2}{M}(1-Mx)\mu.$$

At points outside the z-axis, the current is zero: $j^i = 0$. Therefore, it follows from Maxwell's equations (19) that space-time is shift-free: $\sigma = 0$. As a result, equations (12), (14) and (18) follow

$$\lambda = 0,\ \Psi_2 = \frac{M}{2(1-Mx)}\mu,\ \varepsilon = \frac{M}{4(1-Mx)}.$$

For the μ function, which completely determines the type of metric, the system of equations follows from the NP equations:

$$\partial_x \mu = -\frac{M}{(1-Mx)}\mu,$$

$$\partial_x \phi_2 = 0,$$

$$-\Delta\mu = \mu^2 + \Phi_{22},$$

where $\Delta = n^0\partial_u + n^1\partial_x$. While integrating these equations, we find

$$\mu = \frac{M}{4\sqrt{1-Mx}} e^{-\frac{8\pi^2}{\sigma}\int_0^u I^2 dt},\ \phi_2 = i\sqrt{2}\pi I e^{-\frac{8\pi^2}{\sigma}\int_0^u I^2 dt}.$$

As a result, taking into account (26), we get the metric in the form

$$ds^2 = \frac{1}{\sqrt{1-Mx}} e^{\frac{8\pi^2}{\sigma}\int_0^{t-x} I^2 d\tau}\left(dt^2 - dx^2\right) - (1-Mx)\left(dy^2 + dz^2\right). \quad (31)$$

The solution (31) is a particular solution of the general solution obtained by Misra and Radhakrishna for the null electromagnetic fields with the cylindrical symmetry [8].

The metric has the following properties:

1) Just as in the case when a direct current flows through the plane, when the current is turned off $(I = 0)$, the metric turns into the Taub's static metric (3), and in the case when there is no plane with a current ($I = 0$, $M = 0$), the metric becomes flat.
2) It follows from (27) that the magnetic field of the plane with an alternating current is given by non-zero covariant components of the electromagnetic field tensor:

$$F_{13} = -F_{03} = 2\pi I(t-x).$$

3) In the tetrad (5), the only nonzero Weyl scalar has the form:

$$\Psi_2 = \frac{M^2}{8(1-Mx)^{3/2}} e^{-\frac{8\pi^2}{\sigma}\int\limits_0^{t-x} I^2 d\tau}.$$

This means that the null vector $\boldsymbol{l}$ as well as the null vector $\boldsymbol{n}$ form a shear-free null geodesic congruence and the solution is of Petrov type D. In the literature, space-time of type D is usually interpreted as the gravitational field of insular massive bodies, and the scalar Ψ_2 is associated with their mass, which also confirms the correctness of the conclusion about the connection of the constant M with the surface mass density.

9. The acceleration of the test particle

Consider the case of a weak gravitational field ($Mx \ll 1$, $\left\langle \frac{I^2}{\sigma} \right\rangle \ll 1$, where $\langle\ \rangle$ is the averaging over time). In this case, the g_{00} component of the metric tensor is equal to

$$g_{00} \approx 1 - \frac{4\pi^2}{\sigma}\int\limits_0^{t-x} I^2 d\tau.$$

In such a field, the acceleration of the test sample particle along the x-axis will be equal to

$$\frac{d^2x}{dt^2} \approx -\Gamma^1_{00} \approx -\frac{1}{2}\frac{\partial g_{00}}{\partial x} \approx \frac{4\pi^2}{\sigma} \cdot I^2.$$

The acceleration of the particle is directed from the plane, i.e. the particle experiences the force which repels it from the plane:

$$F \sim \frac{I^2}{\sigma} \tag{32}$$

For the current which changes according to the harmonic law

$$I = I_0 \cos \omega t,$$

the time-averaged acceleration of the particle has the form:

$$\frac{d^2x}{dt^2} \approx \frac{2\pi^2 I_0^2}{\sigma}.$$

The cause of the force (55) is the curvature of space-time by the null Einstein-Maxwell field.

10. Conclusion

Thus, we can draw the following conclusion: if an alternating current flows along the surface of a large flat platform, creating a null electromagnetic field, then a test particle located near this platform, in addition to the gravitational attraction of the platform, will be repelled by the force from the platform, the value of which is directly proportional to the square of the linear current density and inversely proportional to the surface mass density:

$$F \sim \frac{I^2}{\sigma}.$$

References

[1] Taub A H 1951 Empty space-times admitting a three parameter group of motions *Ann. Math.* **53** 472

[2] Bashkov V I 1976 Klassy tochnyh reshenij uravnenij Ejnshtejna [Classes of exact solutions of the Einstein equations] *Itogi nauki i tekhn. Ser. Algebra. Topol. Geom.* **14** 281

[3] Bedran M L, Calvao M O, Paiva F M and Soares D 1997 Taub's plane-symmetric vacuum spacetime reexamined *Phys. Rev.* D **55** 3431
[4] Stephani H, Kramer D, MacCallum M, Hoenselaers D and Herlt E 2003 *Exact Solutions of Einstein Field Equations* (Cambridge: Cambridge University Press)
[5] Jones P, Munoz G, Ragsdale M and Singleton D 2008 The general relativistic infinite plane *Am. J. Phys.* **76** 73
[6] Landau L D and Lifshits E M 1988 *Field theory* (Moscow: Nauka)
[7] Richterek L, Novotný J and Horský J 2002 Einstein-Maxwell Fields Generated from the γ-Metric and Their Limits. *Czechoslov. J. Phys.* **52** 1021
[8] Misra M and Radhakrishna L 1962 Some electromagnetic fields of cylindrical symmetry *Proc. Nat. Inst. Sci.* **28** 632

From Photon to Oganesson: Lie Algebra Realization of the Standard Model Extending over the Periodic Table

Erik Trell

Department of Health, Medicine and Caring Sciences, Linköping University, Se-58183 Linköping, Sweden

E-mail: erik.trell@gmail.com

Abstract. As reported in a series of previous PIRT conferences, a direct SU(3) structural realization of the Standard Model has been developed based upon Marius Sophus Lie's original Norwegian Ph.D. thesis *Over en Classe Geometriske Transformationer* from 1871 (and thus due for a most deserved 150-year anniversary). It elucidates how "the theory of main tangential curves can be brought back to that of rounded curves", anticipating a coherent linear representation of the elementary particles instead of the rotational chosen since they were considered point-like and amorphous when they many years later entered the stage. Under these premises the Standard Model has built a magnificent, undoubtedly true but congested multi-particle system whereas the Lie continuous transformation element, the partial derivative 'straight line of length equal to zero' outlines an isotropic vector matrix lattice of crystallographic Killing root space diagram A_3 form which from the Nucleon and inwards can backtrack the Standard Model geometrically, as well as continue outward iterating to a space-filling solid state $R^3 \times SO(3)$ wave-packet complex tessellating the whole periodic table with electron shells and subshells, isotope spectrum, neutron captures, radiative channels, oxidation states, molecular binding sites etc. in successive layers also including the Lanthanides in the sixth period and the Actinides in the seventh, in which now the concluding Oganesson has been reached in perfectly well-built saturated noble gas shape and condition.

1. Introduction

Developed since the early 70.ies and consolidated by the confirmation of the Higg's field and mechanism in 2012, the Standard Model has reached canonical status and power "to describe the most basic building blocks of the universe" [1] (Figure 1a) and their "greatest equation ever written down" [2] (Figure 1b). However, although mathematically coherent and self-consistent this still "in

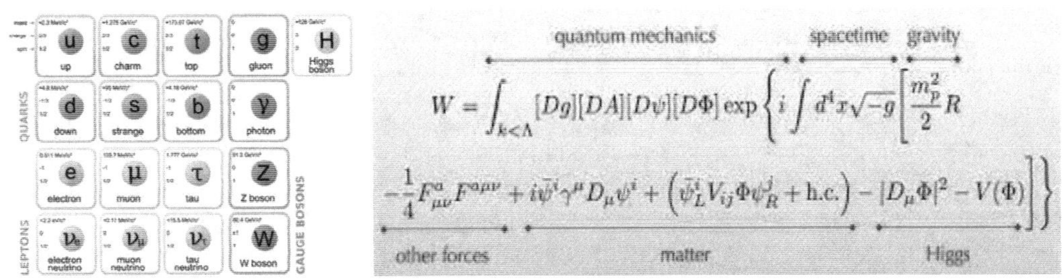

a b

Figure 1. a) Total Standard Model block diagram b) Total Standard Model equation (*Both Wikipedia*)

Journal of Physics: Conference Series **2081** (2021) 012034 doi:10.1088/1742-6596/2081/1/012034

fact...predicts very little about the forces and particles whose behaviour it describes, and information on force strengths, particle mass and so on must be input from experimental measurements" [3]. It is an annoying dilemma, and intensive research is going on, not to replace the mother equation but to amend it. Especially David Tong has emphasized that "if we study the grand equation more there are other patterns that emerge…we should go back to the drawing board and start thinking about different ideas…things like condensed matter physics or quantum information theory" [2].

I have since many years followed a related structural trail based upon Marius Sophus Lie's Norwegian Ph. D. thesis *Over en Classe geometriske Transformationer* from 1871 [4,5], in which he originally, in plain language in order to reach the public and the dissertation committee, carefully presented his continuous infinitesimal transformations and their ties with differential equations and spherical geometry. He got excellent marks and praise in the defense but in fact "not one…had understood a single word of it" [6], so the thesis soon went down to oblivion in both linguistic and physical isolation in the faculty archives until a hundred years later I managed to get a photocopy of it from the Oslo University library (and translated it to English [5]). I realized what a treasure it is, essentially advancing from the very bottom ground "the nature of Cartesian geometry" [4,5], whose fundamental condition is that "the extension in length, width, and depth which constitutes the space occupied by a body, is exactly the same as that which constitutes the body…consequently, there cannot exist a space separate from body, since all spatial extension simply is body" [7].

2. Methods

The Eigenelement of extension is the straight line [8], which is thus a physical structure spanning Euclidean space both as an infinite, basically cubic lattice and its coordinate axes, which remain central in any magnitude down to the limit level where it as the infinitesimal generator of itself ends up in a Lie neighborhood of eight Cartesian segments which determines mirror symmetry to all inclusions and exchanges. Furthermore, "Descartes has among the unlimited manifold of possible coordinate systems chosen a definite one" with "x, y, z...as parameters" [4,5], by which in this connection the Lie groups and transformations in them are bound to Euclidean space. And finally, "Cartesian geometry…translates any geometric theorem into an algebraic one and thus…of the geometry of space a representation of the algebra of three variable quantities", so that Lie groups and their own "reciprocal" algebras share structural element, which in infinitesimal transformations amounts to, "instead of the point, the collection of…imaginary straight lines whose length equals zero", that is, the partial derivatives, where "x, y, z are perceived as parameters, dx, dy, dz on the other hand as direction-cosines" [Ib.]. It is perfectly arranged for an ordinary space realization of the smallest things in a mutually "tangential…curve net" (Figure 2a), where "the Plücker line geometry can be transformed into a sphere geometry" by its "straight lines of length equal to zero turning…into the sphere's rectilinear generatrices" [Ib.] (= infinitesimal generators in present terminology). Their A_2 root diagram over the complex field is duplicated when they join with the central Cartesian coordinate axes in real space to an A_3 form hybrid (Figure 2a) in whose lattice (Figure 2b-d) the infinitesimal continuous transformations

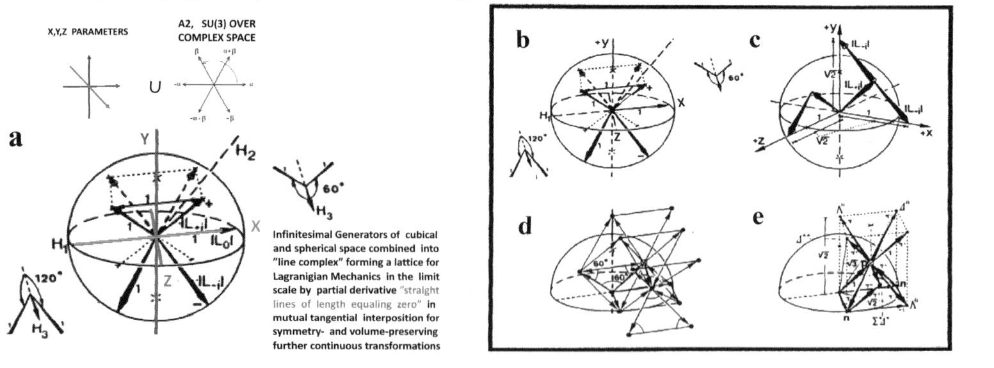

Figure 2. a,b) Cellular automaton hybrid for SU(3), SU(2), U(1) transformations in real space c,d) 90-180° and 60-120° root vector trains e) Cartesian segment lattice cell with major semiaxis projections

are carried out by successively each other "touching" [Ib.] steps of the root vector elements or over the surface of their projections.

3. Results

This "particularly remarkable transformation" system with its "equations that map the two spaces into each other" projects an "infinitesimal sphere" [Ib.] as a differentiable manifold in mutual tangential reciprocity with its Lie neighbourhood. The continuous further transformations there preserving e.g. symmetry and gauge form the respective Lie group as well as algebra. In the elementary particles it is precisely the spherical transformations that replicate the natural events as well as the Standard Model.

3.1. The baryons

Just as in reality, where the transformations occur in response to some reciprocal impact, a virtual "Game of Lie" Cellular Automaton [9] trial can be launched from the Nucleon's dual origin (Figure 2) as either a Neutron from the inner product volume element of the x,y,z cubical infinitesimal generators or a Proton from that of the spherical infinitesimal generators spanning a practically equal-sized volume and commutating in the horizontal plane. The stage is set for stepwise lattice transformations projecting symmetry- and volume-preserving ellipsoids and spheroids from the unit sphere identity. This has been described and illustrated in previous reports [10-21] so is here only exemplified by the basic *u,d,s* supermultiplet baryons (Figures 2 e, 3) and their major semiaxis steps in the lattice with the resulting

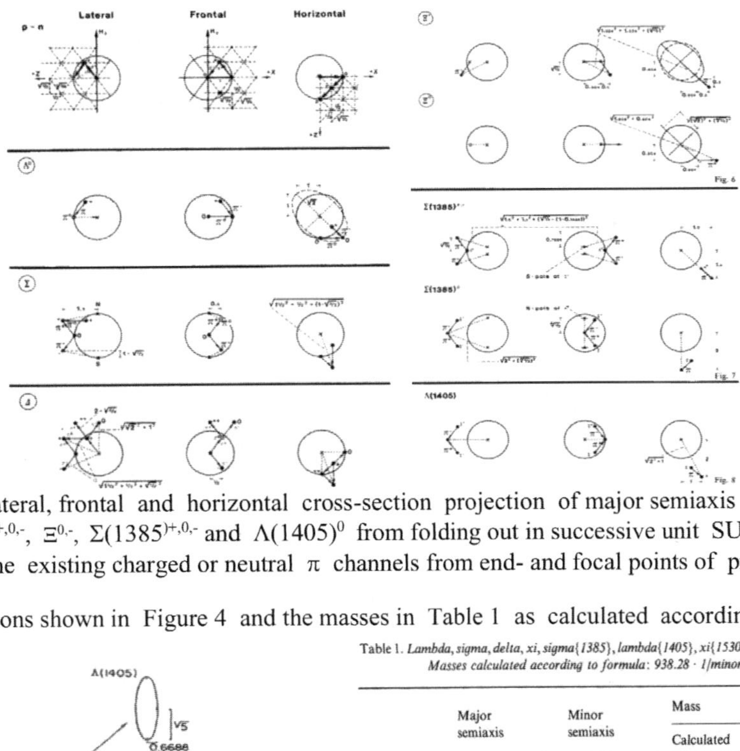

Figure 3. Lateral, frontal and horizontal cross-section projection of major semiaxis of the nucleon, Λ^0, Σ^{+-}, $\Delta^{++,+,0,-}$, $\Xi^{0,-}$, $\Sigma(1385)^{+,0,-}$ and $\Lambda(1405)^0$ from folding out in successive unit SU(3) root vector steps along the existing charged or neutral π channels from end- and focal points of preceding states

shape alterations shown in Figure 4 and the masses in Table 1 as calculated according to the quark

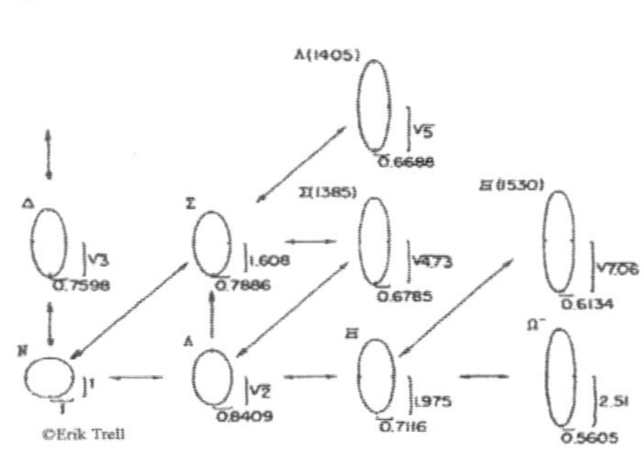

Figure 4. Basic eightfold way multiplet shapes

Table 1. *Lambda, sigma, delta, xi, sigma{1385}, lambda{1405}, xi{1530} and omega hyperons. Masses calculated according to formula: 938.28 · 1/minor semiaxis*

	Major semiaxis	Minor semiaxis	Mass	
			Calculated	Observed
Λ^0	$\sqrt{2}$	$\sqrt[4]{\frac{1}{2}}$	1115.8	1115.6
$\Sigma^{+,0,-}$	1.60804	0.788591	1189.8	1189.4 – 1197
$\Delta^{++,+,0,-}$	$\sqrt{3}$	$\sqrt[4]{\frac{1}{3}}$	1234.8	1230 – 1236
$\Xi^{0,-}$	1.975	0.7116	1318.5	1314.9 – 1321.3
$\Sigma(1385)^{+,0,-}$	$\sqrt{4.71}-\sqrt{4.75}$	0.679 – 0.678	1382.2 – 1385	1383 – 1386
$\Lambda(1405)^0$	$\sqrt{5}$	$\sqrt[4]{\frac{1}{5}}$	1403	1405 ±5
$\Xi(1530)^{0,-}$	$\sqrt{7.06}$	0.6134778	1529.5	1528 – 1534
Ω^-	2.505 – 2.51	0.561 – 0.560*	1673.5 – 1677	1672 – 1674

* Minor semiaxis changed in the transformation (c).

Table 1. Table of same with resulting masses.

pressure formula from the minor semiaxis contraction in inverse proportionality to the radius of the unit Proton. All of this has been thoroughly described in earlier papers for all observed *u,d,s* baryons in extensive and exact correspondence with recorded states and their masses, charge levels and angular momenta [Ib.] so will not be repeated here. The sphere-ellipsoid/spheroid transformation scheme in effect holds a number of interconnecting subgroups, e.g. spatial channel/flavor orientations alluded to in Figure 4, and moreover offers commutating separate degrees of freedom for the *charm* and *bottom* quark classes as well [16,17], and this will be described more in detail to further clarify the model.

While (the majority of) the *u,d,s* baryons are symmetrical and centrally related the *charm* baryons result from the much rarer instances when a high-energy e.g. collision hits directly on a minor semiaxis endpoint of a Proton or Neutron (or other state), compressing only this but leaving the second minor semiaxis unchanged = 1 in the case of a Nucleon impact, and squeezing out the new major semiaxis like a protuberance from the surface. In Figure 5 directly showing this by the projections of Λ_C^+, $\Sigma_C^{++,+,0}$, Ξ_C^+ and Ξ_C^0 along their numerous channels, this in Λ_C^+ leads to the new endpoint from which the length of the major semiaxis back to the origin is $(2^2 + \sqrt{2^2})^{1/2} = 6^{1/2}$ (Figure 5a), and when one of the minor semiaxes through the basis remains unit to preserve gauge reduces the length of the other to $1/6^{1/2}$, the mass according to the quark pressure formula will be $1/(1/6^{1/2}) \times 0.93827 \approx 2.30$ versus real 2.29 GeV. In $\Sigma_C^{++,+,0}$, shown from the Proton surface point (Figure 5 b) the major semiaxis in all varieties is $7^{1/2}$ $\{[(4.5^{1/2})^2 + 1.5^2 + 0.5^2]^{1/2}\}$ in Σ_C^{++}; $[1^2 + (6^{1/2})^2]^{1/2}$ in Σ_C^+ and $[2.5^2 + (0.5^{1/2})^2 + 0.5^2)]^{1/2}$ in Σ_C^0 with mass $= 1/1/7^{1/2} \times 0.93827 \approx 2.48$ versus observed 2.45 GeV in all charge states.

In the same way, Figure 5c,d (enlargeable in the screen) summarizes the plentiful channel spectrum and stages of Ξ_C^+ and Ξ_C^0. Three modes between the observed starting points and terminations are shown, all leading to a major semiaxis length of $7^{1/2}$ and consequential mass value $7^{1/2} \times 0.93827$ GeV = 2.48 GeV, in fine agreement with the observed 2.47 GeV in all states. In Figure 7b, an observed $\Xi^-\pi^+\pi^+$ channel of Ξ_c^+ and two alternatives of Ξ_C^0 are outlined. The $\Xi^-\pi^+\pi^+$ channel is interesting because there have been reports of two peaks of Ξ_C^+, one at 2.47 GeV and the other 75 MeV lower, which fits well with the 2.394 GeV obtained from the $(1.804^2 + 1.804^2)^{1/2} \approx 6.51^{1/2}$ channel. The other two for Ξ_C^0 end up at $7^{1/2}$ and $7.02^{1/2}$ with mass 2.48 and 2.485 respectively.

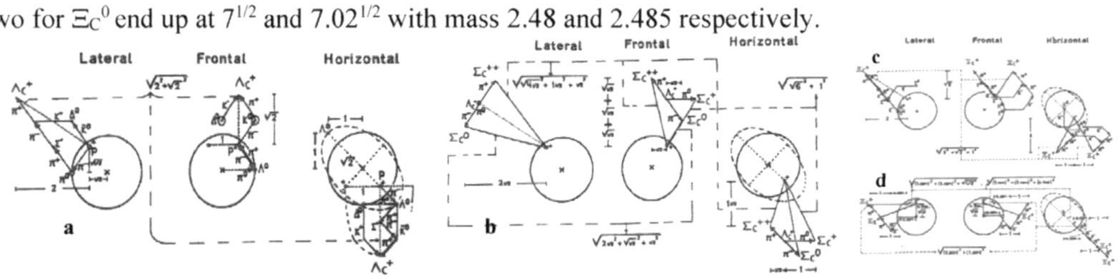

Figure 5. a) The Λ_C^+, b) $\Sigma_C^{++,+,0}$, and c,d) Ξ_C^+ and Ξ_C^0 *charm* baryon resonances and their channels

Also Ω_C^0 can be replicated but is not shown here. It is to be noted that only the ground states of the charmed hadrons and not their '*charmonium*' excited states or extra-Nucleon origins are included in the graphs. However, these can be traced by the same "flavor symmetry" transformations as those of the ordinary (*u,d,s*) baryon/hyperon/resonance series, whereby "the Λ_c and Σ_c spectra look much like the Λ and Σ spectra" and "the Ξ_c and Ω_C spectra ought to be richer than the Ξ and Ω spectra" [22,23].

The *bottom* baryons and their distinct degree of freedom of volume-preserving transformations occur by combining two *charm* major semiaxes to a correspondingly thinned and mass-increased disc, or cap, or torus. The highest observed mass number state is Ω_b^- at 6.07 GeV, obtained by a $6^{1/2} \times 7^{1/2}$ (being the major semiaxis of Λ_c^+ respective $\Sigma_c^{++,+,0}$ and $\Xi_c^{+,0}$) $\times$ 0,93827 $\approx$ 6.08 GeV slightly elliptic disc. When instead a round disc is incorporating the major semiaxis of Λ_c^+, the mass from the compressed minor semiaxis (or/and equivalently from the surface area against the Higgs mechanism) will be 1/1/6 times the Proton mass = 6×0.93827 GeV $\approx$ 5.63 GeV, which is identical with the 5.62 GeV of Λ_b^0. In other respects, too, like the decay modes and the '*bottonium*' spectrum [Ib.], the coincidences are extensive. There are also a few *bottom* baryon states between the Λ_b^0 and Ω_b^- (i.e. $\Sigma_b^{+,0,-}$ and $\Xi_b^{0,-}$) and all with a mass around 5.8 GeV. And there are many corresponding asymmetric disc/cap configuration candidates, e.g. a combination of axes $6^{1/2}$ and $6.5^{1/2}$, giving a mass value of 5.86 GeV, and where the half-integer axis step comes naturally from the root vector lattice, too. There is no need, therefor, to illustrate the *bottom* baryons. And the peaking *top* is not a transformation but a jet in the quark-gluon blowout crescendo.

3.2. The mesons

The mesons appear as transient difference vortices between forming as well as decaying hadrons including themselves to such a multitude that they outnumber the quark-antiquark combinations over the Standard Model flavor range. There are also unflavored mesons, so a more realistic classification is offered by the Lie product group SU(2) × U(1) of their geometrical constitution by which their mass can be approximated as the symmetric area times the antisymmetric tightness between the interchanging particles in the transformation lattice. This has also been reported in further detail before [10-21] so is here only summarized for some basic *u,d,s* mesons in Figure 6 and Table 2.

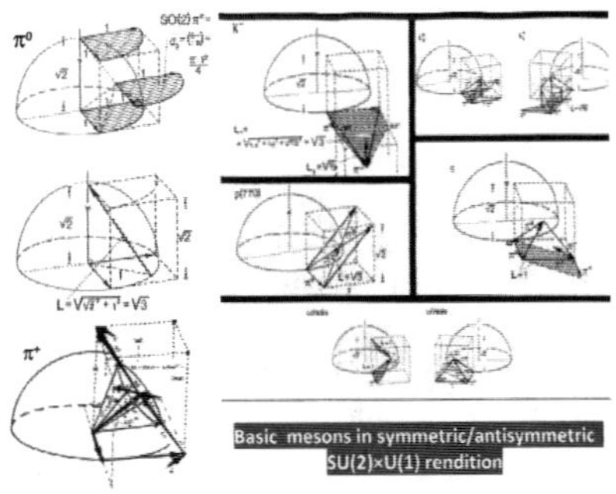

Figure 6. Ground meson reproductions

Table 2
Basic mesons calculated and observed mass numbers (MeV)

π^0	$1/4 \times 938.27 \times 1/3^{1/2}$	135.4	135.0
$\pi^\pm$	$1/6 \times 938.27 \times 1/(5/4)^{1/2}$	139.9	139.6
$K^\pm$	$938.27/4 \times 1/3^{1/2} + 938.27/4 \times 1/3^{1/2} + 938.27/6 \times 1/2^{1/2}$	492.0	492.7
K^0_S	$938.27/(4 \times 1/2^{1/2}) + 938.27/(8 \times 1/2^{1/2})$	497.6	497.67
K^0_L	$938.27/(8 \times 1/2^{1/2}) + 938.27/(8 \times 1/2^{1/2}) + 938.27/(8 \times 1/2^{1/2})$	497.6	497.67
η	$938.27/6 + 938.27/6 + 938.27/4$	547.33	548.8 ± 0.6
$\rho(770)$	$(938.27/1/2^{1/2})/(3/4)^{1/2}$ or $(938.27 \times 2^{-2})/3^{1/2}$	766.1	768.3 ± 0.5
$\omega(783)$	$938.27/4 + 938.27/4 + 938.27/6 + 938.27/6$	781.9	781.95 ± 0.14

Table 2. Computed and observed ground meson masses.

In Figure 7 it is next turned to the *charm* mesons, starting with D^+ and D^0. In D^+ (D^- obtained by reflection) three varieties are discerned (Figure 7 a-c); the longest sequences span a horizontal side length = 4, and in the charged direction = 1, or a length of 2 in both. The distance to the next $D^\pm$ on either side is 0.5. The product mass is thus $1 \times 1/0.5 \times 0{,}93827$ GeV $= 2 \times 0.93827$ GeV ≈ 1.88 (1.877) GeV, which complies very well with the recorded 1.87 (1.869) GeV. In D^0 and its antiparticle this simple plan persists as shown in Figure 7 d. Further example are given by the relatively stable charmed mesons D_s^+ and D_s^{*+}. In the former (D_s^- obtained by charge conjugation), the channels $K^-K^+\pi^+$ and $K^-K^+\pi^+\pi^-\pi^+$ establish a *t* isospin semiaxis of length 2 with the other axis of the weak isospin plane = 3 (Figure 7 e). The distance to the next such plane is $0.5^{1/2}$ and the mass is $3/2 \times 2/2 \times 1/0.5^{1/2} \times 0.93827$ GeV ≈ 1.99 GeV versus the measured 1.97 GeV. The $D_s^{*\pm}$ isodoublet (Figure 7f) has mass 2.11 GeV and a sole $D_s^\pm - \gamma$ mode to a straight vector of length = 1 within an unaltered electromagnetic charge plane which gives a weak isospin plane length = 3. The other sides retain this length so the surface area of the SU(2) plane is $3/2 \times 3/2$, and the distance in the unilateral γ direction to the neighbor = 1. The mass is then $1.5 \times 1.5 \times 0.93827$ GeV ≈ 2.11 GeV.

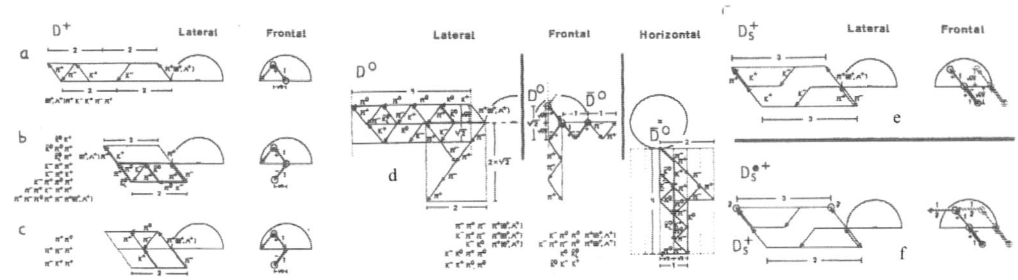

Figure 7. a - c) The D^+ and d) D^0 charmed mesons and their channels e,f) D_s^+ and D_s^{*+}

Also the prototype $B^\pm$ and B^0 *bottom* mesons are retrieved as shown in Figure 8 a, b. There is a close relation to the D mesons by which almost directly or by single pion steps forging the SU(2) plane major semiaxis into the radius of a round or elliptic disc, as sketched at very reduced scale in the D^0-π^+ mode for B^+ (B^- by charge conjugation) in Figure 8 a. It produces a pion semiaxis of length 1 and a D^0 vector of length 4, spanning an area of 1×4. The distance to adjacent $B^\pm$ states is $0.5^{1/2}$ and combined mass expression $4 \times 1/(0.5^{1/2}) \times 0.93827$ GeV ≈ 5.30 GeV in comparison with actual value of 5.28 GeV. In B^0, three varieties are shown (Figure 8 b). In the $D^0/D^0\ \pi^+\pi^-$ modes the elliptic semiaxes are $0.5^{1/2}$ and 4, and 1 and $2(2^{1/2})$, and the distance to the neighbors = 0.5. The resulting mass is $0.5^{1/2} \times 4 \times 1/0.5 = 0.5^{1/2} \times 8 \times 0.93827$, and $1 \times 2(2^{1/2}) \times 1/0.5 = 4(2^{1/2}) \times 0.93827$ GeV, respectively; which in both cases, like in the Λ_c channel ≈ 5.30 GeV versus the real 5.28 GeV. The antiparticle is a degeneracy in the horizontal plane and caused by one semiaxis = 4 and the other = 1, or both = 2, and distance to the neighbor = $0.5^{1/2}$, with mass = $4 \times 2^{1/2} \times 0.93827 \approx 5.30$ GeV, too.

In this connection also the W (Figure 8c) and Z (Figure 8d) gauge vector bosons can be briefly mentioned as the stack of respectively charged and neutral channel vectors contained and released in a colliding e.g. proton/antiproton at what could be called the total ionization energy of the state. In $W^\pm$ this is 80.4 GeV and as earlier reported [16,17] the fact is that when one makes an addition of the mass value of all its vectors (Figure 8c) it amounts to $2 \times [36 \times 0.93827 + (10 \times 2^{1/2}) \times 0.93827]$ GeV ≈ 80.8 GeV. In Z, the result is also within per mille precision; $2 \times [44 + (6 \times 2^{1/2})] \times 0.93827 \approx 90.5$ versus the observed 91.2 GeV (Figure 8d).

And dealing with the exotic aspects of the Standard Model also the gluon octet can be mimicked when coming out in three-jet ejections into the quark-gluon plasma from "break-up vertices" [24] of particles literally rupturing in ultrahigh energy collisions. It is tempting to see these events happening in the triangular outer corners of the bursting particle's eight Cartesian segments to that effect (Figure 8e). And at still higher energies, the whole particle/antiparticle fire ball may be smashed flat so gushing out in a double perpendicular jet the rapidly fragmenting *top* quark from the sides?

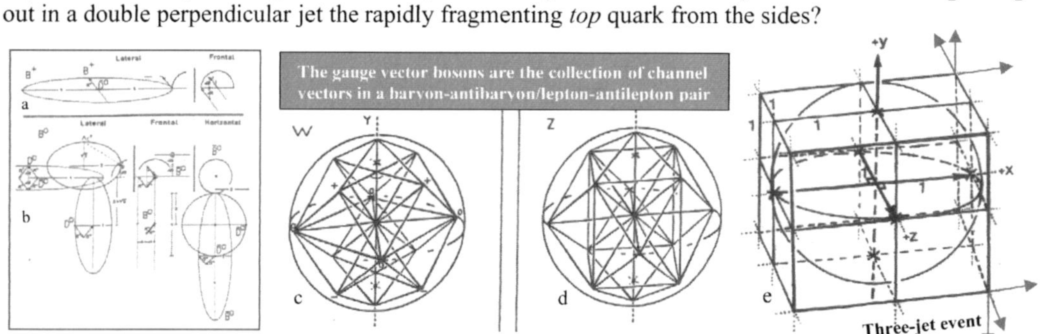

Figure 8. a) The B^+ *bottom* meson b) B^0 c) $W^\pm$ d) Z e) Gluon three-jet events

3.3. The leptons

Figure 9a gives an overview of the Muon and Tau leptons plus Photon and Neutrino events and Figure 9b summarizes electron processes which are extra-Nucleon and thus although included in the Standard

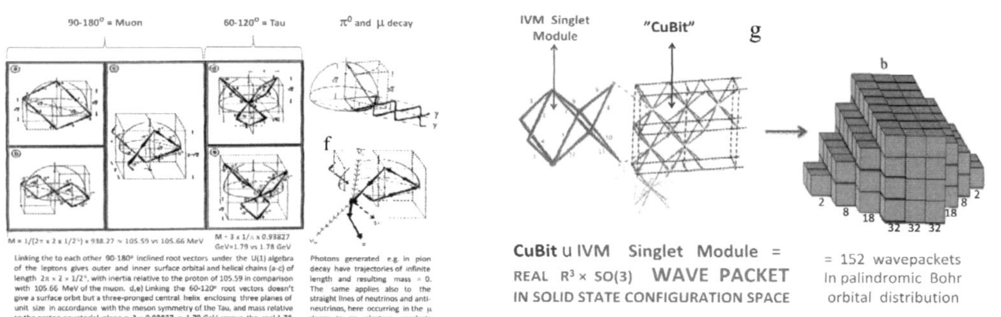

Figure 9. a-f) Muon, Tau, and Photon, Electron Antineutrino and Muon Neutrino g) Electron mesh

Model actually have no locus in it. Figure 2 c and d show the two ways of linking the charged intra-Nucleon root vectors to one-dimensional linear trajectories under the U(1) subalgebra of the leptons.

Joining the 90-180° inclined vectors (Figure 2c) over the surface of or inside the nucleon leads to a single (Figure 9a), duplicated (Figure 9b) or helical (Figure 9c) orbit, whose spherically projected length in all alternatives = $(2\pi \times 2^{1/2})$ and resulting mass in antisymmetric proportionality to the unit radius of the proton $= 1/(2\pi \times 2^{½}) \times 938.27 = 1/(2\pi \times 2\times1/2^{½}) \times 938.27 = 105.59$ MeV in comparison with the measured $\mu^{\pm}$ mass of 105.66 MeV. A linking of the 60-120° root vectors (Figures 2c, 9d, e) which would correspond to the electron (positron by inversion) doesn't lead to an orbit but to a three-pronged closed (Figure 9d) or helical (Figure 9e) propeller-like loop inside the nucleon but leaves no mass there since passing. Instead the three-pronged propeller blades strikingly simulate the Tau, which although classified as a lepton really is governed by the SU(2) part of the weak force SU(2) × U(1) algebra, that is, the mass is directly related to the 938.27 MeV = 0.93827 GeV of the unit Proton equatorial plane of differential area = 1. The area of each Tau blade can be calculated to $1/2^{½} \times 2^{½} = 1$, and there are three blades so the mass would come out as 3×0.93827 GeV ≈ 1.79 GeV. This compares very well with the recorded τ mass of 1.78 GeV and also in other respects, like multitude of production and decay modes, the correspondences are extensive but for matters of brevity only mentioned here as is the case for the remaining leptons; in Figure 9f the Photon (though a boson in the Standard Model) in the decay of π^0, and an Electron Antineutrino and Muon Neutrino in the decay of a Muon to an Electron. Both the Photon and the Antineutrino/Neutrino respectively zigzag and straight trajectories are of infinite length with resulting antisymmetric U(1) mass value zero.

Stepping out of the Nucleon domain (Figures 2e, 9g), the root vector lattice as described in the PIRT-19 conference proceedings [21] takes a space-filling one octahedron-two tetrahedrons isotropic vector matrix (IVM) route where the problem of finding a continuous non-overcrossing path is solved by the chaperon stacking of enveloping space segments as structural wave-packets in palindromic Bohr orbital order to flat-bottom floor modules complemented with a flat-roofed cap to a bilayer, effectively cubical compartment in which like in a ribosome conveyor belt [20] the space-filling geodesics can be hierarchically continued in the global IVM network. Not iterating the detailed description in the previous report [Ib.], a suite of illustrations from it (Figure 10) may recapitulate the process and how the stashing can be used by the combined wave-packet or by its separate constituents in the form of a Neutron and an Electron respectively, filling equal space portions and how this and its combinations allow the reproduction of the elements and their isotopes, here up till $_4$Beryllium^8.

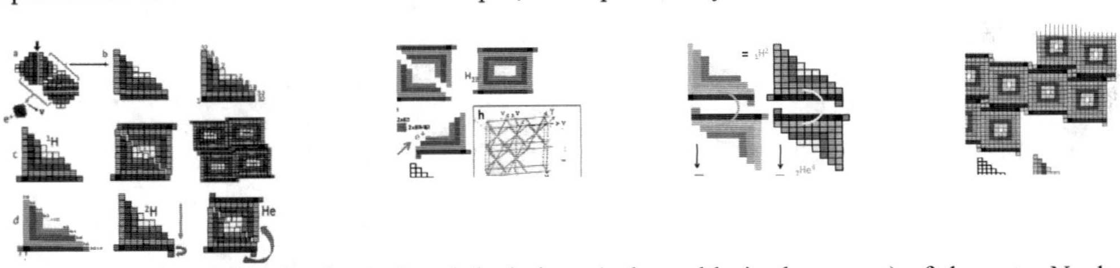

Figure 10. Suite of greatly size-reduced depictions (enlargeable in the screen) of the extra-Nucleon extension of the Standard Model, including the Electron (and Positron by inversion) and Photon and Neutrino production (from breaking-up of wall/lattice sequences). Note two varieties of module stashing; one without central hole (e.g. in Protium, $_1$He^1) an one with one forming a tunnel (from Deuterium and onwards) and that in this case the inertial relation between the unit hub and the total wave length (153 × 12) is 1/1836, that is, the electron/proton mass ratio. The Lego-type building blocks are fine-grained, which in larger assemblies gets blurred even when colors referring to those of the noble gases in raising order is applied to enliven the labor.

Now is only to continue the procedure, which in the previous report had reached Krypton [Ib.] From the third period the fine-grained, Lego-type building blocks used in the first (Figure 10) and second period were replaced by simplified 'Duplo' bricks [25] but after the third period these were modified to 'Triplo' shape, giving better reproduction of marginal layering, binding site variation and space-filling. Like in the elementary particle transformations, computer animation would be most helpful and valuable. This is supported by Figure 11, recapitulating the third and fourth period in Lego-style and Duplo execution plus the introduction of Triplo bricks.

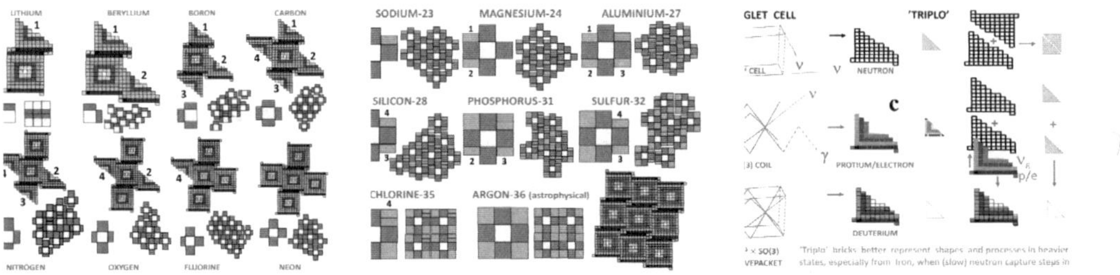

Figure 11. a) Second period in space-filling Lego-style and Duplo reproduction. The binding sites and the saturation of the concluding noble gas are evident. Note symmetric reactivity of Carbon, halogen binding site of Fluorine, and saturation of Neon. b) Duplo reproduction of third period with neutron (grey blocks) excess in stable isotopes c) Construction of 'Triplo' blocks

In the previous PIRT conference the fourth period was reported in Duplo style, and is here updated in Triplo notation in size-reduced figures that can be enlarged in the screen. Like in the previous presentation, (slow) neutron capture setting in from Iron are marked by red arrows in observed beta decay channels. The central parts of the isotope range are also shown from Iron and onwards.

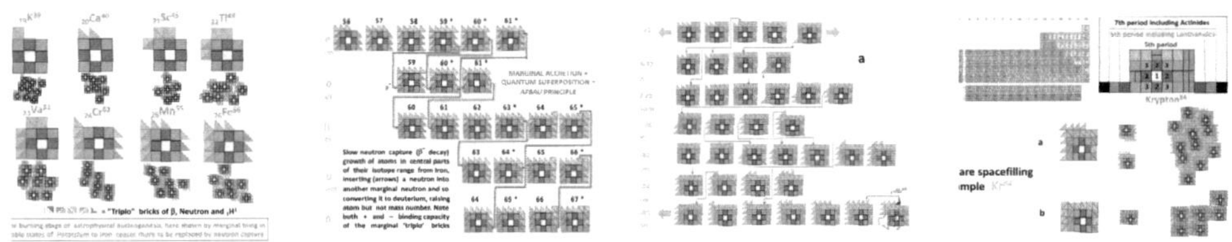

Figure 12. Triplo reproduction of fourth period plus comparison with the standard periodic system. The marginal layering of the present system is discerned, and likewise the absence of a gap in the initial periods.

The fifth period is new and here reproduced in Triplo design. As in all periods the first one or two neutron captures go to the outer shell, then to the flanking and in the terminal states successively returning to saturate the outer.

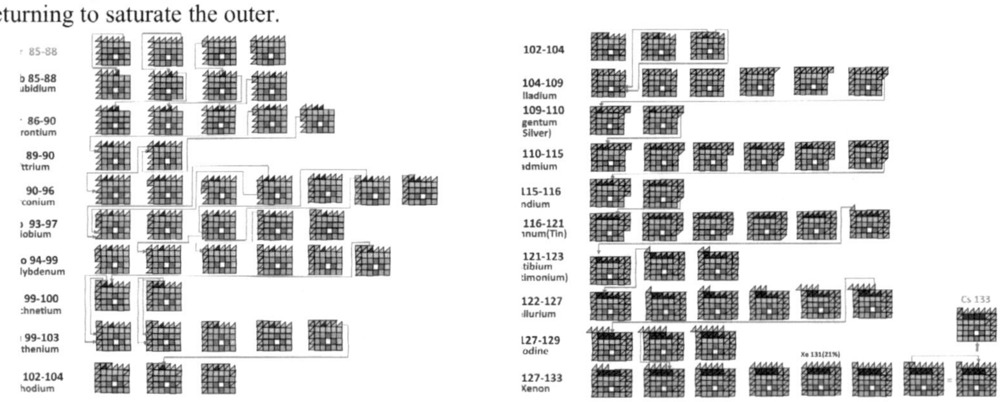

Figure 13. The fifth period with marginal accretion/superposition and observed neutron capture beta decay channels in central parts of the isotope spectrum. The settling of subshells is suggested. Full saturation of Xenon by terminal filling of four alpha blocks in outer layer is also seen.

The marginal growth and accompanying increase of the neutron excess and its implied design function is illustrated both for the fifth and sixth period below

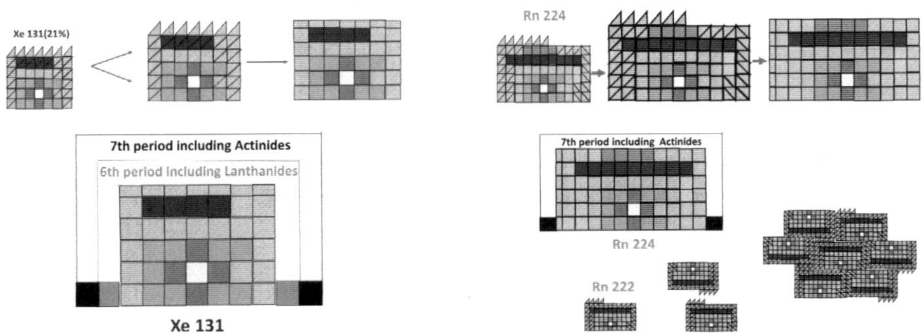

Figure 14. Marginal growth of periodic table in fifth and sixth period

In the sixth period, the lanthanides are incorporated. They show similar properties since mainly their subshells are loaded. One can follow this in the diagram as also the saturation process to Radon.

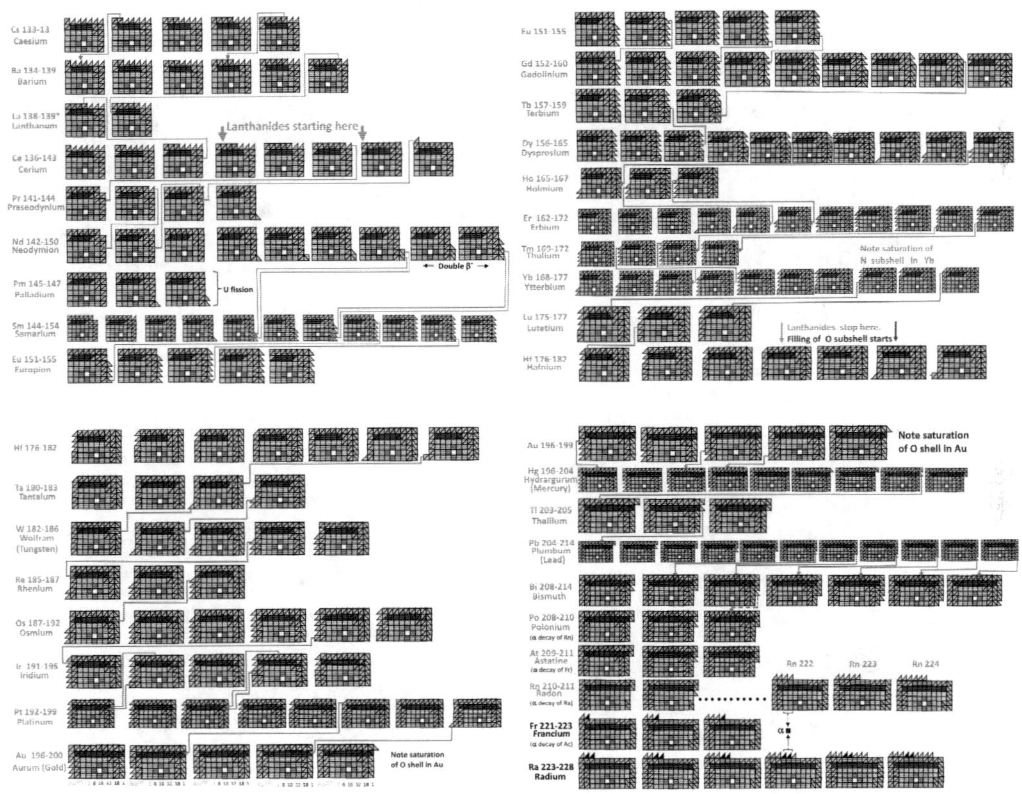

Figure 15. Sixth Period including the Lanthanides. Also alpha decay between Radium and Radon is shown. Many atoms, e.g. Francium are extremely rare and unstable. The highest stable states are Lead and Bismuth between the isotopes of which there are many beta decays. Otherwise the slow Neutron capture process largely stops here to be replaced by downward alpha decay from rapid Neutron capture processes in higher states.

The seventh period includes the actinides and leads to Oganesson.

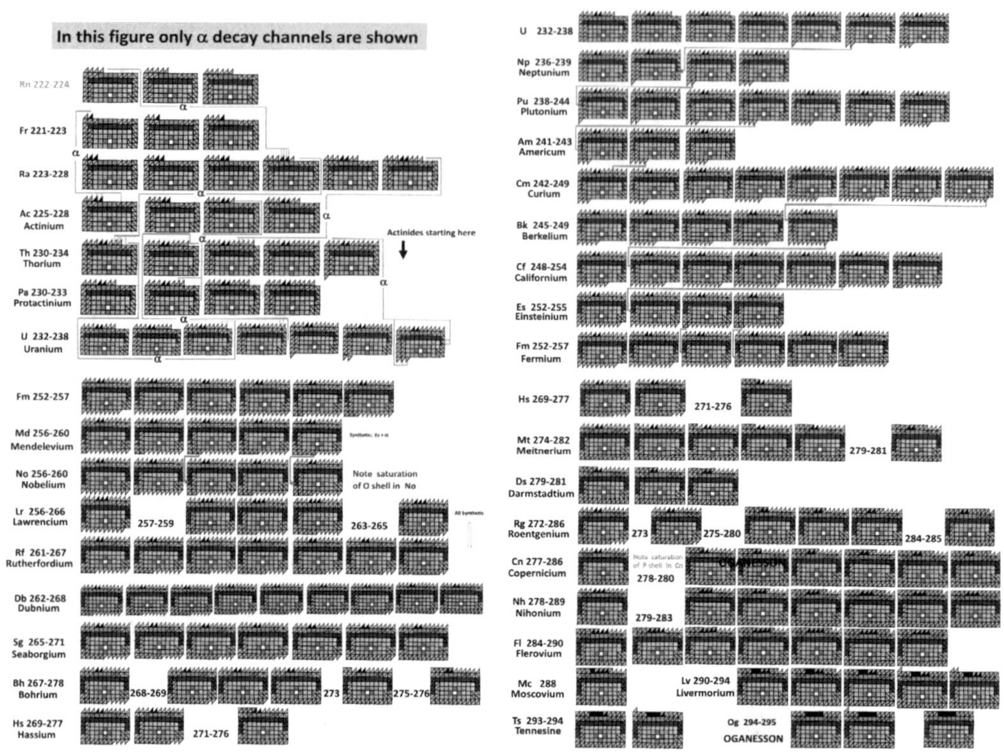

Figure 16. Seventh period culminating with Oganesson. All the atoms in the seventh period are unstable but some Uranium isotopes are long-living. The actinides are included and like the lanthanides mostly engaged in subshell saturation.

With Oganesson the goal is reached. It also forms the perfect full periodic table. It has one higher isotope, Oganesson 295, and that may start an eight period which can be outlined from information in the palindromic Bohr orbital distribution of the atoms.

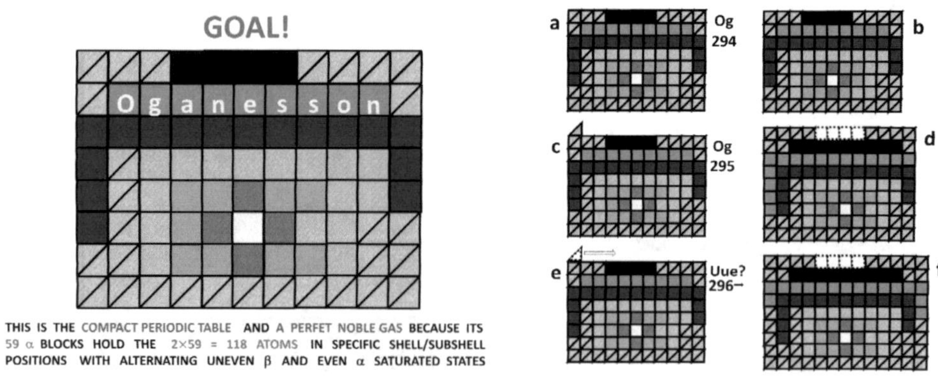

Figure 17. Oganesson and its 295 isotope which may start an eight period with its first, still not confirmed element Ununentium. Projections of full eight period are also shown.

The Bohr palindromic orbital model is not exhausted by the seventh period but has capacity for one more 32-period and a dwindling tail. And nothing is new and the wave-packet with a parallelepiped encasement around an internal coil is reminiscent both of ancient concepts and tilings as well as modern Clifford algebra trivectors and their operations which also have an internal twist.

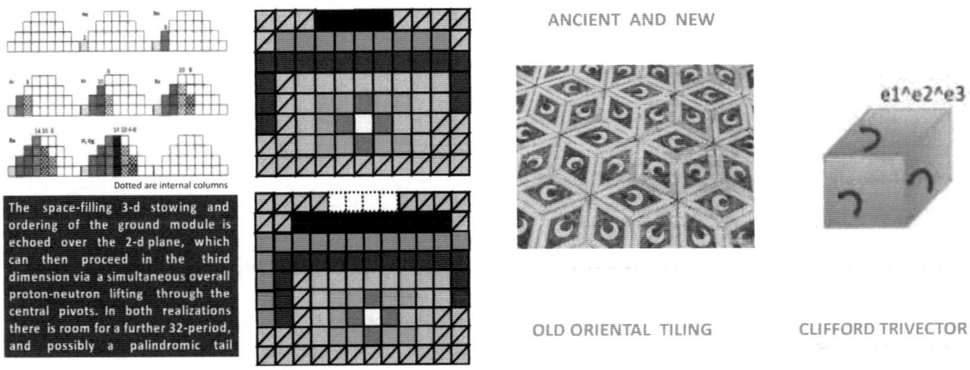

Figure 18. The Bohr palindromic orbital distribution of the number of atoms in the periods is not exhausted with the seventh, but can lodge more. And finally noting that nothing is new under the sun.

4. Discussion

I think that the results are significant and also strongly supporting as well as supported by the Standard Model because showing such remarkable correspondences over the whole range of their mutual comprise and organization. The difference lies in the approach; the Standard Model dealing with primary particles and their mathematical identity as points while the partial derivative "straight line of length equal to zero" [4,5] provides a differential realization of the Lie SU(3) × SU(2) × U(1) product group and algebra which both share at the very core. The transformation between straight and round expands as a phase transition between the inner surface of the Euclidean cube and the outer surface of the perfect sphere as suggested by Figure 8e where one can imagine the attraction from the dark mass of this when inducing in the interstice the leap to an IVM continuation.

In that quantum jump also the spatial frame changes from a unit cube to a parallelepiped of height $2^{1/2}$ (Figure 2e) and that might give a clue to the so far not identified counterpart of the Higg's field and mechanism. This has a proportionality constant of just $2^{1/2}$ and is a scalar, and it is tempting to speculate that stepping out in a new geometry could bring about such effects. However, these are not explained in the Standard Model either so there is a correspondence in that regard as well.

While the Standard Model deals with external objects the present plan would seem to resolve the dilemma that "humanity's best attempts at the ultimate explanation of matter and energy, space and time…suffer from a fundamental weakness…the strings move in a spacetime whose shape has been chosen from the beginning, as if they were actors on a previously constructed stage. A truly fundamental theory…would build the stage itself" [26]. The straight line generates itself as the Eigenvector of the absolute contrast to nothingness, namely extension [7,8], and that it is existentially as well as physically spanning its own infinite space is not a problem but gives a natural explanation of the seemingly increasing acceleration and the utter paradox of an outer border of the Universe. As mentioned in another context; "The immanent endlessness of the straight line as a state rather than a distance brings a relativistic perspective also on the everywhere observer-centered expanding universe impression, which is founded upon radiation from around reaching recording. But in a truly infinite cosmos all emissions from ever so strong sources will be successively absorbed and subjected to relativistic gravitational influences under their way, so that in the long run only some near-exhausted and cumulatively bent and thereby increasingly redshifted photons will remain for the farthest away spectator who is in fact, like we, just on the outermost border as well as in the very midst of a vanishing ray-bubble of the All" [In manuscript].

References

[1] DOE Explains *The Standard Model of Particle Physics*

[2] Tong D 2019 *Quantum Fields: The Real Building Blocks of the Universe*
[3] Shears T 2013 *The Standard Model Philosophical Transactions of the Royal Society*
[4] Lie M S 1871 *Over en Classe geometriske Transformationer* (Kristiania (now Oslo) University: Ph.D. Thesis)
[5] Trell E 1998 Marius Sophus Lie's doctoral thesis Over en Classe geometriske Transformationer *Algebras Groups and Geometries* **15** 395
[6] Stubhaug A 2003 *The mathematician Sophus Lie. It was the Audacity of My Thinking* (Berlin: Heidelberg Springer Verlag)
[7] http://plato.stanford.edu/entries/descartes-physics
[8] Grassmann H 1844 and 1878 Die Ausdehnungslehre von 1844 oder die Lineale Ausdehnungslehre Leipzig Otto Wigand
[9] Gardner M 1970 Mathematical Games: The fantastic combinations of John Conway's new solitaire game "Life" *Scientific American* **223** 120
[10] Trell E 1983 Representation of particle masses in hadronic SU(3) transformation diagram *Acta Phys. Austriaca* **55** 97
[11] Trell E 1990 Geometrical reproduction of (u, d, s) baryon, meson, and lepton transformation symmetries, masses, and channels *Hadronic J.* **13** 277
[12] Trell E 1991 On rotational symmetry and real geometrical representations of the elementary particles with special reference to the N and Δ Series *Phys. Essays* **4** 272
[13] Trell E 1992 Real forms of the elementary particles with a report of the Σ resonances *Phys. Essays* **5** 362
[14] Trell E 1998 The eightfold eightfold way: Application of Lie's true geometriske transformationer to the elementary elementary particles *Algebras Groups and Geometries* **15** 447
[15] Trell E 2008 Elementary particle spectroscopy in regular solid rewrite *AIP Conference Proceedings* **1051** 127
[16] Trell E 2013 3–d realization of τ, and *c* and *b* hadrons in endogenous parity with standard model *Afr. J. Phys.* **3** 24
[17] Trell E 2013 Digital outline of elementary particles via a root space diagram approach *J. Comput.Meth.Sci.Eng.* **13** 245
[18] Trell E 2014 Lie, Santilli, and nanotechnology: From the elementary particles to the periodic table of the elements *AIP Conference Proceedings* **1637** 1100
[19] Trell E, Edeagu S and Animalu A 2017 Geometric Lie Algebra in Matter, Arts and Mathematics with Incubation of the Periodic System of the Elements *AIP Conference Proceedings* **1798** 020162
[20] Trell E, Akpojotor G, Edeagu S and Animalu A 2019 Structural wave-packet tessellation of the periodic table and atomic constitution in real $R^3 \times SO(3)$ configuration space *J. Phys.: Conference Series* **1251** 012047
[21] Trell E 2020 A Space-Frame Periodic Table Representation System Testing Relativity in Nucleosynthesis of the Elements *J. Phys.: Conference Series* **1557** 012006
[22] Beringer J *et al* 2012 Charmed baryons *Particle Data Group* **86** 010001 245
[23] Beringer J et al 2012 Baryon list *Particle Data Group* **86** 010001 1
[24] Ferreres-Solé S and Sjöstrand T 2018 The Space-Time Structure of Hadroniation in the Lund Model (*arXiv:*1808.04619v2 [hep-ph*]*)
[25] Portegeis Zwart S 2018 Computational astrophysics for the future *Science* **36** 979
[26] Cho A 2002 Constructing Spacetime – No Strings Attached *Science* **298** 1166

The Statistical Analysis of Anisotropy Fluctuations of CMB Temperature in Angular Spatial and Temporal Areas

V Vargashkin

Senior research associate, Bauman State Technical University, Moscow, Russia
Associate professor, State University named after I.S. Turgenev, Orel, Russia

E-mail: finstp@bk.ru

Abstract. The changes in the nature of the statistical distributions of the anisotropy of the temperature of CMB in the satellite measurements of the "WMAP" probe were analyzed for the presence of laws in them in changes within the adjacent periods of data accumulation. Statistical distributions of changes in anisotropy between adjacent data accumulation periods were analyzed. The probability of the occurrence of such unidirectional changes at different frequencies of satellite measurements was analyzed and the probability of their occurrence under the influence of non-random factors was estimated.

1. Introduction

In [1] describes a method of searching for frequency independent changes in the anisotropy of the temperature of *CMB* over time in some directions of the celestial sphere. Such changes can be detected using poly-frequency satellite measurement databases such as "*WMAP*" and "*Planck*".

These databases are integrated with a cumulative time of one year for the "*WMAP*" experiments and six months for the "*Planck*" experiments. The changes in anisotropy of temperature over time are differences in temperature anisotropy values for the same cumulative measurement periods, similar in selected measurement frequencies. At the same time, such changes in temperature anisotropy should be observed at selected measurement frequencies in the same directions of the celestial sphere.

It should also be noted that statistically significant time fluctuations in the temperature of *CMB* were recorded directly by the "*WMAP*" probe [2] in 501 directions of the celestial sphere. In the present work the procedure of detection of frequency-independent changes of anisotropy of radiation temperature is considered.

The frequency of such changes in time in different directions of the celestial sphere may exceed their statistically determined frequency due to the randomness of the described phenomenon. In this case, it should be concluded that such changes serve as manifestations of some non-random processes. Since frequency-independent phenomena include gravitational microlensing of electromagnetic radiation, it can be assumed that non-random frequency-independent changes in the anisotropy of the temperature of *CMB* are caused by its gravitational microlensing.

In [3], the described method of searching for frequency-independent changes in temperature anisotropy is used for their possible detection in the satellite measurement database "*Planck*" in its low-frequency frequency band. It has been found that the values of such changes can be within one dozen microkelvins, and the number of frequency-independent changes in anisotropy detected between adjacent pairs of accumulation measurement periods can reach several hundred.

The present work is aimed at detecting frequency-independent jumps in anisotropy of *CMB* using the "*WMAP*" probe database. This base, in particular, contains the results of satellite measurements made [4] on its five frequency bands *K* (23 *GHz*); *Ka* (33 *GHz*); *Q* (41 *GHz*); *V* (61 *GHz*) and *W* (94 *GHz*). The results obtained using the two bands *K* and *Ka* differ significantly in estimating the measurement error of the anisotropy of the radiation temperature from similar estimates for the three bands *Q*; *V* and *W*. As a result, the totality of temperature anisotropy measurement results cannot be considered close measurements. Therefore for the further analysis the database received with use of eight pairs of radiometers was used; *Q*1; *Q*2; *V*1; *V*2 as well as *W*1; *W*2; *W*3 and *W*4, which were grouped according to their alphanumeric designations into the three listed bands.

Estimates of temperature anisotropy are presented in the order range of its absolute values from 10^1 *K* to 10^{-8} *K*, usually with seven significant digits. At the same time, the error in measuring temperature was seriously influenced by foreground radiation, which was the three main influencing factors. These include synchrotron radiation, free radiation and astrophysical dust radiation. The systematic error due to foreground radiation has a frequency dependence, and it was possible to significantly reduce it by introducing corrections precisely for bands *Q*; *V* and *W*.

It should be noted that the published results [2] of measurements of the anisotropy of the radiation temperature are "cleared" of its dipole component due to the movement of the Earth with respect to the flow of the volumetric spectral density of the radiation power. These results are integral data established over a cumulative time of one year.

The procedure for detecting frequency-independent changes in various directions in the celestial sphere is illustrated using the "*WMAP*" probe database to resolve *res* = 9 in the *HEALPix* coordinate system [5]. This resolution corresponds to covering the celestial sphere with numbered pixels, within each of which the anisotropy of temperature is considered constant. The average size of such a pixel in angular terms is 0,1145 angular degrees, or 6.87 angular minutes. Each pixel pulls the body angle $3{,}9947416\times10^{-06}$ *sr*. Thus, the map of the celestial sphere is composed of 3145728 numbered pixels.

2. Analysis of statistical distributions of temperature anisotropy of CMB

The features of the procedure for detecting frequency-independent changes in the anisotropy of *CMB* are considered on the example of data [4] obtained for the first two years of the probe operation. using a pair of radiometers *Q*1 and *V*1, which correspond to the wavelengths of the received radiation, equal to 13 *mm* and 9,1 *mm*, respectively. It is known that gravitational microlensing phenomena for the visible radiation range are also detected by changes in the luminosity of sources using pairs of wavelengths, in particular about 0.7 μm and 0.4 μm, i.e. using red and blue ends of the spectrum. Similarly, the difference of the analyzed temperatures in the respective directions of the celestial sphere allows an estimate of the size of changes in temperature anisotropy, wherein comparing such changes based on the results of measurements at the two indicated wavelengths allows an estimate of the nature of frequency independence of changes in anisotropy.

Figure 1 shows statistical distributions of the values of anisotropy of the radiation temperature *T* measured in all directions of the celestial sphere, respectively, for pairs of radiometers *Q*1 (*a*; *c*) and *V*1 (*b*; *d*). These values, as mentioned above, refer to the second (*a* – c) and first (*b* – *d*) measurement years, respectively. For purposes of clarity, the axis of the statistical relative frequency *f* is constructed on a logarithmic scale of units.

The graphs of Fig.1 also contain numbered vertical lines cutting off values on the abscissa axis, respectively: T_0 (1); $T_0 - 5\cdot\sigma_T$ (2); $T_0 + 5\cdot\sigma_T$ (3); T_{min} (4) and T_{max} (5).

Table 1 contains the characteristics of these distributions used in the further analysis. The first column of the table contains the name of the radiometer pair. The second column contains the year of operation of the probe, starting with the first one from the beginning of the work. The third and fourth columns contain, respectively, the average value of T_0 anisotropy of radiation temperature, as well as its average quadratic deviation σ_T from the average value. Columns five and seven contain minimum T_{min} and maximum T_{max} measured temperature anisotropy values. The sixth and eighth columns contain T_{min} and T_{max} values, respectively, expressed in fractions of ΔT from the fourth column. All values, uniform with temperature, are expressed in millikelvins.

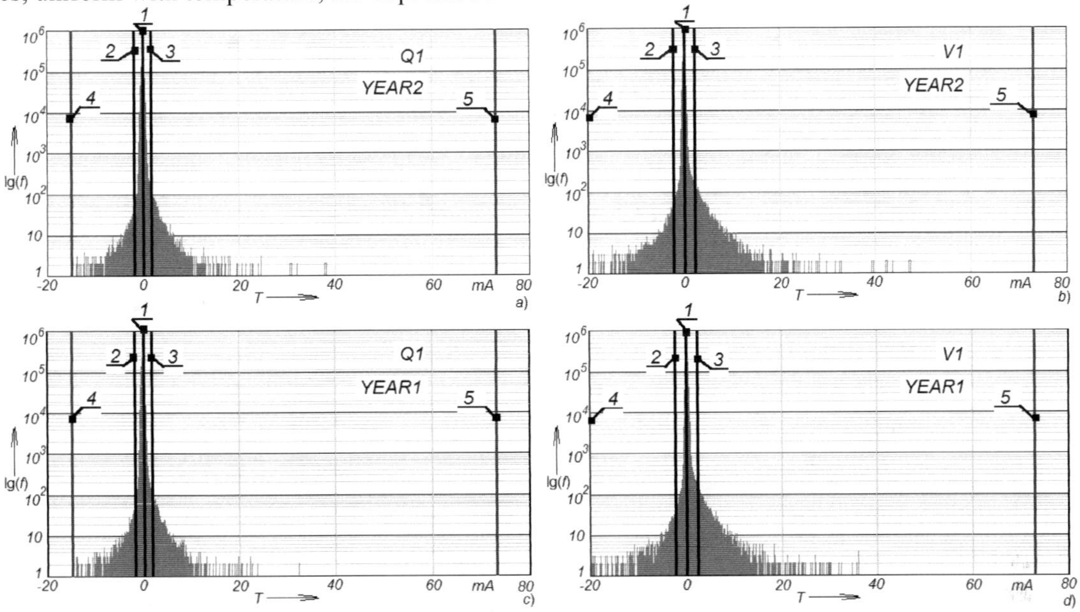

Figure 1. Distributions of measured values of anisotropy of radiation temperature T for pairs of radiometers $Q1$ (a; c) and $V1$ (b; d) relating to the second (a – b) and first (c– d) measurement years.

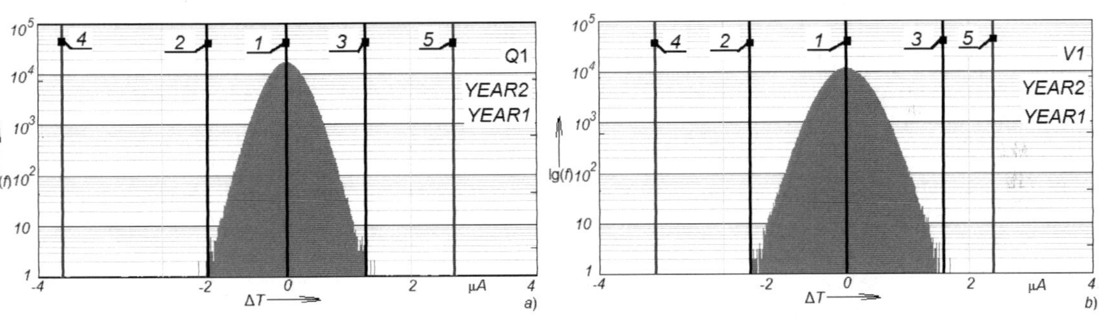

Figure 2. Distributions of measured values of change ΔT in anisotropy of radiation temperature T for pairs of radiometers $Q1$ (a) and $V1$ (b) occurred between the second and first years of measurement.

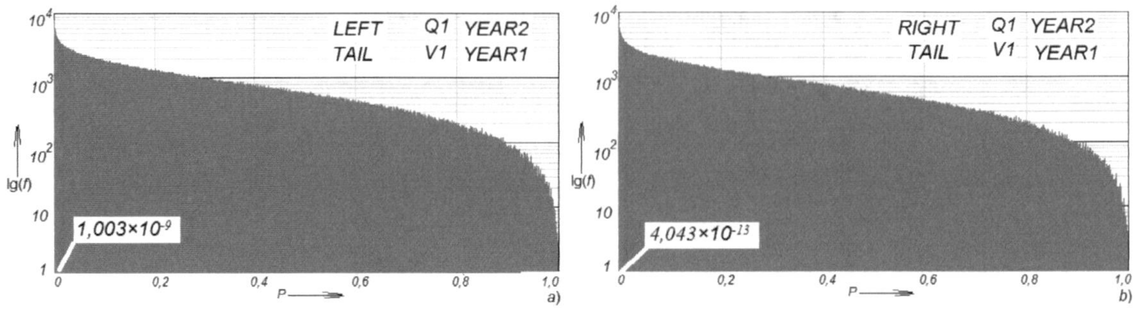

Figure 3. Probability distributions P (1) for negative (a) and negative (b) values of change ΔT in anisotropy of radiation temperature T for pairs of radiometers $Q1$ and $V1$ that occurred between the second and first years of measurement.

Table 1. Characteristics of distributions of anisotropy of temperature of *CMB* measured by pairs of radiometers $Q1$ and $V1$ for the first two years of probe operation.

Radio-meters	Year	T_0, *mK CMB*	σ_T, *mK CMB*	T_{min}, *mK CMB*	T_{min} / σ_T, *mK CMB*	T_{max}, *mK CMB*	T_{max} / σ_T, *mK CMB*
1	2	3	4	6	7	8	9
$Q1$	first	0,01172334	0,4505729	–21,77829	–48,3	72,67414	162
	second	0,01177452	0,4494247	–22,22542	–49,5	73,31463	163
$V1$	first	0,006021815	0,3559442	–14,50349	-40,7	73,17532	206
	second	0,006095985	0,3542433	–41,16442	–116	205,8138	581

Analysis of Figure 1, as well as Table 1, shows that the temperature anisotropy distributions measured on each of the two frequency channels during the two adjacent accumulation periods have sufficiently close sample mean T_0 as well as sample mean square deviations σ_T for each of the accumulation periods to be considered pairs of samples of one of the same general population. In particular, the sample averages T_0 for the two adjacent accumulation periods for the pairs of radiometers $Q1$, as well as $V1$ taken separately, differ by less than one percent, and the corresponding pairs of estimates of average quadratic deviations σ_T – by no more than several percent.

At the same time, the width of the tails of the respective distributions can vary by two or more times or more from one accumulation period to another accumulation period, which indicates the existence of noticeable changes in the anisotropy of the radiation temperature in some directions of the celestial sphere.

Some of these changes should be considered statistically random. For the remainder of these changes, a possible list of non-accidental processes leading to such changes should be identified. In particular, if such changes are found to be frequency-independent, they may be due to the effect of the gravitational microlensing phenomenon on *CMB*.

It should be noted that the frequency independence in the case of microlensing the radiation observed in the optical and radio wave band is subjected to a fairly strict check for proximity to the unit of luminosity ratio in two different frequency bands of observation. The deviation of this deviation from the unit usually does not exceed several percent. However, in the present case of possible microlensing of *CMB*, this ratio can be substantially large. This is evidenced by the fact that when comparing the statistical characteristics of radiation obtained using two different pairs of radiometers, it turns out that their estimates of the average temperature T_0 can differ up to two times, and estimates of the average quadratic deviation σ_T by more than 20%. Such differences may be related, in particular, to features of behavior in the frequency domain of systematic temperature measurement errors caused by foreground radiation.

3. Analysis of statistical distributions of change in temperature anisotropy of CMB

Next, we will go directly to the statistical analysis of changes in the measured temperature values within adjacent pairs of accumulation periods for each of the frequency-differing pairs of radiometers *Q*1 and *V*1 during two adjacent accumulation periods of data starting from the first year of operation of the probe.

Figure 2 (*a*) shows the distribution of changes in the anisotropy of the radiation temperature between the second and first accumulation periods for a pair of radiometers *Q*1, and Figure 2 (*b*) shows the distribution of changes in the anisotropy of the radiation temperature for a pair of radiometers *V*1. For purposes of clarity, the relative frequency axis *f* of these graphs is constructed on a logarithmic scale of units, as in the graphs of Fig.1.

The graphs of Fig.1 also contain numbered vertical lines cutting off values on the abscissa axis, respectively: ΔT_0 (1); $\Delta T_0 - 5\cdot \sigma_{\Delta T}$ (2); $\Delta T_0 + 5\cdot \sigma_{\Delta T}$(3); ΔT_{min} (4) and ΔT_{max} (5).

Table 2 contains the characteristics of these distributions, similar to Table 1. The first column of this table contains the name of the radiometer pair. The second column indicates the years of operation of the probe, starting with the first one from the beginning of operation, for which the temperature difference is calculated. The third and fourth columns contain, respectively, the average value of ΔT_0 change of anisotropy of radiation temperature, as well as its average quadratic deviation $\sigma_{\Delta T}$ from the average value. Minimum ΔT_{min} and maximum ΔT_{max} measured temperature anisotropy change values are entered in columns 5 and 7, respectively. The sixth and eighth columns contain the values of ΔT_{min} and ΔT_{max}, respectively, expressed in fractions of $\sigma_{\Delta T}$ from the fourth column. All values contained in Table 2 and uniform with temperature are expressed, unlike Table 1, in microkelvins.

Table 2. Characteristics of distributions of change of anisotropy of temperature of CMB measured by pairs of radiometers *Q*1 and *V*1 for the first two years of probe operation.

Radio-meters	Year	ΔT_0, μK CMB	$\sigma_{\Delta T}$, μK CMB	ΔT_{min}, μK CMB	ΔT_{min} / $\sigma_{\Delta T}$, μK CMB	ΔT_{max}, μK CMB	ΔT_{max} / $\sigma_{\Delta T}$, μK CMB
1	2	3	4	6	7	8	9
*Q*1	second	0,05117946	0,2525774	3,61435	14,3	2,67885	10,6
*V*1	and first	0,07416961	0,3132958	–3,09434	–0,89	2,37407	7,58

Analysis of Figure 2 as well as Table 2 shows that the distributions of temperature anisotropy changes measured on each of the two frequency channels during the two adjacent accumulation periods have a sufficiently high centering. Estimates of mean values of ΔT_0 change differ from zero value within several hundredths of microkelvin. Estimates of the mean quadratic deviations $\sigma_{\Delta T}$ reach several tenths of the microkelvin, which is three orders of magnitude less than the estimates of the mean quadratic deviation σ_T of the temperature anisotropy itself contained in Table 1. At the same time, at both frequencies of measurement, jumps in temperature anisotropy are observed, which are an order of magnitude higher than their average square deviation and reach microkelvin units.

It should be determined whether such significant changes in temperature anisotropy are frequency independent, i.e. whether they are realized in the tails of the statistical distributions of temperature anisotropy in the same directions of the celestial sphere at two frequencies. The presence of such implementations may serve as a basis for believing the corresponding changes in temperature anisotropy to be statistically non-random, that is, to consider them to be manifestations of non-random processes to be established.

Then we analyze probabilities of variation of anisotropy of radiation temperature measured between the second and the first accumulation periods into tails of statistical distributions of these changes, which were implemented between the second and the first accumulation periods using pairs of radiometers *Q*1 and *V*1 differing in frequency.

4. Analysis of statistical distributions tails of change in temperature anisotropy of CMB

Note that the analyzed tails of temperature anisotropy change distributions correspond to both positive and negative temperature anisotropy change values. Tail bearing negative values of change corresponds to decrease of temperature anisotropy during transition from previous of two adjacent pairs of accumulation periods to subsequent accumulation period. The tail bearing the positive values of the change in anisotropy corresponds to an increase in temperature.

We establish for subsequent reasoning that both tails of the distribution of changes in the anisotropy of temperature, both positive and negative, have the same power, equal to half of the $0{,}5 \cdot N_0$ general distribution of these N_0. Place in ascending order the absolute temperature anisotropy changes measured for each of the two frequencies. We fix at any frequency, in either of the two tails of the distribution of the change in anisotropy, any direction on the celestial sphere, which is characterized by some number $j1$ in the general ascending order of the modulus of change in anisotropy. We will consider this value a boundary for the corresponding tail. Then the probability of accidentally hitting the previously fixed value in the tail in question is equal to the ratio of the width of this tail, i.e. j_1, to the half-volume of $0{,}5 \cdot N_0$ of the general distribution. In the same way, for the same direction in the celestial sphere, the probability of accidentally hitting the corresponding value of the change in anisotropy in the same tail of the distribution obtained by measuring at another frequency is equal to the ratio of the width j_2 of this tail to the half-volume $0{,}5 \cdot N_0$ of the general population.

Then the probability P of claiming that changes in anisotropy measured at two frequencies hit both corresponding tails is randomly described by multiplication:

$$P = \frac{4 j_1 j_2}{N_0^2}. \quad (1)$$

The lower the probability value (1) of accidental ingress of temperature anisotropy values into the tails of distributions at both frequencies, the greater the probability of claiming that such an ingress occurred in a non-random manner.

It is possible to select such a boundary value j_0 of the number limiting the tail of the temperature anisotropy distribution that when the conditions $j_1 < j_0$ and $j_2 < j_0$ are met, the corresponding value P_0 of the probability P

$$P_0 = \frac{4 j_0^2}{N_0^2} \quad (2)$$

cannot be satisfied with any temperature anisotropy value that would accidentally fall into the tail of the distribution at both frequencies. It should then be considered fair to claim that such a hit is the result of a non-random frequency-independent phenomenon, such as gravitational microlensing of *CMB*. It should then be considered fair to claim that such a hit is a consequence of a non-random frequency-independent phenomenon, such as gravitational microlensing of *CMB*.

Figure 3 shows statistical distributions of probability values P (1) calculated in all directions of the celestial sphere for two adjacent annual accumulation periods, starting from the first year of operation of the satellite, for pairs of radiometers $Q1$ and $V1$. Figure 3 (*a*) graph corresponds to negative changes of anisotropy of temperature in tail, and Figure 3 (*b*) graph corresponds to its positive changes. For purposes of clarity, the relative frequency axis f is constructed on a logarithmic scale of units.

Table 3 contains the characteristics of these distributions used in the further analysis. The first column of the table indicates whether the temperature anisotropy change in the temperature distribution is positive or negative. In the second column, the number of $N_0/2$ characterizing the volume of the corresponding distribution is placed. The third column contains the value of the value inverse of $N_0/2$, i.e. the probability P_1 of falling from the first time into the tail of the distribution of the change in temperature anisotropy of any of its values, which is measured at any one of the frequencies. The fifth column contains the value of the value inverse to the square of $N_0/2$, i.e. an estimate of the probability P_2 of falling out from the first time into the tails of the distributions of the change in anisotropy of the temperature of any of its values, which is measured at both frequencies. In the fourth column, that is, between the values P_1 and P_2, the probability P (1) detected by statistical analysis of the database is placed.

Table 3. - Characteristics of probability distributions P (1) measured by pairs of radiometers $Q1$ and $V1$ for the first two years of probe operation.

Tail sign	$N_0/2$	$1/(N_0/2)$	P (1)	$1/(N_0/2)^2$
1	2	3	4	5
positive	1573005	$6,357\times10^{-7}$	$3,641\times10^{-12}$	$4,041\times10^{-13}$
negative	1572723	$6,358\times10^{-7}$	$1,003\times10^{-9}$	$4,043\times10^{-13}$

The analysis of Figure 3 and Table 3 shows that the distributions of positive and negative changes in temperature anisotropy differ slightly in their volumes within a tenth of the percentage, that is, almost half the volume of the total N_0. The values P (1), both for the distribution with positive changes in temperature anisotropy and for the distribution with negative changes, are located inside the segment of the probability scale f limited by the values of P_1 and P_2. At the same time, the obtained result indicates that in the tails of the distribution of both positive and negative values of temperature anisotropy changes measured at both frequencies, in some directions of the celestial sphere, such changes may be contained, in respect of which it is true that their presence in the tails of both distributions is caused by some non-random factor.

Next, we will analyze the estimated number of frequency-independent changes observed in satellite experiments in the temperature anisotropy of *CMB*, which can be observed between adjacent periods of data accumulation and which exceeds their number corresponding to the influence of statistically random factors.

5. Analysis of assumed numbers of occurrence of frequency-independent changes of temperature anisotropy of CMB

Figure 4 shows time diagrams of the amounts of M satellite observations of frequency-independent changes in the temperature anisotropy of *CMB* over nine annual cycles of accumulation of measurement results. Figure 4 (a – b) corresponds to the two observation frequency ranges discussed above with pairs of radiometers $Q1$ and $V1$. These two diagrams contain ten graphs of the change of the number M from time, which for the sake of clarity are depicted with different scales of the magnitude M. To construct these graphs, five values of the number J of frequency-independent changes in temperature anisotropy are selected, which can fall into the tails of the distribution at two measurement frequencies statistically randomly.

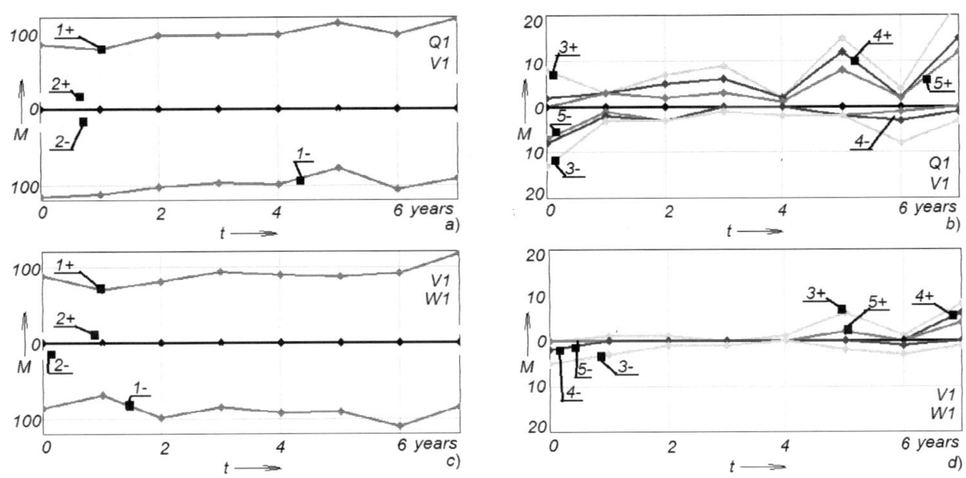

Figure 4. Diagrams of number M dependencies on observation time at different values of factor J for two pairs of radiometers $Q1$ and $V1$ (a - b), as well $V1$ and $W1$ (c - d)

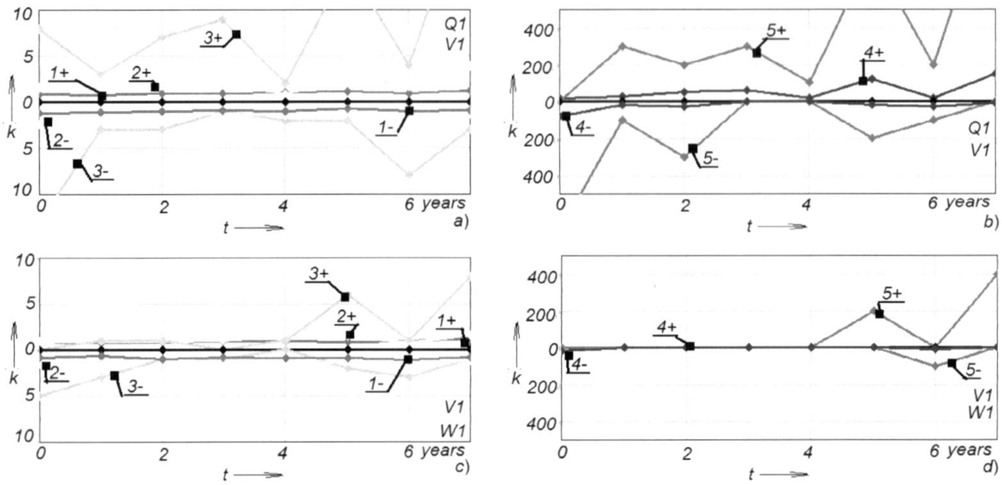

Figure 5. Diagrams of number k dependencies on observation time at different values of factor J for two pairs of radiometers $Q1$ and $V1$ (a - b), as well $V1$ and $W1$ (c - d)

Schedule lines are assigned conditional numbers 1; 2; 3; 4; 5, which correspond to the calculated values of the number J, respectively equal to 100; 10; 1; 0,1; 0,01.

If, for example, at any point in Figure 3 to which $J = 1$ corresponds, the number M of observed changes in anisotropy exceeds one, this means that in the corresponding tails of the distribution of anisotropy of the radiation temperature obtained at each of the two frequencies, there is only one change in anisotropy of temperature, which can be explained from the points of chance. The remaining M - 1 changes arose due to the effect on radiation of some non-random factors.

In the designation of all graphs of Figures 4, except for the numerical designation, there is a sign "+," or "–" for example, "3+," or "3–" The sign "+" corresponds to positive increments of temperature anisotropy in the corresponding tail of the distribution during the transition from the previous period of accumulation of measurement results to the next, and the sign "– " – negative.

To improve the reliability of the calculations, they also included another combination of pairs of satellite radiometers - $V1$ (with a frequency of 9,1 *mm*) and $W1$ (with a frequency of 3,2 *mm*). The diagram of Figure 4 (c – d) for the specified pair of radiometers completely repeats the diagram of Figure 4 (a – b) for the pair of radiometers $Q1$ and $V1$.

In order to compare the numerical values of M and J, Figure 5 shows time diagrams of the number k equal to the ratio M/J. Numeric symbols of graphs of Fig.5 fully correspond to symbols of graphs of Fig.4. At the same time, the greater k, the greater the probability of anisotropy temperature changes in the tailings due to the presence of non-random physical processes in relation to the number of temperature anisotropy changes due to the random combination of temperature anisotropy values.

Table 4. Statistical characteristics of the number M at different values of factor J for two pairs of radiometers $Q1$ and $V1$, as well as $V1$ and $W1$.

Radiometers pairs	Tail sign	J	M_0, *units*	σ_M, *units*	M_{max}, *units*	M_{min}, *units*
1	2	3	4	6	7	8
Q1 *V1*	positive	100	111	21,3	118	78
		10	23,4	10,9	30	8
		1	10,1	7,74	23	2
		0,1	6,71	5,48	15	2
		0,01	4,43	4,42	12	0
	negative	100	114	21,0	116	78
		10	16,7	6,18	24	7
		1	5	4,45	13	1
		0,1	2,71	2,79	8	0
		0,01	2	2.58	7	0
V1 *W1*	positive	100	102	20,7	117	70
		10	12,9	4,21	16	7
		1	2,57	3,27	8	0
		0,1	1,14	2,32	6	0
		0,01	0,86	1,61	4	0
	negative	100	102	19,3	110	69
		10	4,12	14	18	4
		1	2,29	1,76	5	0
		0,1	0,42	0,81	2	0
		0,01	0,143	0,382	1	0

Table 4 contains numerical characteristics of Fig. 4 diagrams. The first column of the table contains the designation of a pair of radiometers. The second column of the table indicates whether the temperature anisotropy change in its distribution is positive or negative. In the third column, the value of factor J. In the fourth column, the mathematical expectation of the M_0 number M is indicated, in the fifth - its average quadratic deviation σ_M, in the sixth and seventh columns - respectively the maximum M_{max} and the minimum M_{min} values of the number M.

Table 5 contains numerical characteristics of Fig. 5 diagrams. Construction of Table 5 is similar to construction of Table 4. Columns of Table 5 from the first to the third coincide with the corresponding columns of Table 4. In the fourth column, the mathematical expectation k_0 of the number k is indicated, in the fifth - its average quadratic deviation σ_k, in the sixth and seventh columns - respectively the maximum k_{max} and the minimum k_{min} of the number k.

Table 5. Statistical characteristics of the factor k at different values of factor J for two pairs of radiometers $Q1$ and $V1$, as well as $V1$ and $W1$.

Radiometers pairs	Tail sign	J	k_0	σ_k	k_{max}	k_{min}
1	2	3	4	6	7	8
Q1 *V1*	positive	100	1,11	0,213	1,78	0,78
		10	2,34	1,09	3,9	0,8
		1	10,1	7,74	23	2
		0,1	67	54	150	20
		0,01	443	442	1200	0
	negative	100	1,15	0,210	1,16	0,78
		10	1,67	0,617	2,4	0,7
		1	5	4,45	13	1
		0,1	22,1	27,9	80	0
		0,01	200	258	700	0
V1 *W1*	positive	100	1,02	0,206	1,17	0,7
		10	1,29	0,421	1,6	0,7
		1	2,57	3,27	8	0
		0,1	11,6	23,15	60	0
		0,01	85,7	161	400	0
	negative	100	1,02	0,196	1,1	0,69
		10	1,4	0,412	1,8	0,8
		1	2,29	1.28	5	0
		0,1	4,39	0,0601	20	0
		0,01	14,3	38,2	100	0

The analysis of the diagrams of Fig.4 and Fig.5, as well of the data of Table 4 and Table 5 shows that with the growth of factor J, the number M increases, but the factor k decreases. In other words, the number of points in the tails of temperature anisotropy changes distributions, which are candidates for frequency-independent changes, increases. However, among them, the number of changes increases disproportionately rapidly, the presence of which can be explained by a random combination of values of changes in temperature anisotropy at two different frequencies. Therefore, at $J = 100$ as well as $J = 10$, it is likely that in the case of the assumption of changes in temperature anisotropy as frequency-independent, these changes are due only to a random combination of their numerical values in the tails of distributions at different measurement frequencies.

On the other hand, when $J = 0.01$, k reaches several hundred. However, in this case, the number M does not exceed several units. As a result, the probability of erroneously considering frequency-independent changes in temperature anisotropy as a result of random combinations of numerical values of temperature anisotropy in the tails of distributions at different measurement frequencies is high.

Summing up this, it can be assumed that the number of candidates for frequency-independent changes in the temperature anisotropy of *CMB* can be several tens of pieces during the annual accumulation period of satellite measurements.

6. Conclusion

The analysis of the anisotropy of the temperature of *CMB* according to the satellite measurements of the "*WMAP*" probe along various directions in the celestial sphere, as well as for various time periods of accumulation of measurement results and for various frequencies of satellite measurements, made it possible to establish a number of features of the change in anisotropy of the temperature of *CMB* in angular spatial as well as time areas.

Tails of statistical distributions of temperature anisotropy can experience changes in length at different measurement frequencies, which exceed several tens of average quadratic deviations of temperature anisotropy from its average value. This makes it possible to set the task of searching for the manifestation of physical laws in changes in the anisotropy of temperature in the angular directions of the celestial sphere, as well as in time.

Analysis of the tails of distributions of changes in the anisotropy of the temperature of *CMB* makes it possible to establish that within the adjacent periods of accumulation of measurement results at various frequencies of satellite measurements, changes in the temperature of anisotropy are observed, reaching three to five microkelvins. This makes it possible to set the task of estimating the probability of such unidirectional changes in the tails of their distributions within the same adjacent periods of accumulation of measurement results in the same directions of the celestial sphere simultaneously at different frequencies.

Analysis of distributions of probabilities of appearance of unidirectional changes of anisotropy of *CMB* results in the same directions of celestial sphere simultaneously at different frequencies makes it possible to conclude that probability of occurrence of such changes calculated on the basis of satellite measurements can exceed by ten times statistical probability of their random appearance. This also makes it possible to search for the effect of the physical laws of occurrence of such changes in temperature anisotropy, as well as to estimate the number of such events within different periods of accumulation of measurement results.

The number of candidates for frequency-independent changes of anisotropy of temperature of *CMB* is estimated at several tens during the annual accumulation period of satellite measurements taking into account resolution of these measurements in 6.87 angular minutes and corresponding number of separately set directions in the celestial sphere equal to 3145728 [2] in the coordinate system *HEALPix* [5].

References

[1] Vargashkin V Ya 2018 The analysis of frequency-independent jumps of CMB according to the «Planck» data *Journal of Physics: Proceeding of XX International Meeting «Physical Interpretations of Relativity Theory»* (London: IOP published house) **1051** 012027

[2] Bennett C L *et al* 2013 Nine-Year Wilkinson Microwave Anisotropy Probe (WMAP) Observations: Final Maps and Results *ApJS* **208** 20B

[3] Vargashkin V Ya 2020 The analysis of CMB anisotropy to temporary domain according to WMAP and Planck probes databases *Journal of Physics: Proceeding of XXI International Meeting «Physical Interpretations of Relativity Theory»* (London: IOP published house) **1557** 012037

[4] Jarosik N *et al* 2003 Implementation And Testing щf ehe Map Radiometers *ApJS* **145** 413

[5] Górski K M. *et al* 2005 HEALPIX - a Framework for High Resolution Discretization, and Fast Analysis of Data Distributed on the Sphere *ApJ* **622** 759

Forces in Schwarzschild, Vaidya and generalized Vaidya spacetimes

Vitalii Vertogradov
Herzen State Pedagogical University of Russia, Moika 48, St. Petersburg 191186, Russia
The SAO RAS, Pulkovskoe shosse 65, 196140, Saint-Petersburg, Russia
E-mail: vdvertogradov@gmail.com

Abstract. In this paper we investigate how the leading term in the geodesic equation in Schwarzschild spacetime changes under the coordinate transformation to Eddington-Finkelstein coordinates. This term corresponds to the Newton force of attraction. Also we consider this term when we add the energy-momentum tensor of the form of the null dust and the null perfect fluid into right-hand side of the Einstein equation. We estimate the value of this force in Vaidya spacetime when the naked singularity formation occurs. Also we give conditions in generalized Vaidya spacetime when this force of attraction is replaced by the force of repulsion.

1. Introduction

In order to describe forces in general relativity one should study the geodesics [1, 2, 3]. In Schwarzschild metric the leading term in geodesic equation is the Newton force of attraction [4]. Under the notion 'the leading term' we mean the Γ^1_{00} component of the Christoffel symbols. In Schwarzschild spacetime this component is given by:

$$\Gamma^1_{00} = \left(1 - \frac{2M}{r}\right)\frac{M}{r^2}\,. \tag{1}$$

Christoffel symbols don't obey the tensor transformation law and if we do coordinate transformation then Christoffel symbols change and depending on the coordinate system, we obtain different inertial forces. However the main goal of this paper is to investigate how the term Γ^1_{00} changes when one does coordinate transformation to Eddington-Finkelstein coordinates.

The Schwarzschild spacetime is the solution of the Einstein equations in the vacuum. However, if we consider the Schwarzschild metric in Eddington-Finkelstein coordinates with the energy-momentum tensor of the null dust in the right-hand side of the Einstein equations then one obtains the Vaidya spacetime. This spacetime is one of the earliest examples of cosmic censorship conjecture violation [5]. Moreover one could consider the mixture of two energy-momentum tensors of type-I (the null dust) and the type-II (the null perfect fluid). In this case one obtains the generalized Vaidya spacetime [6]. The only difference from Schwarzschild spacetime is that in the case of Vaidya spacetime the mass function is the function of the time and in the generalized Vaidya spacetime case the mass function depends upon both and the time and the radial coordinate. The generalized Vaidya spacetime also contains the naked singularities [7, 8, 9, 10].

We investigate the question how these energy-momentum tensors affect the term Γ^1_{00}. We also consider the force expression in the case of the naked singularity formation in order to understand whether it is finite or infinite. Also it is interesting if the attraction can be replaced with repulsion.

The paper is organized as follows: in sec.2 we consider the Schwarzschild spacetime in Eddington-Finkelstein coordinates and see if this transformation changes the term Γ^1_{00}. In sec.3 we consider the Vaidya spacetime and see how the null dust affects the forces and consider these forces in the naked singularity. In sec.4 we consider the generalized Vaidya spacetime and find the condition when the attraction can be replaced with repulsion. Section 5 is the conclusion.

The signature $-,+,+,+$ and the system of units $G = c = 1$ will be used throughout the paper. Also dash and over dot mean the partial derivative with respect to r and v respectively i.e. $M' = \frac{\partial M}{\partial r}\,, \dot{M} = \frac{\partial M}{\partial v}$.

2. Schwarzschild spacetime

Schwarzschild metric in coordinates $\{t\,, r\,, \theta\,, \varphi\}$ has the following form:

$$ds^2 = -\left(1 - \frac{2M}{r}\right)dt^2 + \left(1 - \frac{2M}{r}\right)^{-1} dr^2 + r^2(d\theta^2 + \sin^2\theta d\varphi^2)\,. \tag{2}$$

To obtain this metric in Eddington-Finkelstein Coordinates one should consider the radial null geodesic:

$$\left(\frac{dt}{dr}\right)^2 = \left(1 - \frac{2M}{r}\right)^{-2}\,. \tag{3}$$

Integrating this equation and doing the following transformation from t to new time v:

$$t = v + r + 2M\ln(r - 2M)\,, \tag{4}$$

we obtain the Schwarzschild metric in Eddington-Finkelstein Coordinates:

$$ds^2 = -\left(1 - \frac{2M}{r}\right)dv^2 + 2dvdr + r^2(d\theta^2 + \sin^2\theta d\varphi^2)\,. \tag{5}$$

The geodesic equation for the metric (5) is given by:

$$\begin{aligned}\frac{d^2r}{d\tau^2} = -\frac{M}{r^2}\left(1 - \frac{2M}{r}\right)\left(\frac{dv}{d\tau}\right)^2 + 2\frac{M}{r^2}\frac{dv}{d\tau}\frac{dr}{d\tau} + (r - 2M)\left(\frac{d\theta}{d\tau}\right)^2 +\\ (r - 2M)\sin^2\theta\left(\frac{d\varphi}{d\tau}\right)^2\,.\end{aligned} \tag{6}$$

Comparing the Γ^1_{00} from (1) and (6) we see that it is the same in both coordinate systems. However in comparison to the old coordinates in Eddington-Finkelstein Coordinates we have new force which is proportional to the velocity $\frac{dr}{d\tau}$:

$$2\Gamma^1_{10}\frac{dv}{d\tau}\frac{dr}{d\tau} = -2\frac{M}{r^2}\frac{dv}{d\tau}\frac{dr}{d\tau}\,. \tag{7}$$

As we noted above that Christoffel symbols don't obey the tensor transformation law and one might expect new inertial forces to appear in different frames.

We can conclude that in Eddington-Finkelstein Coordinates the leading term Γ^1_{00} which corresponds to the Newton force of attraction doesn't change. It is useful to write down the energy expression in this metric. For this purpose we should use the Lagrangian which has the following form:

$$\mathcal{L} = g_{ik}\frac{dx^i}{d\tau}\frac{dx^k}{d\tau} = -\left(1-\frac{2M}{r}\right)\left(\frac{dv}{d\tau}\right)^2 + 2\frac{dv}{d\tau}\frac{dr}{d\tau} + r^2\left(\frac{d\theta}{d\tau}\right)^2 + r^2\sin^2\theta\left(\frac{d\varphi}{d\tau}\right)^2 . \tag{8}$$

And we can easily obtain the energy expression from the (8):

$$\varepsilon = \left(1-\frac{2M}{r}\right)\frac{dv}{d\tau} - \frac{dr}{d\tau} = const\,. \tag{9}$$

And the radial geodesic by using (9) and with $\varphi = const\,, \theta = \frac{\pi}{2}$ is given by:

$$\frac{d^2r}{d\tau^2} = -\frac{M}{r^2}\left(\varepsilon\frac{dv}{d\tau} - \frac{dr}{d\tau}\frac{dv}{d\tau}\right) = -\frac{M}{r(r-2M)}\left[\varepsilon^2 - \left(\frac{dr}{d\tau}\right)^2\right] . \tag{10}$$

3. Vaidya spacetime

If we consider Schwarzschild spacetime in Eddington-Finkelstein Coordinates (5) and assume that the mass function M is not a constant but depends on the time v:

$$M = M(v)\,, \tag{11}$$

then one obtains the Vaidya spacetime:

$$\begin{gathered} ds^2 = -\left(1-\frac{2M(v)}{r}\right)dv^2 + 2dvdr + r^2d\Omega^2\,, \\ d\Omega^2 = d\theta^2 + \sin^2\theta d\varphi^2\,. \end{gathered} \tag{12}$$

This is not the solution in empty spacetime anymore. On the right-hand side of the Einstein equation we have the following energy-momentum tensor:

$$T_{ik} = \mu\delta_i^0\delta_k^0\,. \tag{13}$$

As we can see from (13) the only non-vanishing component of the energy-momentum tensor is T_{00}. This corresponds to the case of the null dust.

The energy conditions [11] demands that:

$$\dot{M}(v) > 0\,. \tag{14}$$

Now we can write down the radial geodesic in Vaidya metric:

$$\begin{gathered} \frac{d^2r}{d\tau^2} = -\frac{\left(1-\frac{2M(v)}{r}\right)M(v) + \dot{M}(v)r}{r^2}\left(\frac{dv}{d\tau}\right)^2 + 2\frac{M(v)}{r^2}\frac{dv}{d\tau}\frac{dr}{d\tau} + \\ (r-2M(v))\left(\frac{d\theta}{d\tau}\right)^2 + (r-2M(v))\sin^2\theta\left(\frac{d\varphi}{d\tau}\right)^2 . \end{gathered} \tag{15}$$

If we compare geodesic equations in Vaidya (15) and in Schwarzschild (6) spacetimes then we can see that only the leading term Γ^1_{00} has difference i.e.:

$$\Gamma^1_{00\,vaidya} = \Gamma^1_{00\,sch}(v) + \frac{\dot{M}(v)}{r}\,. \tag{16}$$

If we consider the leading term as the force of attraction then one can see that this force is bigger than in pure Schwarzschild spacetime. It is worth noticing that according to the energy condition (14) the term $\frac{\dot{M}(v)}{r} > 0$ and the leading term (16) doesn't change its sign.

The metric (12) depends upon time hense the energy is not conserved i.e.:

$$\varepsilon(v) = \left(1 - \frac{2M(v)}{r}\right)\frac{dv}{d\tau} - \frac{dr}{d\tau}. \tag{17}$$

Now differentiating ((17)) with respect to τ we obtain the second order geodesic equation which depends on the energy:

$$\varphi = const\,, \theta = \frac{\pi}{2}\,,$$
$$\frac{d^2r}{d\tau^2} = -\frac{\left(1 - \frac{2M(v)}{r}\right)M(v) + 2\dot{M}(v)r}{r^2}\left(\frac{dv}{d\tau}\right)^2 + 2\frac{M(v)}{r^2}\frac{dr}{d\tau}\frac{dv}{d\tau} - \frac{d\varepsilon(v)}{dv}\frac{dv}{d\tau}. \tag{18}$$

And comparing (18) to (15) one can find the energy change law:

$$\frac{\partial\varepsilon(v)}{\partial v} = -\frac{\dot{M}(v)}{r}\frac{dv}{d\tau}. \tag{19}$$

We can conclude that the particle loses its energy despite it falls onto a black hole.

As we mentioned above the Vaidya spacetime contains the naked singularity. It forms during the gravitational collapse at $v = 0\,, r = 0$. The case when the mass function is linear has been considered by P. Joshi [12]. Here we consider the mass function in the form:

$$M(v) = \lambda v^{\beta}\,, \beta \geq 1\,, \lambda > 0\,. \tag{20}$$

In this case the leading term is given by:

$$\Gamma^1_{00} = -\frac{\left(1 - \frac{2\lambda v^{\beta}}{r}\right)\lambda v^{\beta} + \beta\lambda v^{\beta-1}r}{r^2}. \tag{21}$$

We are interested only in the case when $\beta >\geq 1$ because only in this case the naked singularity might form. In the case $\beta < 1$ the result of the gravitational collapse is the black hole. When we have the naked singularity formation then the following condition must be held:

$$\lim_{v\to 0\,, r\to 0}\frac{dv}{dr} = X_0 > 0\,. \tag{22}$$

Moreover X_0 is finite.

Substituting this into (21) we can see that if $\beta \in [1\,, 2)$ then $\lim_{v\to 0\,, r\to 0}\Gamma^1_{00} \to \infty$. When $\beta = 2$ it is a positive real constant and when $\beta > 2$ it is zero.

However we are interested only in gravitationally strong singularities. According to Tipler's deffinition [14] a singularity is termed to be gravitationally strong or simply strong if it destroys by stretching or crushing any object which falls into it. If it does not destroy any object this way then the singularity is termed to be gravitationally weak. Mathematically it is [13]:

$$\xi = \lim_{\tau\to 0}\tau^2 R_{ik}K^iK^k > 0\,, \tag{23}$$

where τ - is the afine parameter and K^i is the tangent vector to outgoing null geodesic congruence.

In our case (23) has the following form:

$$\xi = \lim_{v \to 0} 2\beta(\beta - 1)\lambda v^{\beta-1} X_0^2 \,. \tag{24}$$

And we can see that the singularity is gravitationally strong only in the case $\beta = 1$. Hence we can conclude that when the naked singularity is gravitationally strong the force is infinite and when the force is finite the naked singularity is gravitationally weak.

4. Generalized Vaidya spacetime

When we consider the mixture of two types of the matter field i.e. type-I and type-II [15] then one can obtain the generalized Vaidya spacetime. Type-I is the null dust which has been considered in the previous section. Type-II is the null perfect fluid [6]. So one can write:

$$\begin{aligned} T_{ik}^{(I)} &= \mu L_i L_k \,, \\ T_{ik}^{(II)} &= (\rho + P)(L_i N_k + L_k N_i) + P g_{ik} \,, \end{aligned} \tag{25}$$

here μ is the energy density of the null dust and P and ρ are the pressure and energy density of the null perfect fluid:

$$\begin{aligned} \mu &= \frac{2\dot{M}}{r^2} \,, \rho = \frac{2M'}{r^2} \,, \\ P &= -\frac{M''}{r} \,, L_i = \delta_i^0 \,, \end{aligned} \tag{26}$$

and l^i and n^i are two null vectors:

$$\begin{aligned} N_i &= \frac{1}{2}\left(1 - \frac{2M}{r}\right)\delta_i^0 - \delta_i^1 \,, \\ L_i L^i &= N_i N^i = 0 \,, L_i N^i = -1 \,. \end{aligned} \tag{27}$$

In this case the generalized Vaidya spacetime is given by:

$$ds^2 = -\left(1 - \frac{2M(v,r)}{r}\right) dv^2 + 2dvdr + r^2 d\omega^2 \,. \tag{28}$$

Here the mass function not only the function of the time v but also radial coordinate r.

Weak, strong and dominant energy conditions demand:

$$\begin{aligned} \mu &\geq 0 \,, \\ \rho \geq P &\geq 0 \,. \end{aligned} \tag{29}$$

The radial geodesic equation in this metric has the following form:

$$\begin{aligned} \frac{d^2 r}{d\tau^2} = &-\frac{\left(1 - \frac{2M}{r}\right)(M - M'r) + \dot{M}r}{r^2}\left(\frac{dv}{d\tau}\right)^2 + 2\frac{M - M'r}{r^2}\frac{dv}{d\tau}\frac{dr}{d\tau} + \\ &(r - 2M)\left(\frac{d\theta}{d\tau}\right)^2 + (r - 2M)\sin^2\theta\left(\frac{d\varphi}{d\tau}\right)^2 . \end{aligned} \tag{30}$$

In comparison with (15) we have two changes: I) the inertial force which depends on the velocity $\frac{dr}{d\tau}$ linearly has the extra term $\frac{M'}{r}\frac{dv}{d\tau}\frac{dr}{d\tau}$. II) Our leading term has the following change:

$$\Gamma^1_{00\,generalizedVaidya} = \Gamma^1_{00\,Vaidya}(r,v) - \frac{M'}{r}\,. \tag{31}$$

From (31) one might expect the sign of the force to be changed. But first of all we should find out the explicit form of the mass function. For this purpose let's consider the null perfect fluid with the equation of the state $P = \alpha\rho$ where $\alpha \in [-1\,,1]$. In this case the mass function has the following form [8]:

$$M(v,r) = C(v) + D(v)r^{1-2\alpha}\,, C(v) \geq 0\,, D(v) \geq 0\,. \tag{32}$$

From (30) we can easily see that the only term which can change the force sign is $\frac{(M-M'r}{r^2}$. This term might change the sign for both forces i.e. and for leading term Γ^1_{00} and the inertial force which depends linearly on the velocity $\frac{dr}{d\tau}$. So let's find when this factor can change its sign. For this purpose substituting the mass function (32) into this term:

$$\frac{M - M'r}{r^2} = \frac{C(v) + 2\alpha r^{1-2\alpha}}{r^2}\,. \tag{33}$$

$C(v > 0$ and $D(v) > 0$ these are demanded by the dominant and weak energy condition. So if $\alpha \geq 0$ then (33) doesn't change its sign. The only option is the negative values of α. But in this case we have to violate the strong energy condition and consider the negative pressure.

So the inertial force which is linearly depends on the velocity $\frac{dr}{d\tau}$ changes its sign when the particle crosses the shell of the radias:

$$r = \left(-\frac{C(v)}{2\alpha D(v)}\right)^{\frac{1}{1-2\alpha}}\,. \tag{34}$$

Regarding the leading term Γ^1_{00} things are not so clear. In generall case when (33) changes its sign this term can be still positive. However it can change its sign and attraction will be replaced with repulsion if:

$$\left|\left(1 - \frac{2(C(v) + D(v)r^{1-2\alpha}}{r}\right)(C(v) + 2\alpha r^{1-2\alpha})\right| > \dot{C}(v) + \dot{D}(v)r^{1-2\alpha}\,. \tag{35}$$

That the positive r exists in this case one can see if we consider the mass function which depends only on r. In this case $C(v) = \eta > 0\,, D(v) = \zeta > 0$ and attraction is replaced with repulsion at the radius:

$$r = \left(-\frac{\eta}{2\alpha\zeta}\right)^{\frac{1}{1-2\alpha}} \tag{36}$$

So we can see that in generalized Vaidya spacetime when we consider the negative pressure of the null perfect fluid the force of attraction can be replaced with the force of repulsion.

Now let's find out the law of the energy change. The energy expression in the generalized Vaidya spacetime is:

$$\varepsilon(v) = \left(1 - \frac{2M}{r}\right)\frac{dv}{d\tau} - \frac{dr}{d\tau}\,. \tag{37}$$

Differentiating this with respect to τ one can find:

$$\frac{d^2r}{d\tau^2} = -\frac{\left(1-\frac{2M}{r}\right)(M-M'r)+2\dot{M}r}{r^2}\left(\frac{dv}{d\tau}\right)^2 + 2\frac{M-M'r}{r^2}\frac{dv}{d\tau}\frac{dr}{d\tau} - \frac{d\varepsilon(v)}{dv}\frac{dv}{d\tau}, \tag{38}$$

and comparing this formula to (30) with $\varphi = const\,, \theta = const$ one can find the same law of the energy loss like in the previous section (19):

$$\frac{d\varepsilon(v)}{dv} = -\frac{\dot{M}}{r^2}\frac{dv}{d\tau} \tag{39}$$

5. Conclusion

In this paper we have considered the leading term Γ^1_{00} in Schwarzschild spacetime which is the Newton force of attraction. First of all we have found out that this term doesn't change when we do the coordinate transformation to Eddington-Finkelstein Coordinates. Then the matter has been added to the right-hand side of the Einstein equation in order to track how this term will change. In the begining we considered the energy-momentum tensor of the null dust - so-called Vaidya spacetime. In this case the force of attraction becomes bigger than in Schwarzschild metric and in the case of gravitationally strong naked singularity formation this force is infinite. This force becomes finite in the central naked singularity only in the case when it is gravitationally weak. When we consider the mixture of type-I and type-II of the matter fields (generalized Vaidya spacetime) then the force of attraction can be replaced with repulsion but only if we violate the strong energy condition and consider the negative pressure.

acknowledgments The work was performed within the SAO RAS state assignment in the part "Conducting Fundamental Science Research".

References

[1] Chandrasekhar S 1998 *The mathematical theory of black holes* (New York: Oxford University Press)
[2] Landau L D and Lifshitz E M 1980 *Course of theoretical physics series* vol 2 The classical theory of fields (4th ed. Oxford: Butterworth-Heinemann)
[3] Weinberg S 1972 *Gravitation and cosmology: Principles and applications of the general theory of relativity* (New York: John Wiley & Sons Publ.)
[4] Vladimirov Yu S 2009 *Klassicheskaya teoriya gravitatsii [Classical gravity theory]* (Moscow: Publishing house "'Libroko"')
[5] Papapetrou A 1985 *A Random Walk in Relativity and Cosmology* (Wiley Eastern, New Delhi)
[6] Wang A and Wu Yu 1999 Generalized Vaidya solutions *Gen Relativ. Gravit.* **31** 107
[7] Mkenyeley M D, Goswami R and Maharaj S D 2014 Gravitational collapse of generalized Vaidya spacetime *Phys. Rev. D* **90** 064034
[8] Vertogradov V D 2016 Naked singularity formation in generalized Vaidya space-time *Grav. Cosmol.* **22** 220-223
[9] Vertogradov V D 2018 The eternal naked singularity formation in the case of gravitational collapse of generalized Vaidya spacetime *Int. J. Mod. Phys. A* **33** 17
[10] Vertogradov V D 2020 Some remarks on the naked singularity phenomenon *Physics of Complex Systems* **1** 20-29
[11] Poisson E 2007*A Relativist's Toolkit: The Mathematics of Black-Hole Mechanics* (Cambridge University Press)
[12] Dwivedi I H and Joshi P S 1989 On the nature of naked singularities in Vaidya spacetimes Class. *Quantum Grav.* **6** 1599
[13] Tipler F J 1977 *Phys. Lett. A* **64** 8
[14] Nolan B C 1999 *Phys. Rev. D* **60** 024014
[15] Hawking S W and Ellis G F R 1973 *The Large Scale Structure of Space-Time* (Cambridge University Press)

Method of multiple scales in scalar field cosmology

V M Zhuravlev[1], **S V Chervon** [2,3,4]

[1] Inter-University Department of Space Research, Samara National Research University, 443086 Samara, Russia
[2] Bauman Moscow State Technical University, Moscow, Russia
[3] Ulyanovsk State Pedagogical University, Ulyanovsk, Russia
[4] Kazan Federal University, Kazan, Russia

E-mail: zhvictorm@gmail.com, chervon.sergey@gmail.com

Abstract. In this work, the method of multiple scales is applied to analysis of cosmological dynamics. The method is used to construct solutions to the dynamic equations of the Universe filled with a scalar field in the Friedman-Robertson-Walker metric. A general scheme is described for choosing small dimensionless parameters of the expansion of model functions and applying the method itself to the equations of cosmological dynamics. Solutions are given that are constructed for two different types of a small parameter - a small field value and a small slow roll parameter.

1. Introduction

At present time a scalar field cosmology is actively applied for studying early and later evolution of the Universe. Cosmological dynamic equation of self-interacting scalar field in Friedmann universe is described by the system of ordinary differential equations of the second order with respect to a scale factor or to a scalar field itself [1].

Many attempts were devoted to investigation of the Scalar Cosmology Equations (SCEs) on an inflationary stage by approximate methods (slow-roll and fast oscillation between them) in the first works on inflation by Starobinsky (1980)[2], Guth (1981)[3] Linde (1982)[4], Albrecht and Steinhardt (1982)[5]. Exact solution construction of SCEs is started in the works [6], [7], [8], [9]. More detail about exact solutions and methods of construction of them can be found in the monograph [1]. In the present article we propose new approach for searching approximate solutions of SCEs using method of multiple scales or multi-scale analysis.

Effective methods of asymptotic analysis of models of cosmological dynamics are usually built on the qualitative analysis of dynamical systems [10, 11, 12, 13]. Such methods provide useful information of a general nature and make it possible to understand the limiting states of the Universe evolution. Nevertheless, if it is necessary to obtain more accurate information about the nature of the solutions under certain initial conditions, the methods of asymptotic analysis turn out to be insufficiently effective. For solving problems of analyzing the current state of systems, methods of approximate analysis, built on the expansion of solutions in series in a small parameter, are much more useful. Among such methods, which is rarely used in investigation of cosmological dynamic, is the method of multiple scales or multiple scale analysis.

The method of multiple scales (MoMS) is one of the effective methods for the approximate analysis of the dynamics of various types of objects and processes. The initial field of application of MoMS, most likely, is the mechanics and theory of wave processes [14, 15, 16, 17]. The main idea of the MoMS is reduced to the representation of the dependence of the initial process on time & coordinates to the dependence on a wider range of variables, which are related to the initial variable scale factors and containing small parameters of the model in various degrees. The difference between the method of multiple scales and the standard method of a small parameter is that the introduction of new "slow" or "fast" variables makes it possible to improve the convergence of the series in this small parameter. To control the rate of convergence of series, the MoMS contains a specific tool, which is often called a procedure for eliminating resonances. The meaning of this procedure is that the introduction of additional independent variables ("slow" or "fast") generates a certain arbitrariness in the choice of the balance between the individual terms of the equations obtained as a result of expansions in a series. This arbitrariness allows the selection of dependence expansion elements in a series in such a way that the terms having a power-law form in independent variables (secular terms), which significantly worsen the rate of convergence of the series and reduce the radii of their convergence, are excluded from the series.

In this article, the MoMS is applied to analyzing dynamic of the Universe filled by the self-interacting scalar field in the Friedman-Robertson-Walker space-time. Although this model has been studied quite fully, it can be used as an example to show the features of the application of the MoMS to problems of cosmological dynamics, which can subsequently be usefully applied to more complex models.

The article is organized as follow. In Section 2 we derive cosmological dynamic equation in dimensionless form. Section 3 devoted to obtaining approximate equations for small scalar field by MoMS. In Section 4 we present solutions till the third order. Section 5 contain approximate solution for a scale factor till the second order. In Section 6 we consider approximation based on the first slow roll parameter and in Section 7 we derive equations and find solutions till the second order. In Section 8 we represent the general solution including that for the scale factor. In Section 9 we discuss obtained results.

2. Cosmological dynamic equations in dimensionless form

Let us study a cosmological dynamic of the self-interacting scalar field on the basis of MoMS. Cosmological dynamic equations [1] are:

$$3H^2 = \frac{1}{2}\dot{\phi}^2 + V(\phi), \tag{1}$$

$$2\dot{H} = -\dot{\phi}^2, \tag{2}$$

$$\ddot{\phi} + 3H\dot{\phi} + \frac{\partial V}{\partial \phi} = 0. \tag{3}$$

Here $H = \dot{a}/a = \frac{da}{adt}$ is the Habble parameter, a is a scale factor, ϕ is a scalar field, $V(\phi)$ is the self-interacting potential of the scalar field, t is a cosmic time. Equations of gravitation dynamic (1) and (2) have the field equation (3) as a consequence [1]. Therefore we may exclude this equation from consideration.

To obtain dimensionless variables [11] we set $\xi = H/H_0$, $\eta = \phi/\Phi_0$, $\tau = t/T_0$. Here H_0, Φ_0 and T_0 - some constants of the corresponding dimensions, which should be determined by the choice of the required asymptotic of the the initial system solutions. As a result of substitution in the initial system (1) and (2) of dimensionless variables, we arrive at a system of equations in the following form:

$$3H_0^2\xi^2 = \frac{1}{2}\Phi_0^2\dot{\eta}^2 + \mathcal{V}(\eta, \Phi_0), \tag{4}$$

$$2H_0T_0\dot{\xi} = -\Phi_0^2\dot{\eta}^2. \tag{5}$$

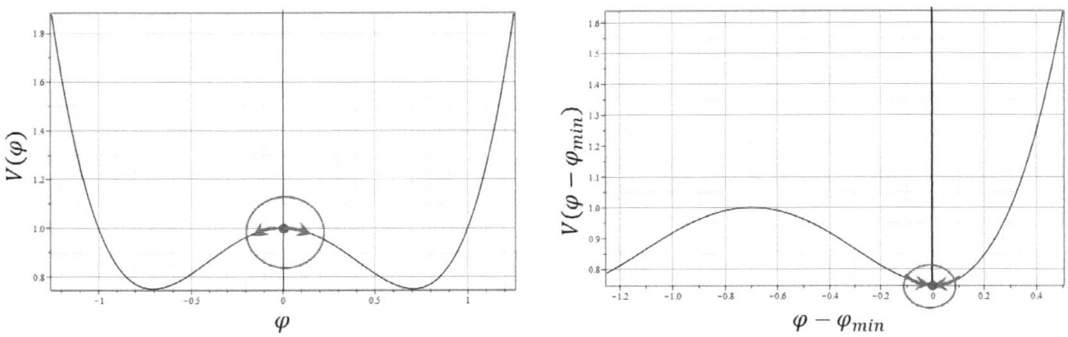

(a) (b)

Figure 1. The positions of the points of the expansion of the potential at the maximum (a) and minimum (b) for the Higgs potential (8).

Equivalently, in another form:

$$\frac{\alpha}{\beta}\dot{\xi} + 3\xi^2 = \frac{1}{H_0^2}\mathcal{V}(\eta, \Phi_0), \tag{6}$$

$$\dot{\xi} = -\frac{1}{2}\beta\dot{\eta}^2. \tag{7}$$

Here $V(\phi) = \mathcal{V}(\eta, \Phi_0)$ and:

$$\alpha = \frac{\Phi_0^2}{T_0^2 H_0^2}, \quad \beta = \frac{\Phi_0^2}{H_0 T_0}.$$

Dimensionless parameters α and β, under given form of the potential $V(\phi)$ and initial conditions, define dynamic of cosmological process. Because the parameter $\gamma = H_0 T_0$ is dimensionless by definition, then Φ_0 is dimensionless as well.

3. Expansion on a scale of the field $\Phi_0 = \sqrt{\varepsilon}$

Let us consider the first option for choosing a small parameter, setting $\gamma = H_0 T_0 = 1$ and a small parameter $\Phi_0 = \sqrt{\varepsilon}$. A small value of the field scale Φ_0 means a small deviation of the field from the point $\phi = 0$, which can be placed by the transformation: $\phi' = \phi - \phi_{min}$ to any other point. In this case, the position of the point $\phi = 0$ can be both at the minimum and maximum, as shown in Fig. 1, as well as at any other point on the ϕ axis. This fact is illustrated by the figures with the positions of the points $\phi = 0$ for the Higgs potential:

$$V = 1 + \phi^4 - \phi^2. \tag{8}$$

To select the field scale of ϕ as a small parameter we have $\alpha = \beta$. Using new variables, we can formally write:

$$\frac{1}{H_0^2}V(\phi) = \mathcal{V}(\eta, \sqrt{\varepsilon}).$$

The assumed smallness of the parameter ε allows, in accordance with the general ideology of multiple scales expansions, to seek solutions to the system (6)-(7) in the form of dependence on the usual τ and slow variables, which will have the following form:

$$\tau_1 = \sqrt{\varepsilon}\tau, \ \tau_2 = \varepsilon\tau, \ \ldots .$$

Hence, for any function we have $F(\tau) = \mathcal{F}(\tau, \tau_1, \tau_2, \ldots)$:

$$\frac{d}{d\tau}F(t) = \left(\frac{\partial}{\partial\tau} + \sqrt{\varepsilon}\frac{\partial}{\partial\tau_1} + \varepsilon\frac{\partial}{\partial\tau_2} + \cdots\right)\mathcal{F}(\tau, \tau_1, \tau_2, \ldots).$$

Further we represent the potential $V(\phi)$ as an expansion series in ϕ:

$$V(\phi) = V_0 + V_1\phi + V_2\phi^2 + \cdots.$$

From here:

$$\mathcal{V}(\eta, \sqrt{\varepsilon}) = v_0 + v_1\eta\sqrt{\varepsilon} + v_2\eta^2\varepsilon + \cdots,$$

where $v_i = H_0^{-2}V_i,\ i = 0, 1, \ldots$ are dimensionless parameters of the potential.

Similarly, the solution for $\xi(t)$ and $\eta(t)$ will be sought in the following form:

$$\xi = \xi_0(\tau, \tau_1, \tau_2, \ldots) + \sqrt{\varepsilon}\xi_1(\tau, \tau_1, \tau_2, \ldots) + \varepsilon\xi_2(\tau, \tau_1, \tau_2, \ldots) + \cdots, \quad (9)$$

$$\eta = \eta_0(\tau, \tau_1, \tau_2, \ldots) + \sqrt{\varepsilon}\eta_1(\tau, \tau_1, \tau_2, \ldots) + \varepsilon\eta_2(\tau, \tau_1, \tau_2, \ldots) + \cdots. \quad (10)$$

Substituting (9) and (10) into (6), with respect to the first three expansion elements we have:

$$\left(\frac{\partial}{\partial\tau} + \sqrt{\varepsilon}\frac{\partial}{\partial\tau_1} + \varepsilon\frac{\partial}{\partial\tau_2} + \cdots\right)\Big(\xi_0(\tau, \tau_1, \ldots) + \sqrt{\varepsilon}\xi_1(\tau, \tau_1, \ldots) + \varepsilon\xi_2(\tau, \tau_1, \ldots) + \cdots\Big) +$$
$$+3\Big(\xi_0(\tau, \tau_1, \ldots) + \sqrt{\varepsilon}\xi_1(\tau, \tau_1, \ldots) + \varepsilon\xi_2(\tau, \tau_1, \ldots) + \cdots\Big)^2 = \quad (11)$$
$$= v_0 + v_1\Big(\eta_0(\tau, \tau_1, \ldots) + \sqrt{\varepsilon}\eta_1(\tau, \tau_1, \ldots)\Big)\sqrt{\varepsilon} + v_2(\eta_0^2\varepsilon + 2\eta_0\eta_1\varepsilon^{3/2}) + v_3\eta_0^3\varepsilon^{3/2} + \cdots.$$

The first four equations of the system (11) in the expansion in orders of $\varepsilon^{1/2}$ have the following form:

$$\varepsilon^0: \quad \frac{\partial\xi_0}{\partial\tau} + 3\xi_0^2 = v_0, \quad (12)$$

$$\varepsilon^{1/2}: \quad \frac{\partial\xi_1}{\partial\tau} + \frac{\partial\xi_0}{\partial\tau_1} + 6\xi_0\xi_1 = v_1\eta_0, \quad (13)$$

$$\varepsilon^1: \quad \frac{\partial\xi_2}{\partial\tau} + \frac{\partial\xi_1}{\partial\tau_1} + \frac{\partial\xi_0}{\partial\tau_2} + 6\xi_0\xi_2 + 3\xi_1^2 = v_1\eta_1 + v_2\eta_0^2, \quad (14)$$

$$\varepsilon^{3/2}: \quad \frac{\partial\xi_3}{\partial\tau} + \frac{\partial\xi_2}{\partial\tau_1} + \frac{\partial\xi_1}{\partial\tau_2} + \frac{\partial\xi_0}{\partial\tau_3} + 6(\xi_0\xi_3 + \xi_1\xi_2) = \quad (15)$$
$$= v_1\eta_2 + 2v_2\eta_0\eta_1 + v_3\eta_0^3$$
$$\ldots$$

Similarly, we find the expansion of (7). After substitution of the series (9) and (10), we find:

$$\left(\frac{\partial}{\partial\tau} + \sqrt{\varepsilon}\frac{\partial}{\partial\tau_1} + \varepsilon\frac{\partial}{\partial\tau_2} + \cdots\right)\Big(\xi_0(\tau, \tau_1, \ldots) + \sqrt{\varepsilon}\xi_1(\tau, \tau_1, \ldots) + \varepsilon\xi_2(\tau, \tau_1, \ldots)\Big) =$$
$$= -\frac{1}{2}\varepsilon\left(\frac{\partial\eta_0}{\partial\tau} + \sqrt{\varepsilon}\frac{\partial\eta_0}{\partial\tau_1} + \varepsilon\frac{\partial\eta_0}{\partial\tau_2} + \sqrt{\varepsilon}\frac{\partial\eta_1}{\partial\tau} + \varepsilon\frac{\partial\eta_0}{\partial\tau_1} + \varepsilon\frac{\partial\eta_2}{\partial\tau_2} + \ldots\right)^2. \quad (16)$$

The first four equations of (7) for the expansion (44) in orders $\varepsilon^{1/2}$ have the following form:

$$\varepsilon^0: \quad \frac{\partial\xi_0}{\partial\tau} = 0, \quad (17)$$

$$\varepsilon^{1/2}: \quad \frac{\partial \xi_1}{\partial \tau} + \frac{\partial \xi_0}{\partial \tau_1} = 0, \tag{18}$$

$$\varepsilon^{1}: \quad \frac{\partial \xi_2}{\partial \tau} + \frac{\partial \xi_1}{\partial \tau_1} + \frac{\partial \xi_0}{\partial \tau_2} = -\frac{1}{2}\left(\frac{\partial \eta_0}{\partial \tau}\right)^2, \tag{19}$$

$$\varepsilon^{3/2}: \quad \frac{\partial \xi_3}{\partial \tau} + \frac{\partial \xi_2}{\partial \tau_1} + \frac{\partial \xi_1}{\partial \tau_2} + \frac{\partial \xi_0}{\partial \tau_3} = -\frac{1}{2}\frac{\partial \eta_0}{\partial \tau}\left(\frac{\partial \eta_0}{\partial \tau_1} + \frac{\partial \eta_1}{\partial \tau}\right), \tag{20}$$

$$\ldots$$

Thus we can solve step by step the systems (12)-(15) and (17)-(20).

4. The solution of the system

From equation (12) and (17) one can find the following solution with respect to ξ_0:

$$\xi_0 = \pm\sqrt{\frac{v_0}{3}} = \pm\Lambda. \tag{21}$$

Hereinafter, the notation is introduced: $\Lambda = \sqrt{v_0/3}$. The solution (39) makes sense if $v_0 \geq 0$. The last inequality is known the consequence of the energy dominance condition. Since $v_0 = \text{const}$, we have: $\xi_0 = \text{const}$. As a result, the solution of (13) and (19) in the first order looks like this:

$$\xi_1 = \pm\frac{v_1}{6\Lambda}\eta_0(\tau_1, \tau_2, \ldots). \tag{22}$$

Moreover, the functions ξ_1 and η_0 depend only on slow variables: $\xi_1 = \xi_1(\tau_1, \tau_2, \ldots)$, $\eta_0 = \eta_0(\tau_1, \tau_2, \ldots)$.

The solution of (14) and (19) in the second order can now be written like this:

$$\xi_2 = \pm\frac{1}{8\sqrt{3v_0}}\Big(4v_0v_1\eta_1(\tau, \tau_1, \ldots) + (4v_0v_2 - v_1^2)\eta_0^2(\tau_1, \tau_2, \ldots)\Big). \tag{23}$$

In this case, from (20) for the function $\eta_1(\tau, \tau_1, \ldots)$ we obtain the equation:

$$\frac{\partial \eta_1}{\partial \tau} + \frac{\partial \eta_0}{\partial \tau_1} = 0.$$

From here we get:

$$\eta_1 = -\tau\frac{\partial \eta_0}{\partial \tau_1} + C_1(\tau_1, \tau_2, \ldots),$$

where $C_1(\tau_1, \tau_2, \ldots)$ is an integration constant wrt τ. Note, the dependence $\eta_0 = \eta_0(\tau_1, \tau_2, \ldots)$ on τ_1 does not defined yet. This can be used to eliminate in the solution for η_1 a term that grows linearly together with τ. Such terms are often called resonant or secular. The presence of such terms leads to a rapid divergence of the series in ε with increasing τ. By setting $\eta_0 = \eta_0(\tau_2, \ldots)$, we eliminate the dependence of η_1 on τ. This procedure is called the resonance elimination procedure. In this case, ξ_2 will also not depend on τ, but will depend on slow variables.

The third order scale factor equation can now be written in the following form:

$$\xi_3 = \frac{1}{6}\sqrt{\frac{3}{v_0}}\Big(v_1\eta_2 + 2v_2\eta_0\eta_1 + v_3\eta_0^3 - 6\xi_1\xi_2\Big). \tag{24}$$

In this case, the following relations are also fulfilled:

$$\xi_2(\tau_1, \tau_2, \ldots) = -\tau_1\frac{\partial \xi_1}{\partial \tau_2} + \xi_1^{(0)}(\tau_2, \ldots).$$

The last relations define dependence ξ_2 on slow variable τ_1.

From the formal point of view, the linear dependence of $\xi_2(\tau_1, \tau_2, \ldots)$ on τ_1 can also be considered as a 'resonant' term leading to an accelerated divergence of the series. Therefore, it can also be eliminated by setting $\xi_1 = \xi_1(\tau_2, \ldots)$. In the next orders, one can continue to apply further a similar procedure.

5. Solution for the scale factor

Since ξ_0, ξ_1 and ξ_2 do not depend on the variable τ, it is possible to find out the character of the cosmological expansion. From definition of ξ up to $\varepsilon^{3/2}$ we have:

$$H = H_0\Big(\xi_0 + \sqrt{\varepsilon}\xi_1 + \varepsilon\xi_2 + \varepsilon^{3/2}\xi_3 + \ldots\Big) = H_0\frac{\dot{a}}{a} = \frac{H_0}{a_0 + \sqrt{\varepsilon}a_1 + \varepsilon a_2 + \varepsilon^{3/2}a_3 + \ldots}\Bigg(\frac{\partial a_0}{\partial \tau} + \\ + \sqrt{\varepsilon}\left(\frac{\partial a_0}{\partial \tau_1} + \frac{\partial a_1}{\partial \tau}\right) + \varepsilon\left(\frac{\partial a_0}{\partial \tau_1} + \frac{\partial a_1}{\partial \tau_1} + \frac{\partial a_2}{\partial \tau} + \ldots\right) + \ldots\Bigg).$$

Using this expansion we arrive at the system of equations for calculating the evolution of the scale factor $a = a_0 + \sqrt{\varepsilon}a_1 + \varepsilon a_2 + \ldots$. Thus, we have:

$$\xi_0 = \frac{\partial \ln a_0}{\partial \tau},$$
$$\xi_1 = \frac{1}{a_0}\left(\frac{\partial a_0}{\partial \tau_1} - \frac{a_1}{a_0}\frac{\partial a_0}{\partial \tau} + \frac{\partial a_1}{\partial \tau}\right),$$
$$\xi_2 = -\frac{\partial a_0}{\partial \tau}\left(\frac{a_2}{a_0^2} - \frac{a_1^2}{a_0^3}\right) - \left(\frac{\partial a_0}{\partial \tau_1} + \frac{\partial a_1}{\partial \tau}\right)\frac{a_1}{a_0^2} + \left(\frac{\partial a_0}{\partial \tau_2} + \frac{\partial a_1}{\partial \tau_1} + \frac{\partial a_2}{\partial \tau}\right)\frac{1}{a_0}.$$

Using the solution for ξ_0 (39) we find:

$$a_0 = A_0(\tau_1, \tau_2, \ldots)e^{\tau\sqrt{v_0/3}}.$$

Thus, in the zero order, evolution is an inflationary expansion with the Hubble parameter $H = H_0\Lambda$ at the sign of $+$ in the exponent. It is this option that is discussed below.

In the first order, the equation for a_1 is reduced to the form:

$$\frac{\partial a_1}{\partial \tau} - \sqrt{\Lambda}a_1 = A_0 e^{\Lambda\tau}\left(\frac{v_1}{6\Lambda}\eta_0 - \frac{\partial \ln A_0}{\partial \tau_1}\right).$$

The condition for the absence of resonance in this equation is the vanishing of the right side of this equation, which leads to the equation:

$$A_0\frac{v_1}{6\Lambda}\eta_0 - \frac{\partial A_0}{\partial \tau_1} = 0.$$

Hence, under the condition $\xi_1 = \text{const}$ (from (22) this means that $\eta_0 = \text{const}$ too) it follows:

$$A_0 = A_{00}\exp\left(\frac{v_1}{6\Lambda}\eta_0\tau_1\right). \tag{25}$$

Integration constant A_{00} may depend on $\tau_2, \ldots$. Note that the sign of the exponent in (25) is defined both the sign of v_1 (i.e., by the sign of the coefficient of the Taylor series of the expansion of the potential at zero) and the sign of the field itself in the zero order. Consequently, with the opposite signs of η_0 and v_1, the inflation rate will decrease in comparison with the zero order.

The solutions for the scale factor obtained in the first two orders can be represented as follows:

$$a = e^{\Lambda\tau}\left(A_{00}\exp\left(\frac{v_1}{6\Lambda}\eta_0\tau\sqrt{\varepsilon}\right) + \sqrt{\varepsilon}A_{10}\right) + O(\varepsilon), \tag{26}$$

where A_{10} is the constant of integration.

Considering η_0 as a constant in (26) and expanding the exponent $\exp\left(\frac{v_1}{6\Lambda}\eta_0\tau\sqrt{\varepsilon}\right)$ in Taylor series in $\sqrt{\varepsilon}$ one can obtain the next form of the solution

$$a = e^{\Lambda\tau}\left(A_{00} + \sqrt{\varepsilon}\left(\tilde{\eta}\tau + A_{10}\right)\right) + O(\varepsilon), \tag{27}$$

where $\tilde{\eta} = \frac{v_1}{6\Lambda}\eta_0$. The solution in the form (27) corresponds to exponent power-law inflation, considered, for example, in the work [18] (p.4, formula (14)).

6. Expansion on the slow roll parameter

Let us now consider the results of the analysis of equations in the case of choosing the scales in accordance with the relations: $\Phi_0 = 1$, $H_0T_0 = \varepsilon^{-1}$, where $\varepsilon \ll 1$ is a new small parameter. In dimensionless form, the slow roll parameter

$$p = -\frac{\dot{H}}{H^2},$$

will be like follows:

$$p = -\frac{1}{H_0T_0}\frac{\dot{\xi}}{\xi^2} = -\varepsilon\frac{\dot{\xi}}{\xi^2},$$

that is, it is a quantity of the first order of smallness in ε. This means that $\varepsilon = (H_0T_0)^{-1}$ can be interpreted as a slow roll scale.

Passing to the dimensionless equations (6), we find in this case:

$$\alpha = \varepsilon^2, \quad \beta = \varepsilon.$$

With this choice of parameters, the equations (6) take the following form:

$$\varepsilon\dot{\xi} + 3\xi^2 = \frac{1}{H_0^2}\mathcal{V}(\eta), \tag{28}$$

$$\dot{\xi} = -\frac{1}{2}\varepsilon\dot{\eta}^2. \tag{29}$$

This system is singular, since in two equations the small parameter stands at the highest derivative. To eliminate this drawback, we introduce a new time variable, setting $\theta = \tau\varepsilon^{-1} = H_0t$. For the new variable, the equations will look like this:

$$\dot{\xi} + 3\xi^2 = \frac{1}{H_0^2}\mathcal{V}(\eta), \tag{30}$$

$$\dot{\xi} = -\frac{1}{2}\dot{\eta}^2. \tag{31}$$

Hereinafter dot over the function means the derivative wrt new variable θ.

An unclear element of the last system of equations is the term on the right-hand side of (30) associated with the potential. The crux of the problem is how the values of the $\mathcal{V}(\eta)$ function relate to H_0^2. There are two possibilities. The first is that the values of $H_0^{-2}\mathcal{V}(\eta)$ are of order 1. The second is that $T_0^2\mathcal{V}(\eta)$ is the order 1. In the first case, the equations do not contain the small parameter ε, and in the second, the term with the potential in the right-hand side of (30) is of order ε^2. In the general case, it can be assumed that the functional dependence of the potential on η can be split into two parts. The first part is of order 1, according to the case of $H_0^{-2}\mathcal{V}$. The second part, corresponding to the case when $T_0^2\mathcal{V}$, is of order 1. Thus, we assume:

$$H_0^{-2}\mathcal{V} = V_0(\eta) + \varepsilon^2 U(\eta),$$

where $\mathcal{V}_0(\eta)$ and $U(\eta)$ are functions of order 1.

7. Model's equations and their solutions

Let us study the case $V_0 = v_0 = \text{const}$. Under this condition v_0 can be considered as cosmological constant. We will construct an expansion in ε, assuming:

$$\xi = \xi_0 + \varepsilon\xi_1 + \dots, \quad \eta = \eta_0 + \varepsilon\eta_0 + \dots.$$

The dependence of the expansion coefficients on slow variables will be determined by the variables $\theta_i = \varepsilon^i\theta,\ i = 0, 1, 2, \dots$. For this expansion, we restrict ourselves to only the first two orders. As a result, we have the following system of equations in the orders of the equation (30):

$$\dot{\xi}_0 + 3\xi_0^2 = v_0, \tag{32}$$

$$\dot{\xi}_1 + \frac{\partial\xi_0}{\partial\theta_1} + 6\xi_0\xi_1 = 0, \tag{33}$$

$$\dot{\xi}_2 + \frac{\partial\xi_1}{\partial\theta_1} + \frac{\partial\xi_0}{\partial\theta_2} + 6\xi_0\xi_2 + 3\xi_1^2 = U(\eta_0). \tag{34}$$

Similarly, for the equation (31) we have:

$$\dot{\xi}_0 = -\frac{1}{2}\dot{\eta_0}^2, \tag{35}$$

$$\dot{\xi}_1 + \frac{\partial\xi_0}{\partial\theta_1} = -\dot{\eta}_0\left(\dot{\eta}_1 + \frac{\partial\eta_0}{\partial\theta_1}\right), \tag{36}$$

$$\dot{\xi}_2 + \frac{\partial\xi_1}{\partial\theta_1} + \frac{\partial\xi_0}{\partial\theta_2} = -\dot{\eta}_0\left(\dot{\eta}_2 + \frac{\partial\eta_1}{\partial\theta_1} + \frac{\partial\eta_0}{\partial\theta_2}\right) - \frac{1}{2}\left(\dot{\eta}_1\frac{\partial\eta_0}{\partial\theta_1}\right)^2. \tag{37}$$

7.1. *Zero order*

The zero-order solution for ξ_0 is constructed using the formal replacement:

$$\xi_0 = \frac{1}{3}\frac{d\ln u}{d\theta}. \tag{38}$$

For new function $u(\theta)$ we have the equation:

$$\ddot{u} = 3v_0 u.$$

General solution for u has the following form:

$$u = C_1 e^{\mu\theta} + C_2 e^{-\mu\theta}.$$

Here $\mu^2 = 3v_0$. Solution for ξ_0 is:

$$\xi_0 = \frac{\mu}{3}\left[\frac{1 - Ce^{-2\mu\theta}}{1 + Ce^{-2\mu\theta}}\right].$$

Here $C = C_2/C_1$. Since $\dot{\xi}_0 < 0$, then there are restrictions on the choice of C. For the function ξ_0 to be monotonically decreasing, it is necessary that $C < 0$. The value of C determines in this case the position of the cosmological singularity. If we assume that the instant of time of the cosmological singularity is chosen equal to $t = 0$, then $C = -1$. As a result, the solution for ξ_0 can be written as follows:

$$\xi_0 = \frac{\mu}{3\tanh(\mu\theta)}. \tag{39}$$

From here:

$$\dot{\xi}_0 = -\frac{\mu^2}{3\sinh^2(\mu\theta)} < 0, \tag{40}$$

$$a_0 = A_0 \sinh^{1/3}(\mu\theta). \tag{41}$$

In this formula $a_0(\theta)$ is a scale factor in zero-order, A_0 - the constant of integration. Note, that such type of solution for arbitrary exponent α instead of 1/3 was presented in [1] (p.22, formula (2.28).

Using (40), the zero-order field equation (35) can be reduced to the following form:

$$\dot{\eta}_0 = s\frac{\sqrt{2}\mu}{\sqrt{3}\sinh(\mu\theta)}, \quad s = \pm 1.$$

From here we find:

$$\eta_0 = f_0 + \int \frac{\mu d\theta}{\sinh(\mu\theta)} = f_0 + \frac{s\sqrt{2}}{\sqrt{3}}\int \frac{dz}{z^2-1} = f_0 + \frac{s}{\sqrt{6}}\ln\left(\frac{\cosh(\mu\theta)-1}{\cosh(\mu\theta)+1}\right). \tag{42}$$

7.2. First order

In order zero, the solution for ξ_0 does not contain arbitrary constants that might depend on slow variables. Therefore, from (33) in the first order for ξ_1 we have the following equation:

$$\dot{\xi}_1 = -\frac{2\mu}{\tanh(\mu t)}\xi_1.$$

From here we find:

$$\xi_1 = \frac{q_1(\tau_1,\ldots)}{\sinh^2(\mu\theta)}, \tag{43}$$

where $q_1 = q_1(\theta_1,\ldots)$ is the integration constant depending on slow variables.

Unlike ξ_0, the solution for η_0 contains a constant of integration, which may depend on slow variables. As a result, for η_1 we have from (36) the following equation:

$$\dot{\eta}_1 = -\frac{1}{\dot{\eta}_0}\left(\dot{\xi}_1 + \dot{\eta}_0\frac{\partial f_0}{\partial\theta_1}\right) = \frac{2sq_1}{\sqrt{3}}\frac{\cosh(\mu\theta)}{\sinh^2(\mu\theta)} - \frac{\partial f_0}{\partial\theta_1}. \tag{44}$$

From here:

$$\eta_1 = -\frac{2sq_1}{\sqrt{3}\mu}\frac{1}{\sinh(\mu\theta)} - \frac{\partial f_0}{\partial\theta_1}\theta + f_1(\theta_1,\ldots). \tag{45}$$

The second term in (45) is linear in θ and it is 'resonant'. To exclude it, one need to put

$$\frac{\partial f_0}{\partial\theta_1} = 0, \quad f_0 = f_0(\theta_2,\ldots).$$

Thus, the solution for η_1 is:

$$\eta_1 = -\frac{2sq_1}{\sqrt{3}\mu}\frac{1}{\sinh(\mu\theta)} + f_1(\theta_1,\ldots). \tag{46}$$

7.3. Second order

The equation (34) for the scale factor in the second order, taking into account the first two: (39) and (43), can be written as follows:

$$\dot{\xi}_2 + \frac{\partial q_1}{\partial \theta_1}\frac{1}{\sinh^2(\mu\theta)} + \frac{2\mu}{\tanh(\mu t)}\xi_2 + 3\frac{q_1^2}{\sinh^4(\mu\theta)} = U(\eta_0). \tag{47}$$

The solution of (47) depends on the form of the function $U(\eta_0)$. Therefore, to obtain its explicit form, it is necessary to select a certain type of this dependence. However, to clarify the presence of resonance terms in this equation, it is sufficient to consider only some components of the Taylor series of the function $U(\eta_0)$ at zero. The eigenfunction of the equation (47), up to a constant factor, coincides with $\sinh^{-2}(\mu\theta)$. Consequently, only the Taylor series component quadratic in η_0 is of interest to eliminate resonances, since $\eta_0 \sim \sinh^{-1}(\mu\theta)$. Denoting this term by $u_2\eta_0^2$, we find an equation for $q_0(\tau_1)$, which is equivalent to the condition for the absence of resonances in the following form:

$$\frac{\partial q_1}{\partial \theta_1} = u_2. \tag{48}$$

From here:

$$q_1 = u_2\theta_1 + q_{10}.$$

As a result, the final solution for ξ_2 can be represented as follows:

$$\xi_2 = \frac{1}{\sinh^2(\mu\theta)}\left(q_2 + 3\frac{(u_2\theta_1 + q_{10})^2}{\sinh(\mu\theta)} + \int\limits_0^\theta \Big(U(\eta_0(z)) - u_2\eta_0^2(z)\Big)\sinh^2(\mu z)dz\right). \tag{49}$$

Here q_2 is the constant of integration, depending on slow variables.

The field equation in the second order (37) now has the following form:

$$\frac{1}{\dot{\eta}_0}\left(\dot{\xi}_2 + \frac{u_2}{\sinh^2(\mu\theta)}\right) - \frac{2su_2}{\sqrt{3}\mu}\frac{1}{\sinh(\mu\theta)} + \frac{\partial f_1}{\partial \theta_1} + \frac{\partial f_0}{\partial \theta_2} = -\dot{\eta}_2.$$

This equation is easy to integrate, but like the equation for ξ_2, it depends on the form of the function $U(\eta)$. Therefore, to construct solutions in the second and subsequent orders, it is necessary to clarify the general ideas about the form of the self-interacting potential. In the first two orders, the form of the function does not matter, and the entire dynamics is determined by the initial conditions, as well as the value of the constant v_0, which plays the role of the cosmological constant.

8. General solution

Combining the obtained solutions in the first two orders, we find:

$$\xi = \frac{\mu}{3\tanh(\mu\theta)} + \varepsilon\frac{u_2\theta_1 + q_{10}}{\sinh^2(\mu\theta)} + O(\varepsilon^2), \tag{50}$$

$$\eta = f_0 + \frac{s}{\sqrt{6}}\ln\left(\frac{\cosh(\mu\theta) - 1}{\cosh(\mu\theta) + 1}\right) - \varepsilon\left(\frac{2s(u_2\theta_1 + q_{10})}{\sqrt{3}\mu}\frac{1}{\sinh(\mu\theta)} - f_1(\tau_1, \dots)\right) + O(\varepsilon^2) \tag{51}$$

Accordingly, the solution for the scale factor is:

$$a = \Big(A_0\mu + \varepsilon\Big(A_1 - A_0(u_2\theta_1 + q_{10})\frac{1}{\tanh(\mu\theta)}\Big)\Big)\sinh^{1/3}(\mu\theta) + O(\varepsilon^2). \tag{52}$$

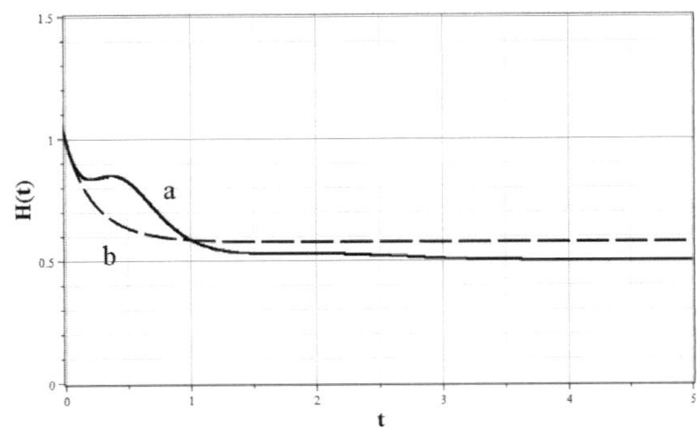

Figure 2. Comparison of the Hubble parameter variation curve obtained numerically (a) and using the expansion in a series in ε (b) for the Higgs potential in zeroth order (8).

Here A_0 and A_1 are integration constants.

To illustrate the nature of the obtained approximation in the zeroth order, let us compare the numerical solution of the problem with the Higgs potential (8) and the solution obtained using the MoMS. The corresponding curves are shown in Fig. 2. It can be seen that the zero order in the considered model for the Hubble parameter is universal, i.e. depends only on the values of the self-interacting potential at zero and does not depend on its specific form. Higher-order corrections require certain calculations related to the satisfaction of the initial conditions, which is beyond the scope of this work.

9. Conclusions

The calculations presented in this work demonstrate the general principles of using the method of multiple scales in scalar field cosmology. As has been shown, dynamical systems used to describe cosmological dynamics contain, as a rule, implicitly several dimensionless constants that determine the type of solutions. In the considered model, such dimensionless constants are the field scale Φ_0 and the slow roll parameter: $p = (H_0 T_0)^{-1}$. The use of MoMS allows one to obtain approximate solutions for physical conditions corresponding to the smallness of these parameters, with a correction of the convergence rate. In this case, the information about the solution itself obtained with the help of MoMS supplements the general information that can be obtained by numerical methods and methods of qualitative analysis of dynamical systems. The results obtained indicate the possibility of using this approach in cosmology for more complex models than that considered in this work.

10. Acknowledgments

This research was partly funded by RUSSIAN FOUNDATION OF BASIC RESEARCH grant number 20-02-00280 a. V.M. Zhuravlev was supported within the framework of the project 0777-2020-0018, financed from the funds of the state assignment to the winners of the competition of scientific laboratories of educational institutions of higher education subordinated to the Ministry of Education and Science of Russia. S.V. Chervon was supported by the Program of Competitive Growth of Kazan Federal University.

References

[1] Chervon S, Fomin I , Yurov V, Yurov A 2019 *World Scientific Publisher* **13**
[2] Starobinsky A A 1980 *Phys. Lett. B* **91** 99
[3] Guth A H 1981 *Phys. Rev. D* **23** 347
[4] Linde A D 1982 *Phys. Lett. B* **108** 389

[5] Albrecht A and Steinhardt P J 1982 *Phys. Rev. Lett.* **48** 1220
[6] Ivanov G G 1981 *Gravitation and Theory of Relativity* (Kazan university publishing house) **18** 54
[7] Lucchin F, Matarrese S 1985 *Phys. Rev. D* **32** 1316
[8] Muslimov A G 1990 *Class. Quant. Grav.* **7** 231
[9] Barrow J D 1990 *Phys. Lett. B* **235** 40
[10] Zhuravlev V M 2016 *Space, Time and Fundamental Interactions* **4** 39
[11] Zhuravlev V M, Chervon A V 2020 *Universe* **6** 195
[12] Lowski M S, Hrycyna O, Stachowski A 2014 *Int. J. Geom. Methods Mod. Phys.* **11** 1460012
[13] Paliathanasis A, Tsamparlis M, Basilakos S, Barrow J D 2015 *Phys. Rev. D* **91** 123535
[14] Bhatnagar P L 1979 *Nonlinear waves in one-dimensional dispersive systems* (Oxford)
[15] Nayfen A H 1981 *Introduction to perturbation techniques* (John Wiley & Sons)
[16] Dodd R K , Eilbeck C J, Gibbon J D, Morris H C 1982 *Soliton and nonlinear wave equations* (Academic Press Inc. [Harcourt Brace Jovanovich Publishers])
[17] Eliseev V V 1999 *Mechanics of elastic bodies* (SPb: Publishing house of SPbSTU) p. 341
[18] Fomin I V, Chervon S V 2020 *J.Phys.Conf.Ser.* **1557** 012020

The principle of materiality of space and the theory of fundamental fields

V M Zhuravlev [1]

[1] Inter-University Department of Space Research, Samara National Research University, 443086 Samara, Russia

E-mail: zhvictorm@gmail.com

Abstract. The work formulates the principle of materiality of space and on its basis a brief critical analysis of the general ideology of the Special and General Theories of Relativity is carried out. The connection of the new principle with the previously developed Topological Theory of Fundamental Fields (TTFF) is considered. A method of constructive implementation of the principle of materiality in the framework of the physical theory of fundamental fields is considered. General equations of the dynamics of markers of material points of physical space are derived and their physical meaning is established.

1. Introduction

In papers [1, 2, 3, 4, 5, 6, 7, 8] there is a new approach for describing the dynamic of matter and its structures, including electromagnetic and gravitational fields, which will be further called **topological theory of fundamental fields** or short **TTFF**. One of the features of this approach is the possibility, based on geometric and topological ideas about the structure of space, to obtain a completely adequate unified description of not only gravity and electromagnetism, but also quantum phenomena, including elementary particles. This geometric approach is an alternative to the geometric approach of General Relativity (GR) [9]. In contrast to GR, topological and geometric ideas in TTFF were initially applied not to the gravitational field, but to the electromagnetic field, and first of all, to the concept of an electric charge. But on the basis of a developed ideology, this theory served as the basis for a new description of the gravitational field. Although the general construction of the developed theory is not yet closed ([5]), nevertheless, a number of important problems of modern physics find a fairly clear explanation within the framework of this theory. These include, for example, the topological interpretation of electric and baryon charges, which explains their discreteness. The geometric interpretation is given to the concept of mass and the concept of the wave function of particles, which reduces Born's probabilistic postulate to the principle of geometric averaging.

The main problem of TTFF, to which special attention was paid in papers [5, 8], is the absence of a physical ideology in it, with the help of which one can describe geometrodynamics of a **physical material hypersurface** $\mathcal{V}^3$ (**PMH**), embedded in the ambient space $\mathcal{W}^4$ per unit of a larger number of dimensions. The geometry of space in TPFF is simpler than in GR, and is determined by the embedding of a three-dimensional physical hypersurface into a

[1] Present address: Department of Theoretical Physics, Ulyanovsk State University, Leo Tolstoy str. 42, Ulyanovsk, Russia

Euclidean space of dimension 4. Therefore, all the properties of such a hypersurface at any given moment of time t are determined by only one function of height - $\mathcal{F}(\mathbf{x}, t)$ But until the physical nature of the hypersurface itself and its environment is known, it is not possible to fully formulate the principles of describing its dynamics. It should be noted that a similar problem stood at the first stages of the creation of general relativity. This was due to the fact that there was no direct data on what determines the properties of the curvature of space-time. The solution was found in the form of postulating the principle of least action in the form of the Hilbert-Einstein variational principle. This principle was involved in general relativity on the basis of some indirect ideas concerning the invariant description of the dynamics of matter and material fields in pseudo-Riemannian space-time, provided that they should transform into the equations of Newton's theory of gravitation in the limit to flat space-time. The last requirement is a particular consequence of the correspondence principle, which plays an important role in modern physics.

In TTFF, the solution to the formulated problem also has to be sought, relying on some indirect data. For example, the passage to the limit to the properties of electromagnetic fields in vacuum can serve as such indirect information. In the case of weak fields, electromagnetic waves in empty space propagate at a fixed speed - the speed of light and are described by the d'Alembert equations. In TTFF, such a transition should also take place. By analogy with general relativity, there should also be a passage to the limit to Newton's theory of gravitation. Since the TTFF claims to describe quantum phenomena as well, then there must be a passage to the limit in it and to quantum mechanics. All these passage to the limit were found in the above papers. All of them are a consequence of the mathematical properties of describing the dynamics of PMH within the framework of the proposed interpretation of electric charge, mass, etc. Due to this interpretation, all the attributes of matter necessary for modern physics appear in theory in the form of mathematical connections between the introduced fields and objects of geometry and topology.

At the same time, in order to substantiate the TTFF, it is required to carry out a fundamental analysis of the modern geometric theory of space-time - GR. This analysis is necessary for several reasons. The first of these reasons concerns the need to explain why the method chosen in GR to describe the geometry of space-time and its relationship with the material objects in it cannot be considered physically adequate to the general concepts of matter as such. With the mathematical flawlessness of general relativity, it contradicts the basic concepts of the properties of matter. The second reason is that such an analysis can provide the missing elements for describing the dynamics of space, which are necessary to complete the whole concept of TTFF.

Partially a preliminary analysis of the justification for the need to revise the entire concept of general relativity was carried out in the paper [7]. In this work, we will conduct a more detailed study of the difficulties of SR and GR, which will give us the opportunity to formulate a special principle for constructing theories of matter and fields, including the concept of space and time. Formalization of this principle, in turn, will make it possible to obtain general information about how it is necessary to formulate the equations of the dynamics of space as a hypersurface embedded in the ambient space of four dimensions. The need for such an approach is due to the fact that the entire concept of TTFFT is based on a set of mathematical identities that allow one to construct from a unified standpoint the entire set of equations for the dynamics of particles with mass, electric and baryon charges, which are involved in electromagnetic and gravitational interactions. However, the theory lacks a dynamic principle that would single out, among all the possible variants of the geometric dynamics of a three-dimensional hypersurface in the ambient four-dimensional space, the only one that corresponds to the dynamics being realized in reality. From a physical point of view, this problem can be formulated as the problem of the absence of any knowledge about what physical properties the physical hypersurface possesses as a material object and what properties the "physical environment" that surrounds it in the ambient space of four dimensions possesses. This formulation becomes possible only after the analysis of the admissible general description of the dynamics of material objects from the point of view of markers is carried out.

At the first stage in the proposed work, after discussing the main difficulties of SRT and GRT, the principle of materiality of PMH is formulated, which is absent in SR and GR. This principle is a revision of a number of remarks concerning problems with the materiality of space-time in general relativity and special relativity, which was given in the papers [15, 13, 14, 16], equations for the fundamental potential are derived in general form - functions of the FMH height, written in terms of the marker transfer equation. These equations are the main goal of this work, which represent a tool that allows one to formulate specific models of FMG dynamics also in terms of marker dynamics.

2. Fundamental disadvantages of SRT and GRT

One of fundamental disadvantages of the Special relativity (SR) and General relativity (GR) ([7]) is the immateriality of physical space. Formally, at least within the framework of GR, sometimes space-time is thought of as a material object, on the natural basis that it is endowed with physical properties of curvature. The problem is that no direct measurements for fixing any elements of this space, as material objects, are provided in SR and GR. This fact can be formulated approximately as follows: points of the space-time of GR cannot be marked with any physical marker. However, in the absence of such a procedure, space-time cannot be interpreted as a material object. Thus, in SR and GR there is matter and an intangible object - space-time, which nevertheless in some way interact with each other.

The absolute space of classical mechanics is also not material, but this space has no other properties, except for the obvious property of extension. This means that in reality, distances in space can be measured exclusively between material bodies or parts of the same material body. In the latter case, one speaks of length. However, it is not very convenient to remember the set of individual distances and lengths between the gigantic set of individual material objects and their component parts - points. It is convenient to combine all these distances into a common consistent scheme - a space that has three dimensions. According to the observations of classical mechanics, a consistent scheme is a mathematical construction - a Euclidean three-dimensional space. Therefore, the absolute Euclidean space of classical mechanics is not a material object, but a convenient way to describe the set of distances between material bodies. It is important to state that the absolute space of classical mechanics itself, being a mathematical construction, does not have any effect on material bodies, which is just postulated in Newton's first law and Galileo's principle of relativity.

Newton's corpuscular theory of light did not contradict the idea of the immateriality of space and time. Individual light particles were some localized material objects, and in this sense, the distances to them were included in the general scheme of classical mechanics. However, the contradiction to this fact of the immateriality of space and time was the wave theory of light, which was first formulated clearly by Huygens. If light is waves, then some medium in which these waves propagate must fill the entire space as a whole. In this case, it is possible to check the properties of the model of space as a Euclidean three-dimensional space, considering changes in the characteristics of the waves that have passed through it. However, in this case, there is a problem with the need to distinguish the space itself from the environment in it. This problem became acute after Maxwell created the general theory of electromagnetism. This medium, which was called ether, had to be discovered and its properties established.

As you know, in the Michelson-Morley experiments, the ether could not be detected. Actually SR appeared as a theory that tries to build mechanics and electrodynamics without the presence of ether - a luminiferous medium. In SR, with the help of Lorentz transformations, it is possible to eliminate from the equations of electrodynamics the terms arising from the transition from one inertial frame of reference to another (as in acoustics), moving uniformly relative to the first frame. However, the environmental problem remained unchanged. If electromagnetic waves are a material object that carries recorded energy, then the question of what oscillates at the observation point when an electromagnetic wave passes through it remains unanswered. As a result, we have to admit that the electromagnetic field itself is a specific form of matter, which does not have mass and other natural properties of matter.

A completely analogous situation exists with any other field, in particular, the gravitational field. As a result, the fields are singled out into a separate category of matter, which does not have any material embodiment, but has a number of measurable and observable characteristics, i.e. fields have a measurable effect on matter.

But if the properties of the electromagnetic field are to some extent similar to matter in the sense that, according to the quantum theory, it has corpuscular properties, then the situation with the gravitational field turns out to be much more complicated. Gravitation does not show any corpuscular properties. The forces of interaction of this field with matter are many times less than the electromagnetic one. But nevertheless, it has a significant effect on bodies of large mass and determines the entire dynamics of celestial bodies. However, the gravitational field in the area between the bodies does not manifest itself in any way. To explain this behavior of the gravitational field at the end of the 19th century, the general idea was proposed that the gravitational field is a manifestation of the properties of space itself. If there is no environment, then only the space itself can be endowed with the necessary properties. It was this idea that formed the basis of Einstein's theory of gravitation in the form of general relativity, although a more radical idea was put forward by Clifford [11].

How does SR solve the problem of the absence of an environment as a material object that would be a carrier of electromagnetic waves and a repository of other fields? For this, in SR space-time is endowed with special properties, consisting in the fact that it affects the movements of material bodies, although in itself, as noted above, it is not a material object. This influence consists in the fact that, without any physical reason, the transition from one inertial frame of reference to another leads to a change in the scales of length and time. This change is detected by comparing the standards of length and time in close contact with the help of light signals, i.e. is a physically detectable phenomenon. It is important to emphasize that the reason for the change in the standards of length and time under these conditions cannot be any physical mechanism, since, according to the first postulate of the SR, all physical laws are the same in all inertial reference frames. Thus, a fundamental contradiction of SR appears - an intangible object - space-time, has quite measurable physical properties.

In general relativity, the non-material space-time is endowed with an even greater number of physical properties - the components of the curvature tensor of the four-dimensional pseudo-Riemannian space-time. The curvature of non-material space-time in general relativity affects the motion of material bodies and, moreover, determines the observed properties of material bodies - their interaction with each other using the forces of gravity. The result of including both material and non-material objects in the theory is the presence of a number of paradoxes in SR and GR. An example is the paradox of the SR twins. In general relativity, this is, for example, the Unruh effect, as well as an intractable problem with the energy of the gravitational field [12].

2.1. Kinematic paradoxes of SR and GR

All explanations of paradoxes in SR and GR are usually based on proving the mathematical consistency of the theory itself, and the irrationality of some of the conclusions of these theories usually refers to the inability to understand reality from the point of view of the practical experience of a person living in the macrocosm of classical mechanics. For example, the paradox of twins requires for its explanation the fact that the frames of reference in which the twins are located are not "symmetric". If we compare the frames of reference moving uniformly and rectilinearly relative to each other, then all the physical laws in them are the same and there is no way to indicate the mechanism of occurrence of the discrepancy in the clock speed and the change in the length of the rulers in them. Actually, this is the twins paradox. For twins, the problem is that aging is a physical process that takes place in the cells of a living organism. Since all physical processes in both frames of reference are exactly the same, the faster aging of the twin remaining on Earth is not associated with any physical mechanism. From a formal point of view, the asymmetry between twins can be found if we take into account the presence of a mandatory segment of the path of the departing twin astronaut, on which the spacecraft accelerates. In the frame of reference associated with the departing twin, inertial forces arise, and in the frame of reference associated with the Earth, there are no inertial forces. Therefore, it is logical to assume that the discrepancy in the readings of the clocks and the length of the rulers arises precisely because of the presence of inertial forces in the frame of reference of the departing twin. However, inertial forces, generally speaking, are the kinematic effect of the formal transformation of coordinates from one frame of reference to another. Therefore, this asymmetry is irrational from a physical point of view. The forces of inertia in the accelerated frame of reference are manifested only in the emergence of reaction forces of the support acting on the astronaut, which ensure its acceleration. Literally, the departing cosmonaut is "pressed" against the wall of the spacecraft, which creates asymmetry from the point of view of material forces. However, it is difficult to imagine that it is the impact of the spacecraft wall on the astronaut that causes changes in the course of his biological clock. With such an explanation, it can be assumed that any action of the reaction force of the support should lead to a change in the lengths of the standards and hours. But even in this case, it remains unclear why the Lorentz transformations do not contain any references to the presence of inertial forces. With a "mathematical" explanation of this paradox, they usually stop at the very fact of the presence of asymmetry in frames of reference, considering that the intangible space-time itself "somehow" manages to influence all clocks and rulers, regardless of their device. A number of problems of this kind were discussed in the paper of Brillouin [10].

The "muon effect" is often cited as evidence that scale contractions in a moving frame of reference take place in reality. This effect consists in the fact that decaying muons, being born in the upper layers of the atmosphere, have time to fly to the surface of the Earth. Since their lifetime in a laboratory experiment in a stationary frame of reference is approximately $2\cdot 10^{-6}$пïS, even at the speed of light, these particles could not travel to decay a distance greater than 600 m. The distance from the region where these particles are born to the Earth's surface is tens of kilometers. However, it should be borne in mind that today there is no real theory of the muon decay process. Therefore, it can be assumed that there is a real physical mechanism that is not associated with the Lorentz transformations, which explains the fact of slowing down of muon decay. This is all the more important, since, as already noted, the physical processes in the own frame of reference and the frame of reference associated with the Earth should be the same for a muon.

3. The principle of materiality

Considering the above analysis as a program for the formation of the concept of three-dimensional space as a material object, we introduce a special principle of materiality, which will be the starting point in the creation of a consistent physical theory of fields and particles. The essence of this principle can be summarized as the following statement: **Any object that has a physically measurable (detectable with the help of instruments) effect on material bodies must itself be a material object.**

For the practice of physical research of observed material objects, this principle does not provide anything particularly new. However, in relation to the concepts of field, space and time, this principle leads to important consequences in relation to their physical properties. For space and time, this principle has at least two ways of consistent implementation.

The first method is demonstrated by classical mechanics, according to which space and time do not possess any properties of materiality and serve as a formal mathematical model for describing lengths and distances on the one hand, and periods and durations of phenomena on the other. It is important to emphasize here once again that distances in the absolute space of classical mechanics can be measured and compared only to certain material objects. At the same time, it is postulated that material objects are something completely different from immaterial space and time. Only the distance from one material body to another has physical meaning, although formally the entire space is covered by a coordinate grid, which reflects a consistent way of conveniently generalizing the set of measurements of distances between material objects. Therefore, in classical mechanics, the methods for measuring length and time are the same in all space and do not depend on the choice of an inertial frame of reference. At the same time, the limitation of the area of ??operation of this principle only by inertial reference frames is explained by the possibility of the appearance in non-inertial reference frames of forces that change the length of the rulers and the speed of the clock due to obvious physical mechanisms. In a non-inertial frame of reference, for example, support forces must always arise, forcing the standard to accelerate. Support reaction forces acting on elastic bodies cause their relative lengthening or contraction. These facts were discussed by Brillouin in paper [10]. In contrast to this, in SR, the change in the length of the standards occurs during the transition from one inertial frame of reference to another, although the first postulate of this theory is the sameness of all the laws of physics in these frames of reference. Therefore, the change in the scales of length and time in SRT cannot be associated with any real physical mechanism of interaction of material bodies.

Another way of interpreting space, corresponding to the outlined new theory, is to recognize space as a material object with measurable physical properties. In this case, the points of this space are recognized as material objects, and any change in the distances between these points must be caused and explained by specific physical reasons. As a way of describing the geometric properties of such a material space, a model of embedding of a three-dimensional smooth hypersurface into an enclosing four-dimensional space is proposed $\mathcal{W}^4$. This model means that the properties of a three-dimensional material hypersurface differ in a measurable way from the properties of a four-dimensional enclosing space. Since this model includes a new mathematical object - four-dimensional ambient space and separately time, then the same initial dilemma arises with a single description of them. It is necessary to make a choice, either we believe that this space is only a formal way of calculating the lengths and distances in it, or we believe that this space is also material. In the first case, we can rely on our experience in measuring distances in classical mechanics, since the absolute space of this theory is part of the ambient space. In this case, the most natural model for calculating the lengths in this four-dimensional space is the four-dimensional Euclidean space $\mathcal{W}^4$, at all points of which the length of the standards remains unchanged. Moreover, in order for such a procedure to be implemented, the presence of material objects in $\mathcal{W}^4$ is required. As in classical mechanics, in TTFF distances can be measured only between material objects. Such material objects in TTFF are points of a three-dimensional physical material hypersurface (PMH).

In the second case, it is necessary to indicate the model of materialization of this space, which will determine its geometry as a material object. However, in the absence of any precise information about this space, it is rather difficult to do this. Due to this, the most logical choice is the assumption that the ambient four-dimensional space $\mathcal{W}^4$ is an analogue of the absolute space of classical mechanics, but having one more dimensions. This approach was adopted earlier in the discussed new theory of TTFF.

However, in contrast to classical mechanics, in which the very practice of a physical experiment was based on the possibility of decomposing matter into separate material points, in TTFF it is necessary to restrict ourselves, at least for now, to distances only between the points of the physical material hypersurface itself. This creates certain difficulties in attempts to formulate the theory of ambient space, relying on standard classical techniques. Since we have no idea of ??how a material hypersurface is arranged in a small one, we can still formally rely on mathematical axioms, according to which a hypersurface in Euclidean space can be represented as a set of mathematical points, which can be formally associated with some physical parameters, for example, density masses, energy density, speed, etc. In this case, the PMH is represented as a continuous and even smooth hypersurface $\mathcal{V}^3 \in \mathcal{W}^4$ with physical quantities distributed on it. As a result, we get an object consisting of many points, the physical meaning of which is not fully defined, but which can serve as a good model for describing the dynamics of matter. It should be emphasized that these points are not directly related to the material points of classical or quantum mechanics. The objects of these theories are particles, which are treated in the TTFF as extended FMG regions, distinguished in a certain way ([6, 2, 3, 4, 5, 8]).

Suppose $\mathcal{W}^4$ as Euclidean space, a Cartesian coordinate system of the following form can be introduced: $\mathbf{X} = (x_1, x_2, x_3, x_4) = (\mathbf{x}, u)$, where one of the coordinates is highlighted, in this case $u = x^4$, and orthogonal hyperplane $\mathcal{P}^3$ with Cartesian coordinates $\mathbf{x} = (x_1, x_2, x_3)$. Then any hypersurface $\mathcal{V}^3 \in \mathcal{W}^4$ can be mathematically singled out unambiguously using one height function $\mathcal{F}(\mathbf{x}, t)$ using the equation:

$$u = \mathcal{F}(\mathbf{x}, t). \tag{1}$$

It is in this form that the physical material hypersurface (PMH) was described in the papers [5, 8]. However, this selection implies that in $\mathcal{W}^4$ it is possible not only formally, but also physically, to single out the hyperplane $\mathcal{P}^3$. As pointed out in [5, 8], such hyperplane in $\mathcal{W}^4$ stands out by us in the form of an intuitive image based primarily on the rectilinear propagation of light in a vacuum. Since light does not pass through dense matter in a straight line or does not pass through individual material objects at all, the idea of the flatness of our world intuitively continues into the area of space occupied by these objects. Thus, the hyperplane $\mathcal{P}^3$ - is a set of our intuitive ideas in relation to the PMH about the location of material objects relative to each other, which can be interpreted as the choice of a physical frame of reference.

Following this analysis and the general idea underlying the hypothesis of Clifford and Einstein that matter and material fields, gravitational and electromagnetic, should be explained by the properties of the geometry of some space and time, it is logical to relate them to the geometric and topological properties of the physical hypersurface V^3. But now this method should be based not on formal considerations such as choosing the Ricci curvature tensor as the density of the Lagrangian function of the gravitational field of the scalar, but on the very ideology of materiality of the physical hypersurface. Such an ideology, first of all, should reflect the very meaning of defining the materiality of an object as an object that can be observed in a physical experiment.

4. A method for describing the properties of a material hypersurface

The most obvious property of material objects is their observability, i.e. the ability to track them using one or another direct physical measurements. This means that each point of a material object can be associated with a set of physical markers that reflect measurable properties that distinguish one point from others. Such markers in hydrodynamics are called Lagrangian markers and represent the coordinates of each material point at a certain fixed moment in time. It is these Lagrangian markers that are the most general way to describe any moving material medium at subsequent times. In this regard, to describe the distribution of matter in the ambient four-dimensional space $\mathcal{W}^4$ it is also necessary to introduce a set of markers $E^a(\mathbf{x}, t)$, that uniquely highlight each point of matter located in it at a point with a coordinate $\mathbf{X} = (X_1, X_2, X_3, X_4)$ in time moment t. As noted, we will assume that $\mathcal{W}^4$ is a Euclidean space and therefore the coordinates $\mathbf{X}$ can be chosen as Cartesian without loss of generality.

Suppose that a material object has the form of a hypersurface V^d of dimension d, embedded in the ambient space $\mathcal{V}^D$ of dimension $D > d$. The physical material hypersurface distinguishes the object under study, and the enveloping space is necessary for introducing the extension property into the theory of the standard, comparing with which we can get an idea of the change in dimensions and distances on the physical hypersurface. Geometric markers $e^a(\mathbf{X}, t)$, $a = 1, \ldots$ will further mean the functions of the coordinates of the ambient space and time, which are associated with each point of the hypersurface $V^d \subset \mathcal{V}^D$ and allow one point to be uniquely distinguished from other points of this hypersurface. By analogy with hydrodynamics, the value of the marker function can be the value of the coordinate of a material point in $\mathcal{V}^D$ at one selected moment in time $t = 0$ conduct the measurement $e_0^a(\mathbf{X})$. Based on this, the number of geometric markers $e^a(x_1, x_2, \ldots, x_D)$, $a = 1, \ldots, D$ should be equal to D independent functions. In this case, all markers of points of a material object should be divided into two classes. The first class is **global markers**, highlighting the physical material hypersurface in the ambient space as a whole. The number of such markers must be equal to $M = D - d$. Indeed, the material hypersurface of dimension d in the Euclidean space $\mathcal{V}^D$ at time moment t can be specified using a system of M algebraic equations:

$$e_a(\mathbf{X}, t) = C_a = \text{const}, \quad a = d + 1, \ldots, D.$$

$C^a$is a collection of a certain set of real constants that distinguishes a particular hypersurface among all possible hypersurfaces with other values of global markers.A collection of functions $e_a(\mathbf{X}, t)$ can be considered as M markers, which will be called **global markers**. Their values are determined at the moment $t = 0$ using one or another measuring procedure and remain unchanged at any point of the PMH for any changes in its geometry.

The second class of markers should highlight the points of the FMG itself. These markers reflect the movement of points of the FMH with specified physical properties along the PMH itself when its geometry changes. These markers will be called **local or geometric markers.** This means that local markers must be associated with such transformations of the functional dependence of global markers on $\mathbf{X} \to e_a(\mathbf{X}, t)$, that do not change the geometry of the PMH, but only reduce to redesignating its points.

To describe the motion of the hypersurface V^d with time, it is now sufficient to introduce into the theory the equation of transfer of markers:

$$\frac{\partial e^a}{\partial t} + U^i(\mathbf{x}, t)\frac{\partial e^a}{\partial x^i} = 0, \quad a = 1, \ldots, D; \quad i = 1, \ldots, D. \tag{2}$$

The field $\mathbf{U}(\mathbf{X}, t)$ with components $U^i(\mathbf{X}, t)$ will be called the marker transfer rate field. Along the integral curves of this field, i.e. integral curves of the system of equations:

$$\frac{dx^i}{dt} = U^i(\mathbf{X}, t), \quad \frac{de^a}{dt} = 0,$$

Geometric marker values remain constant. Therefore, the field $\mathbf{U}(\mathbf{X}, t)$ of the marker transfer rate should play a fundamental role in the theory of PMH dynamics, since it is associated with all the main elements of the marker movement and, as a consequence, the entire material hypersurface. In this regard, the equation (2) is a mathematical expression of the second postulate of materiality. Namely: **The main mathematical method for describing the dynamics of a material hypersurface is the equation (2), and the field $\mathbf{U}(\mathbf{X}, t)$ of the transfer of markers is the fundamental field for describing the dynamics of a material object V^d.**

5. Three-dimensional material hypersurface in $\mathcal{W}^4$

In accordance with the main provisions of the TTFF, the physical material hypersurface $\mathcal{V}^3$ has dimension $d = 3$ and is embedded in the ambient Euclidean space $\mathcal{W}^4$ of dimension $d = 4$: $\mathcal{V}^3 \in \mathcal{W}^4$. In this case, among the four markers, one of the markers E^4, which we will further denote as the function ϕ, is a global marker, and the other three are local.

Since the function $\phi(\mathbf{X}, t)$ is a global marker, the FMG can be distinguished into $\mathcal{W}^4$ using one algebraic equation:

$$\phi(\mathbf{X}, t) = \phi_0 = \text{const.} \tag{3}$$

However, in TTFF, to describe the geometry of the FMH and the physical properties of material objects that we identify as particles of matter, not the ϕ function, but the height function $\mathcal{F}(\mathbf{x}, t)$, is used, with the help of which the PMH is distinguished in accordance with equation:

$$\mathcal{F}(\mathbf{x}, t) = u = x^4, \quad \mathbf{x} = (x_1, x_2, x_3). \tag{4}$$

here u - the coordinate of $\mathcal{W}^4$, corresponding to the orthogonal direction to some distinguished hyperplane $\mathcal{P}^3 \subset \mathcal{W}^4$, on which the Cartesian coordinates are given $\mathbf{x} = (x_1, x_2, x_3)$. It is this function, called the fundamental potential in the TTFF, that plays the most important role in the description of the fundamental fields - electromagnetic and gravitational. In particular, the solution of the problem of a topological-geometric description of the properties of electric charges, electromagnetic and gravitational fields in the TTFF is based on the fact that the function of the height of the PMH $\mathcal{F}$ is associated with local markers $\mathbf{e}$ on each simple topological cell ([2, 3, 4, 5] Appendix) using the equation:

$$\mathcal{F} = \mathcal{F}_0 + \frac{\varepsilon}{2}|\mathbf{e}(\mathbf{x}, t)|^2. \tag{5}$$

Here $|\mathbf{e}|^2 = R^2(\mathbf{x}, t) = (e^1)^2 + (e^2)^2 + (e^3)^2$, and $\mathcal{F}_0$ - the function value $\mathcal{F}$ at its extremum belonging to a given cell, and $\varepsilon = \pm 1$ depending on whether the extremum is a maximum or a minimum.

Marker transfer equation (2) should, by definition, look like:

$$\frac{\partial e^a}{\partial t} + U^i(\mathbf{X}, t)\frac{\partial e^a}{\partial x^i} = 0, \quad a = 1, \ldots, 4; \quad i = 1, \ldots, 4. \tag{6}$$

However, in papers [6, 2, 3, 4, 5, 8] only local markers were considered, the transfer equation of which has the following form:

$$\frac{\partial e^a}{\partial t} + V^\alpha(\mathbf{x}, t)\frac{\partial e^a}{\partial x^\alpha} = 0, \quad a = 1, \ldots, 3; \quad \alpha = 1, \ldots, 3, \tag{7}$$

where the field $V^\alpha(\mathbf{x}, t)$ was specified on the hyperplane $\mathcal{P}^3$ and associated with an electromagnetic field. In particular, it was shown that by introducing geometric averaging with density $|J|$, the role of which is played by the Jacobian of the transformation $\mathbf{x} = (x^1, x^2, x^3) \to \mathbf{e} = (e^1, e^2, e^3)$, it is possible to construct all the basic equations of classical and quantum physics, as well as the equations of the classical theory of gravity in an extended form. It was also shown [3, 8], that a simple modification of the theory of gravitation in the classical form makes it easy to explain the phenomenon called dark matter in modern physics. In fact, this means that the concept of markers is a very effective tool for solving a number of problems in modern physics. Therefore, we can assume that the equations (6) for local markers should be reduced to the form of equations (7).

The equation for the global marker can now be constructed from the definition of the height function. Since on FMG the global marker takes some constant value, then based on this we can assume that the global marker $\phi(\mathbf{x}, t, u)$ is some function of the following form:

$$\phi = \Phi(u - \mathcal{F}(\mathbf{x}, t)). \tag{8}$$

Differentiating this equation with respect to $\mathbf{x}$, t and u, and then, excluding the derivative of the function $\Phi'(\xi)$, from the obtained relations, we arrive at the following system of equations for the function $\phi(\mathbf{x}, t, u)$:

$$\frac{\partial \phi}{\partial x^i} = -\frac{\partial \mathcal{F}}{\partial x^i}\frac{\partial \phi}{\partial u}, \quad i = 0, 1, 2, 3. \tag{9}$$

This system of equations was considered in [14, 7] and describes a special class of waves - rivertons. Convolving this system with respect to the index i with an arbitrary continuous vector field $u^i(\mathbf{x}, t, u)$, we arrive at the following equation for $\phi(\mathbf{x}, t, u)$:

$$u^0\frac{\partial \phi}{\partial t} + u^\alpha \frac{\partial \phi}{\partial x^\alpha} + (\mathbf{u}, \nabla \mathcal{F})\frac{\partial \phi}{\partial u} = 0, \quad (\mathbf{u}, \nabla \mathcal{F}) = u^i \mathcal{F}_{,i}. \tag{10}$$

This equation can be given the following form:

$$\frac{\partial \phi}{\partial t} + U^\alpha(\mathbf{x}, t)\frac{\partial \phi}{\partial x^\alpha} + U^4(\mathbf{x}, t)\frac{\partial \phi}{\partial u} = 0. \tag{11}$$

Here:

$$U^\alpha = \frac{u^\alpha}{u^0}, \quad U^4 = \frac{(\mathbf{u}, \nabla \mathcal{F})}{u^0}. \tag{12}$$

The field $\mathbf{U}$, defined on $\mathcal{W}^4$, with the components U^i (12) is the field for transferring the global marker to $\mathcal{W}^4$. It follows from the last relation that the component U^4 of the field $\mathbf{U}$can be represented as follows:

$$U^4 = U^\alpha \frac{\partial \mathcal{F}}{\partial x^\alpha} + \frac{\partial \mathcal{F}}{\partial t}.$$

Therefore, this component U^4 of the field $\mathbf{U}$ is determined by the derivatives of the height function $\mathcal{F}$ and the components of the field $\mathbf{V} = (U^1, U^2, U^3, 0)$, tangent to the hyperplanes $\mathcal{P}^3$. The case $U^4 = 0$, equivalent to the execution of the equation:

$$\frac{\partial \mathcal{F}}{\partial t} + V^\alpha \frac{\partial \mathcal{F}}{\partial x^\alpha} = 0,$$

defines the transfer equation of the height function and, as a consequence, local markers.

Note that the relations (12) can be slightly changed, proceeding from the fact that the relation (8) can be generalized, assuming:

$$\phi = \Phi(u - \mathcal{F}(\mathbf{x}, t, \phi)). \tag{13}$$

This generalization is connected with the fact that, solving the equation (3) with respect to u, we have the relation:

$$u = \mathcal{F}(\mathbf{x}, t, \phi),$$

which is equivalent to (4) for some specific constant value $\phi(\mathbf{x}, t, u) = C = \text{const}$, which is what the PMH selects. With this generalization, the global marker can be a multivalued function of coordinates in $\mathcal{W}^4$, which extends the admissible structure of the PMH.

Differentiating (13) with respect to $\mathbf{x}, t$ and u and excluding the derivative $\Phi'(\xi)$, we have the same equation (9):

$$\frac{\partial \phi}{\partial x^i} = -A_i \frac{\partial \phi}{\partial u}, \quad A_i = \left. \frac{\partial \mathcal{F}(\mathbf{x}, t, \phi)}{\partial x^i} \right|_{\phi = \text{const}}, \quad i = 0, 1, 2, 3, \tag{14}$$

with the difference that the function $\mathcal{F}(\mathbf{x}, t, \phi)$ contains a dependence on ϕ, which is not affected by differentiation with respect to coordinates and time. Thus, all the relations (10),(11) remain valid in this generalized case, and in the equation (12) it is necessary to take into account that the derivatives with respect to x^i on the right-hand side are calculated under the condition $\phi = \text{const}$.

6. Equation of geometrodynamics of the second order

The resulting transfer equation for the global marker contains the vector $\mathbf{U}$ with its arbitrary distribution on the distinguished hyperplane $\mathcal{P}^3$. This means that the introduction of a global marker does not yet solve the problem of distinguishing among all possible types of FMG dynamics in $\mathcal{W}^4$ the one that corresponds to the real state of affairs. For such a possibility to arise, it is necessary to obtain some consequences of the equations derived in the previous section, which can be related to some inherently clear dynamic FMG models. Such consequences can be second order equations for ϕ, which can be obtained from equations (9) or (14), and which can be interpreted by comparing with some form of dynamics of classical or mathematical physics.

The calculations will be carried out for the generalized system (14). Instead of the time t we introduce the variable $x^0 = ct$, where c - is a constant that further plays the role of the speed of light. This is useful for comparison with existing classical theories. Let us reduce the equations of the system (14) with the components g^{ij} of the diagonal matrix $\hat{g}$, which has the form:

$$\hat{g} = \text{diag}(1, -1, -1, -1), \tag{15}$$

which imitates the Minkowski space-time metric. As a result, we get the following equation:

$$g^{ij} \frac{\partial \phi}{\partial x^j} = g^{ij} A_j(\phi, \mathbf{x}) \frac{\partial \phi}{\partial u}, \quad i = 0, 1, 2, 3.$$

Calculating now the divergence from the right and left sides of the last equation, we arrive at the following second-order equation:

$$\overline{\Diamond}\phi = \frac{\partial}{\partial u}\left(|\mathbf{A}(\phi, \mathbf{x})|^2 \frac{\partial \phi}{\partial u} \right) + \overline{\Diamond} F \frac{\partial \phi}{\partial u}, \tag{16}$$

where

$$\overline{\Diamond} = g^{ij} \frac{\partial^2}{\partial x^i \partial x^j} = \frac{\partial^2}{\partial x^{0 2}} - \sum_{\alpha=1}^{3} \frac{\partial^2}{\partial x^{\alpha 2}}$$

- is the d'Alembert operator in the coordinate space-time, the action of which on the function $F(\mathbf{x}, t, \ phi)$ is carried out under the condition $\phi = const$. In addition, the designation has been introduced:

$$|\mathbf{A}(\mathbf{x}, t, \phi)|^2 = g^{ij} A_i A_j = \Big(A_0(\phi, \mathbf{x}) \Big)^2 - \sum_{\alpha=1}^{3} \Big(A_\alpha(\phi, \mathbf{x}) \Big)^2.$$

Note that from the formal point of view, the choice of the constant nondegenerate matrix $\hat{g}$, over which the convolution is performed in the equation (16), is arbitrary. For example, as a matrix $\hat{g}$, one could choose any non-degenerate matrix with signature $(+,+,+,+)$ or $(+,+,-,-)$. In this case, instead of the standard d'Alembert operator, the equation (16) would have an operator with the corresponding signature of the second derivatives. In particular, for the signature $(+,+,+,+)$ it would be the Laplace operator in four-dimensional space. This choice obviously corresponds to a different version of the FMG dynamics, which could be realized in some other situation than the one that is realized for our Universe.

Regardless of the choice of the matrix g_{ij}, the equation (16) is the equation of the dynamics of the global marker, which can be compared to some equations of classical physics in order to understand its interpretation from the experimental point of view. First of all, it should be noted that the reduced general form of the equation (16) is a general and natural consequence of the original postulate that the FMG is a three-dimensional hypersurface in $\mathcal{W}^4$. Because of this, the equation contains two functional parameters $\Lambda = |\mathbf{A}|^2$ and $P = \overline{\Diamond}\mathcal{F}$, connected exclusively with the height function $\mathcal{F}$:

$$\Lambda = |\mathbf{A}|^2 = \left(\left(\frac{\partial \mathcal{F}(\mathbf{x}, t, \phi)}{\partial x^0} \right)^2 - \sum_{\alpha=1}^{3} \left(\frac{\partial \mathcal{F}(\mathbf{x}, t, \phi)}{\partial x^\alpha} \right)^2 \right) \Bigg|_{\phi=\text{const}}, \tag{17}$$

$$P = \overline{\Diamond}\mathcal{F} = \frac{\partial^2 \mathcal{F}}{\partial x^{0\,2}} - \sum_{\alpha=1}^{3} \frac{\partial^2 \mathcal{F}}{\partial x^{\alpha\,2}}.$$

Hence it follows that the geometrodynamics of the FMG is determined by two functions Λ and P, which must be related to the physical quantities that are measured in the experiment. In fact, if Λ and P are represented in the form of measurable distributions of some physical parameters of matter, then the relations (17) can be considered as equations for the fundamental potential - the height function $\mathcal{F}$. From this point of view, the choice of the matrix $\hat{g}$ in the form (15) is only a way to bring such equations for $\mathcal{F}$ to a form close to the form of the wave equations of classical physics. The meaning of this choice lies in the assumption that in the limit when the functions Λ and P are small, which can apparently be interpreted as the absence of matter in space, the processes of change in $\mathcal{F}$ and ϕ should be are similar to the propagation of waves with a certain fixed speed c, which coincides with the speed of light.

7. Interpretation of the equations of geometrodynamics

Note that from the point of view of the measurements that we can make as an element of the FMG, the equations (16) taking into account (17) are not very useful, since we do not have, at least not yet, information about the ambient space $\mathcal{W}^4$. Because of this, the equations (17), which relate to the height function and do not contain dependence on the variable u, are of particular interest. The function $\mathcal{F}(\mathbf{x}, t, \phi)$ itself formally depends on the value of ϕ on the PMH. But this dependence is purely parametric and does not actually affect the functional dependence $\mathcal{F}(\mathbf{x}, t, \phi)$ with respect to $\mathbf{x}$ and t. Moreover, if the dependence is known $\mathcal{F}(\mathbf{x}, t, \phi)$ with respect to $\mathbf{x}$ and t, then the dependence $\phi(\mathbf{x}, t, u)$ will be determined directly from (8), or from the solution of the equation (16) or (17) with additional boundary or initial conditions. Thus, to clarify the dynamics of the PMH, it is necessary to focus on the interpretation of the (17) equations from the point of view of a physical experiment.

The first thing to notice is that the equation:

$$\overline{\Diamond}\mathcal{F} = P \tag{18}$$

has an obvious similarity with the equation of vibrations of a thin three-dimensional membrane in four-dimensional Euclidean space under the action of the "force" P, referred to the unit volume of the PMH, and applied to the membrane along the u axis, orthogonal to the hyperplane $\mathcal{P}^3 \in \mathcal{W}^4$. Indeed, the equation of vibrations of a thin two-dimensional membrane has a similar form, but with the dimension of the coordinate space less than one. For a two-dimensional membrane, the P function is the ratio of the external pressure applied to the membrane to the density of the membrane material. As a result, we can interpret the equation (18) as the equation of vibrations of an elastic three-dimensional membrane $\mathcal{V}^3$ in $\mathcal{W}^4$, under the action of some "force" P.

Since to calculate the function $\mathcal{F}(\mathbf{x}, t, \phi)$, one equation (18) is sufficient in the presence of boundary and initial conditions, the second equation of the system (17):

$$\Lambda = \left(\left(\frac{\partial \mathcal{F}(\mathbf{x}, t, \phi)}{\partial x^0} \right)^2 - \sum_{\alpha=1}^{3} \left(\frac{\partial \mathcal{F}(\mathbf{x}, t, \phi)}{\partial x^\alpha} \right)^2 \right) \Bigg|_{\phi=\mathrm{const}} . \tag{19}$$

must be a consequence of (18). However, there is no simple link between (18) and (19), since there is no simple link between the functions P and Λ. Such a connection can appear if we use a formal factorization of these equations, similar to the relations (14) for the function ϕ. This possibility will be discussed later. For the function Λ there is a formal interpretation as the Lagrangian of the linear part of the equation (18). However, this interpretation does not make much physical sense yet.

8. General principles of the dynamics of a physical hypersurface

Interpreting the equation (18) as the equation of vibrations of a three-dimensional membrane in four-dimensional Euclidean space under the action of the "forces" described by the function P allows us to analyze the meaning and properties of P from the point of view of known facts. The first important fact that must be taken into account is the fulfillment of the law of conservation of energy in our Universe on all scales observed today. In particular, this means that the equation (18) itself should most likely be autonomous. The condition of autonomy guarantees that no external factors interfere with the dynamics of the system, which, as a rule, lead to a violation of the laws of conservation of momentum and energy. The latter means that the force acting on the PMH should depend only on the parameters of the PMH itself and its derivatives:

$$P = P(\mathcal{F}, \mathcal{F}_{,\alpha}, \mathcal{F}_{,\alpha,\beta}, \ldots).$$

In the simplest case, we can assume that the function $P(\mathcal{F}, \mathcal{F}_{,\alpha}, \mathcal{F}_{,\alpha,\beta}, \ldots)$ explicitly depends only on the function $\mathcal{F}$ itself. In this case, the equation (18) is the nonlinear Klein-Gordon equation, which occurs in various physical applications. In particular, in the work [16], exact solutions of the nonlinear Klein-Gordon equations in the class of rivertons were found. The solutions obtained in this work are multivalued, which seems to be quite useful for using in the TTFF, in which the multivalued structures of the fundamental hypersurface are associated with baryons [5].

Similar reasoning can be applied to the function Λ, assuming that the equation (19) must also be autonomous. In this case, the function Λ must be a function of only $\mathcal{F}$ and its derivatives:

$$\Lambda = \Lambda(\mathcal{F}, \mathcal{F}_{,\alpha}, \mathcal{F}_{,\alpha,\beta}, \ldots).$$

Unfortunately, until the direct connections of the equations (18) and (19) with experimentally measurable quantities have not been established, it is difficult to give any instructions for the functional choice of P as a function of $\mathcal{F}$ and its derivatives. However, we can establish some useful properties of this function by considering some limiting situations that should be realized in reality.

As an example of the "simplest" state of pmh, consider the equations for the fundamental potential corresponding to the situation: $P = 0$. In this case, the complete equation (16) turns into an equation of the following form:

$$\Diamond \phi = \frac{\partial}{\partial u} \left(|\mathbf{A}(\phi, \mathbf{x})|^2 \frac{\partial \phi}{\partial u} \right).$$

Accordingly, the equations (18) and (19) take the following form:

$$\Diamond \mathcal{F} = 0, \tag{20}$$

$$\left(\left(\frac{\partial \mathcal{F}(\mathbf{x}, t, \phi)}{\partial x^0} \right)^2 - \sum_{\alpha=1}^{3} \left(\frac{\partial \mathcal{F}(\mathbf{x}, t, \phi)}{\partial x^\alpha} \right)^2 \right) \Bigg|_{\phi=\text{const}} = \Lambda. \tag{21}$$

The first equation of this system is the equation of free vibrations of a three-dimensional thin membrane in four-dimensional space. The phase velocity of waves arising due to edge effects is determined by the number c, which is related to the relations (15). This speed in the experiment is obviously equivalent to the speed of light. The function Λ in this case is some equivalent of the energy density of such waves, corresponding to each specific solution for $\mathcal{F}$. In this case, such a situation can be correlated with PMH oscillations in the absence of complex topological structures, which are compared in the TTFF with elementary particles [5].

However, the equality to zero of P is obviously a very special case of possible situations. Although the set of solutions to the D'Alembert equation (20) contains an extensive subset of multivalued solutions ([13, 14]) - rivertones, nevertheless, solutions from this set alone can hardly give something similar Wheeler's handles [5]. However, in order for such a possibility to appear, there is an extended version of using the results of [13, 14], described in the work [16]. In this last work, using the theory of rivertons, multivalued solutions of first-order quasilinear equations, an extensive set of multivalued solutions of multidimensional nonlinear Klein-Gordon equations of general form was constructed. those. equations of the form:

$$\Diamond \mathcal{F} = F(\mathcal{F}),$$

where $F(\mathcal{F})$ - the quite function general.

9. Rivertons and multivalued solutions of equations of geometrodynamics

The idea of constructing solutions of multidimensional nonlinear Klein-Gordon equations using rivertons is reduced to the following. By rivertons we mean solutions of a system of quasilinear equations of the following form:

$$\frac{\partial \psi}{\partial x^\alpha} = B_\alpha(\psi) \frac{\partial \psi}{\partial t}, \quad \alpha = 1, \ldots, n. \tag{22}$$

The general solution to this system [13, 14] can be written as follows:

$$H(\psi, t + U(\mathbf{B}, \mathbf{x})) = 0, \tag{23}$$

where:

$$U(\mathbf{B}, \mathbf{x}) = \sum_{\alpha=1}^{n} B_\alpha(\psi) x^\alpha.$$

Simple differential consequence (22) is the equation:

$$\Delta \psi = \frac{\partial}{\partial t} \left(|\mathbf{B}|^2 \psi_t \right). \tag{24}$$

Since the components of the vector field $\mathbf{B}(\psi)$ are arbitrary functions of ψ, they can be chosen in such a way that the condition is satisfied:

$$|\mathbf{B}|^2 = c^{-2}, \tag{25}$$

where c - constant from relations (16). As a result, the equation (24) becomes the equation (20). Note that in this case the equation (19) with $\Lambda = 0$ will be automatically satisfied for the functions ψ. This follows directly from the equations (22) and the choice of the functions $\mathbf{B}(\psi)$ in accordance with (25).

In order to obtain in a similar way the solutions of equations of the (18) type, it is necessary to consider functions of the following general form as $\mathcal{F}$:

$$\mathcal{F} = W(\psi, \psi_t, \psi_{tt}, \ldots), \tag{26}$$

where ψ - riverton, and W - some differentiable function of its arguments, the form of which will be determined by the functional form of the function P as a function of $\mathcal{F}$ and its derivatives.

For simplicity, consider the case $\mathcal{F} = W(\psi, \psi_t)$. Let us introduce additional notation $u = \psi_t$, $v = \psi_{tt}$. Differentiate (26) with respect to independent variables x^α and t. As a result, we find:

$$\mathcal{F}_{,\alpha} = W_\psi \psi_{,\alpha} + W_u \psi_{t,\alpha} = W_\psi B_\alpha u + W_u \frac{\partial}{\partial t}(B_\alpha u),$$
$$\mathcal{F}_{,t} = W_\psi u + W_u u_t.$$

using the last relation, we find:

$$\mathcal{F}_{,\alpha} = B'_\alpha F(\psi, u) + B_\alpha \mathcal{F}_t.$$

Where:

$$F(\psi, u) = u^2 \frac{\partial W}{\partial u};$$

From the last relation, after some transformations ([16]) follows:

$$\Delta \mathcal{F} - R\mathcal{F}_{tt} = R'' uF + \Omega^2 \Big(u^2 F_{,u} - 2uF\Big) + R' \Big(uF_{,\psi} + F_{,u} u_t + u\mathcal{F}_t\Big). \tag{27}$$

The notation is introduced here:

$$R = \sum_{\alpha=1}^{n} B_\alpha^2(\psi), \quad \Omega^2 = \sum_{\alpha=1}^{n} B_\alpha'^2(\psi), \quad B'_\alpha = \frac{d}{d\psi} B_\alpha.$$

In the case of a special choice $R = \text{const} = c^{-2}$ the equation (27) turns into the equation:

$$\Delta \mathcal{F} - R\mathcal{F}_{tt} = \Omega^2 \Big(u^2 F_{,u} - 2uF\Big),$$

which turns into the Klein-Gordon equation if the function $W(\psi, u)$ satisfies the equation:

$$u^2 F_{,u} - 2uF = u^2 \frac{\partial}{\partial u}\left(u^2 \frac{\partial W}{\partial u}\right) - 2u^3 \frac{\partial W}{\partial u} = u^4 \frac{\partial^2 W}{\partial u^2} = \frac{1}{\Omega^2(\psi)} G(W). \tag{28}$$

If $W(\psi, u)$ is the solution of this simple differential equation, then the function $\mathcal{F}(\mathbf{x}, t) = W\Big(\psi(\mathbf{x}, t), \psi_t(\mathbf{x}, t)\Big)$, where $\psi(\mathbf{x}, t)$ - is one of rivertons, is a solution to the nonlinear Klein-Gordon equation:

$$\Delta \mathcal{F} - c^{-2} \mathcal{F}_{tt} = G(\mathcal{F})$$

for an arbitrary choice of the function $G(W)$. This equation is standalone and can serve as a variant of the (18) equation. Since the choice of the function $G(W)$ is not predetermined in advance, it is possible to select the properties of feasible solutions based on the choice of the function $G(W)$. Note that in [16] more general options for constructing solutions of equations of the Klein-Gordon type and nonlinear telegraph equations based on the general dependence (26) were described. Although this approach has not been studied well enough from the point of view of constructing a general model of FMG dynamics based on the equation (18), nevertheless, in the absence of direct connections between the FMG geometry parameters and ordinary experimental data, this approach seems to be quite natural for solving the problems posed. However, solving such problems is beyond the scope of this article.

10. Equations of the dynamics of fundamental fields and local markers

The previous constructions are a natural consequence of the principle of materiality and its embodiment in the form of dynamics of PMH point markers. As shown in the papers [2, 3, 4, 5], all the basic equations of the dynamics of particles and physical fields, including electrodynamics, the theory of gravity and mechanics in the form of a modified Newtonian theory and quantum mechanics, follow from the equations of the dynamics of local markers. To clarify the role played by the global marker equation in the TTFF, it is useful to consider the basic principles of deriving the equations of modern physical theories from the set of marker transfer equations (7). Using the conclusions [2, 3, 4, 5], we will show that all these equations are, in fact, a consequence of only the equations (7) in some of their natural interpretation.

The first step is to derive the differential density conservation equation $J = |\det\left(\partial e^a/\partial x^\alpha\right)|$, where $e^a = e^a(\mathbf{x})$ - local markers on PMH. Indeed, (7) implies the equation for J:

$$\frac{\partial J}{\partial t} + \operatorname{div}\left(\mathbf{V}J\right) = 0, \tag{29}$$

where $\mathbf{V}$ - vector translating markers field with components V^α. Since markers should play a fundamental role in the theory, the TTFF suggests that the conserved density J can be considered as the mass density of matter ρ_m, based on the formulas:

$$\rho_m = m_0 J I(\mathbf{e}), \quad M = m_0 \int\limits_V J I(|\mathbf{e}|) dV, \tag{30}$$

where the integral is taken over the volume of the coordinate space $\mathcal{P}^3$, which is occupied by the considered element of matter, M is the mass of the material body, and m_0 is a certain dimensional factor. The function $I(|\mathbf{e}|)$ is an invariant associated with the metric of the marker space, which determines the deviations from Newton's law of gravitation of the "dark matter" type. See [2, 3, 4, 5, 8]. The presence of the conserved density J allows one to introduce into the theory geometric averaging for any function $f(\mathbf{x})$ according to the rule:

$$\overline{f(t)} = \frac{1}{\mu} \int\limits_{V_0} f(\mathbf{x}) J dV, \quad \mu = \int\limits_V J I(|\mathbf{e}|) dV. \tag{31}$$

where, like in (30), the integral is taken over the volume of the topological cell $V_0 \in \mathcal{P}^3$. In particular, you can introduce the average coordinates of a material particle, assuming:

$$X^\alpha = \overline{x^\alpha} = \int\limits_{V_0} x^\alpha J dV. \tag{32}$$

The velocities and accelerations associated with the average coordinates are as follows:

$$U^\alpha = \frac{dX^\alpha}{dt} = \int\limits_{V_0} V^\alpha J dV, \tag{33}$$

$$A^\alpha = \frac{d^2 X^\alpha}{dt^2} = \int\limits_{V_0} \left(\frac{\partial}{\partial t} V^\alpha + V^\beta \frac{\partial}{\partial x^\beta} V^\alpha\right) J dV.$$

If we formally represent the transfer field of markers $\mathbf{V}$ as a set of vortex $\gamma_0\mathbf{A}$ and potential $\nabla\chi$ parts:

$$\mathbf{V} = -\gamma_0\mathbf{A} + \nabla\chi,$$

where γ_0 is a dimensional constant, then the last equation in (33) can be considered as Newton's equation of the averaged motion of a charged particle in an averaged electromagnetic field with a vector potential $\mathbf{A}$ [2]:

$$\frac{d^2X^\alpha}{dt^2} = \gamma_0\overline{\mathbf{E}} - \gamma_0[\overline{\mathbf{V}} \times \overline{\mathbf{B}}] - \nabla_{\mathbf{X}}\overline{U} + \mathbf{F}_q. \tag{34}$$

Here $\overline{\mathbf{E}}$ and $\overline{\mathbf{B}}$ is average strength of electric and induction of magnetic fields:

$$\overline{\mathbf{E}} = -\frac{1}{c}\frac{\partial \overline{A^\alpha}}{\partial t} - \frac{\partial\Phi}{\partial X^\alpha}, \quad \overline{\mathbf{B}} = \mathrm{rot}_X\overline{\mathbf{A}},$$

$\mathbf{F}_q$ is a quantum addition to the forces acting on a particle and $U(\mathbf{X},t)$ is an additional potential:

$$U(\mathbf{x},t) = -\frac{1}{2}|\mathbf{V}|^2 - \chi_t - \gamma_0 c\Phi, \tag{35}$$

ăŠpresumably playing the role of averaged gravitational forces ([2]). For a given potential $U(\mathbf{x},t)$, the ratio (35) can be considered as the Jacobi equations:

$$\chi_t + \frac{1}{2}\Big(|\nabla\chi|^2 + 2\gamma_0(\mathbf{A},\nabla\chi) + \gamma_0^2|\mathbf{A}|^2\Big) + \gamma_0 c\Phi + U = 0. \tag{36}$$

for fuction χ, which in this case is the action function for a particle in an electromagnetic field with potentials Φ and $\mathbf{A}$ and in a scalar field with potential U. This also implies that the function

$$\Psi = \sqrt{|J|}e^{i\chi/\hbar}, \tag{37}$$

where $\hbar$ - Planck's constant, satisfies the Schr?dinger equation

$$i\hbar\frac{\partial\Psi}{\partial t} = \frac{1}{2}\Big(-i\hbar\nabla - \gamma_0\mathbf{A}\Big)^2\Psi + \Big(\gamma_0\Phi - U_G\Big)\Psi, \tag{38}$$

where Φ is the potential of the electric field and

$$U_G = U - \frac{\hbar^2}{2}\frac{\Delta|J|}{|J|}. \tag{39}$$

Thus, averaging (31) automatically leads to the equations of quantum theory and classical mechanics in the form of averaged equations of motion with a geometric interpretation of the wave function (37).

The field part of the theory is based on identities connected by local markers $e^a(\mathbf{x})$. The basic principle, according to which equations of fundamental fields arise in TTFF, is that all fundamental fields are functions of the markers themselves and their derivatives. Let $e^a(\mathbf{x})$ be markers as functions of coordinates x^α, $\alpha = 1,2,3$ on the hyperplane $\mathcal{P}^3 \in \mathcal{W}^4$ distinguished in the ambient space and time t. The set of values ?of the functions $e^a(\mathbf{x})$ can be considered as some coordinate space $\mathcal{E}^3$ with the metric induced by the mapping: $\mathcal{P}^3 \to \mathcal{E}^3$. This space $\mathcal{E}^3$ will be called the marker space. Two obvious differential identities hold in the marker space:

$$\frac{\partial e^a}{\partial e^a} = 3, \quad \frac{\partial}{\partial e^a}\left(\frac{e^a}{|\mathbf{e}|^3}\right) = 4\pi\delta(\mathbf{e}), \tag{40}$$

where $\mathbf{e} = (e^1, e^2, e^3)$ - the radius vector on a Cartesian marker space map, $|\mathbf{e}|^2 = (e^1)^2 + (e^2)^2 + (e^3)^2$ and $\delta(\mathbf{e})$ - Dirac delta function with a carrier at the origin of the Cartesian map $\mathcal{E}^3$. After converting to coordinates on the identities (40) turn into the following equations:

$$\frac{\partial}{\partial x^\alpha} g^\alpha = 4\pi G \zeta(|\mathbf{e}|)\rho_m, \quad \frac{\partial}{\partial x^\alpha} D^\alpha = 4\pi \rho_e, \tag{41}$$

where the following notation is introduced:

$$g^\alpha = \frac{4\pi m_0 G}{3} I(|\mathbf{e}|)|J| e^a \frac{\partial x^\alpha}{\partial e^a}, \quad D^\alpha = \frac{|J|}{|\mathbf{e}|^3} e^a \frac{\partial x^\alpha}{\partial e^a},$$

$$\rho_e = \sum_{k=1}^{N} \varepsilon_k \delta(\mathbf{x} - \mathbf{x}_k), \quad \zeta(R) = 1 + \frac{e^a}{3} \frac{\partial \ln I(\mathbf{e})}{\partial e^a}$$

The factor G - Newton's constant of gravity, function ρ_m is mass density given by the ratio (30), ρ_e - electric charge density, which is the density of point charges associated with critical points of the fundamental potential $\mathcal{F}$. See papers [6, 2, 3, 4, 5, 8]. As a result, the identities (41) take the standard form of the Poisson equations for the gravitational field strength $\mathbf{g}$ (with components g^α) and Maxwell's first equation for the electric induction field $\mathbf{D}$ (with components D^α). Another pair of equations is obtained directly from (7) using some transformations, the details of which can be found in [4, 8]. This pair of equations can be represented in this form:

$$\frac{\partial \mathbf{D}}{\partial t} = -\mathrm{rot}\Big([\mathbf{D} \times \mathbf{V}]\Big) - 4\pi \rho_e \mathbf{V}, \tag{42}$$

$$\frac{\partial \mathbf{g}}{\partial t} = -\mathrm{rot}\Big([\mathbf{g} \times \mathbf{V}]\Big) - 4\pi G \zeta(R) \rho_m \mathbf{V}. \tag{43}$$

Here $\mathbf{V}$ - markers transfer field. The last pair of equations are analogs of the equations of induction of electromagnetic and gravitational fields. The other two Maxwell equations actually relate the magnetic and electric fields and require the absence of magnetic charges. In this theory, the magnetic field strength looks like this:

$$\mathbf{H} = -\frac{1}{c}[\mathbf{D} \times \mathbf{V}] + \nabla \Phi_H, \tag{44}$$

where Φ_H - an initially arbitrary potential that leaves the equation (42) unchanged, calculated from the equation of the absence of magnetic charges:

$$\mathrm{div}\mathbf{H} = 0.$$

The equation (43) is the equation of induction of the gravitational field, which is similar to the equation of induction of electromagnetic fields, which is an extension of Newton's theory of gravitation, which consists in introducing the gravitational field strength and, as a consequence, the gravimagnetic field into the theory of vortex fields $\mathbf{Z} = [\mathbf{g} \times \mathbf{V}] + \nabla \Phi_g$ with a potential that also, by analogy with (44), find from the equation of the absence of gravimagnetic charges:

$$\mathrm{div}\mathbf{Z} = 0.$$

The entire set of the presented equations forms a complete system of equations for describing the fundamental electromagnetic and gravitational fields, which in TTFF turn out to be closely related, since the field strengths $\mathbf{D}$ and $\mathbf{g}$ differ only in scalar factors from each other and contain in their notation the fundamental field $\mathbf{K}$:

$$\mathbf{K} = e^a \frac{\partial x^\alpha}{\partial e^a}, \quad \mathbf{D} = \frac{|J|}{|\mathbf{e}|^3}\mathbf{K}, \quad \mathbf{g} = \frac{4\pi m_0 G}{3} I(|\mathbf{e}|)|J|\mathbf{K}. \tag{45}$$

Thus, from the system (7) all the basic equations of the modern physical picture of the world follow. In TTFF, this picture is supplemented by a topological interpretation of the electric charge, which provides an explanation of the charge properties of elementary particles, including electric and baryon charges.([2, 5]). There is one flaw in this scheme, which is that for the function $R = |\mathbf{e}|$, which plays an important role in the theory, there is no equation describing the variation of this function with time. However, since R, according to (5), is associated with the fundamental potential $\mathcal{F}$, the missing equation is the equation for $\mathcal{F}$, the role of which is played by the equation (18) with some interpretation of the function "external pressure".

11. Conclusion

The General Theory of Relativity paved the way for a completely new approach in physics - the description of the properties of matter and fields, based on the properties of non-Euclidean space-time geometry. In fact, thanks to general relativity, concepts such as the theory of the Grand Unification or the theory of "everything" appeared in physics. However, in the process of implementing such unifying ideas, a number of problems of the general concept of general relativity emerged, which served as the basis for many attempts to construct its generalizations that would solve at least some of these problems. From a mathematical point of view, the whole concept of general relativity looks quite adequate. Nevertheless, problems, for example, the energy of the gravitational field, deprive general relativity of completeness and do not provide an opportunity to adequately relate to quantum theory. In this work, it was demonstrated that the main disadvantage of the path chosen in SRT and GRT to the geometrization of physics is the endowment of non-material objects with physical properties, in particular, Einstein's space-time itself. Proceeding from this, the problem of energy seems to be natural and requires a solution for at least some real theory of matter.

The main conclusion of this work can be an indication that in order to construct a theory that would correctly include electromagnetic and gravitational fields, quantum particles with electric and baryon charges, and even "dark matter", it is necessary to initially involve the principle of materiality of space. This principle excludes from the theory non-material objects that can be implicitly endowed with physical properties, which leads to various paradoxes with the outwardly consistency of the theory from a mathematical point of view.

In addition to formulating the principle of materiality of space, the paper proposes a method for its general implementation, which would make it possible to initially formulate the geometric theory in terms of material particles and objects. This approach is closely related to the TTFF [6, 2, 3, 4, 5, 8], which provides a deeper justification for this theory. In this work, the TTFF problem of deriving the FMG dynamics equation was formally solved, relying on general ideas about fundamental and geometric markers. Nevertheless, the theory cannot be considered complete, since it lacks ideas for a general description of the P function in the geometrodynamics equation (18) based on experimental data. The problem is that there are no such experimental data at this time. The paper proposes a way to derive the general form P, relying on the selection of this function, proceeding from the analysis of solutions to this equation in the class of rivertones. However, such an analysis is beyond the scope of this work.

12. Acknowledgments

The work was carried out within the framework of the project 0777-2020-0018, financed from the funds of the state assignment to the winners of the competition of scientific laboratories of educational institutions of higher education subordinate to the Ministry of Education and Science of Russia and partly within the framework of the project RFBR 20-02-00280.

References

[1] Zhuravlev V M 2000 *Izvestiay vuzov, Fizika* **2** 134
[2] Zhuravlev V M 2014 *Space, Time and Fundamental Interactions* **4** 6
[3] Zhuravlev V M 2014 *Space, Time and Fundamental Interactions* **4** 25
[4] Zhuravlev V M 2015 *Space, Time and Fundamental Interactions* **3** 44
[5] Zhuravlev V M 2015 *Space, Time and Fundamental Interactions* **4** 1
[6] Zhuravlev V M 2011 *Gravitation and Cosmology* **17** 201
[7] Zhuravlev V M 2016 *Space, Time and Fundamental Interactions* **2** 5
[8] Zhuravlev V M 2017 *Gravitation and Cosmology* **23** 95
[9] Einstein A 1979 *Albert Einstein and the theory of gravity*, ed E Kuranskiy (Moscow: Mir) pp. 146-196
[10] Brillouin L 1970 *Relativity reexamined* (New York and London: Academic Press)
[11] Cliiford V 1979 *Albert Einstein and the theory of gravity.*, ed E Kuranskiy (Moscow: Mir) pp. 36-37
[12] Fok V A 1959 *The theory of space, time and gravitation* (London: Permagon Press)
[13] Zhuravlev V M 2013 *Theor. and Math. Phys.* **174** 236
[14] Zhuravlev V M 2016 *Theor. and Math. Phys.* **186** 371
[15] Zhuravlev V M 2013 *Space, Time and Fundamental Interactions* **4** 56
[16] Zhuravlev V M 2018 *Theor. and Math. Phys.* **186** 371

AUTHOR INDEX

Institute of Physics
Dirac House, Temple Back
Bristol BS1 6BE UK

ISSN: 1742-6588
ISBN 978-1-7138-5052-6

CURRAN ASSOCIATES INC.
proceedings
.com